Advances in Planar Filters Design

Related titles on electromagnetic waves:

Dielectric Resonators, 2nd Edition Kajfez and Guillon
Electronic Applications of the Smith Chart Smith
Fiber Optic Technology Jha
Filtering in the Time and Frequency Domains Blinchikoff and Zverev
HF Filter Design and Computer Simulation Rhea
HF Radio Systems and Circuits Sabin
Microwave Field-Effect Transistors: Theory, design and application, 3rd Edition Pengelly
Microwave Semiconductor Engineering White
Microwave Transmission Line Impedance Data Gunston
Optical Fibers and RF: A natural combination Romeiser
Oscillator Design and Computer Simulation Rhea
Radio-Electronic Transmission Fundamentals, 2nd Edition Griffith, Jr
RF and Microwave Modeling and Measurement Techniques for Field Effect Transistors Jianjun Gao
RF Power Amplifiers Albulet
Small Signal Microwave Amplifier Design Grosch
Small Signal Microwave Amplifier Design: Solutions Grosch
2008+ Solved Problems in Electromagnetics Nasar
Antennas: Fundamentals, design, measurement, 3rd Edition Blake and Long
Designing Electronic Systems for EMC Duff
Electromagnetic Measurements in the Near Field, 2nd Edition Bienkowski and Trzaska
Fundamentals of Electromagnetics with MATLAB®, 2nd Edition Lonngren *et al.*
Fundamentals of Wave Phenomena, 2nd Edition Hirose and Lonngren
Integral Equation Methods for Electromagnetics Volakis and Sertel
Introduction to Adaptive Arrays, 2nd Edition Monzingo *et al.*
Microstrip and Printed Antenna Design, 2nd Edition Bancroft
Numerical Methods for Engineering: An introduction using MATLAB® and computational electromagnetics Warnick
Return of the Ether Deutsch
The Finite Difference Time Domain Method for Electromagnetics: With MATLAB® simulations Elsherbeni and Demir
Theory of Edge Diffraction in Electromagnetics Ufimtsev
Scattering of Wedges and Cones with Impedance Boundary Conditions Lyalinov and Zhu
Circuit Modeling for Electromagnetic Compatibility Darney
The Wiener–Hopf Method in Electromagnetics Daniele and Zich
Microwave and RF Design: A systems approach, 2nd Edition Steer
Spectrum and Network Measurements, 2nd Edition Witte
EMI Troubleshooting Cookbook for Product Designers Andre and Wyatt
Transmission Line Transformers Raymond Mack and Jerry Sevick
Electromagnetic Field Standards and Exposure Systems Grudzinski and Trzaska
Practical Communication Theory, 2nd Edition Adamy
Complex Space Source Theory of Spatially Localized Electromagnetic Waves Seshadri
Electromagnetic Compatibility Pocket Guide: Key EMC facts, equations and data Wyatt and Jost
Antenna Analysis and Design Using FEKO Electromagnetic Simulation Software Elsherbeni, Nayeri and Reddy
Scattering of Electromagnetic Waves by Obstacles Kristensson
Adjoint Sensitivity Analysis of High Frequency Structures with MATLAB® Bakr, Elsherbeni and Demir
Developments in Antenna Analysis and Synthesis vol 1 and vol 2 Mittra

Advances in Planar Filters Design

Edited by
Jiasheng Hong

The Institution of Engineering and Technology

Published by SciTech Publishing, an imprint of The Institution of Engineering and Technology, London, United Kingdom

The Institution of Engineering and Technology is registered as a Charity in England & Wales (no. 211014) and Scotland (no. SC038698).

First published 2019

The Institution of Engineering and Technology
Michael Faraday House
Six Hills Way, Stevenage
Herts, SG1 2AY, United Kingdom

www.theiet.org

British Library Cataloguing in Publication Data
A catalogue record for this product is available from the British Library

ISBN 978-1-78561-589-4 (hardback)
ISBN 978-1-78561-590-0 (PDF)

Typeset in India by MPS Limited

Contents

Preface

This edited book aims to present recent advances in planar filters design. It covers a wide range of different design types, technologies, and applications for wireless, microwave, communications and radar systems. The book is based on the latest development in this area and draw on expertise from academic professionals in the field.

The book editor would like to express his sincere thanks to all the contributors for their valuable contributions, without which editing this book would not have been possible. To recognize their contributions, their names are listed as follows (in alphabetical order):

J. Ross Aitken, Institute of Sensors, Signals and Systems, School of Engineering and Physical Sciences, Heriot-Watt University, Edinburgh, UK. Airborne & Space Systems Division, Leonardo MW Ltd, Edinburgh, UK

Ahmed Basti, XLIM UMR7252, CNRS/Université de Limoges, Limoges, France

Stéphane Bila, XLIM UMR7252, CNRS/Université de Limoges, Limoges, France

Vicente E. Boria, Department of Communications, Universitat Politècnica de València, Valencia, Spain

Ludovic Carpentier, CNES, DSO/RF/HNO, Toulouse, France

Laetitia Estagerie, CNES, DSO/RF/HNO, Toulouse, France

Zhang-Cheng Hao, The State Key Laboratory of Millimeter Waves, School of Information Science and Engineering, Southeast University, Nanjing, China

Jiasheng Hong, Institute of Sensors, Signals and Systems, School of Engineering and Physical Sciences, Heriot-Watt University, Edinburgh, UK

Feng Huang, School of Electronic and Optical Engineering, Nanjing University of Science and Technology, Nanjing, China

Roberto Gómez-García, Signal Theory and Communications Department, University of Alcala, Spain

Hervé Leblond, Thales Alenia Space, Toulouse, France

Juseop Lee, College of Informatics, Korea University, South Korea

Yo-Shen Lin, Department of Electrical Engineering, National Central University, Taoyuan, Taiwan

Haiwen Liu, School of Information Engineering, East China Jiaotong University, Nanchang, China. School of Electronics and Information Engineering, Xi'an Jiaotong University, Xi'an, China

Jorge D. Martínez, Department of Electronic Engineering, Universitat Politècnica de València, Valencia, Spain

Matthew A. Morgan, National Radio Astronomy Observatory, Central Development Lab, Charlottesville, USA

Aurélien Périgaud, XLIM UMR7252, CNRS/Université de Limoges, Limoges, France

Dimitra Psychogiou, Department of Electrical, Computer, and Energy Engineering, University of Colorado, Boulder, USA

Miguel Ángel Sańchez-Soriano, Department of Physics, Systems Engineering and Signal's Theory, University of Alicante, Alicante, Spain

Jhih-Ying Shao, Department of Electrical Engineering, National Central University, Taoyuan, Taiwan

Stefano Sirci, Institute of Telecommunications and Multimedia Applications, Universitat Politècnica de València, Valencia, Spain

Baoping Ren, School of Information Engineering, East China Jiaotong University, Nanchang, China

Serge Verdeyme, XLIM UMR7252, CNRS/Université de Limoges, Limoges, France

Jianpeng Wang, School of Electronic and Optical Engineering, Nanjing University of Science and Technology, Nanjing, China

Xuedao Wang, School of Electronic and Optical Engineering, Nanjing University of Science and Technology, Nanjing, China

Zhou Zhou, Institute of Sensors, Signals and Systems, School of Engineering and Physical Sciences, Heriot-Watt University, Edinburgh, UK

The book editor would also wish to acknowledge the Newton Mobility Grant from the Royal Society (UK) and the Zijin Visiting Professorship from Nanjing University of Science and Technology (China), which are helpful for him to edit this book. His sincere gratitude extends to the editorial team at the IET publisher including, in alphabetical order, Paul Deards, Nicki Dennis, Amber Thomas, and Olivia Wilkins for their great support throughout the produce of this new book.

Jiasheng Hong

Chapter 1
Introduction

Jiasheng Hong[1]

Radio frequency (RF) spectrum is a valuable recourse for the transformative technologies such as the connectivity and Internet-of-Things. As a result, communications including emerging 5G wireless and radar systems are facing extremely complex radio environments. Advanced RF technology is an enabler for controlling the spectrum of RF signals and tackling interference issues in RF ecosystems. To this end, planar filters, which, in a broad sense, exhibit a low profile including microstrip filters, substrate-integrated waveguide (SIW) filters and acoustic-wave filters based on surface-mounted technology, find applications where size, weight, and/or cost are of importance. For example, significant filtering requirements exist for planar and drop-in receiver filters at IF (intermediate frequency) bands in the up- or down-converter of wireless communication systems including the satellite payloads. Due to the position of the IF filters in the RF chain, absolute insertion losses can be easily compensated by means of the power amplification stages located before and after the filter. However, sharp rejection and flat pass band insertion loss are required which lead to high-Q resonators. Lossy and predistorted design theory for filters can provide the required rejection and flatness avoiding the use of high-Q materials, which are more expensive, and allowing the required compactness. Filter requirements become even more challenging when operational frequencies increase such as in millimeter-wave and terahertz wireless systems for high data rate communications; or when flexible reconfigurations of RF transceiver's operational frequencies and/or bandwidths are required, or when multiple functions such as multi-bands, balance-to-unbalance (balun) transformation or diplexing need to be integrated.

The book is organized as follows.

In Chapter 2, several types of planar millimeter-wave and terahertz filters and their design methods are described. The millimeter-wave SIW filter using high-order resonating mode is introduced at first, and then a fast and accurate design method, which uses the space mapping method with different types of coarse-models, for the complex SIW components and the W-band filter is described. As followed, terahertz filters using the silicon deep reactive ion-etching process are introduced in detail.

[1]Institute of Sensors, Signals and Systems, School of Engineering and Physical Sciences, Heriot-Watt University, UK

Chapter 3 deals with planar SIW coaxial resonator filters. This type of planar filter is proposed as a direct translation of the classical combline waveguide filter to a substrate integrated technology, which presents important advantages in terms of compactness, and can easily accommodate magnetic and electric couplings in a single-layer structure suitable for batch production using low-cost PCB fabrication procedures. Additionally, the structure can withstand moderate power levels both in continuous and in pulsed signal conditions. Advanced topologies like dual-mode filters and singlet sections can also be implemented based on the same principle.

Chapter 4 presents planar lossy filters for satellite transponder applications. Different design approaches are compared considering the same specifications and the same technology. A classical filter is designed and fabricated first. Afterwards, lossy filters formed on the one hand by a transversal network with non-uniform Q resonators and, on the other hand, by an in-line network with resistive cross couplings are designed and fabricated for comparison. Finally, absorptive lossy filters are designed and fabricated. Such filters integrate the property of attenuating reflected wave that is often a requirement for protecting the receiver subsystem.

In Chapter 5, a type of microstrip extracted-pole lossy filter is presented, which is realized by introducing resistive cross couplings into a microstrip extracted-pole filter to achieve a flat passband. The high selectivity is achieved by introducing two transmission zeros using two extracted poles, which can be adjusted independently. In the first part of the chapter, microwave filter design from a system perspective is briefly introduced. Microwave lossy filter and the extracted-pole filter techniques are then described. After that, the design and the microstrip implementation of extracted-pole lossy filters are discussed in detail.

It is found in many circumstances that the reactive termination presented by conventional filters in their stop bands is harmful to overall system performance, as a consequence of the standing waves that develop between the filter and neighboring components. Even out-of-band, these interactions can have a negative impact on dynamic range, stability, power capacity, and a host of other subtle performance parameters, especially when filters are combined with nonlinear devices such as mixers, multipliers, or compressed-mode amplifiers. In such situations, it is advantageous to develop a filter that blocks stop-band energy by absorption rather than by reflection. To this end, reflectionless filter topology is introduced and discussed in Chapter 6.

This is followed by Chapter 7, in which a simple and very effective way to realize narrowband and wideband absorptive bandstop filters is presented. Analytical design equations are established to allow the direct synthesis of this type of planar filter according to the desired specifications.

Chapter 8 presents a type of miniature bandpass filter deploying acoustic-wave lumped-element resonators. It introduces coupling matrix modeling of two-terminal one-port-type acoustic-wave resonators and a simple coupling matrix-based model for the representation of low- and high-frequency spurious resonances. The coupling matrix modeling of the most fundamental element, i.e., the so-called *acoustic-wave-lumped-element resonator*, is then discussed. Afterwards, RF design methodologies are described with several design examples including quasi-elliptic type bandpass filters with high out-of-band isolation levels; bandpass filters with continuous analog RF tuning and transfer functions that can be designed to exhibit either one or multiple passbands as well as for the realization of flat-in-band and RF diplexers.

The main purpose of Chapter 9 is to introduce the recent advances in frequency tunable and reconfigurable SIW filter designs based on the use of frequency-tunable SIW resonators. It first deals with a frequency-tunable bandpass filter design demonstrated with a K-band frequency-tunable filter of this type. It then describes bandstop filter designs using the frequency-tunable SIW resonators. Bandstop filter topologies with and without inter-resonator couplings are discussed. The concepts and theories for the design of multifunction filters based on using the bandstop filter topologies containing the inter-resonator couplings are introduced.

Chapter 10 presents high-temperature superconducting (HTS) dual-band filters based on multimode resonators for compact size and high performance. It includes several compact HTS dual-band bandpass filters. To achieve steep sideband and flat passband, a specific dual-mode resonator is utilized to design high-order HTS dual-band bandpass filter with controllable bandwidths and multi-transmission zeros.

Chapter 11 deals with balun bandpass filters. It introduces a coupled line based balun topology and a method to design a class of planar-type balun bandpass filters. To extend the application of this simple and effective method, a dual-mode balun diplexer is also presented. In addition, several planar balun bandpass filters based on different transition structures are described.

Chapter 12 addresses the design of millimeter wave SIW diplexer with relaxed fabrication tolerances. The equivalence between SIW and dielectric-filled waveguide is outlined to introduce a method of designing SIW components entirely in dielectric-filled waveguide and translating the final design to SIW. The design of the Ka-band SIW diplexer is demonstrated. To investigate the diplexer's handling of fabrication errors, simulated results of a tolerance analysis are discussed.

In general, each of these chapters is self-sustaining for easy reading. The book, supplementing the 2011 book on microstrip filters [1], will be suitable for R&D engineers, specialists, research students, and academics working on the topic of RF/microwave filters and related system applications, and other specialists in RF/microwave engineering.

Reference

[1] J. Hong, Microstrip Filters for RF/Microwave Applications, Second Edition, John Wiley & Sons, INC, Hoboken, New Jersey, 2011.

Chapter 2

Planar millimeter-wave and terahertz filters

Zhang-Cheng Hao[1]

2.1 Introduction

As the demand for high integrity wireless services with a high data rate is increasing rapidly, millimeter-wave and terahertz wireless systems are drawing more and more attentions from both academic and industrial sectors because they can offer many advantages such as very wide available frequency spectrum and small system size. Millimeter-wave and terahertz filters play as one of key components for rejection of the interference from outside operating frequency band and improving the signal over noise ratio of the millimeter-wave and terahertz system. Many efforts have been made to develop novel techniques and technologies for high performance millimeter-wave and terahertz filters. Although having an expensive cost, an complicated assembling process and a three-dimensional (3-D) topology, the conventional waveguide filters based on the computer numerical control (CNC) milling, micromachining, microelectromechanical systems (MEMS), silicon deep reactive ion-etching (DRIE), thick SU-8 processes well as the 3-D printing process can offer excellent selectivity, low insertion loss, high channel-to-channel isolation with high quality factor [1–17]. The up-to-date research shows that a very small insertion loss can be achieved for a waveguide bandpass filter with a center operating frequency around 650 GHz. Due to its excellent performance, the waveguide filter is popularly adopted in the military arena in which the performance is the major requirement. On the other hand, with the rapidly developing millimeter-wave private consuming wireless system, filters are required with low cost, supporting massive-volume production and easy integration ability. That leads to the planar millimeter-wave filter becoming a hot topic in microwave and millimeter-wave research area. Usually, the planar millimeter-wave and terahertz filters are designed on or inside a low loss substrate. It can be integrated with active devices by using mounting or wire assembling process. Since the wavelength inside the substrate is much smaller than that in the air, high accuracy manufacturing process is required for the planar millimeter-wave and terahertz implementation. The

[1]The State Key Laboratory of Millimeter Waves, School of Information Science and Engineering, Southeast University, China

Integrated-Circuit (IC) process such as the CMOS, GaAs, and GeSi process can provide a few nanometers fabrication accuracy. It has been adopted to develop millimeter-wave and sub-terahertz bandpass filters [18–31]. Although the IC-based filters can be seamlessly integrated with active devices for system application, the millimeter-wave and terahertz filters based on the IC process have a relatively large insertion loss due to the pretty small space between metallic layers of the IC process. To obtain a balance among the insertion loss, the integration capability and the cost, planar filters using conventional substrate and commercial printed circuit board process are adorable for modern consuming wireless system. Among them, the substrate-integrated waveguide (SIW) filters are quite attractive due to their attractive merits including relatively high quality factor, small insertion loss, small size, little electromagnetic leaking, high integration capability, and low cost [32–62]. In this chapter, several types of planar millimeter-wave and terahertz filters and their design methods are described. The millimeter-wave SIW filter using high-order resonating mode is introduced at first, and then a fast and accurate design method, which uses the space mapping (SM) method with different types of coarse-models, for the complex SIW components and the W-band filter is described. As followed, terahertz filters using the silicon deep reactive ion-etching process are introduced in detail.

2.2 Compact planar millimeter-wave filter using multiple-mode SIW cavities

Usually, planar filters are designed by using its fundamental mode. Each cavity has only one resonating mode. Then high selectivity filter requires more cavities and larger size. To reduce filter size and improve its selectivity, multiple-mode resonator is a good choice for the filter design. Theoretically, a resonator has N resonators modes resonating at around the same frequency can be used to design an N-order filter. A few multiple-mode SIW filters have been reported in recent years, including the dual-mode SIW filter and diplexer [34,38,39,41,42,51,54] and the triple mode SIW filters [44,45,48–50,55]. Although multiple modes can be obtained by etching additional resonators such as slots or complementary split ring resonators on the broad surface of a SIW cavity [55], an increased radiation loss due to unwanted radiations is unavoidable, which reduces the quality factor in somehow. In order to reduce the radiation loss, the high-order resonances of a SIW cavity may be a good choice for the design of a low insertion loss filter.

A triple-mode circular SIW cavity is shown in Figure 2.1 [45], where a metallic via that has a diameter of r_{add} is inserted at the center of a circular SIW cavity. It also can be treated as a flat substrate integrated coaxial cavity. The metallic via that is used to form the circular SIW cavity has a diameter of d_{via}. Then, by properly adjusting the diameter of the metallic center via, the resonant frequency of the dominant mode of the SIW circular cavity increases and become close to that of the two degenerated modes, i.e., TM_{110}-like mode. At the same time, the size of the triple-mode cavity remains almost the same as the circular dual-mode SIW resonator.

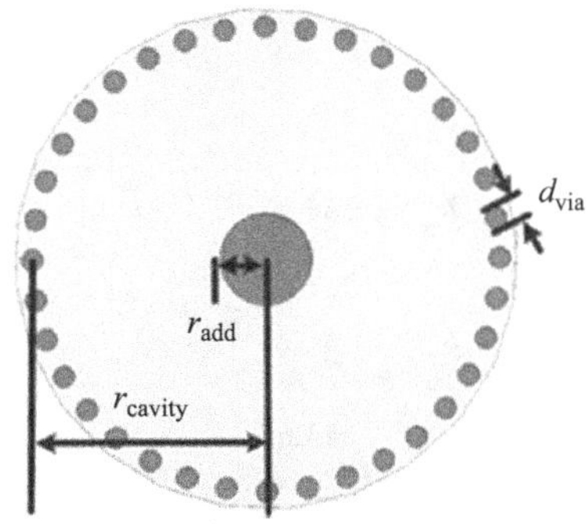

Figure 2.1 Structure of a triple-mode circular SIW resonator. Reprinted with permission from Reference [45]; © 2013 IEEE

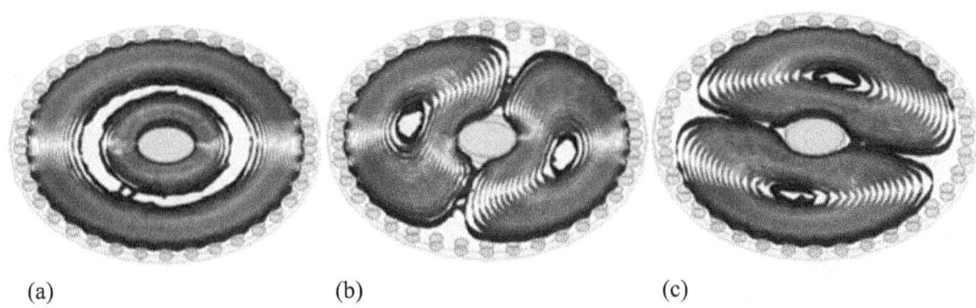

Figure 2.2 Electric field distributions of three resonating modes of the circular SIW cavity shown in Figure 2.1. (a) TM_{020}-like mode, (b) and (c) two degenerated TM_{110}-like modes. Reprinted with permission from Reference [45]; © 2013 IEEE

At the center of the circular SIW cavity, the electric field strength of the fundamental mode reaches its maximum, and that of the TM_{110}-like mode reaches its minimum. Thus, the inserted metallic via affects little on the electromagnetic field distribution of the TM_{110}-like mode. The electric field distributions of the fundamental mode and two TM_{110}-like modes are shown in Figure 2.2(a), (b), and (c), respectively. The electric field of these three modes is normal to the substrate surface, while that of TEM mode in the traditional coaxial resonator is parallel to the substrate surface. To avoid the resonator working with the TEM mode, the following requirement should be satisfied:

$$\pi\left(r_{\text{cavity}} + r_{\text{add}}\right) > 2h_{\text{substrate}} \tag{2.1}$$

where $h_{\text{substrate}}$ is the thickness of substrate, which is usually much smaller than the equivalent width and length of the SIW cavity.

As an example, considering a triple-mode SIW cavity resonator with $r_{\text{cavity}} =$ 3.68 mm and $d_{\text{via}} = 0.3$ mm fabricated on the 0.254 mm thick Rogers 5880 substrate,

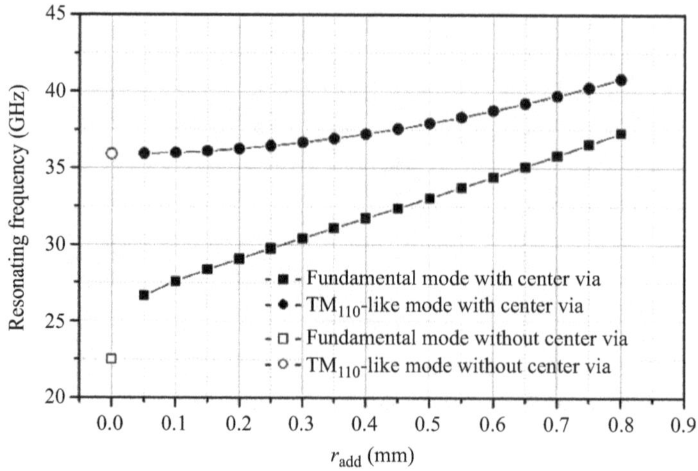

Figure 2.3 Resonant frequencies of the triple-mode SIW cavity resonator with $r_{cavity} = 3.68$ mm and $d_{via} = 0.3$ mm fabricated on the 0.254 mm thick Rogers 5880 substrate. Reprinted with permission from Reference [45]; © 2013 IEEE

its resonant frequencies versus r_{add} are shown in Figure 2.3. These resonant frequencies are obtained by using the high-frequency structure simulator (HFSS) software. As can be clearly observed, the electromagnetic field of the fundamental mode is greatly changed and the corresponding resonant frequency is rapidly increased when a metallic via is inserted at the center of the circular SIW cavity. At the same time, the resonant frequency of the second-order resonant modes remains almost the same. For example, when the diameter of the inserted metallic via increases from 0.0 mm to 0.8 mm, the resonating frequency of the fundamental mode varies from 22.5 to 36.0 GHz, while the resonating frequency of the TM_{110}-like mode varies from 36.0 to 40.0 GHz. Compared to the variation of the fundamental mode resonating frequency, the variation of the TM_{110}-like mode is pretty small. The difference between the resonant frequencies of fundamental mode and TM_{110}-like modes becomes smaller as the diameter of the inserted via increases. Thus, a triple-mode SIW cavity resonator can be created by selecting proper size of the SIW cavity and the inserted via.

The unloaded quality factor Q_u of the circular SIW cavity with or without center inserted metallic via is estimated by the full-wave simulator and listed in the Table 2.1, where the resonating frequency of the three resonating modes of the triple-mode cavity can be adjusted to be closed to that of the original dual-mode cavity with a high unloaded quality factor.

The triple-mode circular SIW cavity can be used to design planar millimeter-wave filters. Figure 2.4(a) shows a three-pole filter using one triple-mode circular SIW cavity, and Figure 2.4(b) shows a six-pole filter using two triple-mode circular SIW cavities. Compared to the conventional single-mode filter, the size of multiple-mode filter is significantly reduced. Those filters are designed to be

Table 2.1 Unloaded quality factor Q_u of the circular SIW cavity with or without center inserted metallic via ($r_{cavity} = 3.68$ mm, $r_{add} = 0.55$ mm, $\sigma = 5.8 \times 10^7$ S/m)

	Dual-mode cavity (without center-inserted metallic via)		**Triple-mode cavity (with center-inserted metallic via)**	
	Resonating frequency (GHz)	**Q_u**	**Resonating frequency (GHz)**	**Q_u**
Fundamental mode	–	–	33.0755	383.0906
TM_{110}-like mode	34.2083	410.9381	37.0539	420.4989

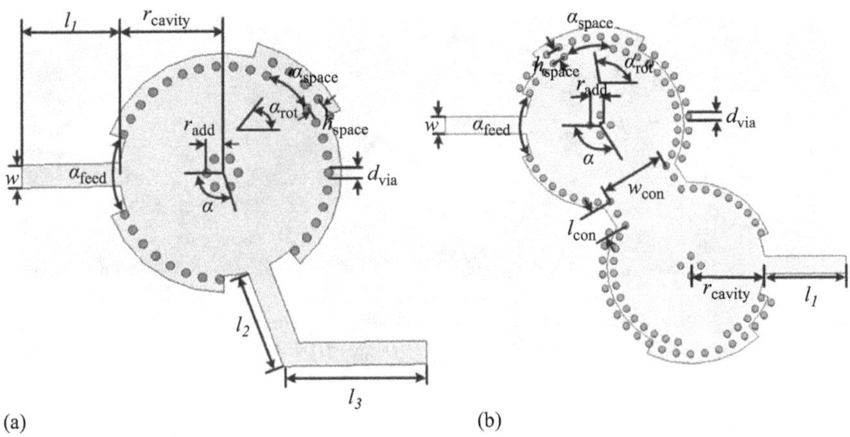

Figure 2.4 Planar millimeter-wave filters using the triple-mode circular SIW cavities. Reprinted with permission from Reference [45]; © 2013 IEEE

operated around 35 GHz. The inserted via is replaced by a circular via-array to avoid unnecessary too larger holes in the implementation. The angle α between input and output ports can be used to adjust the location of transmission zeros. Two three-pole filters (Filters I and II) and one six-pole filter (Filter III) are designed and their dimensions are listed in Table 2.2.

The resonant frequencies of two degenerated TM_{110}-like modes are finely tuned by the perturbation at the wide side of the circular SIW cavity and asymmetric layout of coupling structure. Three transmission zeros are generated. That is due to the cross coupling among input/output ports, the fundamental and TM_{110}-like modes. If the angle between the input and output ports of the SIW cavity is less than 90°, the transmission zeros are found to be located between the resonating frequencies of the dominant mode and TM_{110}-like modes. Thus, the angle should be larger than 90° to guarantee the transmission zeros locate at the upper stopband.

Table 2.2 Geometries of the designed three-pole and six-pole triple-mode filters

	r_{cavity} (mm)	r_{add} (mm)	d_{via} (mm)	A (deg)	α_{feed} (deg)	α_{rot} (deg)	α_{space} (deg)	h_{space} (mm)	W (mm)	l_1 (mm)	l_2 (mm)	l_3 (mm)	w_{con} (mm)	l_{con} (mm)
Filter I	3.68	0.55	0.3	110	22	51.5	28	0.5	0.8	3.32	3.32	5	–	–
Filter II	3.6	0.575	0.3	125	22.5	62	27.1	0.6	0.8	3.4	3.9	5	–	–
Filter III	3.5	0.46	0.3	122.5	25	85	37.4	0.5	0.8	7.0	–	–	2.75	1.4

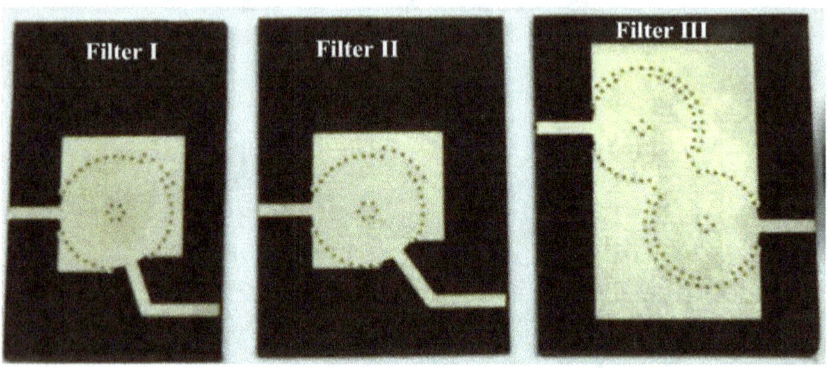

Figure 2.5 Photographs of fabricated triple-mode filters. Reprinted with permission from Reference [45]; © 2013 IEEE

As the angle increases, the cross-coupling between input/output ports becomes weaker. That leads to one transmission zero shifts up while the other two transmission zeros remain almost the same. Then an optimal angle between the input and output ports is used to achieve a wide upper stopband by adjusting the location of one transmission zero.

The commercial printed-circuit-board (PCB) process is used to fabricate those designed filters, and their photos are shown in Figure 2.5. An Antitsu test fixture 3680V and Agilent E8363B vector network analyzer (VNA) are used to measure their response. The simulated and measured *S*-parameters of the filters with one or two triple-mode SIW cavity resonators are shown in Figures 2.6 to 2.8, respectively. Generally, measured results agree well with simulated results. The insertion loss of the filters with one triple-mode SIW cavity resonator is around 1.55 dB, which is 0.3 dB larger than the simulated result. It may be caused by the slight impedance mismatch between the feeding section and the measurement fixture, and the increased conductor loss at millimeter-wave frequency. It can be seen from Figures 2.6 and 2.7, when the angle between the coupling ports of single triple-mode cavity filter is adjusted to 125°, a wider upper stopband can be achieved while the bandwidth of those two filters keeps unchanged, i.e., around 15%. Two triple-mode SIW cavities

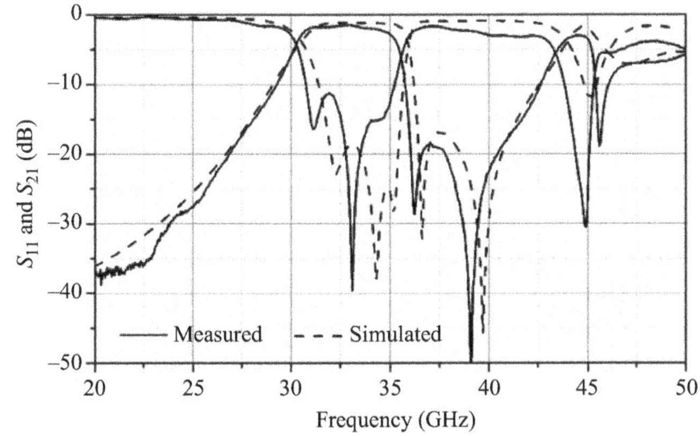

Figure 2.6 Measured and simulated results for one triple-mode cavity filter with $\alpha = 110°$. *Reprinted with permission from Reference [45]; © 2013 IEEE*

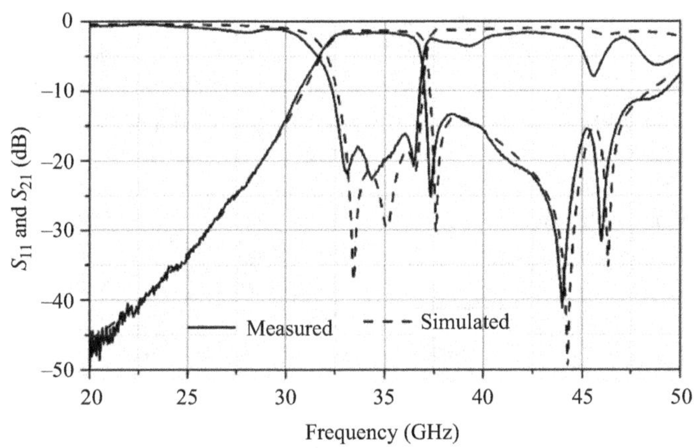

Figure 2.7 Measured and simulated results for one triple-mode cavity filter with $\alpha = 125°$. *Reprinted with permission from Reference [45]; © 2013 IEEE*

with $\alpha = 122.5°$ are cascaded to design a sixth-order filter, whose measured and simulated results are presented in Figure 2.8. Six transmission zeros can be found at the upper stopband band, and the lower and upper stopband become shaper. The measured insertion loss in the passband is better than 1.8 dB, 0.5 dB larger than the simulated result. That shows a compact high performance planar millimeter-wave filter can be designed by using multiple-mode cavities.

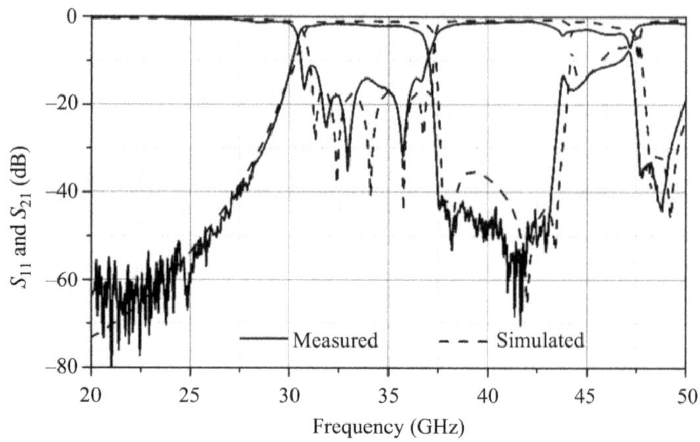

Figure 2.8 Measured and simulated results for the six-pole triple-mode filter with $\alpha = 122.5°$. Reprinted with permission from Reference [45]; © 2013 IEEE

2.3 Design SIW components using the space-mapping method

The merits of low cost, high Q-factor, compact size, and ease of manufacture make the SIW technology attractive for the design of planar filters and other high performance components, especially in millimeter-wave frequency band. However, because the SIW components use a large amount of metallic vias, which require high-density meshing associated with a large number of unknowns, very long computation time, up to a few days with a high performance workstation, is required for full-wave simulations of a complicated SIW structure. Thus, high efficient and high accurate design method is desirable for developing SIW components. Over past years, a number of design methods have been proposed to design microwave and millimeter-wave components [63–81], including the equivalent network [63,64], the coupling matrix method [69,70], the semi-analytical model [65], the hybrid optimization method [66,67], and the fictitious reactive load concept [68], and the SM method [71–81]. Different to traditional design methods that directly utilize the full-wave simulation and parameter derivatives to force the responses to satisfy design specifications, the SM method or aggressive space mapping (ASM), which combines the speed of the circuit simulation and the accuracy of the full-wave simulation, has been verified as a universal method for the design of various microwave and millimeter-wave components [71–84]. Circuit simulation and CAD tools uses empirical models and analytical solutions, the optimal can be converged with less accuracy within a few seconds. On the contrary, the full-wave simulation requires solving density matrix based on the Maxwell equation. It usually needs a long computing time up to a few hours. Especially for large scale problems, the computing time may be as long as a few days. The SM method utilizes the circuit simulation to obtain an initial design and the full-wave simulation as verifications. Then a surrogate algorithm is constructed to exchange or update design parameters between the circuit-mode and the full-wave simulation model. Most of the optimization processes are carried on with the circuit-simulation. In the implementation of the SM method, the circuit-model is usually named as the "coarse-model" and the full-wave simulation model is named as the "fine-model". Since the full-wave simulation model is generally based on a conventional Maxwell-equitation based algorithm, the surrogate algorithm, used to exchange parameters between above two types of models, and the circuit-model, used to provide the initial design and map the full-wave simulation, are critical for the SM method. A high-accuracy coarse-model can improve efficiently the convergence and accuracy of the SM method. Due to its high efficiency and high accuracy, the SM method has been adopted for designing SIW components with hybrid optimization algorithms [82–84]. In this section, two types of coarse-models are introduced in detail for designing SIW components. Specially, examples of designing of a complicated X-band SIW multiplexer and W-band SIW filters are introduced with two different design modes, respectively.

2.3.1 Design compact SIW multiplexers using the space-mapping method with equivalent rectangular-waveguide coarse-model

Because the operating principle of a conventional SIW is similar to that of a dielectric rectangular waveguide, the circuit-model of a rectangular waveguide can be used as the coarse-model of a SIW in the space-mapping method. As an example, the design of a complex SIW component using the equivalent rectangular-waveguide coarse-model is presented in the following [82].

Figure 2.9 illustrates the structures of a compact four-contiguous-channel SIW Multiplexer (MUX), which comprises the SIW channel bandpass filters, the grounded-coplanar-waveguide to SIW (GCPW-SIW) transitions and a compact SIW power-combiner. Different to conventional L-shape or T-shape SIW multiplexer in which the neighboring channels have a space around $\lambda_g/2$, where the λ_g is the guided-wavelength of the SIW, the topology in Figure 2.9 in much more compact though its structure is more complex. If a multiple-way broadband SIW power-combiner is adopted in Figure 2.9, an SIW multiplexer with a large number of output channels can be extended, as shown in the configuration of Figure 2.10. The H-plane SIW bandpass filter is chosen for the SIW multiplexer because it has no critical geometries to be optimized and its in-line structure is suitable for the proposed configuration. The space among SIW channel filters and the output ports of the SIW power-combiner is $Li0$ ($i = 1, \ldots, 4$). As depicted in Figure 2.9, the GCPW has been used for the input/output port, which has a width of W50 and a gap of G50. The GCPW-SIW transition has a width of W_{ti} ($i = 0, \ldots, 4$), a length of L_{ti} ($i = 0, \ldots, 4$), and a gap of Gti ($i = 0, \ldots, 4$) $= (W_{\text{siw}} - W_{\text{ti}}$ ($i = 0, \ldots, 4$))/2. The cavity lengths of channel filters are $L_{i,j}$ ($i = 1, \ldots, 4, j = 1, \ldots, 5$), and the widths of the coupling windows between cavities are $W_{i,j}$ ($i = 1..4$, $j = 1, \ldots, 6$), where the subscript i denotes the ith channel filter and j represents the jth coupling window of the ith channel filter. The SIW multiplexer is designed with a 1.57-mm-thick

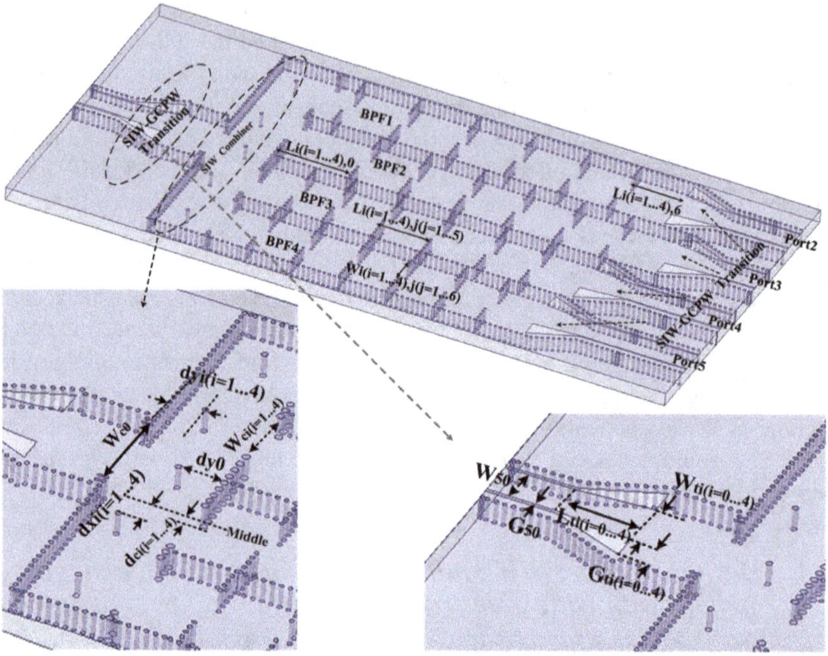

Figure 2.9 Structures of a compact four-contiguous-channel SIW Multiplexer. Reprinted with permission from Reference [82]; © 2015 IEEE

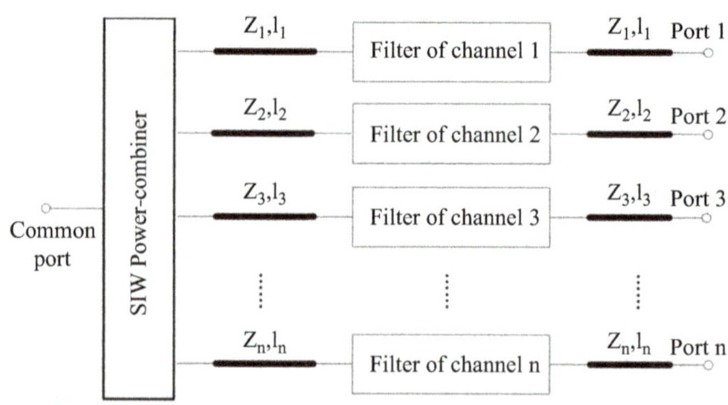

Figure 2.10 Generalized configuration of a compact multiple-channel SIW Multiplexer. Reprinted with permission from Reference [82]; © 2015 IEEE

Taconic TLY-5A substrate having a relative dielectric constant of 2.2 and a loss tangent of 0.0009 at 10 GHz. The SIW power-combiner consists of a number of metallic matching vias having a 0.4-mm diameter and coupling windows with different offsets and widths. All other metallic vias have a radius of 0.15 mm, and the pitch between metallic vias is 0.6 mm. A coupling window having a width of W_{c_0} is used to help obtain a good matching. As depicted in Figure 2.9, the space between the sidewall of the SIW and the metallic matching vias is d_{yi} $(i = 1, \ldots, 4)$, and the offsets between the centers of the metallic matching vias and the corresponding SIW are *dxi* $(i = 1, \ldots, 4)$. The coupling windows have widths of W_{c_i} $(i = 1, \ldots, 4)$, and the offsets between the coupling windows and the corresponding SIW are d_{ci} $(i = 1, \ldots, 4)$. The specifications of the demonstrated Ku-band contiguous-channel SIW multiplexer include channeled passbands of 18.0–19.0 GHz, 19.0–20.0 GHz, 20.0–21.0 GHz, and 21.0–22.0 GHz and a return loss of the input port that is smaller than −20 dB from 180.0 to 22.0 GHz.

Since there is almost no frequency space among adjacent channels, the filtering performance of each channel is pretty sensitive to other geometries of the SIW multiplexer. As shown in Figure 2.9, there are over 80 variables involved in the design. Obviously, it is impossible to accomplish the design using the direct full-wave optimization method. To overcome this challenge, the complicated SIW multiplexer is divided a few separated parts at first, and each of them can be designed independently within a short computation time, and then fulfilled the final design using the ASM method. The design process can be summarized as follows:

Step (1): Designing a wideband SIW power-combiner using the full-wave simulator HFSS. In this step, the multiple-way SIW power-combiner is decomposed into a few power-dividing blocks that can be designed with the desired couplings. Once all of the blocks are accomplished, they are directly integrated together to implement the wideband SIW power-combiner. Then, this power-combiner will remain unchanged in the aftermentioned design and is represented by its *S*-parameters in the coarse model of the SIW multiplexer.

Step (2): Designing the SIW channel filters with the SM. In this step, the equivalent circuits of the metallic via-iris in the SIW are extracted with full-wave simulations at first; then, curve-fitting code is developed in MATLAB® to establish the relationship between the geometries of the SIW iris and its equivalent circuits. This relationship is used to establish the coarse model of each channel filter whose fine mode is set up in HFSS. Using the SM, all of the channel filters can be designed with the required specifications with a few full-wave simulations (less than 3 in the study).

Step (3): Designing the SIW multiplexer. In this step, the aforementioned SIW wideband power-combiner and the channel filters are integrated together using SIW transmission lines, as shown in Figure 2.9. GCPW-SIW transitions are integrated at the input/output ports of the SIW multiplexer as well. Then, the coarse and fine models for the SIW MUX are established and optimized to meet the required specifications with the ASM.

The SIW power-combiner has a symmetrical structure, and its design is based on the simple full-wave tuning. To reduce the full-wave simulation cost, the SIW power-combiner is divided into four types of key blocks, as shown in Figure 2.11. For a SIW multiplexer having an odd number of output channels, blocks II, III, and IV are used in the design. Block III is used for the center channel, block IV is used

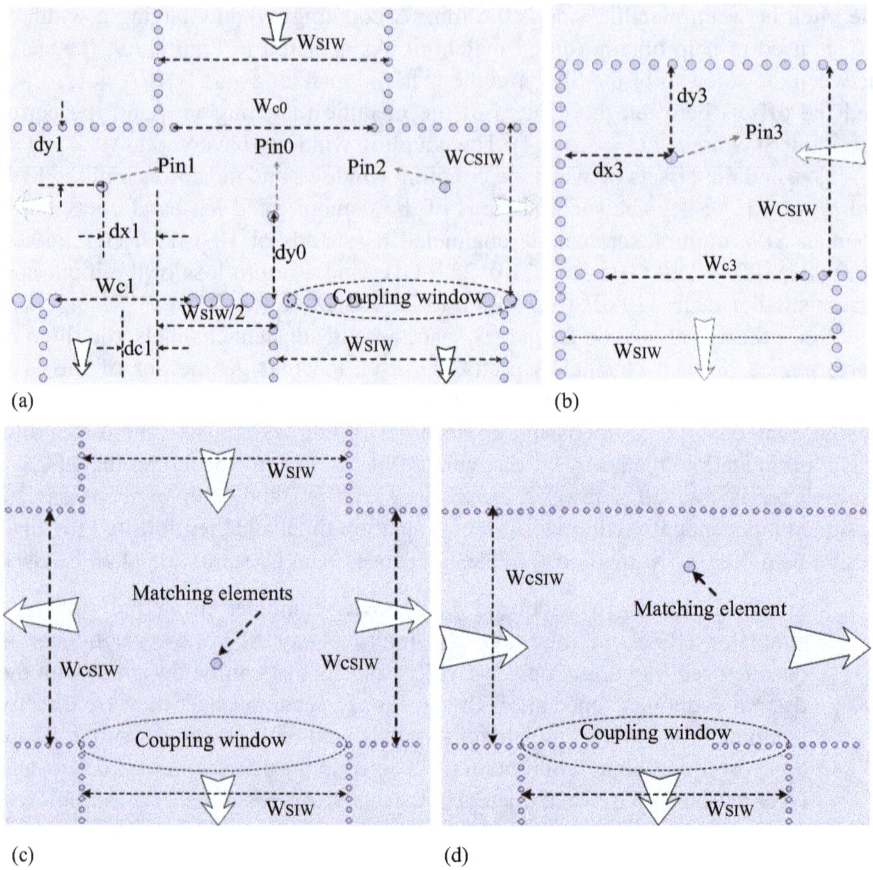

Figure 2.11 Power-dividing components for the compact power-combiner: (a) type I; (b) type II; (c) type III; and (d) type IV. Reprinted with permission from Reference [82]; © 2015 IEEE

for the middle channels, and block II is used for the first and last channels. Alternately, for an SIW multiplexer having an even number of output channels, blocks I, II, and IV are adopted. A typical implementation of the compact SIW power-combiner is shown in Figure 2.9 (the case of an even number of output ports), where only blocks I and II are used.

The SIW power-combiner is designed to have a phase shifting of approximately 180° between neighboring output ports. The width of the input/output SIW, i.e., W_{SIW} in Figure 2.11, is chosen to support the operating frequency, and the width of the connecting SIW, i.e., W_{cSIW}, is chosen to support the phase shifting between neighboring output ports. Once the W_{SIW} and W_{cSIW} are determined, each power-dividing block is individually designed with the desired coupling for the output channels through a short-time full-wave tuning at first, and they are then integrated together to implement the compact power-combiner for the SIW multiplexer. During this process, the size of the coupling window and the position of metallic matching vias denoted as Pin1, 2 and 3 in Figure 2.11(a) and (b) are adjusted to obtain a good return loss with the desired coupling, i.e., −6 dB for a four-way power-combiner, with the help of HFSS. After all of the SIW power-dividing blocks are designed, they are integrated together to implement the proposed compact SIW power-combiner.

The SM method is adopted to design SIW channel filters of the SIW multiplexer. To this end, coarse and fine models are established for the SIW filter. An equivalent rectangular waveguide (RW) model coarse model, which reflects the intrinsic physical meanings of the SIW filter, is adopted for reducing the iterations and improving the accuracy of ASM, whose width can be obtained from (2.2) [85],

$$\bar{a} = \frac{a_{\text{rec}}}{W_{\text{SIW}}} = \xi_1 + \frac{\xi_2}{\dfrac{S_{vp}}{D} + \dfrac{\xi_1 + \xi_2 - \xi_3}{\xi_3 - \xi_1}} \tag{2.2}$$

where $\bar{a}$ is a normalized coefficient, a_{rec} is the width of the corresponding RW, and W_{SIW} is the width of SIW. S_{vp} is the pitch between the vias and D is the diameter of the via. ξ_1, ξ_2, and ξ_3 are defined as

$$\begin{aligned}
\xi_1 &= 1.0198 + \frac{0.3465}{\dfrac{W_{\text{SIW}}}{S_{\text{vp}}} - 1.0684} \\
\xi_2 &= -0.1183 - \frac{1.2729}{\dfrac{W_{\text{SIW}}}{S_{\text{vp}}} - 1.2010} \\
\xi_3 &= 1.0082 - \frac{0.9163}{\dfrac{W_{\text{SIW}}}{S_{\text{vp}}} + 0.2152}
\end{aligned} \tag{2.3}$$

The via iris is another important structure that needs to be modeled in the coarse model. To improve the accuracy of the coarse model, an equivalent circuit shown in Figure 2.12, which is composed by two series capacitors and one shunt

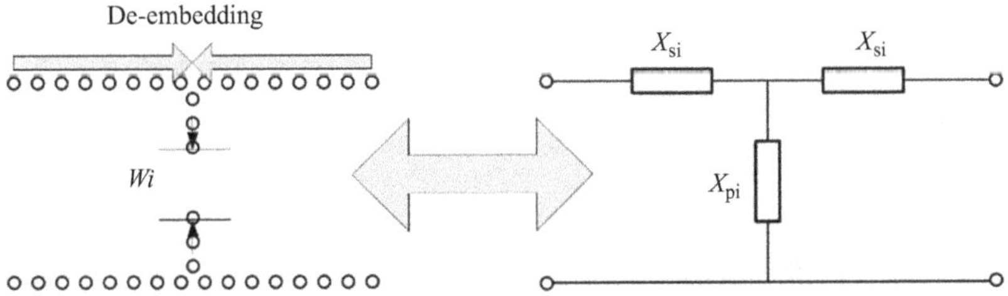

Figure 2.12 Equivalent circuit of the metallic-via iris in SIW. Reprinted with permission from Reference [82]; © 2015 IEEE

inductor, is utilized to model the metallic-via iris in the SIW, whose element values can be extracted by using (2.4) [86],

$$
\begin{aligned}
j\frac{X_{si}}{Z_0} &= \frac{1 - S_{12} + S_{11}}{1 - S_{11} + S_{12}} \\
j\frac{X_{pi}}{Z_0} &= \frac{2S_{12}}{(1 - S_{11})^2 - S_{12}^2}
\end{aligned}
\tag{2.4}
$$

where S_{11} and S_{12} are the full-wave simulated *S*-parameters of the via-iris shown on the left side of Figure 2.12. To remove the effects from the addition SIW transmission lines, a de-embedding process is carried out when exporting the *S*-parameters from the HFSS [87]. By changing the width of the via-iris, i.e., the W_i, a number of equivalent circuits can be extracted with full-wave simulations. To express the values of series capacitors and shunt inductor in terms of the width of the via-iris, a curve-fitting program is then developed in MATLABR$_{2010b}$ [88] to express X_{si} and X_{pi} with two functions of *W*, respectively.

The coarse model shown in Figure 2.13(a) is established in the advanced design system (ADS) [89] circuit simulator, where the values of the inductors and capacitors are expressed by functions of *W*, and the width of the RW is obtained after evaluation of (2.2).

The fine model for the SIW channel filter is shown in Figure 2.13(b), which is established in a full-wave simulator HFSS. It should be pointed out that although the equivalent RW in (2.2) can be used to model the transmission characteristics of the SIW, it cannot be used to model the SIW discontinuities with high accuracy. Hence, the SIW structure composed by plenty of metallic vias, as shown in Figure 2.13(b), is adopted as the fine model for the SIW channel filter. To illustrate the design process in detail, a Ku-band double-terminated SIW channel filter is designed as an example.

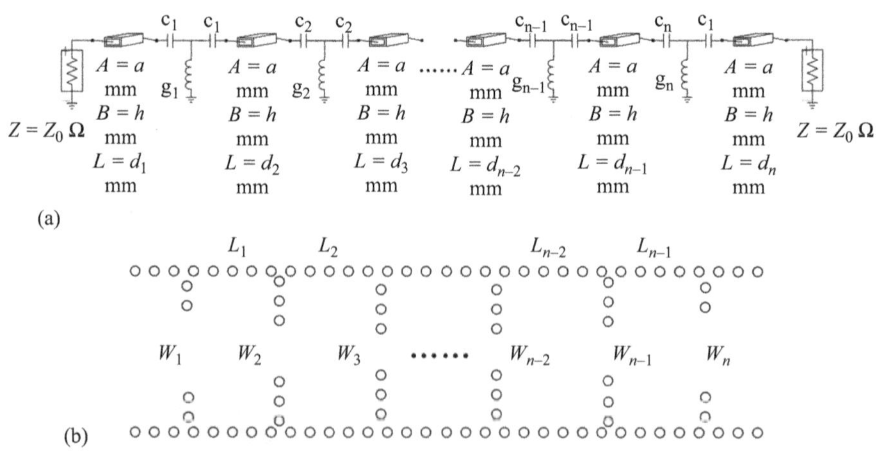

Figure 2.13 Coarse and fine models of SIW channel filter: (a) coarse model and (b) fine model. Reprinted with permission from Reference [82]; © 2015 IEEE

It uses five SIW cavities, and has a 20-dB return-loss passband from 20 to 21 GHz. This filter is designed with a 1.57-mm-thick Taconic TLY-5A substrate, which has a relative dielectric constant of 2.2, the metallic vias for the sidewall and the coupling windows have a diameter of 0.3 mm, and the width of the SIW is 8.79 mm. The pitch between adjacent vias is 0.6 mm. The SIW channel filter has 11 geometries needed to be optimized. However, because of the symmetry of the filter topology, only six geometrical variables are used in the fine model for the optimization, which are the width of via-irises, i.e., $W_i(i = 1, ..., 3)$, and the length of the SIW cavities, i.e., $L_i(i = 1, ..., 3)$. Correspondingly, three sets of inductors and capacitors represented by $W_{ci}(i = 1, ..., 3)$ and $L_{ci}(i = 1, ..., 3)$ three lengths of the equivalent RW represented by were employed in the coarse mode for optimization.

The relationship in (2.5) among Wi and the elements of the equivalent circuit in Figure 2.12 is established by using a curve-fitting process with full-wave simulated results. Then the well-known low-pass prototype, which uses K-invertors for traditional RW filters with a closed-form formula [90], is adopted for obtaining the initial values geometries of the SIW filter.

$$\begin{aligned} &Xs = p_1 \times w^3 + p_2 \times w^2 + p_3 \times w + p_4 \\ &p_1 = -0.00007631, p_2 = -0.00004758, p_3 = 0.01216, p_4 = -0.0201 \\ &Xp = a_1 \times e^{-\left(\frac{w - b_1}{c1}\right)^2} + a_2 \times e^{-\left(\frac{w - b_2}{c2}\right)^2} \\ &a_1 = 13.97, a_2 = 0.2197, b_1 = 10.47, b_2 = 4.591, c_1 = 3.197, c_2 = 1.983 \end{aligned} \tag{2.6}$$

Figure 2.14 shows the responses for the initially synthesized SIW filter. Apparently, the return loss and the operation passband need to be optimized for achieving the desired specifications. This is accomplished by using the SM optimization algorithm. All the simulations are carried out with an 8-processor Dell workstation Precision T5600, which has a memory of 64 GB, and two 2.4-GHz Intel Xeon E52609 CPUs. The coarse and fine models have the same number of

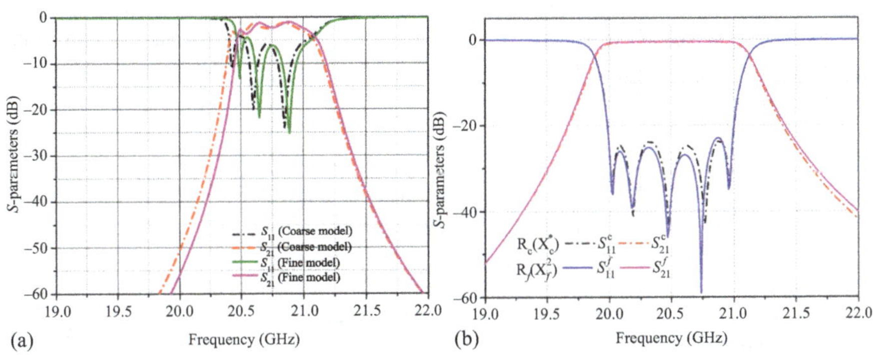

Figure 2.14 Response of the SIW channel filter: (a) initial response and (b) responses of the Rc(Xc) and Rf(Xf) of the second iteration. Reprinted with permission from Reference [82]; © 2015 IEEE*

optimization variables. The SM is then used to optimize the SIW filter to achieve the desired specifications, and it is summarized as follows:

Step (1) Optimize the coarse-model to obtain Xc^* and the optimal response $Rc(Xc^*)$.

Step (2) Simulate the fine model with Xf and obtain the fine-mode response $Rf(Xf)$. For the first full-wave simulation, Xf is set as Xc^*.

Step (3) Obtain the corresponding optimization vector $Xc^{(c)}$ through a parameter-extraction procedure, in which the Gradient optimization method in ADS is used to extract $Xc^{(c)}$ and the following error function is used for the objective:

$$\underset{Xc^{(c)}}{\text{Min.}} \, \|Rf(Xf) - Rc(Xc^{(c)})\| \tag{2.6}$$

where $\|\bullet\|$ is the 2-norm.

Step (4) Update Xf with a parameter-mapping procedure, in which $Xf^{(i+1)} = Xf^{(i)} + h^{(i)}$ and $h^{(i)}$ is computed using a quasi-Newton Step as follows:

$$B^{(i)}h^{(i)} = -f^{(i)} \tag{2.7}$$

where $f^{(i)} = Xf^{(i)} - Xc^{(i)}$ and $B^{(i)}$ is the Broyden matrix. For the first iteration, $B^{(1)}$ is an identity matrix with a dimension of $m \times m$ (m is the dimension of the vector Xc). In the ith iteration, $B(i)$ is updated using Broyden's update process [79].

The objective of the ASM is to minimize the function of $\|Xf^{(i)} - Xc^{(*)}\|$or $\|Rf(Xf^{(i)}) - Rc(Xc^{(*)})\|$. If this condition is not reached, then go back to step (2) for the next iteration.

The design specifications are achieved through two iterations. Figure 2.14(b) shows the responses of the coarse and fine models, i.e., $Rc(Xc^*)$ and $Rf(Xf)$, where a good agreement is obtained. The variables and geometries of the coarse and fine models are listed in Table 2.3. Because a high-accuracy coarse model extracted from full-wave simulated results is adopted, small variations can be found between the initial (Xc^*) and final geometries ($Xf^{(2)}$).

Once the multiple-way SIW power-combiner and all SIW channel filters are designed with the desired specifications, the SM method can be used to design the SIW multiplexer, which integrates SIW power-combiner, channel filters, and SIW-GCPW transitions together. Figure 2.15 shows the coarse model established in the circuit simulator ADS for the SIW multiplexer, where the SNP component represents

Table 2.3 Variables and geometries of the coarse and fine models of the SIW filter (unit: mm)

	L_1	L_2	L_3	W_1	W_2	W_3
Synthesized	5.0583	5.6081	5.6733	3.502	2.006	1.7
$Xc^* = Xf^{(0)}$	5.0549	5.6017	5.6593	3.9313	2.4989	2.2392
$Xf^{(2)}$	5.0406	5.602	5.665	3.9774	2.5409	2.27

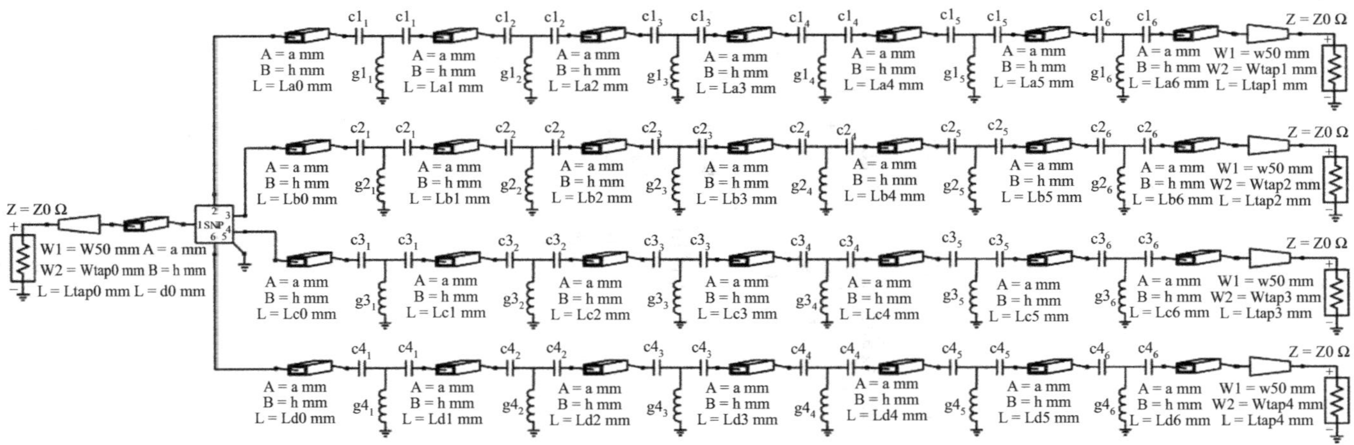

Figure 2.15 A coarse model for the four-channel SIW multiplexer. Reprinted with permission from Reference [82]; © 2015 IEEE

the n-port ($n = 2,3,\ldots$) S-parameter of the SIW power-combiner, which is not involved in the optimization. Because the field distributions of the GCPW are similar to those of the microstrip line, microstrip taper lines are used in the coarse model to model the transitions from the SIW to the GCPW. The microstrip taper line has a length of L_{tap_i} ($i = 0,\ldots,4$). One of its ends is connected to a 50-Ω microstrip line. Hence, its width, *W50*, is determined by the 50-Ω microstrip lines. A 50-Ω load and equivalent rectangular waveguides with width of W_{tap_i} ($i = 0,\ldots,4$) are connected with input/output ports, respectively. The fine model is established by using the full-wave simulator HFSS, as shown in Figure 2.9. To reduce the computation time, perfect conductors are used for the metallic material in the fine model.

Because the fine model of the SIW multiplexer uses a few hundred metallic vias, each full-wave simulation requires approximately one day of computation time with the computer mentioned above. On the contrary, the coarse model requires less than a few seconds for each simulation. 63 variables are optimized in the SM process for the coarse and fine models. All of the channel filters use the geometries of the predesigned BPFs, which use the design procedure mentioned above as the initial geometries, and an initial space of L_{ij} ($i = 1,\ldots,4$, $j = 0,6$) is chosen as one-half the guided wavelength at the center frequencies of each channel, the initial lengths of the GCPW-SIW transition, L_{ti} ($i = 0,\ldots,4$), are chosen as a quarter-wavelength of the center frequencies, and its width, W_{ti} ($i = 0,\ldots,4$), is determined by the impedance of the equivalent rectangular waveguide at the input/output ports. The initial responses of the SIW multiplexer are shown in Figure 2.16. Although a bad return loss and bad filtering performance are observed in Figure 2.16, multiplexing performance can still be watched. Similar to the design process of the SIW BPF, the SM is used in the design to achieve the desired specifications. To this end, the coarse-model is first optimized to meet the design specifications, whose variables are denoted as Xc^*. Then, the fine-model is simulated by HFSS with the geometries of Xc^*. Figure 2.17 shows the responses of the

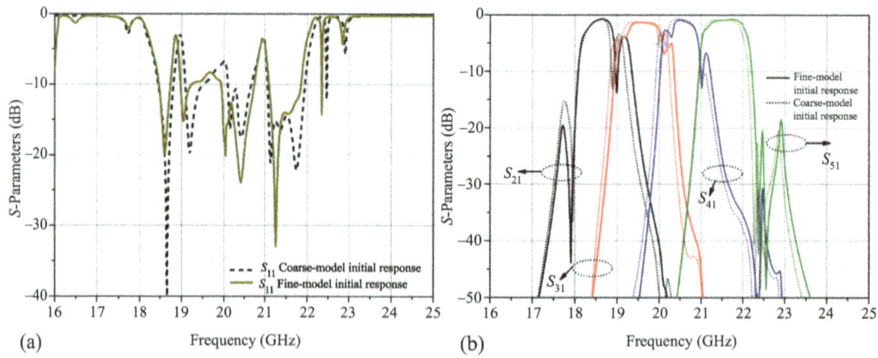

Figure 2.16 Initial responses of the Ku-band SIW multiplexer by direct integrating all the predesigned blocks together. (a) return-loss and (b) insertion loss. Reprinted with permission from Reference [82]; © 2015 IEEE

coarse and fine models of the SIW multiplexer, i.e., Rc(Xc^*) and Rf($Xf(0) = Xc^*$). Because an accurate coarse model is established, Rf($Xf(0) = Xc^*$) has good multiplexing performance. However, its return loss must be improved to satisfy the design specifications.

Following the procedures mentioned above, the SIW multiplexer is designed to achieve the desired specifications with SM which uses two iterations and three full-wave simulations. The final fine-model results of Rf($Xf^{(2)}$) and the optimal coarse-model responses of Rc(Xc^*) are shown in Figure 2.18, where a good return loss, good filtering performance and high channel selectivity can be observed. In addition, it can be found through the ASM optimization that the final fine-model responses Rf($Xf^{(2)}$) are well matched with the optimal coarse-model responses Rc(Xc^*) except for a small deviation in a very narrow frequency band at approximately 18.25 GHz. The variables and geometries, i.e., Xc^* and the $Xf^{(2)}$, of the

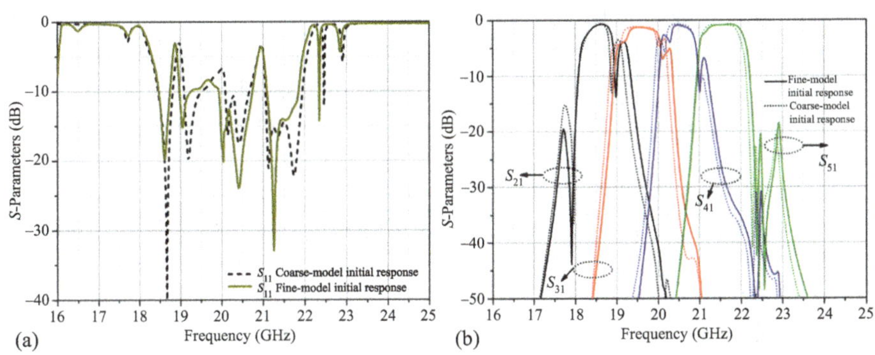

Figure 2.17 The responses of Rc(Xc^*) and Rf($Xf^{(0)} = Xc^*$) of the Ku-band SIW multiplexer: (a) return loss and (b) insertion loss. Reprinted with permission from Reference [82]; © 2015 IEEE

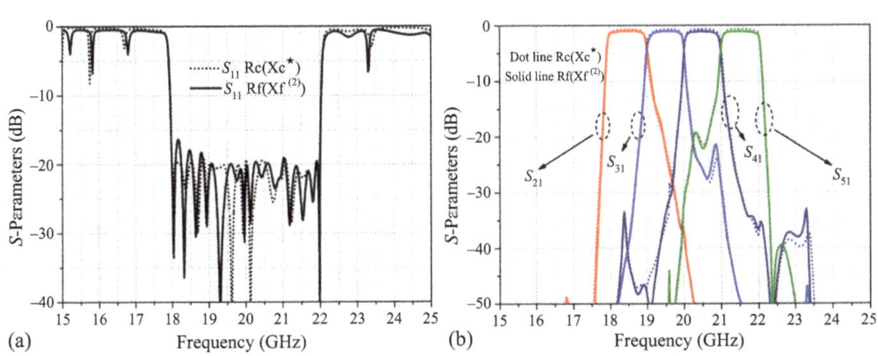

Figure 2.18 The responses of Rc(Xc^*) and Rf($Xf^{(2)}$) of the Ku-band SIW multiplexer: (a) return loss and (b) insertion loss. Reprinted with permission from Reference [82]; © 2015 IEEE

Table 2.4 Variables and geometries of the coarse- and fine-models of the SIW multiplexer (unit: mm)

	L10	*L11*	*L12*	*L13*	*L14*	*L15*	*L16*	*W11*	*W12*	*W13*	*W14*	*W15*
Initial	2.442	6.222	6.456	6.555	6.456	5.741	1.821	4.98	2.974	2.64	2.64	2.974
$Xc^* = Xf^{(0)}$	2.749	6.241	6.594	6.417	6.136	5.546	1.663	3.804	2.572	2.604	2.906	3.246
$Xf^{(2)}$	2.6402	6.258	6.582	6.412	6.136	5.515	1.58	3.806	2.63	2.658	2.93	3.292
	W16	*L20*	*L21*	*L22*	*L23*	*L24*	*L25*	*L26*	*W21*	*W22*	*W23*	*W24*
Initial	4.514	7.427	5.508	5.983	6.061	5.983	5.347	1.812	3.964	2.732	2.43	2.43
$Xc^* = Xf^{(0)}$	4.61	8.349	5.597	6.015	5.639	5.551	5.279	1.719	3.94	2.366	2.582	3.022
$Xf^{(2)}$	4.584	8.386	5.607	6.011	5.65	5.526	5.295	1.804	3.958	2.378	2.614	3.038
	W25	*W26*	*L30*	*L31*	*L32*	*L33*	*L34*	*L35*	*L36*	*W31*	*W32*	*W33*
Initial	2.732	4.238	4.98	4.842	5.58	5.645	5.58	5.01	5.833	3.374	2.55	2.27
$Xc^* = Xf^{(0)}$	2.948	4.274	4.706	4.448	5.477	5.551	5.927	5.079	6.056	3.778	2.83	2.246
$Xf^{(2)}$	3	4.256	4.794	4.367	5.471	5.55	5.965	5.057	6.087	3.804	2.86	2.252
	W34	*W35*	*W36*	*L40*	*L41*	*L42*	*L43*	*L44*	*L45*	*L46*	*W41*	*W42*
Initial	2.27	2.55	4.002	3.141	4.697	5.221	5.282	5.221	4.682	2.911	3.344	2.45
$Xc^* = Xf^{(0)}$	3.236	3.1	3.736	4.521	4.994	5.279	5.292	5.283	4.711	2.797	3.98	2.31
$Xf^{(2)}$	3.32	3.12	3.77	4.559	4.989	5.289	5.297	5.297	4.699	3.155	4.038	2.33
	W43	*W44*	*W45*	*W46*	*L00*	*Wt0*	*Lt0*	*Wt1*	*Lt1*	*Wt2*	*Lt2*	*Wt3*
Initial	2.168	2.168	2.450	3.868	3.321	2.723	3.852	2.745	2.611	2.822	2.723	2.833
$Xc^* = Xf^{(0)}$	2.164	2.574	2.748	3.886	3.186	2.307	4.0272	2.266	3.261	2.165	3.307	2.433
$Xf^{(2)}$	2.172	2.58	2.762	3.884	3.658	3.012	3.599	3.123	2.171	3.08	3.5	3.08
	Lt3	*Wt4*	*Lt4*									
Initial	3.456	2.785	3.733									
$Xc^* = Xf^{(0)}$	4.151	2.458	3.995									
$Xf^{(2)}$	3	3.5	3.9									

coarse and fine models are listed in Table 2.4. Apparently, although the variation of the geometries, i.e., $\left|Xf^{(2)} - Xf^{(0)}\right|$, is small, one or two frontal cavities and the connecting SIWs having a length of Li_0 $(i = 0, \ldots, 4)$ have a significant effect on the multiplexer performance.

Experiments are used to verify the designed SIW multiplexer, and the photograph of the fabricated SIW is shown in Figure 2.19. Measured results which verified full-wave simulated results are shown in Figure 2.20, where a relative dielectric constant of 2.17 provided by the substrate manufacturer is adopted for the simulation.

2.3.2 Design W-band bandpass filters using the space-mapping method with coupling-matrix coarse-model

Section 2.3.1 presents a fast and efficient design method for the designing of complicated SIW multiplexer, where an equivalent rectangular waveguide is adopted to represent the conventional SIW. However, as the developing of the substrate integrated technique, cavities having irregular shapes such as circular, triangular, or

Figure 2.19 Photograph of the fabricated SIW multiplexer. Reprinted with permission from Reference [82]; © 2015 IEEE

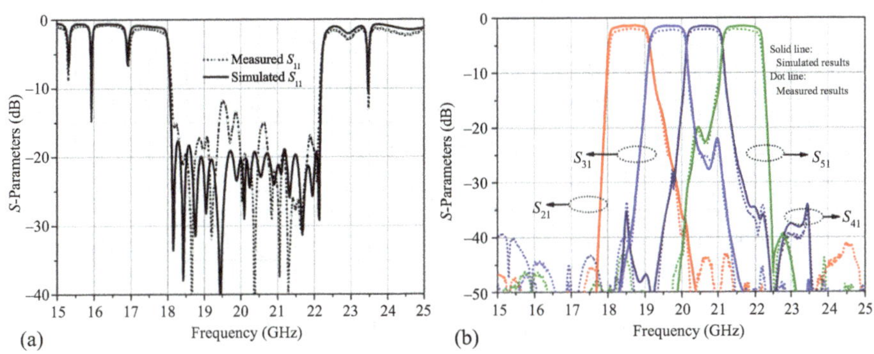

Figure 2.20 Measured and simulated results for the fabricated prototype: (a) return loss and (b) insertion loss. Reprinted with permission from Reference [82]; © 2015 IEEE

even hexagonal substrate integrated cavities, are also adopted for the design of SIW filters. The electromagnetic field distribution of those cavities is different to the rectangular waveguide. Hence, the equivalent rectangular waveguide cannot be used to model the substrate integrated structure any more. This section introduces a flexible model based on the coupling-matrix for modeling the substrate integrated resonators in the SM [53].

A W-band SIW cascaded trisection (CT) filter is designed as a demonstration. This filter has a center frequency of 80 GHz, a bandwidth of 2 GHz and an in-band reflection coefficient smaller than −20 dB. A transmission zero located at 82.5 GHz is desired to improve the filter selectivity. Figure 2.21 shows the configurations and coupling topologies of the CT-filter, in which the CT cell is used to produce the desired transition zero [91,92]. The coupling matrix synthesizing

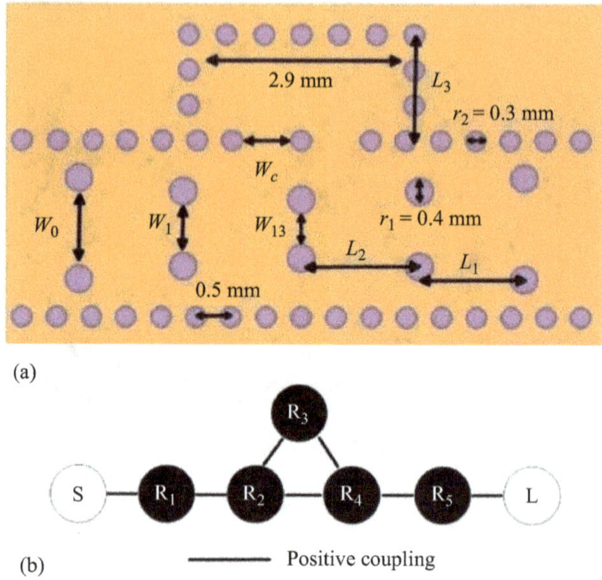

Figure 2.21 (a) Configurations of the W-band SIW CT-filter and (b) coupling topologies of the W-band SIW CT-filter. Reprinted with permission from Reference [53]; © 2016 IEEE

procedure is used to synthesize the normalized coupling matrix for the given specification [92]. The normalized coupling matrix is presented in (2.8) and the external quality factor Q_e is 29.175.

$$M = \begin{bmatrix} 0.006 & 0.0278 & 0 & 0 & 0 \\ 0.0278 & 0.001 & 0.0188 & 0.00705 & 0 \\ 0 & 0.0188 & -0.009 & 0.0188 & 0 \\ 0 & 0.00705 & 0.0188 & 0.001 & 0.0278 \\ 0 & 0 & 0 & 0.0278 & 0.006 \end{bmatrix} \tag{2.8}$$

The initial size of the TE_{10}-like mode based SIW cavity should be determined before the design, and it can be evaluated according to the following formula [93] when setting the resonating frequency as the center frequency of 80 GHz:

$$f_0 = \frac{c_0}{2\sqrt{\varepsilon_r}}\sqrt{\frac{1}{a_{\text{eff}}^2} + \frac{1}{l_{\text{eff}}^2}} \tag{2.9}$$

a and l are the width and length of the TE_{10}-like mode based SIW cavity, respectively. d and p are the diameter of metallized via and pitch between adjacent via-holes, c_0 is the light velocity in vacuum, and ε_r is the relative dielectric constant of the substrate. a_{eff} and l_{eff} are the width and length of the equivalent cavity, respectively.

According to (2.9), the geometries of the SIW cavity are chosen as: $a = 2.5$ mm, $d = 0.3$ mm, $p = 0.5$ mm, $\varepsilon_r = 2.2$, and $l = 1.675$ mm. Then coupling coefficients in (2.8) can be extracted based on the designed SIW cavity and the coupling coefficient k can then be calculated by the following relation [91]:

$$k = \frac{f_1^2 - f_2^2}{f_1^2 + f_2^2} \tag{2.10}$$

where f_1 and f_2 can be obtained by extracting the Eigen mode of two coupled resonators in HFSS. They stand for high and low resonant frequencies, respectively. Figure 2.22(a), (b), and (c) represents the positive coupling between cavity resonators, i.e., SIW cavity 1, 2 and 2, 3, respectively. By adjusting the width, i.e., W_c, of the via irises, coupling coefficients that against the W_c were obtained. And the desired coupling coefficient can be obtained through a curve-fitting program developed in the MATLAB.

The external quality factor can be obtained by full-wave analyzing of a SIW transmission line fed SIW cavity with via-irises. By changing the width W_0 in

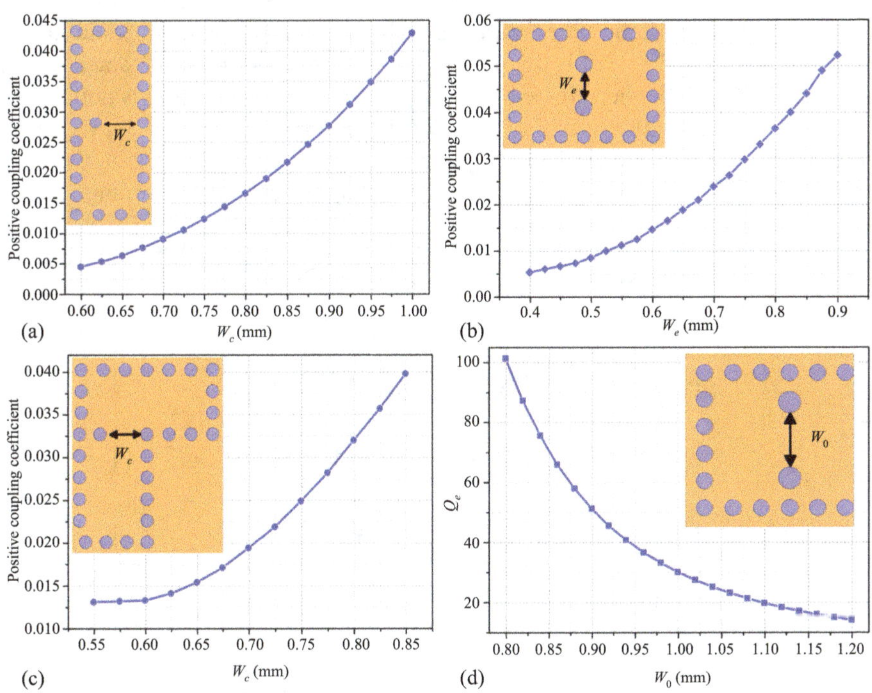

Figure 2.22 (a) Coupling coefficient of the lateral coupling structure; (b) coupling coefficient of the broadside coupling structure; (c) coupling coefficient for the coupled CT cell and SIW cavity; and (d) external quality factor of the input/output structure. Reprinted with permission from Reference [53]; © 2016 IEEE

Figure 2.22(d), the external quality factor Q_e is calculated by [91]:

$$Q_e = \frac{2\pi f_0 \tau_d(f_0)}{4} \tag{2.11}$$

where $\tau_d(f_0)$ represents the group delay value at center frequency $f_0 = 80$ GHz. The extracted Q_e and its corresponding fitted curve are presented in Figure 2.22(d). Figure 2.23 shows the full-wave response of the W-band SIW CT filter with extracted initial geometries. Apparently, large discrepancies between full-wave simulated and desired responses can be observed. And optimizations need to be carried out for achieving desired specifications. It should be mentioned that in order to model accurately the coupling coefficient and the loss for W-band operation, a practical copper model used in the commercial PCB process is adopted as metallic material in the simulation instead of the perfect conductor. Traditionally, the copper in the commercial PCB process has a conductivity of 58,000,000 S/m. However, as the frequency increases, its conductivity decreases rapidly [53]. Moreover, being associated with the wavelength of the operating frequency, the surface roughness of the commercial PCB process at W-band is much larger than that at microwave frequency. Hence, an equivalent conductivity of 6,000,000 S/m is used in the simulation for W-band application for accurate predictions with previous experiences.

The SM is used to fulfill the design. To this end, two models named as the coarse- and fine-model need to be established at first. The fine-model is built in the full-wave simulator HFSS, as shown in Figure 2.21. As mentioned above, seven geometrical variables represented as $X_f = [L_1, L_2, L_3, W_0, W_1, W_e, W_{13}]$ in the fine-model needs to be optimized for archiving desired specifications. Accordingly, same

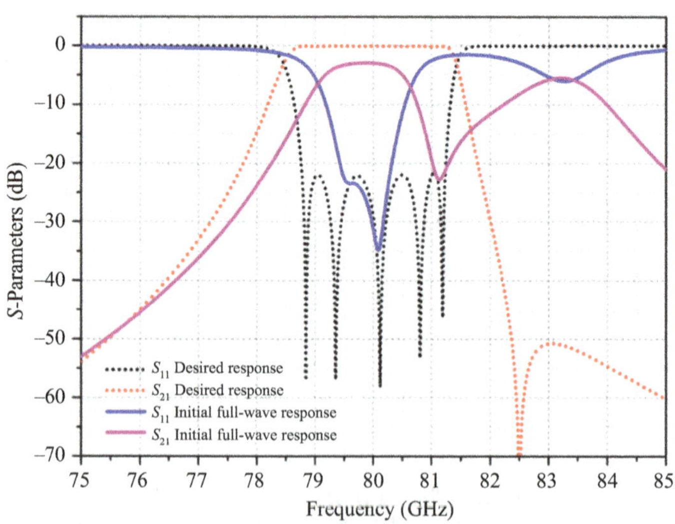

Figure 2.23 The initial full-wave response and the desired specification of the W-band SIW CT-filter (where the W-band equivalent conductivity of the copper used in the PCB process are included). Reprinted with permission from Reference [53]; © 2016 IEEE

number of variables should be used in the coarse-model as well for the implementation of the SM. The offset resonating frequencies of SIW cavities, external quality factors and the coupling matrixes as optimization variables in the coarse-model, and it is denoted as $X_c = [f_1, f_2, f_3, Q_e, M_{12}, M_{23}, M_{24}]$. The reason for using the coupling matrix theory based circuit as the coarse-model is that the coupling matrix theory is independent to the physical structures of the designed filter. It is well known that the modeling accuracy of the coarse-model affects seriously the convergence of the SM. To improve the design efficiency, the commercial circuit simulators such as ADS and AWR Microwave Office are usually adopted to establish the physical structure related coarse-model owning to their powerful circuit simulator and optimization algorithms. However, for some special cases such as the topologies shown in Figure 2.21, it is impossible to find a circuit model in the commercial circuit simulators to model the special physical structure. Hence, the coupling matrix based coarse-model is more generalized than other types of coarse-model for filter design [92]. In order to utilize the powerful circuit simulator and optimization algorithms, the coarse-model in Figure 2.24 is developed in the ADS for representing the coupling matrix response, which uses phase shifters and the RLC circuits to represent losses, coupling coefficients and resonating frequencies of the SIW cavities [53]. To model the loss of the filter in the coarse-model, the resonators in Figure 2.24 are modeled with resistors of R, which can be evaluated as $R = Z^*Q_u/Q_e$ [91], where Z is the terminal impedance at the I/O ports, i.e., 50 Ω, Q_e is the external quality factor, and Q_u is the unloaded quality factor of resonators, respectively. The Q_u is obtained from an eigenmode analysis by using the full-wave simulator HFSS for a SIW cavity with a resonating frequency of 80 GHz.

At first, the optimal coarse-model and fine-model variables are obtained from (2.8) and the corresponding geometries in Figure 2.22, respectively, as:

$$X_c^* = [79.96\,\text{GHz}, 79.95\,\text{GHz}, 80.46\,\text{GHz}, 29.175, 0.0278, 0.0188, 0.00705];$$
$$X_f^0 = [1.45\,\text{mm}, 1.6\,\text{mm}, 1.5\,\text{mm}, 1.0\,\text{mm}, 0.75\,\text{mm}, 0.67\,\text{mm}, 0.53\,\text{mm}]; \tag{2.12}$$

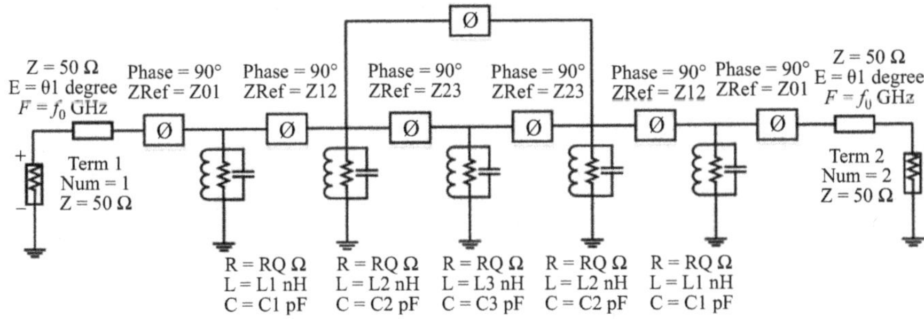

Figure 2.24 Coarse-model of the W-band SIW CT-filter. Reprinted with permission from Reference [53]; © 2016 IEEE

To initially set up the mapping relation among variables of the coarse- and fine-models, a manual calibration process is then adopted in the design by adjusting the SIW cavities with small perturbation [94,95], and the calibrated fine-model variables are denoted as:

$$X_f^a = [1.464\,\text{mm}, 1.616\,\text{mm}, 1.515\,\text{mm}, 1.01\,\text{mm}, 0.735\,\text{mm}, 0.677\,\text{mm}, 0.519\,\text{mm}] \tag{2.13}$$

The parameter extracting process in Section 2.3.1 is then used to extract $X_c^{\,0}$ and $X_c^{\,a}$ by fitting the coarse-model response to the fine-model response, and they are:

$$\begin{aligned} X_c^0 &= [81.471\,\text{GHz}, 81.629\,\text{GHz}, 80.528\,\text{GHz}, 31.155, 0.0451, 0.0148, 0.0143]; \\ X_c^a &= [80.991\,\text{GHz}, 81.257\,\text{GHz}, 79.735\,\text{GHz}, 28.996, 0.0416, 0.0152, 0.0137]; \end{aligned} \tag{2.14}$$

Once $X_f^{\,0}$, $X_f^{\,a}$, $X_c^{\,0}$, and $X_c^{\,a}$ are obtained, iterations following the description in Section 2.3.1 are used to complete the design.

Table 2.5 shows variations of parameters in the SM optimization of the W-band SIW CT filter. The final geometries of the W-band SIW CT-filter were obtained through four iterations. As shown in Figure 2.25(a), desired specifications were achieved except a little bit frequency shifting of the transmission zero. The lower transmission zero resulted in a higher selectivity for the designed filter. For experiments which use the WR10 waveguide as testing interface, a fin-line SIW-waveguide transition is designed and integrated with the filter [53,96]. The full-wave response of the filter that integrated the fin-line SIW-waveguide transition is shown in Figure 2.25(b). Although the in-band reflection coefficient becomes a little bit worse, it is acceptable for experiments. Experiments were used to verify the design, and the results are shown in Figure 2.26. The measured insertion loss is 3.89 dB at 80 GHz. It includes the loss from two fin-line SIW-waveguide transitions. The in-band reflection coefficient is smaller than −13 dB. Due to the increased dielectric loss at W-band, the measured insertion loss is 0.39 dB worse than that of the simulation.

Table 2.5 Parameter variations of the iteration for the SIW CT-filter

Initial		**first**		**fourth**	
x_c^*	x_f^0 **(mm)**	x_c^1	x_f^1 **(mm)**	x_c^4	x_f^4 **(mm)**
79.96 GHz	1.45	81 GHz	1.496	79.785 GHz	1.521
79.95 GHz	1.6	81.26 GHz	1.672	79.802 GHz	1.685
80.46 GHz	1.5	79.74 GHz	1.501	80.678 GHz	1.493
29.175	1	29	1.01	29.587	1.002
0.0278	0.75	0.0416	0.677	0.0279	0.659
0.0188	0.67	0.0152	0.742	0.0172	0.688
0.00705	0.53	0.0137	0.408	0.0076	0.437

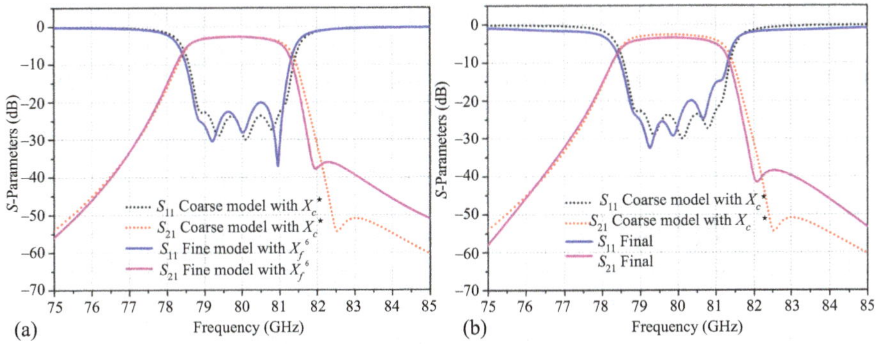

Figure 2.25 (a) The final full-wave response of the W-band SIW CT-filter and (b) the final response of the W-band SIW CT-filter integrated with fin-line SIW-waveguide transitions. Reprinted with permission from Reference [53]; © 2016 IEEE

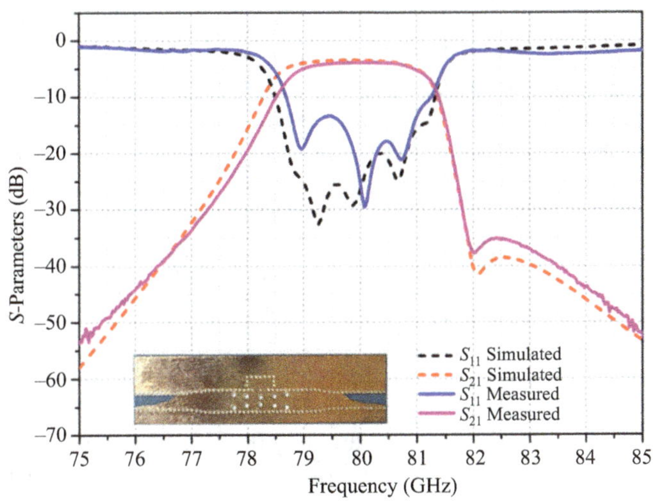

Figure 2.26 Measured and simulated responses of the W-band SIW CT-filter integrated with fin-line SIW-waveguide transitions. Reprinted with permission from Reference [53]; © 2016 IEEE

2.4 Developing planar terahertz filters using the deep reactive ion-etching (DRIE) process

The terahertz technology has recently attracted many attentions from researchers for its potential applications in short-range communications, medical imaging, and security scanning. Waveguide component is popularly used in the terahertz system because it can provide excellent performance [1–9,11–13]. In recent years, numerous efforts have been made in developing terahertz waveguide components. The SU-8 photoresist micromachining technology has been studied and proposed in [4,6,7], and single-band and dual-band bandpass filters with center frequencies around 300 GHz have been reported with impressive results by utilizing this technology [9]. Canonical waveguide filters, including E- and H-plane five-pole bandpass filters with operating frequency around 600 GHz, have been reported in [5] by using the DRIE micromachining process which is one type of special MEMS processes, where key fabrication parameters were analyzed. By utilizing the Bosch DRIE process, a 350–460 GHz frequency band waveguide was reported with a low average insertion loss of 0.086 dB/mm in [97]. A lowpass filter having a cutoff frequency of 275 GHz was developed with a high selectivity and low insertion loss in [98] by using the UV-LIGA process. A two-pole Butterworth bandpass filter having center operating frequency of 380 GHz was studied in [3], where an insertion loss of 2.7 dB was reported with a 3 dB frequency bandwidth of 25 GHz. A 140 GHz waveguide bandpass filter was investigated in [2] with an unloaded quality factor of 815. Those achievements show that micromachining technologies, including DRIE, UV-LIGA, and SU-8 process, can provide enough fabrication tolerance for terahertz components.

Theoretically, in order to reduce the insertion loss, high quality factor cavities are favorable for the design of terahertz filters. Waveguide cavities are attractive for the terahertz components due to its high quality factor and high power-handling capability. However, different waveguide cavity may have different quality factor characteristics. Investigations reveal that in-regular shape cavities having relatively large footprint may offer a higher unloaded quality factor. Moreover, multiple-mode cavity may provide a high unloaded quality factor as well. It functions as multiple single-mode cavities with a compact size, and can not only offer a higher quality factor that of its single-mode counterpart, but also increase the filter order as well as its selectivity with compact footprint. Generally, cavities having higher unloaded quality factor may require a relatively larger circuit size. For a given manufacturing tolerance, that means those cavities may have more fabrication reliability than the lower unloaded quality factor cavity.

A number of cavities operated at around 400 GHz with single- or dual-mode are shown in Figure 2.27, and their unloaded quality factors. The quality factors shown in Table 2.6 are extracted by using the HFSS. The conductivity is set to 3.3e7 S/m. Table 2.6 shows the comparison of the Q_u of different cavities with

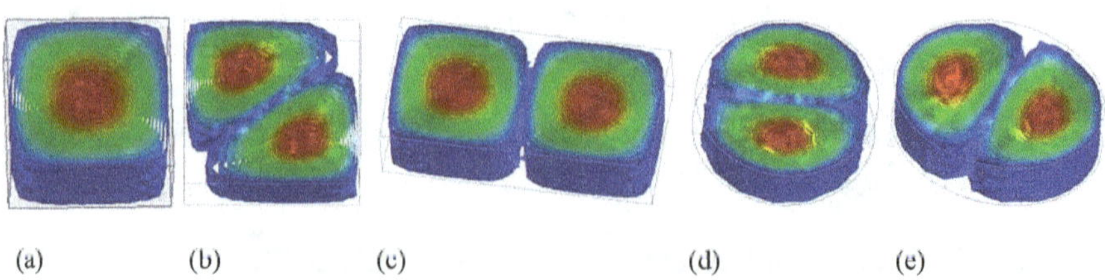

Figure 2.27 (a) single-mode square cavity; (b) dual-mode square cavity; (c) high-order rectangular cavity; (d) single-mode circular cavity; and (e) single-mode elliptic cavity

Table 2.6 Comparison of the unloaded quality factor

Cavity shape	Operating mode	Dimension (μm)	Q_u and frequency
Square cavity 1	TE_{10}	530*530	983@400 GHz
Square cavity 2	TE_{201}/TE_{102}	838*838	1,211@400 GHz
Rectangular cavity	TE_{201}	530*1060	1,127@400 GHz
Circular cavity	Dual-TM_{110}	$r = 463$	1,240@400 GHz
Elliptic cavity	Dual-Quasi-TM_{110}	$r = 440$	1235@393 GHz
		$AR = 1.1$	1246@412 GHz

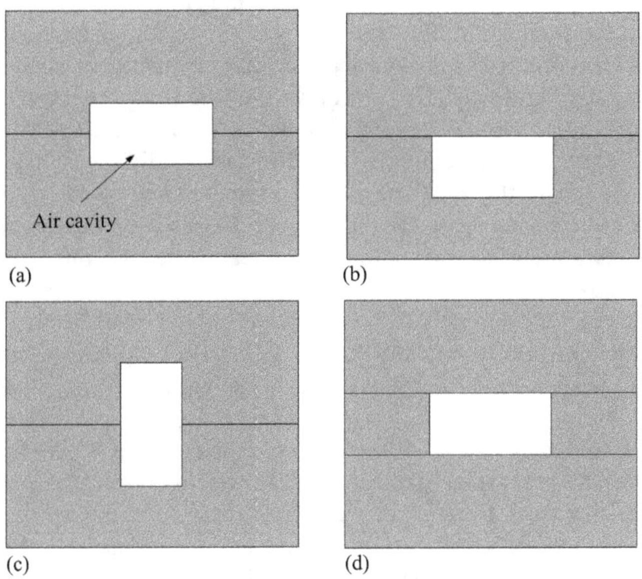

Figure 2.28 Four different layer structures of the silicon based rectangular waveguides. Reprinted with permission from Reference [8]; © 2015 IEEE

different modes at around 400 GHz. The square cavity with fundamental mode has the lowest Q_u and the smallest size. To the dual-mode cavities, the circular and the elliptic cavities have the higher Q_u than the square and the rectangular cavities. Sizes of the dual-mode cavities are larger than single mode cavity at the same frequency. Therefore, the better tolerance and insertion loss will be achieved if the dual-mode circular or elliptic cavities are used for the terahertz components.

The deep reactive ion-etching process has been adopted usually for the fabrication of waveguide terahertz components. In order to integrating or measuring the fabricated terahertz components, waveguides with standard size are usually used as input/output interfaces. By using DRIE process, a certain depth needs to be etched on silicon wafers. There are several different structures to form the waveguide, as shown in Figure 2.28, regarding etching depth and the stack way of wafers.

In [99] and [100], waveguide components are combined by two identical halves in the H-plane, as shown in Figure 2.28(a). These two halves are formed by etching the half height of the standard waveguide. Roughness surfaces are usually existed on the top and the bottom surfaces of the waveguide, which result in an increased insertion loss. In addition, a small misalignment of these two halves in the fabrication may increase the insertion loss as well. In [3,101], the height of the standard waveguide is etched in one silicon wafer. After bonded with another wafer, waveguide will be formed, as shown in Figure 2.28(b). In this way, no misalignment exists, and the top surface of the waveguide is much smooth. For example, measured insertion losses of 0.4 dB/mm in the frequency band of 325–440 GHz and 0.15 dB/mm @ 600 GHz are reported in [3] and [101], respectively. To reduce the loss from the fabrication, the terahertz waveguide can also be split along its E-plane with less current disturbance and fabricated by using the DRIE process, as shown in Figure 2.28(c) and [5]. To remove the loss from the roughness surface as much as possible, another topology shown in Figure 2.28(d) that consists of three layers, while the waveguide is etched in the middle layer, may be a good solution. The height of the middle layer is the standard waveguide height. The top and the bottom layers are glass or silicon layer. In this way, the roughness in the top and the bottom surface are eliminated. Meanwhile, the misalignment error could be avoided.

A fabricated terahertz WR-2.2 waveguide having a cross section size of 560 μm × 280 μm is shown in Figure 2.29. The topology shown in the Figure 2.28(d) is adopted in the DRIE process. The high resistivity silicon with the thickness of 280 μm is chosen as the middle layer. The top and bottom layers are both glass with the thickness of 550 μm. The silicon and glass wafers are all double side polished. The processing steps are as follows: firstly, the thermal oxidation mask layer thicker than 3 μm on the middle silicon wafer is formed. Then ICP-etching process is used to etch 280 μm for the designed structures. After etching, the bottom layer is bonded with the silicon layer by anodic bonding technology. For metallization, the side walls of the etched structures are sputtered with 0.3 μm gold, then a 2.5 μm gold layer is electroplated for the further metallization, while the skin depth for gold is 0.124 μm at 400 GHz. Next, the top glass wafer is bonded with the middle layer by the Au-Si eutectic bonding technique to form the waveguide or cavities. The bonding interfaces of the top and bottom glass wafers are electroplated with gold already. Then the wafer is divided into pieces by dicing saw. At last, in case the energy leaks into the silicon [97]. Every side surface is metalized

Figure 2.29 Photograph of the lateral view of the fabricated waveguide. Reprinted with permission from Reference [8]; © 2015 IEEE

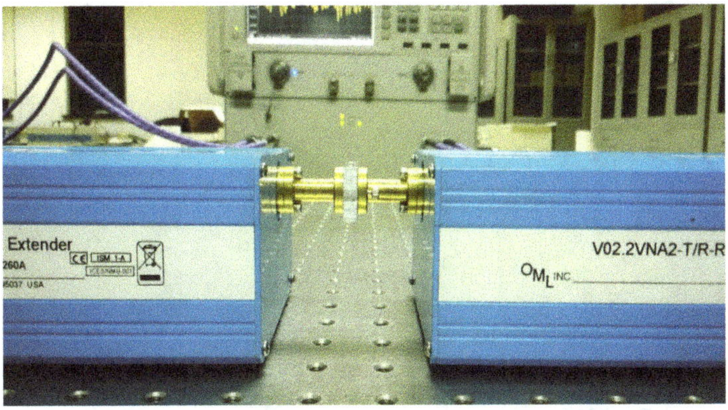

Figure 2.30 Photograph of the measurement setup of the WR-2.2 waveguide. Reprinted with permission from Reference [8]; © 2015 IEEE

by spurting with 0.3 μm gold for single piece. After fabrication, the size of the pieces are 6 mm × 6 mm × 1.38 mm or 6 mm × 8 mm × 1.38 mm to measure the performance of the fabricated terahertz waveguide, a test fixture is fabricated by using the CNC milling process. The keysight VNA N5245A shown in Figure 2.30 with two OML WR-2.2 frequency extenders are used to measure the terahertz waveguide. The trough-reflect-line (TRL) calibration is used during the measurement. The flanges (UG-387) on the extenders are used and fixed by screws to connect the test fixture.

Figure 2.31 shows the measured results of the straight waveguide with a length of 8 mm. As can be seen, the insertion loss is approximately changing monotonically and the return loss is better than 13 dB within 320–500 GHz. The ripple of the return loss is caused by the mismatch between the flange interface and the WR-2.2 waveguide under test. Figure 2.32 shows the detail of the interface. The measured insertion loss is 1.86–0.92 dB from 325–500 GHz. The approximately unit length loss can be achieved using the insertion loss divided by the length of the waveguide, which is about 0.144 dB/mm at 400 GHz. It should be mentioned, although a 2.5 μm gold layer is electroplated for the metallization, surface roughness still exits inner the waveguide and has an impact on the insertion loss. For the surface roughness analysis, a number of models have been designed for different structures. Among them, the Huray model is a good choice to model the roughness of the gold layer [102]. Figure 2.33 shows the effect of Hall–Huray surface ratio on the insertion loss. Nodule radius and layer thickness are set to 0.5 and 2.5 μm, respectively. With the increasing of Hall–Huray surface ratio, the insertion loss will increase and present a rising tendency. And this change is almost linear to the increasing of surface ratio.

Misalignment between the WR-2.2 waveguide under test and flanges of extenders is one of sources for increasing the measured loss. Misalignment may be existed in all three directions. Figure 2.34 is the enlarged view of two waveguide

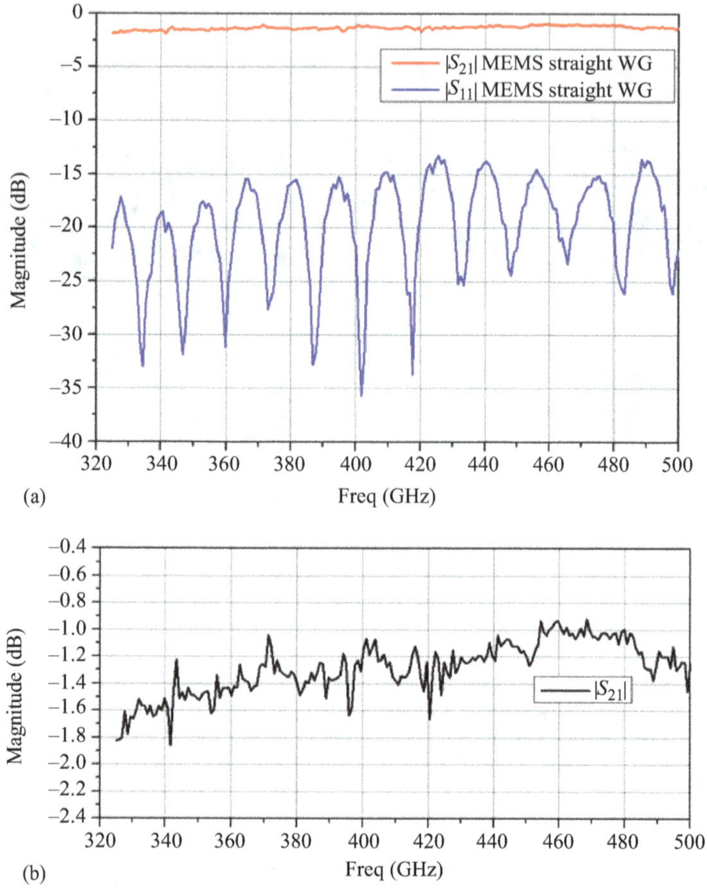

Figure 2.31 Measured results of the 8 mm straight WR-2.2 waveguide: (a) wide frequency response and (b) enlarged view. Reprinted with permission from Reference [8]; © 2015 IEEE

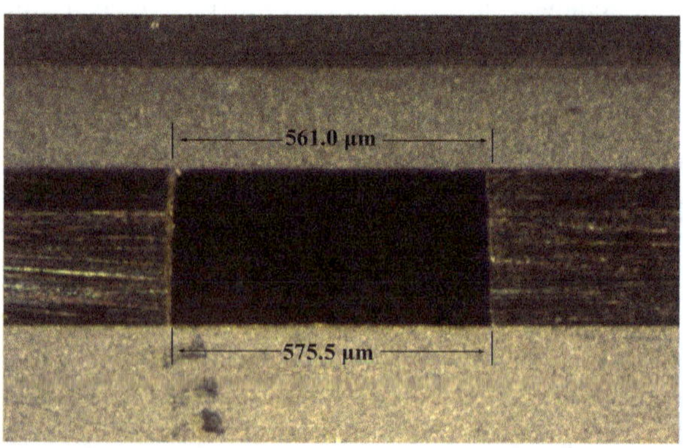

Figure 2.32 Measurement of the size of the input/output interface of the waveguide. Reprinted with permission from Reference [8]; © 2015 IEEE

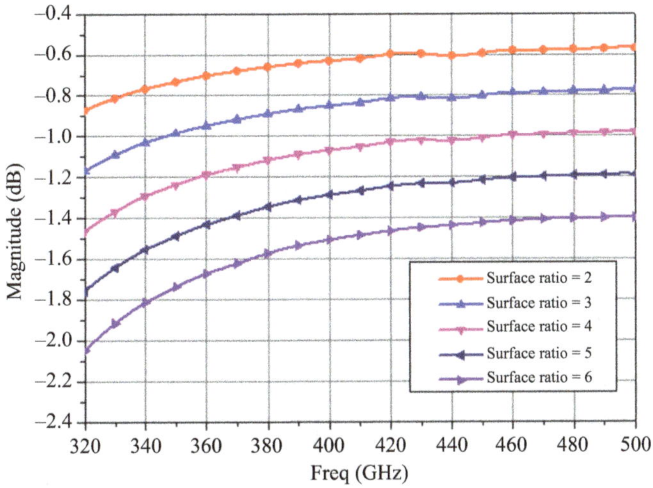

Figure 2.33 The effect of the Hall–Huray surface ratio to the insertion loss. Reprinted with permission from Reference [8]; © 2015 IEEE

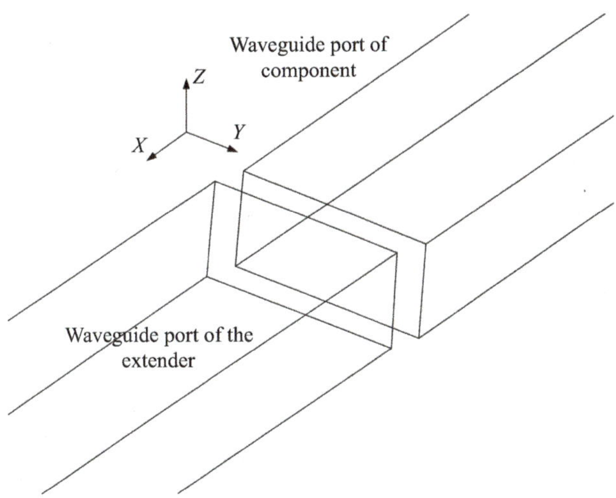

Figure 2.34 Enlarged view of the connection of two waveguide ports. Reprinted with permission from Reference [8]; © 2015 IEEE

ports which are used to model the match condition of measurement. The shift in the x-direction causes an air gap between two ports. The shifts in the y- or z-direction lead to an H- or E-plane leakage, respectively. In the y-direction, the farther away from the center of the waveguide, the weaker the E-field. Therefore, measured results are less affected with y-direction misalignment, which may be ignored. As shown in Figure 2.35, the gap or the shift in the x-direction has a great impact on

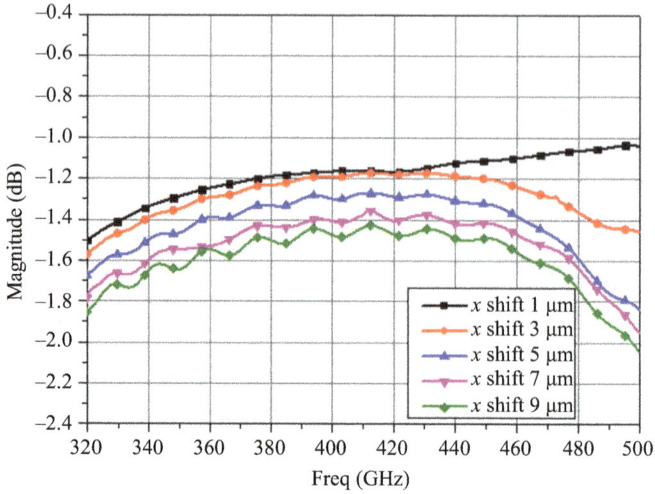

Figure 2.35 The effect of the shift in the x-direction (air gap) to the insertion loss. Reprinted with permission from Reference [8]; © 2015 IEEE

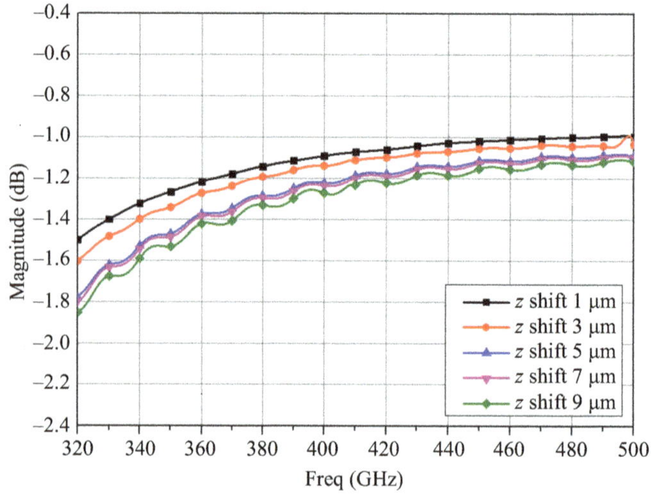

Figure 2.36 The effect of the shift in the z-direction to the insertion loss. Reprinted with permission from Reference [8]; © 2015 IEEE

insertion loss. With the effect of the air gap, the loss at upper frequency band is larger than lower frequency band. In the z-direction, the shift mainly affects the lower frequency band, while the upper frequency loss changes a little, as shown in Figure 2.36. In addition, the ripple will become larger with the increasing of any type of shift. In those simulations, the Hall–Huray surface ratio is set as 4. Based on

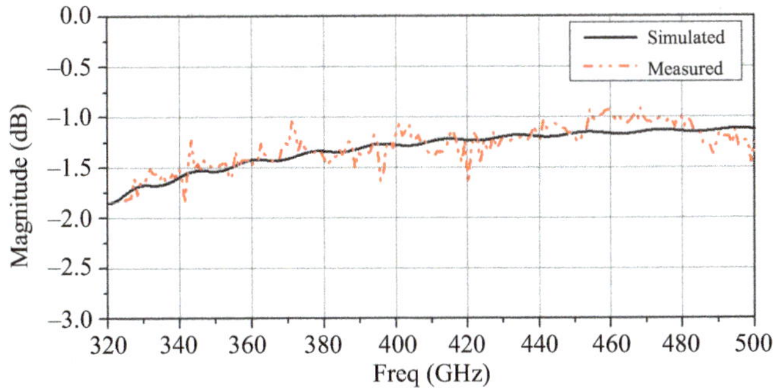

Figure 2.37 Comparison of the re-simulated results and the measured results of the straight waveguide. Reprinted with permission from Reference [8]; © 2015 IEEE

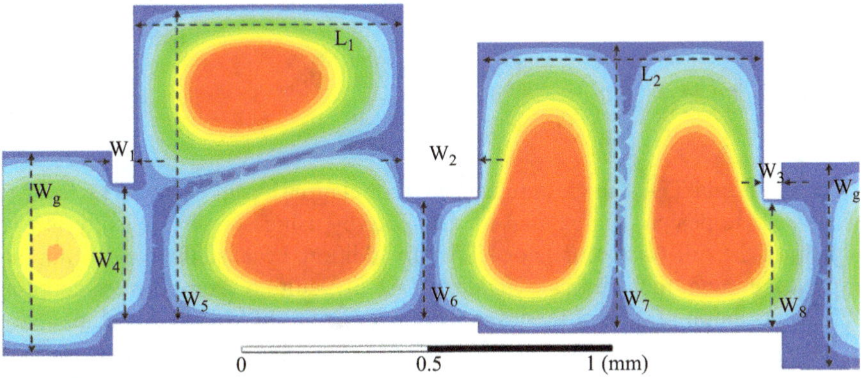

Figure 2.38 Structure and electric-field distribution of a terahertz filter using two dual-mode cavities

the analysis mentioned above, simulated results including in 0 μm x-direction shift and 10 μm z-direction shift is presented in Figure 2.37, and the Hall–Huray surface ratio is set to 4. The simulated and measured results are in good agreement with each other for the adopted DRIE process. Hence, the simulated model including the surface roughness, dimension errors, and the misalignment could be used to analysis the performance of fabricated terahertz components with specified process.

Shown in Figure 2.38 is a terahertz filter using two dual-mode rectangular cavities. Because each dual-mode cavity functions as two equivalent single-mode cavities, this functions as a four-order filter with a compact size and can reduce efficiently the insertion loss of the filter. More importantly, it can provide multiple transmission zeros at designated frequencies, and then greatly improves the filter selectivity.

The dual-mode bandpass prototype shown in Figure 2.38 is designed with a center frequency of 400 GHz and a 20 dB return loss of 20 GHz (5% fractional bandwidth). Two transmission zeros are predestinated at 376 and 423 GHz, respectively, to improve the selectivity. Each dual-mode cavity shown in Figure 2.38 contains two orthogonal modes, and its input/output aperture simultaneously couples to both Tem0n and TEp0q modes. Since those two orthogonal modes have opposite directions in the cavity, they can be designed to cancel each other at stopband. Then each dual-mode cavity produces two transmission poles and one transmission zero that can be located at the left or right side of the passband by adjusting dimensions of the cavity [86,103]. Hence, once the operated two orthogonal modes, i.e., TEm0n and TEp0q, are chosen, the resonating frequency f_0 can be estimated by [91]

$$f_0 = \frac{c_0}{2}\sqrt{\left(\frac{m}{a}\right)^2 + \left(\frac{n}{l}\right)^2} = \frac{c_0}{2}\sqrt{\left(\frac{p}{a}\right)^2 + \left(\frac{q}{l}\right)^2} \tag{2.15}$$

where c_0 is the velocity of the light, and a and l are the width and length of the dual-mode cavity, respectively. Then the initial choice of the ratio a/l for obtaining two orthogonal modes is

$$\frac{a}{l} = \sqrt{\frac{m^2 - p^2}{q^2 - n^2}} \tag{2.16}$$

If TE_{201} and TE_{102} modes are chosen, the relationship in equation (2.16) then can be used to design desired dual-mode cavity by using eigenmode analysis with the help of the full-wave simulator HFSS. In addition, because there exist multiple couplings among four resonant modes, it is difficult to extract cross-coupling coefficients among resonant modes in different cavities. To avoid this complexity, the four-pole dual-mode filters are divided into two separated blocks, and each of them only includes one dual-mode cavity with input/output apertures. Coupling-matrixes are then synthesized for those blocks with specified passband of 390–410 GHz and a transmission zero located at 376 or 423 GHz, respectively. A coupling coefficient extraction process in [91] is used to obtain initial geometries for those blocks. To obtain required coupling coefficients, resonating frequencies of two orthogonal modes are split by adjusting a and l. The coupling aperture, for example, W_4 in Figure 2.38, is adjusted to obtain desired external quality factor. Finally, those two blocks are integrated together by using a coupling aperture with geometries of W_2 and W_6 in Figure 2.38. The HFSS is then used to finalize the design by slightly adjusting geometries of the prototype, and final geometries shown in Figure 2.38 are: $W_1 = 62$ μm, $W_2 = 202$ μm, $W_3 = 50$ μm, $W_4 = 382$ μm, $W_5 = 867$ μm, $W_6 = 339$ μm, $W_7 = 793$ μm, $W_8 = 361$ μm, $L_1 = 737$ μm, $L_2 = 785$ μm. The input/output waveguide has a dimension of 560 μm × 280 μm. Simulated results of the designed prototype are shown in Figure 2.39, where a perfect conductor is used. Although the simulated return loss (< 18 dB from 398 to 403 GHz) is a little bit smaller than 20 dB in a narrow frequency range, the filter is good enough for the fabrication.

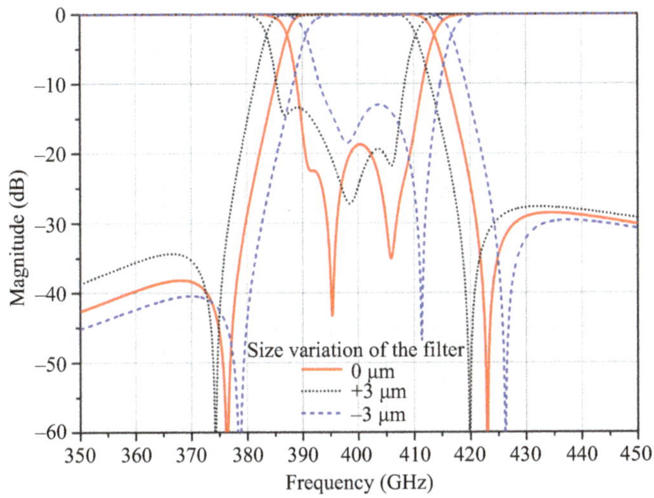

Figure 2.39 Tolerance simulation for the designed prototype with varied geometries

At the terahertz frequency, the fabrication tolerance seriously affects filter's performance. Hence, it needs to be investigated to predict the filter performance before fabrication. The influence of the size variation is investigated by supposing that all geometries shown in Figure 2.38 vary simultaneously from −3 to + 3 μm. As illustrated in Figure 2.39, varied geometries lead operating frequency shifting. However, a good return loss can still be obtained within a fabrication tolerance of 3 μm for the designed prototype. As discussed above, another important tolerance factor is the surface roughness. As the operating frequency increases, the surface roughness plays an important role in modeling the insertion loss and bandwidth of terahertz filter. This is because the surface roughness causes a surface resistance, which then leads to a low quality factor of terahertz resonator.

Since the quality of the surface metallization and the scale of the surface roughness are depended on the adopted micromachining process, each DRIE process has its own surface roughness model. To this end, we use the equivalent conductivity to model the quality of the surface metallization, and use the Huray surface model to model the surface roughness, as studied above. In the simulation, the equivalent conductivity and the nodule radius are varied with a fixed Hall–Huray surface ratio of 4.5. Full-wave simulated effects of the surface roughness are illustrated in Figure 2.40. As the surface roughness increases, i.e. the equivalent conductivity decreases or the nodules radius increases, the insertion loss dramatically increases, and the operation bandwidth becomes narrower as well.

A designed prototype has been fabricated by using a three-layer glass–silicon–glass DRIE micromachining process. As shown in Figure 2.41, the designed prototype is etched inside a silicon wafer that has been polished to have a thickness of 280 μm. Since this step uses a deep through etching process, vertical walls, which

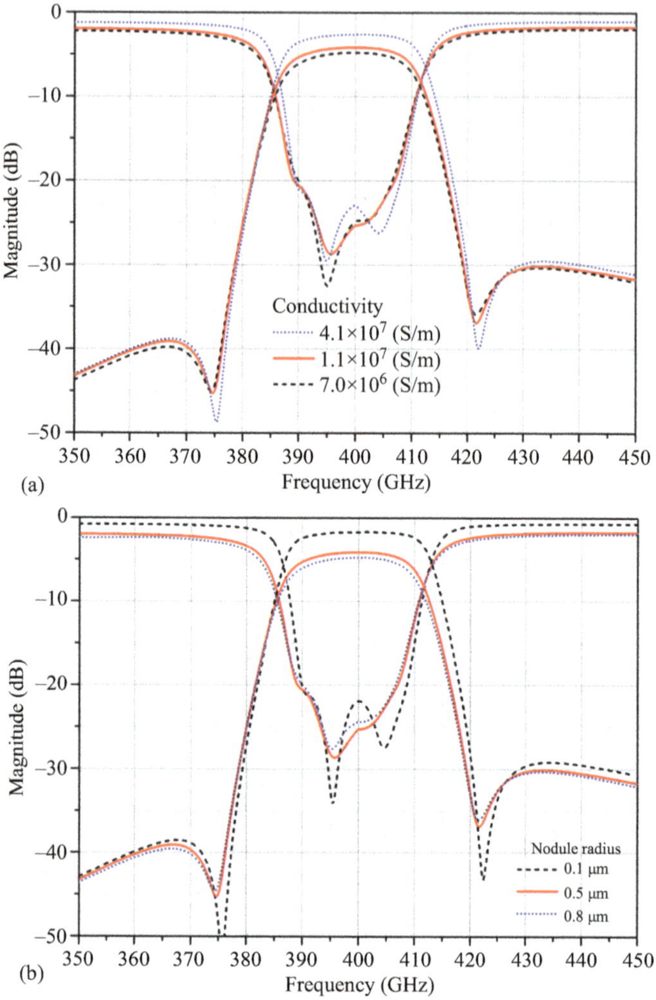

Figure 2.40 Simulated results with different roughness for the designed terahertz filter: (a) varied equivalent conductivity (Hall–Huray surface ratio = 4.5, nodule radius = 0.5 μm) and (b) varied roughness (Equivalent conductivity = 1.1×10^7 S/m)

are sputtered with gold and have very small deviation, are obtained for the sidewall of the filter. Next, to implement metallic top- and bottom-surfaces of the prototype, two 500 μm glass wafers are deposited with gold on one of their surfaces, respectively. Gold to silicon bonding is used to bond glass and silicon wafers together to realize the designed prototype. Because glass wafers only function as metallized cover, this process avoids the alignment error of the inter-wafer bonding. The designed filter has a length of 1,844 μm, and an extra WR2.2 waveguide having a length of 4,156 μm has been used in the fabrication for easy measurement.

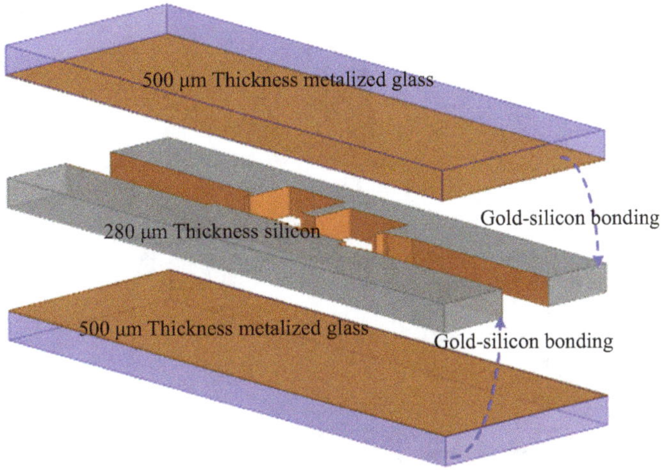

Figure 2.41 Schematic of the adopted micromachining process

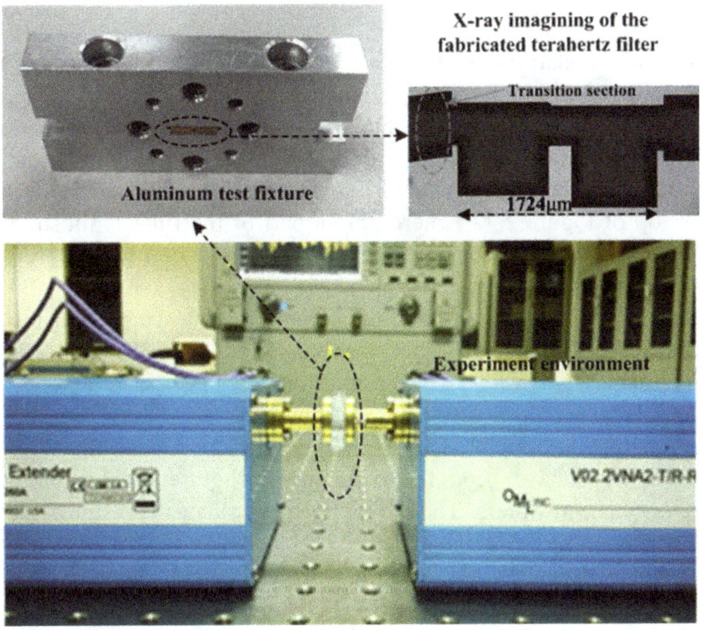

Figure 2.42 Measurement setup for the fabricated prototype

The fabricated prototype is measured by using an Agilent VNA equipped with two 325 to 500 GHz OML frequency extenders. TRL calibration was adopted in the measurement. Figure 2.42 shows the setup of the measurement. An aluminum test fixture is used to assemble the fabricated prototype, as shown at the left top of

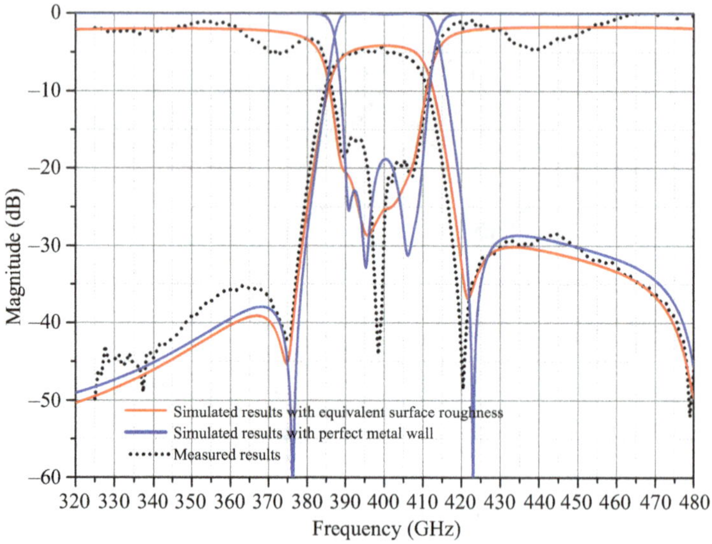

Figure 2.43 Measured and simulated results of the fabricated prototype

Figure 2.42. The standard flange interface UG387 is adopted for the test fixture for connecting with the VNA by using alignment dowels and screws. Because the designed filter adopts a topology in which input/output interface of are misaligned, a transition shown in the right top of Figure 2.42 is used to keep input/output ports inline for easy measurement. The X-ray imagining of the fabricated filter is shown at the right top of Figure 2.42, where only a part of the filter is illustrated due to the limitation of the X-ray scope. Measured and simulated results are shown in Figure 2.43. Measured results have an insertion loss of 4.3 dB at 400 GHz. It includes losses from additional 4,200 μm long WR2.2 waveguide at input/output ports. The prototype has a 10 dB return loss bandwidth from 387 to 410.5 GHz, i.e., 5.875% fractional bandwidth. Two transmission zeros can be observed at 375 and 420 GHz, respectively. It greatly improves the filter selectivity. Because bonded wafers have a thickness up to 1,280 μm, sidewalls of input/output ports of the fabricated prototype have been found having a slightly tilted angle when the component is cut from wafers. These slightly tilted sidewalls result in a small gap between the filter and VNA in the experiment. It leads to un-wanted terahertz-wave leaking, which results in noticeable ripples of the S_{11} and additional losses of the S_{21}, as shown in Figure 2.43. Nevertheless, a good performance is still obtained for the fabricated prototype. To explain effects from the misalignment gap and surface roughness, an equivalent conductivity of 1.1×10^7 S/m, a Hall–Huray surface ratio of 4.5 and a nodule radius of 0.5 μm are used in the simulation. Then the insertion loss is below 3.0 dB at 400 GHz if additional WR2.2 waveguides are excluded. In general, measured results are well agreed with simulated results, and it has much high selectivity. Considering the fabrication tolerance and measurement errors mentioned above, it can be concluded that a high selectivity terahertz filter can be

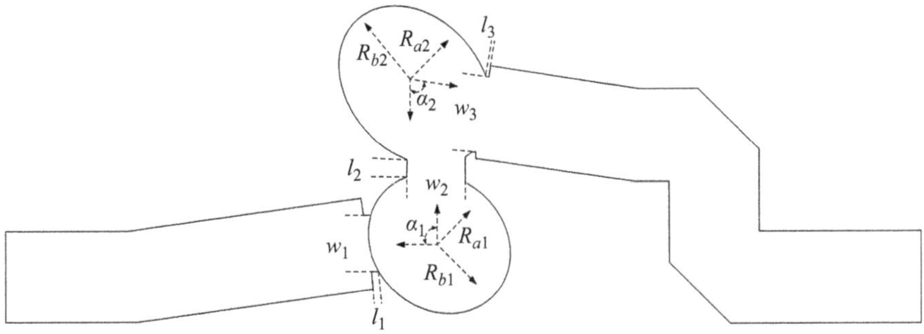

Figure 2.44 Schematic topology and the structure of the bandpass filter. Reprinted with permission from Reference [8]; © 2015 IEEE

Table 2.7 Parameters of the proposed filter (Micrometer)

Parameter	Ra_1	Rb_1	Ra_2	Rb_2	α_1	α_2
Value	405	470	415	519	48	42
Parameter	w_1	w_2	w_3	l_1	l_2	l_3
Value	368	373	467	50	123	54

implemented by using adopted dual-mode topology with a small size and low insertion loss.

Shown in Figure 2.44 is a prototype of a terahertz filter using two dual-mode elliptic cavities, whose geometries are presented in Table 2.7. In Figure 2.44, R_a, R_b are the length of minor axis and major axis, respectively. *Ratio* is the ratio of the major axis and minor axis. α is the angle between the input and output ports. *slot_w* and *slot_l* are the width and the length of coupling aperture. α is set to 140° and the center frequency f_o is set to 400 GHz. Since the length of the major and minor axis of elliptic cavity is different, the elliptic cavity has some characteristics that circular cavity dose not has. The circular cavity is a special case of elliptic cavity when the length of major and minor axis is the same. The position of the transmission zero for the fundamental mode circular cavity can be adjusted by changing the angle between the input port and output port, as demonstrated in [41]. In the case of the elliptic cavity, the position of the transmission zero can be adjusted by changing the ratio of the major and minor axis.

Transmission zeros are generated due to the input and output port are all coupled to the quasi-TM_{110} modes in the elliptic cavity. By simultaneous exciting those two modes, one of the mode may change the sign at a certain frequency. Therefore, a node will be generated at the output port, and a transmission zero will appear [104,105]. With the changing of the ratio of the major and minor axis, the fields in the cavity are squashed and stretched. The node will appear at the different position while the operating frequency is fixed. The transmission performance of a

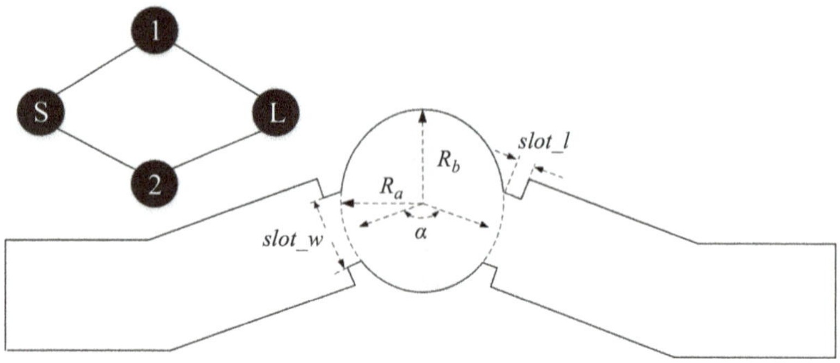

Figure 2.45 Schematic topology and the structure of a single elliptic cavity

Table 2.8 Comparison of the dimensions of different cases (micrometer)

	Ratio	***α***	$\boldsymbol{R_a}$	**slot_w**	**slot_l**
Case 1	1.3	140°	385	360	100
Case 2	1.2	140°	395	360	92
Case 3	1.1	140°	410	360	91

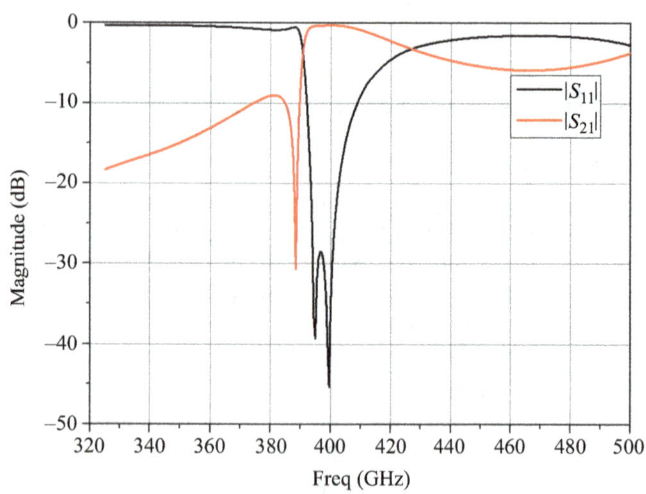

Figure 2.46 Simulated results of the first case

single elliptic cavity shown in Figure 2.45 is investigated with geometries illustrated in Table 2.8. The location of the transmission zero can be adjusted from the lower sideband to the upper sideband as the *Ratio* is decreasing monotonously, as demonstrated in Figures 2.46–2.48. Meanwhile, R_a and *slot_l* need to be fine-tuned to ensure the center frequency is 400 GHz.

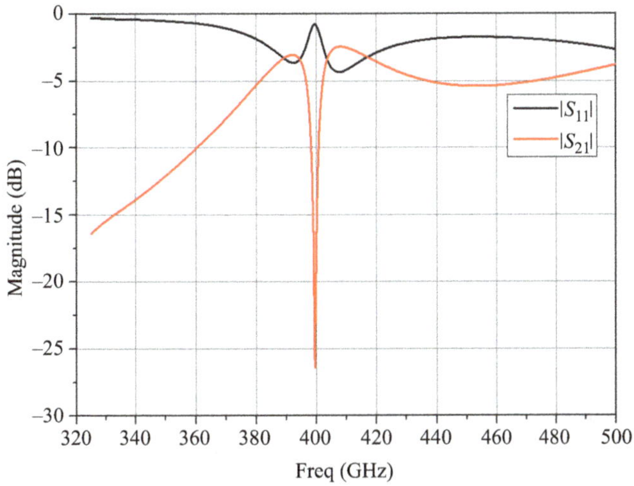

Figure 2.47 Simulated results of the second case

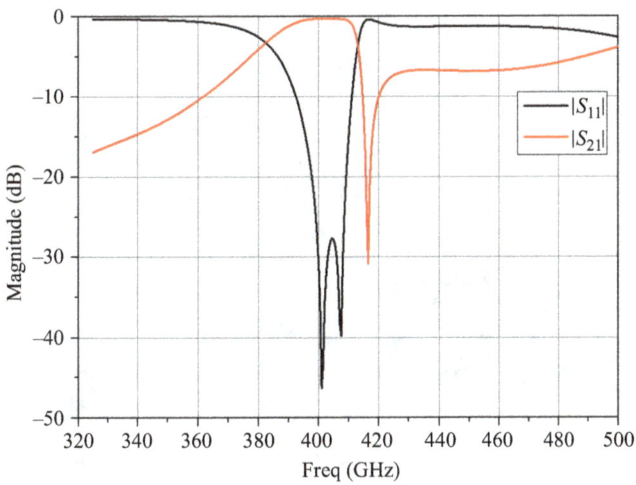

Figure 2.48 Simulated results of the third case

The position of the transmission zeros against the *Ratio* is shown in Figure 2.49. Three curves represent the different angle α, 120°, 140°, and 160°, respectively. The transmission zeros are almost unchanged while the *Ratio* is 1. Meanwhile, the position of the transmission zeros change slowly while the *Ratio* is larger than 1.3. With the increasing of the angle α, the slope of the curves is decreasing. And the bandwidth of the filter is getting larger and the stopband rejection is getting worse while α is increasing. On the other hand, small α results in the difficulty of the fabrication because the input and the output ports are too close to each other. Therefore, α is chose as 140° for the implementation of a cascaded filter.

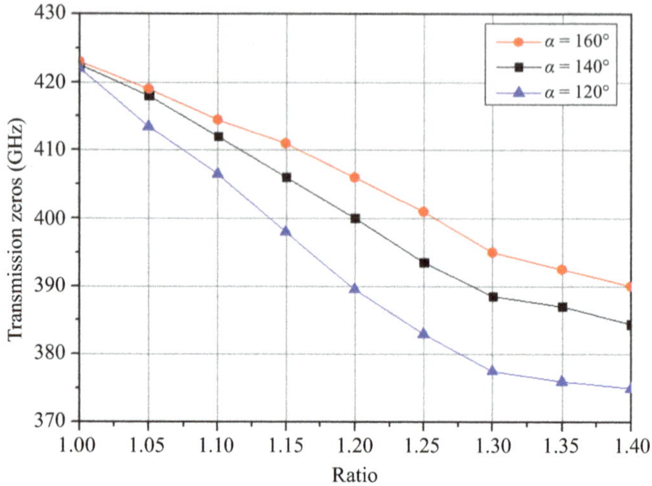

Figure 2.49 Varying of the transmission zeros with respect to the Ratio

Figure 2.50 X-ray photograph of a terahertz bandpass filter using elliptic cavities. Reprinted with permission from Reference [8]; © 2015 IEEE

Based on the above analysis, a dual-mode bandpass filter using two cascaded elliptic cavities is designed and demonstrated, as illustrated in Figure 2.50. Those two elliptic cavities have different AR. Therefore, a transmission zero on each side of the filter passband can be obtained. The other transmission zeros at the upper frequency band are caused by the higher modes in the elliptic cavities. The elliptic cavity filter has a center frequency of 400 GHz, and its optimized dimensions are listed in Table 2.7. The bended waveguides on both side of the cavity are used to make sure the input and output ports are align on the same axis for easy measurement.

The simulated and measured results of the proposed filter are shown in Figure 2.51. Simulated results are obtained by using a full-wave simulation with HFSS. The simulation is carried out with smooth gold and the conductivity is set to 3.3e7 S/m. The simulated 3 dB bandwidth is 8.05%, i.e., from 383.9 to 416.1 GHz,

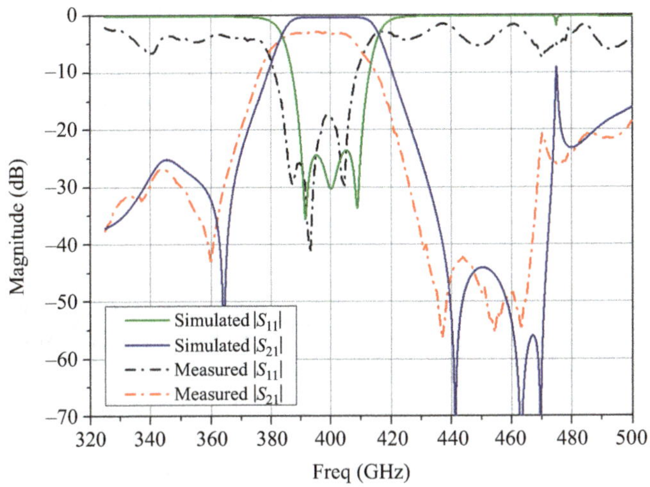

Figure 2.51 Comparison of the simulated results and the measured results of the bandpass filter. Reprinted with permission from Reference [8]; © 2015 IEEE

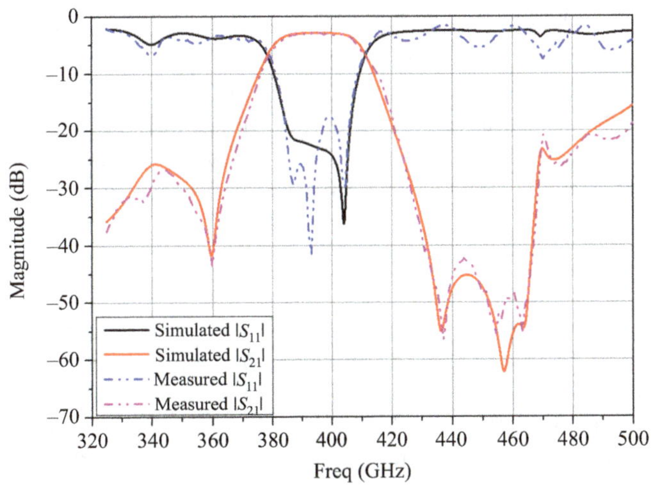

Figure 2.52 Comparison of the re-simulated results and the measured results of the bandpass filter. Reprinted with permission from Reference [8]; © 2015 IEEE

while the measured 3 dB bandwidth is 7.52%, i.e., from 380.2 to 409.9 GHz. A little frequency shift happens due to the taper angle of the lateral wall, as discussed in the design of the rectangular dual-mode cavity filter. The insertion loss is 0.39 dB and 2.84 dB for the simulated and the measured results, respectively. According to the above analysis, by considering the surface roughness and taper angle 1.46° of the lateral walls in the simulation model, the prototype has been re-simulated. The surface ratio and nodule radius are set to 4 and 0.5 μm, respectively. And the shift in the *z*-direction is set to 5 μm. The re-simulated and measured results are shown in Figure 2.52, where those two curves match very well with each other. It should be mentioned that the insertion loss of the prototype includes the loss comes from bended waveguides at the both side of the cavities.

2.5 Summary

In this chapter, various techniques for developing millimeter-wave and terahertz planar filters have been introduced. For the design of a high performance millimeter-wave planar filter, the SIW technique shows the merits of low insertion loss and good multilayered integration capability. Two efficient design methods based on the SM technique have been introduced in detail for the design of complicated SIW components. The method which combines the equivalent circuit of the SIW and the SM technique is capable of designing various complicated SIW components such as multiplexer, Butler matrix, and other waveguide-based components which use hundreds or thousands of metallic vias. The method which combines the coupling matrix and the SM method can be used to narrow or middle bandwidth filters with any types of topologies. For the terahertz design, it has been shown that the high-order mode cavity can provide higher quality factor than its fundamental-mode counterpart, as well as a lower insertion loss. In addition, multiple-mode cavity can reduce efficiently the component size and improve significantly the selectivity of the terahertz filter. For the terahertz filter implementation, the analysis of the DRIE process is carried out, which is useful to predict accurately the measured response of the terahertz filter. A number of examples with experimental results have been presented and discussed in detail to illustrate the design and measuring process of planar millimeter-wave and terahertz filters.

References

[1] X. Shang, M. L. Ke, Y. Wang, and M. J. Lancaster, Micromachined WR-3 waveguide filter with embedded bends, *Electronics Letters*, vol. 47, no. 9, pp. 488–490, 2011.

[2] X. H. Zhao, J. F. Bao, G. C. Shan, *et al.*, D-Band Micromachined Silicon Rectangular Waveguide Filter, *IEEE Microwave and Wireless Components Letters*, vol. 22, no. 5, pp. 230–232, 2012.

[3] J. Hu, S. Xie, and Y. Zhang, Micromachined Terahertz Rectangular Waveguide Bandpass Filter on Silicon-Substrate, *IEEE Microwave and Wireless Components Letters*, vol. 22, no. 12, pp. 636–638, 2012.

[4] X. Shang, M. Ke, Y. Wang and M. J. Lancaster, WR-3 Band Waveguides and Filters Fabricated Using SU8 Photoresist Micromachining Technology, *IEEE Transactions on Terahertz Science and Technology*, vol. 2, no. 6, pp. 629–637, 2012.

[5] C. A. Leal-Sevillano, T. J. Reck, C. Jung-Kubiak, *et al.*, Silicon Micromachined Canonical E-Plane and H-Plane Bandpass Filters at the Terahertz Band, *IEEE Microwave and Wireless Components Letters*, vol. 23, no. 6, pp. 288–290, 2013.

[6] X. Shang, Y. Tian, M. J. Lancaster, and S. Singh, A SU8 Micromachined WR-1.5 Band Waveguide Filter, *IEEE Microwave and Wireless Components Letters*, vol. 23, no. 6, pp. 300–302, 2013.

[7] Q. Chen, X. Shang, Y. Tian, J. Xu, and M. J. Lancaster, SU-8 micromachined WR-3 band waveguide bandpass filter with low insertion loss, *Electronics Letters*, vol. 49, no. 7, pp. 480–482, 2013.

[8] J.-X. Zhuang, Z.-C. Hao, and W. Hong, Silicon Micromachined Terahertz Bandpass Filter With Elliptic Cavities, *IEEE Transactions on Terahertz Science and Technology*, vol. 5, no. 6, pp. 1040–1047, 2015.

[9] X. Cao, Z. Tang, and J. Bao, Design of a Dual-Band Waveguide Filter Based on Micromachining Fabrication Process, *IET Microwaves, Antennas & Propagation*, vol. 10, no. 4, pp. 459–463, 2016.

[10] Y. Li, B. Pan, C. Lugo, M. Tentzeris, and J. Papapolymerou, Design and Characterization of a W-Band Micromachined Cavity Filter Including a Novel Integrated Transition From CPW Feeding Lines, *IEEE Transactions on Microwave Theory and Techniques*, vol. 55, no. 12, pp. 2902–2910, 2017.

[11] M. Stickel, P. Kremer, and G. V. Eleftheriades, A Millimeter-Wave Bandpass Waveguide Filter Using a Width-Stacked Silicon Bulk Micromachining Approach, *IEEE Microwave and Wireless Components Letters*, vol. 16, no. 4, pp. 209–211, 2017.

[12] J. Wang, B. Zhang, X. Wang, J. Duan, and W. Wang, Dual-Mode Band-Pass Filters Made by SU-8 Micromachining Technology for Terahertz Region, *Electronics Letters*, vol. 53, no. 11, pp. 730–732, 2017.

[13] J.-Q. Ding, S.-C. Shi, K. Zhou, D. Liu, and W. Wu, Analysis of 220-GHz Low-Loss Quasi-Elliptic Waveguide Bandpass Filter, *IEEE Microwave and Wireless Components Letters*, vol. 27, no. 7, pp. 648–650, 2017.

[14] X. Shang, M. Lancaster, and Y.-L. Dong, W-Band Waveguide Filter Based on Large TM120 Resonators to Ease CNC Milling, *Electronics Letters*, vol. 53, no. 7, pp. 488–490, 2017.

[15] J.-Q. Ding, S.-C. Shi, K. Zhou, Y. Zhao, D. Liu, and W. Wu, WR-3 Band Quasi-Elliptical Waveguide Filters Using Higher Order Mode Resonances, *IEEE Transactions on Terahertz Science and Technology*, vol. 7, no. 3, pp. 302–309, 2017.

[16] M. D. Auria, W. J. Otter, J. Hazell, *et al.*, 3-D Printed Metal-Pipe Rectangular Waveguides, *IEEE Transactions on Components, Packaging and Manufacturing Technology*, vol. 5, no. 9, pp. 1339–1349, 2015.

[17] X. Shang, P. Penchev, C. Guo, *et al.*, W-Band Waveguide Filters Fabricated by Laser Micromachining and 3-D Printing, *IEEE Transactions on Microwave Theory and Techniques*, vol. 64, no. 8, pp. 2572–2580, 2016.

[18] G. Wolf, G. Prigent, E. Rius, *et al.*, Band-Pass Coplanar Filters in the G-Frequency Band, *IEEE Microwave and Wireless Components Letters*, vol. 15, no. 11, pp. 799–801, 2005.

[19] B. Dehlink, M. Engl, K. Aufinger, and H. Knapp, Integrated Bandpass Filter at 77 GHz in SiGe Technology, *IEEE Microwave and Wireless Components Letters*, vol. 17, no. 5, pp. 346–348, 2007.

[20] S. Sun, J. Shi, L. Zhu, S. C. Rustagi, and K Mouthaan, Millimeter-Wave Bandpass Filters by Standard 0.18-μm CMOS Technology, *IEEE Electron Device Letters*, vol. 28, no. 3, pp. 220–222, 2007.

[21] C.-Y. Hsu, C.-Y. Chen, and H.-R. Chuang, 70 GHz Folded Loop Dual-Mode Bandpass Filter Fabricated Using 0.18um Standard CMOS Technology, *IEEE Microwave and Wireless Components Letters*, vol. 18, no. 9, pp. 587–589, 2008.

[22] L.-K. Yeh, C.-Y. Hsu, C.-Y. Chen, and H.-R. Chuang, A 24-60-GHz CMOS On-Chip Dual-Band Bandpass Filter Using Trisection Dual-Behavior Resonators, *IEEE Electron Device Letters*, vol. 29, no. 12, pp. 1373–1375, 2008.

[23] C.-Y. Hsu, C.-Y. Chen, and H.-R. Chuang, A 60-GHz Millimeter-Wave Bandpass Filter Using 0.18-um CMOS Technology, *IEEE Electron Device Letters*, vol. 29, no. 3, pp. 246–248, 2008.

[24] L. Nan, K. Mouthaan, Y.-Z. Xiong, J. Shi, S. C. Rustagi, and B-L Ooi, Design of 60- and 77-GHz Narrow-Bandpass Filters in CMOS Technology, *IEEE Transactions on Circuits and Systems-II: Express Briefs*, vol. 55, no. 8, pp. 738–742, 2008.

[25] S.-C. Chang, Y.-M. Chen, S.-F. Chang, *et al.*, Compact Millimeter-Wave CMOS Bandpass Filters Using Grounded Pedestal Stepped-Impedance Technique, *IEEE Transactions on Microwave Theory and Techniques*, vol. 58, no. 12, pp. 3850–3858, 2010.

[26] C.-Y. Hsu, C.-Y. Chen, and H.-R. Chuang, A 77-GHz CMOS On-Chip Bandpass Filter With Balanced and Unbalanced Outputs, *IEEE Electron Device Letters*, vol. 31, no. 11, pp. 1205–1207, 2010.

[27] Q. Xu, X. J. Bi, and G. Wu, Ultra-compacted sub-terahertz bandpass filter in 0.13 mm SiGe, *Electronics Letters*, vol. 48, no. 10, pp. 570–571, 2012.

[28] C.-L. Yang, M.-C. Chiang, H.-C. Chiu, and Y.-C. Chiang, Design and Analysis of a Tri-Band Dual-Mode Chip Filter for 60-, 77-, and 100-GHz Applications, *IEEE Transactions on Microwave Theory and Techniques*, vol. 60, no. 4, pp. 989–997, 2012.

[29] K. Ma, S. Mou, and K. S. Yeo, Miniaturized 60-GHz On-Chip Multimode Quasi-Elliptical Bandpass Filter, *IEEE Electron Device Letters*, vol. 34, no. 8, pp. 945–947, 2013.

[30] Y. Yang, X. Zhu, E. Dutkiewicz, and Q. Xue, Design of a Miniaturized On-Chip Bandpass Filter Using Edge-Coupled Resonators for Millimeter-Wave

Applications, *IEEE Trasactions on Electron Devices*, vol. 64, no. 9, pp. 3822–3828, 2017.

[31] V. N. R. Vanukuru and V. K. Velidi, Compact Millimeter-Wave CMOS Wideband Three-Transmission-Zeros Bandstop Filter Using a Single Coupled-Line Unit, *IEEE Transactions on Circuits and Systems-II: Express Briefs*, vol. 64, no. 9, pp. 1022–1026, 2017.

[32] M. Ito, K. Maruhashi, K. Ikuina, T. Hashiguchi, S. Iwanaga, and K. Ohata, A 60-GHz-Band Planar Dielectric Waveguide Filter for Flip-Chip Modules, *IEEE Transactions on Microwave Theory and Techniques*, vol. 49, no. 12, pp. 2431–2436, 2001.

[33] S. T. Choi, K. S. Yang, K. Tokuda, and Y. H. Kim, A V-band Planar Narrow Bandpass Filter Using a New Type Integrated Waveguide Transition, *IEEE Microwave and Wireless Components Letters*, vol. 14, no. 12, pp. 545–547, 2004.

[34] X. Chen, Z.-C. Hao, W. Hong, and T. Cui, Planar Asymmetric Dual-Mode Filters Based on Substrate Integrated waveguide (SIW), *IEEE Microwave Symposium Digest (IMS)*, June 12–17, 2005.

[35] J.-H. Lee, S. Pinel, J. Papapolymerou, J. Laskar, and M. M. Tentzeris, Low-Loss LTCC Cavity Filters Using System-on-Package Technology at 60 GHz, *IEEE Transactions on Microwave Theory and Techniques*, vol. 53, no. 12, pp. 3817–3824, 2005.

[36] D. Stephens, P. R. Young, and I. D. Robertson, Millimeter-Wave Substrate Integrated Waveguides and Filters in Photoimageable Thick-Film Technology, *IEEE Transactions on Microwave Theory and Techniques*, vol. 53, no. 12, pp. 3832–3838, 2005.

[37] L. Rigaudeau, P. Ferrand, D. Baillargeat, *et al.*, LTCC 3-D Resonators Applied to the Design of Very Compact Filters for Q-Band Applications, *IEEE Transactions on Microwave Theory and Techniques*, vol. 54, no. 6, pp. 2620–2627, 2006.

[38] X. Chen, W. Hong, T. Cui, Z. C. Hao, and K. Wu, Symmetric dual-mode filter based on substrate integrated waveguide (SIW), *Electrical Engineering - ELECTR ENG*, vol. 89, no. 1, pp. 67–70, 2006.

[39] J.-H. Lee, S. Pinel, J. Laskar, and M. M. Tentzeris, Design and Development of Advanced Cavity-Based Dual-Mode Filters Using Low-Temperature Co-Fired Ceramic Technology for V-Band Gigabit Wireless Systems, *IEEE Transactions on Microwave Theory and Techniques*, vol. 55, no. 9, pp. 1869–1879, 2007.

[40] T.-M. Shen, C.-F. Chen, T.-Y. Huang, and R.-B. Wu, Design of Vertically Stacked Waveguide Filters in LTCC, *IEEE Transactions on Microwave Theory and Techniques*, vol. 55, no. 8, pp. 1771–1779, 2007.

[41] H. J. Tang, W. Hong, J.-X. Chen, G. Q. Luo, and K. Wu, Development of Millimeter-Wave Planar Diplexers Based on Complementary Characters of Dual-Mode Substrate Integrated Waveguide Filters With Circular and Elliptic Cavities, *IEEE Transactions on Microwave Theory and Techniques*, vol. 55, no. 4, pp. 776–782, 2007.

[42] R. Q. Li, X.H. Tang, and F. Xiao, Substrate Integrated Waveguide Dual-Mode Filter Using Slot Lines Perturbation, *Electronics Letters*, vol. 46, no. 12, 2010.
[43] X.-P. Chen and K. Wu, Self-Packaged Millimeter-Wave Substrate Integrated Waveguide Filter With Asymmetric Frequency Response, *IEEE Transactions on Components, Packaging and Manufacturing Technology*, vol. 2, no. 5, pp. 775–782, 2012.
[44] M. Rezaee and A. R. Attari, Realisation of New Single-Layer Triple-Mode Substrate Integrated Waveguide and Dual-Mode Half-Mode Substrate-Integrated Waveguide Filters Using a Circular Shape Perturbation, *IET Microwaves, Antennas & Propagation*, vol. 7, no. 14, pp. 1120–1127, 2013.
[45] X.-C. Zhu, W. Hong, K. Wu, *et al.*, Design and Implementation of a Triple-Mode Planar Filter, *IEEE Microwave Wireless Component Letter*, vol. 23, no. 5, pp. 243–245, 2013.
[46] S. W. Wong, K. Wang, Z.-N. Chen, and Q.-X. Chu, Electric Coupling Structure of Substrate Integrated Waveguide (SIW) for the Application of 140-GHz Bandpass Filter on LTCC, *IEEE Transactions on Components, Packaging and Manufacturing Technology*, vol. 4, no. 2, pp. 316–322, 2014.
[47] S. W. Wong, K. Wang, Z.-N. Chen, and Q.-X. Chu, Design of Millimeter-Wave Bandpass Filter Using Electric Coupling of Substrate Integrated Waveguide (SIW), *IEEE Microwave and Wireless Components Letters*, vol. 24, no. 1, pp. 26–28, 2014.
[48] X.-P. Chen and Ke Wu, Substrate Integrated Waveguide Filters: Practical Aspects and Design Considerations, *IEEE Microwave Magazine*, vol. 15, no. 7, pp. 75–83, 2014.
[49] X.-P. Chen and Ke Wu, Substrate Integrated Waveguide Filters: Design Techniques and Structure Innovations, *IEEE Microwave Magazine*, vol. 15, no. 6, pp. 121–133, 2014.
[50] X.-P. Chen and Ke Wu, Substrate Integrated Waveguide Filter: Basic Design Rules and Fundamental Structure Features, *IEEE Microwave Magazine*, vol. 15, no. 5, pp. 108–116, 2014.
[51] T. V. Duong, W. Hong, Z. C. Hao, W. C. Huang, J. X. Zhuang, and M. H. Nguyen, A New Class of Selectivity-Improved mm-Waves Dual-Mode Substrate Integrated Waveguide Filters, *IEEE Asia-Pacific Microwave Conference 2015*, Nanjing, China, vol. 1, pp. 1–3, 6–9, 2015.
[52] S.-W. Wong, R. S. Chen, K. Wang, Z.-N. Chen, and Q.-X. Chu, U-Shape Slots Structure on Substrate Integrated Waveguide for 40-GHz Bandpass Filter Using LTCC Technology, *IEEE Transactions on Components, Packaging and Manufacturing Technology*, vol. 5, no. 1, pp. 128–134, 2015.
[53] Z.-C. Hao, W.-Q. Ding, and W. Hong, Developing Low-Cost W-Band SIW Bandpass Filters Using the Commercially Available Printed-Circuit-Board Technology, *IEEE Transactions on Microwave Theory and Techniques*, vol. 64, no. 6, pp. 1775–1786, 2016.
[54] T.-V.Duong, W. Hong, Z.-C. Hao, W.-C. Huang, J.-X. Zhuang, and V.-P. Vo, A Millimeter Wave High-Isolation Diplexer Using Selectivity-Improved

Dual-Mode Filters, *IEEE Microwave and Wireless Components Letters*, vol. 2, no. 2, pp. 104–106, 2016.

[55] Z. Liu, G. Xiao, and L. Zhu, Triple-Mode Bandpass Filters on CSRR-Loaded Substrate Integrated Waveguide Cavities, *IEEE Transactions on Components, Packaging and Manufacturing Technology*, vol. 6, no. 7, pp. 1099–1105, 2016.

[56] A. Vosoogh, A. A. Brazález, and P.-S. Kildal, A V-Band Inverted Microstrip Gap Waveguide End-Coupled Bandpass Filter, *IEEE Microwave and Wireless Components Letters*, vol. 26, no. 4, pp. 261–263, 2016.

[57] K. Wang, S.-W. Wong, G.-H. Sun, Z. N. Chen, L. Zhu, and Q.-X. Chu, Synthesis Method for Substrate-Integrated Waveguide Bandpass Filter With Even-Order Chebyshev Response, *IEEE Transactions on Components, Packaging and Manufacturing Technology*, vol. 6, no. 1, pp. 126–135, 2016.

[58] Y. Li, L.-A. Yang, H. Zou, H.-S. Zhang, X.-H. Ma, and Y. Hao, Substrate Integrated Waveguide Structural Transmission Line and Filter on Silicon Carbide Substrate, *IEEE Electron Device Letters*, vol. 38, no. 9, pp. 1290–1293, 2017.

[59] Y. Li, L.-A. Yang, S.-B. Wang, L. Du, and Y. Hao, Design of 3D filtering antenna for the application of terahertz, *Electronics Letters*, vol. 53, no. 1, pp. 7–8, 2017.

[60] Y. Li, L.-A. Yang, L. Du, K. Zhang, and Y. Hao, Design of Millimeter-Wave Resonant Cavity and Filter Using 3-D Substrate-Integrated Circular Waveguide, *IEEE Microwave and Wireless Components Letters*, vol. 27, no. 8, pp. 706–708, 2017.

[61] M. S. Sorkherizi and A. A. Kishk, Self-Packaged, Low-Loss, Planar Bandpass Filters for Millimeter-Wave Application Based on Printed Gap Waveguide Technology, *IEEE Transactions on Components, Packaging and Manufacturing Technology*, vol. 7, no. 9, pp. 1419–1431, 2017.

[62] K. Zhou, C.-X. Zhou, and W. Wu, Resonance Characteristics of Substrate-Integrated Rectangular Cavity and Their Applications to Dual-Band and Wide-Stopband Bandpass Filters Design, *IEEE Transactions on Microwave Theory and Techniques*, vol. 65, no. 5, pp. 1511–1524, 2017.

[63] E. G. Cristal and G. L. Matthaei, A Technique for the Design of Multiplexers Having Contiguous Channels, *IEEE Transactions on Microwave Theory and Techniques*, vol. 12, no. 1, pp. 88–93, 1964.

[64] J. D. Rhodes and R. Levy, Design of General Manifold Multiplexers, *IEEE Transactions on Microwave Theory and Techniques*, vol. 27, no. 2, pp. 111–123, 1979.

[65] M. Guglielmi, Simple CAD Procedure for Microwave Filters and Multiplexers, *IEEE Transactions on Microwave Theory and Techniques*, vol. 42, no. 7, pp. 1347–1352, 1994.

[66] A. A. Kirilenko, S. L. Senkevich, V. I. Tkachenko, and B. G. Tysik, Waveguide Diplexer and Multiplexer Design, *IEEE Transactions on Microwave Theory and Techniques*, vol. 42, no. 7, pp. 1393–1396, 1994.

[67] L. Accatino and M. Mongiardo, Hybrid Circuit-Full-Wave Computer-Aided Design of a Manifold Multiplexers without Tuning Elements, *IEEE Transactions on Microwave Theory and Techniques*, vol. 50, no. 9, pp. 2044–2047, 2002.

[68] J. R. Montejo-Garai, J. A. Ruiz-Cruz, and J. M. Rebollar, Full-Wave Design of H-Plane Contiguous Manifold Output Multiplexers Using the Fictitious Reactive Load Concept, *IEEE Transactions on Microwave Theory and Techniques*, vol. 53, no. 8, pp. 2628–2632, 2005.

[69] C.-F. Chen, T.-M. Shen, T.-Y. Huang, and R.-B. Wu, Design of Multimode Net-Type Resonators and Their Applications to Filters and Multiplexers, *IEEE Transactions on Microwave Theory and Techniques*, vol. 59, no. 4, pp. 848–856, 2011.

[70] X. Shang, Y. Wang, W. Xia amd M. J. Lancaster, Novel Multiplexer Topologies Based on All-Resonator Structures, *IEEE Transactions on Microwave Theory and Techniques*, vol. 61, no. 11, pp. 3838–3845, 2013.

[71] J. W. Bandler, R. M. Biernacki, S. H. Chen, P. A. Grobelny, and R. H. Hemmers, Space Mapping Technique for Electromagnetic Optimization, *IEEE Transactions on Microwave Theory and Techniques*, vol. 42, no. 12, pp. 2536–2544, 1994.

[72] J. W. Bandler, R. M. Biernacki, S. H. Chen, R. H. Hemmers, and K. Madsen, Electromagnetic Optimization Exploiting Aggressive Space Mapping, *IEEE Transactions on Microwave Theory and Techniques*, vol. 43, no. 12, pp. 2874–2882, 1995.

[73] M. H. Bakr, J. W. Bandler, R. M. Biernacki, S. H. Chen, and K. Madsen, A Trust Region Aggressive Space Mapping Algorithm for EM Optimization, *IEEE Transactions on Microwave Theory and Techniques*, vol. 46, no. 12, pp. 2412–2425, 1998.

[74] M. H. Bakr, J. W. Bandler, and N. Georgieva, An Aggressive Approach to Parameter Extraction, *IEEE Transactions on Microwave Theory and Techniques*, vol. 47, no. 12, pp. 2428–2439, 1999.

[75] J. W. Bandler, Q. S. Cheng, D. M. Hailu, and N. K. Nikolova, A Space-Mapping Design Framework, *IEEE Transactions on Microwave Theory and Techniques*, vol. 52, no. 11, pp. 2601–2610, 2004.

[76] K.-L. Wu, Y.-J. Zhao, J. Wang, and M. K. Cheng, An Effective Dynamic Coarse Model for Optimization Design of LTCC RF Circuits with Aggressive Space Mapping, *IEEE Transactions on Microwave Theory and Techniques*, vol. 52, no. 1, pp. 393–402, 2004.

[77] J. W. Bandler, Q. S. Cheng, S. A. Dakroury, *et al.*, Space Mapping: The State of the Art, *IEEE Transactions on Microwave Theory and Techniques*, vol. 52, no. 1, pp. 337–361, 2004.

[78] M. A. Ismail, D. Smith, A. Panariello, Y. Wang, and M. Yu, EM-Based Design of Large-Scale Dielectric-Resonator Filters and Multiplexers by Space Mapping, *IEEE Transactions on Microwave Theory and Techniques*, vol. 52, no. 1, pp. 386–392, 2004.

[79] J. M. Ros, P. S. Pacheco, H. E. Gonzalez, *et al.*, Fast Automated Design of Waveguide Filters Using Aggressive Space Mapping with a New Segmentation Strategy and a Hybrid Optimization Algorithm, *IEEE Transactions on Microwave Theory and Techniques*, vol. 53, no. 4, pp. 1130–1142, 2005.

[80] D. Gorissen, L. Zhang, Q.-J. Zhang, and T. Dhaene, Evolutionary Neuro-Space Mapping Technique for Modeling of Nonlinear Microwave Devices, *IEEE Transactions on Microwave Theory and Techniques*, vol. 59, no. 2, pp. 213–229, 2011.

[81] F. Feng, C. Zhang, V.-M.-R. Gongal-Reddy, Q.-J. Zhang, and J. Ma, Parallel Space-Mapping Approach to EM Optimization, *IEEE Transactions on Microwave Theory and Techniques*, vol. 62, no. 5, pp. 1135–1148, 2014.

[82] Z.-C. Hao, X. Huo, W. Ding, and W. Hong, Efficient Design of Compact Contiguous-Channel SIW Multiplexers using the Space-Mapping Method, *IEEE Transactions on Microwave Theory and Techniques*, vol. 63, no. 11, pp. 3651–3662, 2015.

[83] Z.-C. Hao, M. He, and W. Hong, Design of a Millimeter-Wave High Angle Selectivity Shaped-Beam Conformal Array Antenna using Hybrid Genetic/ Space Mapping Method, *IEEE Antennas and Wireless Propagation Letters*, vol. 15, pp. 1208–1212, 2016.

[84] Z.-C. Hao and M. He, Developing Millimeter-Wave Planar Antenna With a Cosecant Squared Pattern, *IEEE Transactions on Antennas and Propagation*, vol. 65, no. 10, pp. 5565–5570, 2017.

[85] L. Yan, W. Hong, K. Wu, and T.-J. Cui, Investigations on the Propagation Characteristics of the Substrate Integrated Waveguide Based on the Method of Lines, *Proceedings of the IEE.—Microwave Antennas & Propagation*, vol. 152, no. 1, pp. 35–42, 2005.

[86] J. C. Richard, M. K. Chandra, and R. M. Raafat, Microwave Filters for Communication Systems: Fundamentals, *Design and Applications*. New York, NY, USA: Wiley, 2007.

[87] ANSYS HFSS 19, https://www.ansys.com/en-gb/products/electronics/ansys-hfss/

[88] MATLAB, https://uk.mathworks.com/products.html?s_tid=gn_ps

[89] Advanced Design System (ADS), https://www.keysight.com/en/pc-1297113/advanced-design-system-ads?nid=-34346.0&cc=GB&lc=eng&cmpid=zzfindeesof-ads

[90] G. Matthaei, L. Young, and E. M. T. Jones, Microwave Filters, *Impedance Matching Networks and Coupling Structures*, 1st ed. New York, NY, USA: McGraw-Hill, 1964.

[91] J. S. Hong and M. J. Lancaster, Microstrip Filter for *RF/Microwave Applications*. New York, NY, USA: Wiley, 2001.

[92] R. J. Cameron, Advanced Coupling Matrix Synthesis Techniques for Microwave Filters, *IEEE Transactions on Microwave Theory and Techniques*, vol. 51, no. 1, pp. 1–10, 2003.

[93] Y. Cassivi, L. Perregrini, P. Arcioni, M. Bressan, K. Wu, and G. Conciauro, Dispersion Characteristics of Substrate Integrated Rectangular Waveguide, *IEEE Microwave and Wireless Components Letters*, vol. 12, no. 2, pp. 333–335, 2002.

[94] F. Feng, C. Zhang, V.-M.-R. Gongal-Reddy, Q.-J. Zhang, and J. Ma, Parallel Space-Mapping Approach to EM Optimization, *IEEE Transactions on Microwave Theory and Techniques*, vol. 62, no. 5, pp. 1135–1148, 2014.

[95] Y. L. Zhang, T. Su, Z. P. Li, and C. H. Liang, A Hybrid Computer Aided Tuning Method for Microwave Filters, *Progress in Electromagnetic Research*, vol. 139, pp. 559–575, 2013.

[96] Z. C. Hao and J. Wang, Investigations on the Rectangular Waveguide to SIW Transition for Terahertz Applications, *Wiley, Microwave and Optical Technology Letters*, vol. 59, no. 7, pp. 1546–1553, 2017.

[97] P. L. Kirby, D. Pukala, H. Manohara, I. Mehdi, and J. Papapolymerou, Characterization of Micromachined Silicon Rectangular Waveguide at 400 GHz, *IEEE Microwave and Wireless Components Letters*, vol. 16, no. 6, pp. 366–368, 2006.

[98] J. R. Stanec and N. S. Barker, Fabrication and Integration of Micromachined Millimeter-Wave Circuits, *IEEE Microwave and Wireless Components Letters*, vol. 21, no. 8, pp. 409–411, 2011.

[99] Y. Li, P. L. Kirby, O. Offranc, and J. Papapolymerou, Silicon Micromachined W-Band Hybrid Coupler and Power Divider Using DRIE Technique, *IEEE Microwave and Wireless Components Letters*, vol. 18, no. 1, pp. 22–24, 2008.

[100] H. J. Tang, W. Hong, G. Q. Yang, and J. X. Chen, Silicon Based THz Antenna and Filter with MEMS Process, in *Proceedings of the International Workshop on Antenna Technology*, Mar. 2011, pp. 148–151.

[101] K. Leong, K. Hennig, C. Zhang, *et al.*, WR-1.5 Silicon Micromachined Waveguide Components and Active Circuit Integration Methodology, *IEEE Transactions on Microwave Theory and Techniques*, vol. 60, no. 4, pp. 998–1005, 2012.

[102] S. H. Hall, S. G. Pytel, P. G. Huray, *et al.*, Multi-GHz, Causal Transmission Line Modeling Methodology with a Hemispherical Surface Roughness Approach, *IEEE Transactions on Microwave Theory and Techniques*, vol. 55, no. 12, pp. 2614–2624, 2007.

[103] M. Guglielmi, O. Roquebrun, P. Jarry, E. Kerherve, M. Capurso, and M. Piloni, Low-Cost Dual-Mode Asymmetric Filters in Rectangular Waveguide, *IEEE MTT-S Int. Microwave Symp. Dig.*, vol. 3, pp. 1787–1790, 2001.

[104] U. Rosenberg and S. Amari, Novel Design Possibilities for Dual-Mode Filters Without Intracavity Couplings, *IEEE Microwave and Wireless Components Letters*, vol. 12, no. 8, pp. 296–298, 2002.

[105] M. Guglielmi, P. Jarry, E. Kerherve, O. Roquebrun, and D. Schmitt, A New Family of All-Inductive Dual-Mode Filters, *IEEE Transactions on Microwave Theory and Techniques*, vol. 49, no. 10, pp. 1764–1769, 2001.

Chapter 3

Advances in planar coaxial SIW resonator filters design

Stefano Sirci[1], Jorge D. Martínez[2], Miguel Ángel Sánchez-Soriano[3], and Vicente E. Boria[4]

3.1 Introduction

During the last years, substrate-integrated waveguide (SIW) technology has evolved into a well established and successful choice for implementing microwave and mm-wave filters with low losses and high selectivity, while keeping a monolithic integration with other planar technologies. Additionally, SIW components can be fabricated using different batch manufacturing processes as PCB or low-temperature co-fired ceramic (LTCC), thus enabling mass production and low cost [1].

More recently, increased interest has grown on the development of more compact SIW structures, mostly due to the inherently larger size of SIW devices compared to other common planar technologies. Different strategies have been followed in order to achieve this goal. A common approach has consisted on bisecting the structure at quasi-perfect magnetic walls (i.e. as in folded [2], half-mode [3] and quarter-mode SIW bandpass filters (BPFs) [4]). Other techniques have relied on loading the SIW cavities with complementary split-ring resonators [5] or rods of dielectric material [6].

Furthermore, advanced filtering responses including transmission zeros (TZs) at finite or imaginary frequencies are highly demanded in numerous applications for implementing high selectivity or linear phase filters. The introduction of positive and negative cross-couplings between resonators has been the most prominent technique, based on signal cancellation along the multiple paths between the filter input and output ports [7]. Thus, magnetic couplings have been broadly used in SIW filters by means of post-wall irises between adjacent resonators, while electric coupling structures have been broadly researched by several authors using single [8,9] and double-layer structures [10].

[1]Institute of Telecommunications and Multimedia Applications (iTEAM), Universitat Politècnica de València, Spain
[2]Department of Electronic Engineering, Universitat Politècnica de València, Spain
[3]Department of Physics, Systems Engineering and Signal's Theory, University of Alicante, Spain
[4]Department of Communications, Universitat Politècnica de València, Spain

In this context, a SIW coaxial filter was recently proposed as a direct translation of the classical combline waveguide filter to a substrate-integrated technology. This solution was proposed for the first time in [11], and extensively studied for implementing compact filters in several works [12,13].

Coaxial SIW filters present important advantages in terms of compactness, and can easily accommodate magnetic and electric couplings in a single-layer structure suitable for batch production using low-cost PCB fabrication procedures. Additionally, the structure can withstand moderate power levels both in continuous and pulsed signal conditions. Lastly, advanced topologies like dual-mode filters and singlet sections can also be implemented based on the same principle.

3.2 Coaxial SIW technology

In this section we present, analyse and study a novel structure for implementing moderate-to-high *Q*-factor cavity resonators, based on a translation of the well-known 3D coaxial resonator concept to a SIW scheme.

The most important advantages of such configuration are the high miniaturisation degree, and the straightforward accessibility to the capacitance section of the coaxial resonator, but keeping fabrication and integration easiness. As it will be shown in the following sections, its external location enables the designers to implement the control of the loading capacitance in very practical ways, allowing to create out-of-phase couplings between non-adjacent resonators.

3.2.1 Study of a coaxial SIW cavity

As it is well known, coaxial or combline filters have been extensively used in waveguide and planar technologies, especially for the compact implementation of higher-order filters. In its simplest form, a combline resonator in waveguide technology consists of a conductive cavity with a metal rod placed within it symmetrically [14,15]. The latter is short circuited at one end located at the bottom of the metallic cavity, while the other end is spaced some distance from the top cover of the cavity. The inner conductive rod can be rectangular, or can have other cross section in shape (depending on the manufacturing process).

The enormous similarity between waveguide and SIW technologies allows us to easily implement such topology in SIW technology. Indeed, the layout and the main design parameters of a coaxial SIW resonator are shown in Figure 3.1.

Thus, a coaxial SIW resonator consists on a square or circular SIW cavity where a plated via hole has been inserted at the centre. The inductive section of the combline resonator has been obtained by the plated via hole, which is connected to ground at one of its ends. At the other side, a metal disk (square with a semi-side length or circular with radius, both named r_p) having a size much bigger than the inner post diameter d_v is connected. Between this disk and the top metal plane a small air gap s_p is inserted, so generating a high capacitance towards ground (termed loading capacitance C_l), which represents the capacitive section of the coaxial resonator. In particular, the fringing fields across the air gap generate the integrated

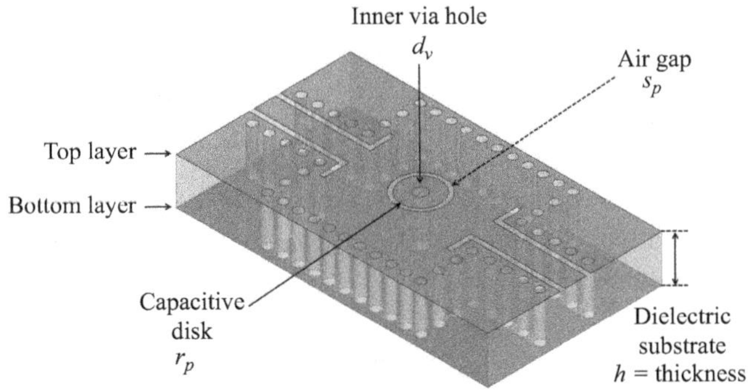

Figure 3.1 3D view of a coaxial SIW resonator. © 2015 IEEE

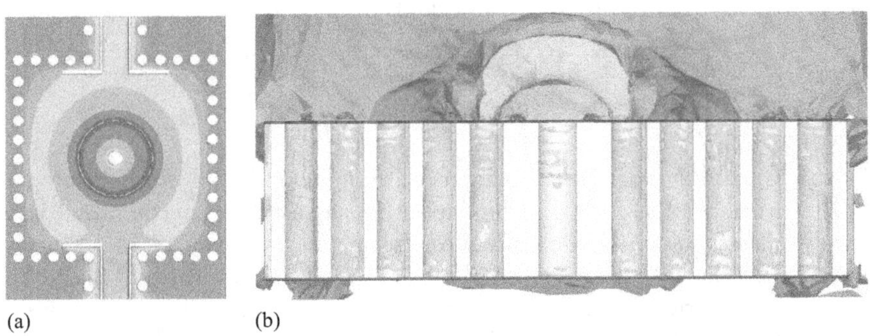

Figure 3.2 3D E-field distribution for a coaxial SIW resonator: (a) top view and (b) in the air box surrounding the structure

C_l of a typical combline configuration, which can be clearly seen by observing the electric field distribution around the air gap, as it is shown in Figure 3.2.

3.2.2 Filter synthesis

As Figure 3.1 shows, in a coaxial SIW resonator, the inner via is short-circuited at the bottom metal layer and open-ended at the top. Therefore, the resonator can be seen as a piece of circular-square coaxial line of length h and characteristic admittance Y_0 embedded into the dielectric substrate [12]. The external square conductor is composed by the via hole walls that form the SIW cavity, while the coaxial inner circular conductor is the central plated via hole. Such combline resonator can be modeled as a TEM-mode transmission line short-circuited at one end and terminated with a capacitor on the other side. The TEM-mode resonant frequency is given by the condition $B(\omega) = 0$, taking into account that the susceptance B of the coaxial resonator is expressed as

$$B(\omega) = \omega C_l - \frac{1}{Z_0} \cot \beta h \tag{3.1}$$

where β is the TEM-mode propagation constant, Z_0 is the coaxial resonator characteristic impedance, and θ_0 is the resonator electrical length at the desired centre frequency f_0. It should be remarked that the length of the coaxial transmission line h corresponds to the substrate laminate thickness, so that it becomes now a key design parameter, increasing the structure design flexibility.

As it has been demonstrated in [16], for a circular inner conductor of diameter d_v and a external square contour of side l_{SIW}, the admittance Y_0 of this coaxial guide can be well approximated by

$$Y_0 = \left[\frac{60}{\sqrt{\varepsilon_r}}\ln\left(1.079\frac{l_{\text{siw}}}{d_v}\right)\right]^{-1} \text{for } l_{\text{SIW}} \gg d_v \tag{3.2}$$

where ε_r is the dielectric substrate permittivity. On the other hand, as it has been explained in [17], the loading capacitance C_l established between the top metal layer and the circular patch of radius r_p can be estimated by using

$$C_l = \frac{2\pi r_p(1+\varepsilon_r)}{\ln\left(1+\dfrac{s_p}{r_p}\right)}\int_0^{\infty}\left[J_0(\gamma r_p) - J_0(\gamma(r_p+s_p))\right]\frac{J_1(\gamma r_p)}{\gamma}d\gamma \tag{3.3}$$

where J_n is the n-th order Bessel function of the first kind. It is worth mentioning that (3.3) is only valid for a circular capacitive patch neglecting the metal layer thickness. Thus, an optimisation of the patch structure by means of 3D EM simulations, which includes the metal layer thickness t, is usually needed to finely tune C_l. Hence, the fabrication process selected for implementing such SIW components plays a fundamental role in setting C_l, since metal layer t relies on the properties of the manufacturing procedure.

Let us now consider the design of an all-pole direct-coupled BPF using shunt resonators and frequency-invariant admittance inverters, whose equivalent circuit is shown in Figure 3.3. Given a filter response with centre frequency $\omega_0 = \sqrt{\omega_L\omega_H}$ and bandwidth $\Delta\omega$, being ω_H and ω_L the upper and lower cutoff frequencies, C_l and Y_0 may be obtained by mapping the resonator B and the admittance inverters of the BPF with those of the low-pass prototype. The former condition can be expressed as

$$B(\omega) = \omega'\left(\frac{b\dfrac{\Delta\omega}{\omega_0}}{\Omega_C}\right) \tag{3.4}$$

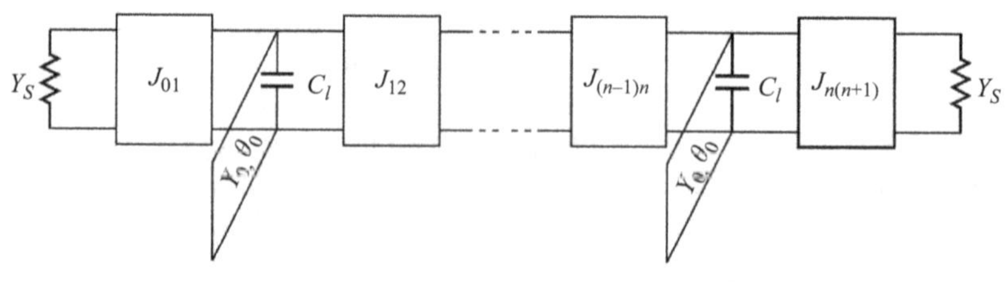

Figure 3.3 BPF prototype using shunt resonators and frequency-invariant admittance inverters. © 2012 IEEE

where ω' and $\Omega_C = 1$ rad/s are the angular frequency and cut-off frequency of the low-pass prototype, respectively. Now, substituting (3.1) into (3.4) for the corresponding pass-band edge frequencies ($\omega' = \pm 1 \Rightarrow \omega = \{\omega_H, \omega_L\}$), we can obtain

$$C_l = \left(b\frac{\Delta\omega}{\omega_0}\right)\frac{\cot\theta_H + \cot\theta_L}{\omega_H \cot\theta_L - \omega_L \cot\theta_H} \tag{3.5}$$

$$Y_0 = \left(b\frac{\Delta\omega}{\omega_0}\right)\frac{\omega_H + \omega_L}{\omega_H \cot\theta_L - \omega_L \cot\theta_H} \tag{3.6}$$

where $\theta_H = \beta_H h$ and $\theta_L = \beta_L h$, and β_H and β_L are the phase constants at the pass-band edge frequencies. Then, the admittance inverters $J_{n,n+1}$ can be computed from the low-pass prototype coefficients $(g_0, g_1, \cdots, g_{n+1})$ using the well-known expressions presented in [18].

In this context, to design a BPF based on coaxial SIW resonators, the synthesis procedure usually starts by choosing the resonator b at the desired centre angular frequency ω_0. Using the definition of the susceptance b for resonators having zero susceptance at ω_0 proposed in [18], and taking into account (3.1), the slope parameter for the proposed coaxial SIW topology can be particularised as

$$b = \frac{1}{2}\left[\omega_0 C_l + \frac{\theta_0}{Z_0 \sin^2\theta_0}\right] \tag{3.7}$$

Note that the b must be conveniently chosen as a trade-off between compactness, physical feasibility of synthesised values of C_l and Z_0, and resonator Q-factor, as it will be shown in the next section. Then, the corresponding coaxial Z_0 may be expressed as

$$Z_0 = \left(\frac{\cot\theta_0 + \theta_0 \csc^2\theta_0}{2b}\right) \tag{3.8}$$

The latter is used to extract the ratio between the outer cavity side l_{SIW} and the inner hole d_v as

$$\frac{l_{\mathrm{SIW}}}{d_v} = 0.9268 \cdot e^{\frac{Z_0\sqrt{\varepsilon_r}}{60}} \tag{3.9}$$

The d_v is usually set to the minimum diameter permitted by the fabrication technology, so that the cavity size for a given Y_0 level is reduced. Then, the SIW cavity side l_{SIW} is calculated from the synthesised Y_0 using (3.9).

Next, given the required C_l, the patch dimensions s_p and r_p may be obtained numerically from (3.3), and then optimised by means of 3D electromagnetic (EM) simulations including metal thickness. In addition to this, s_p has to be kept as small as possible to implement strongly loaded configurations and reduce radiation losses due to the isolating gaps. Once the resonators are designed, the input/output and inter-resonators couplings will be adjusted. The external quality factor Q_e and coupling coefficients $k_{i,i+1}$ are extracted from the synthesised $J_{i,i+1}$ values, and then, using full-wave EM simulations, it is possible to obtain the desired Q_e and

$k_{i,i+1}$ by adjusting the coupling mechanisms between coaxial SIW resonators that will be presented in Sections 3.3.1 and 3.3.2. Finally, a fine-tuning procedure using full-wave EM simulations of the whole filter structure should be accomplished. From a practical point of view, when adjusting the coaxial SIW resonator centre frequencies, it is recommended to perform optimisation directly on C_l values, since a fine control of the f_0 value is easily attained.

3.2.3 EM performance of coaxial SIW resonators

An SIW structure is inherently a dielectric-filled synthesised rectangular waveguide with periodic metallised via holes or slot arrays. So, the substrate properties do play a very important role for the physical and electrical performance of any microwave SIW filter. For instance, the intrinsic sources of losses in SIW structures are due to the metal layers (conductor loss), the dielectric substrate (dielectric loss), and the power leakage through the lateral sidewalls. In particular, if the longitudinal spacing of the via holes on the sidewalls becomes too large, the EM field is not confined anymore inside the structure and the SIW tends to radiate throughout the surrounding dielectric. The aforementioned statements are completely valid for the coaxial SIW topology, even though there are other parameters that strongly affect its EM performance. In such structures, the fundamental mode is the TEM-mode, meanwhile in a standard SIW the fundamental mode is the TE_{101}-mode. That means that the EM field distributions are completely different among them. Indeed, the EM field of the TEM-mode is mainly concentrated around the integrated loading capacitance, with a high current that flows along the inner plated via hole, and a strong electric field across the isolating gap. As it can be observed in Figure 3.2, the EM field strength progressively decreases when getting close to lateral sidewalls.

Hence, key design parameters that control EM performance are the inner via diameter d_v, the gap spacing s_p and the patch perimeter p_p. Let us analyse the effects of these parameters on the unloaded Q-factor (Q_u) and resonant frequency for the coaxial SIW resonator of Figure 3.4. In this context, let us consider a single resonator having centre frequency 4 GHz as benchmark component. Figure 3.4(a) depicts a top view of the resonator showing its main parameters. The device is implemented in a 2.54 mm-thick Rogers TMM10i dielectric substrate ($\varepsilon_r = 9.8 \pm 0.245$, $\tan\delta = 2 \cdot 10^{-3}$). The resonator b is set to 37 mS that gives: $C_l = 1.25$ pF (i.e. $r_p = 2$ mm, $s_p = 0.125$ mm), $Z_0 = 40.8\,\Omega$ (i.e. $d_v = 0.8$ mm, $w_{siw} \times l_{siw} = 6.3 \times 6.3\ \text{mm}^2$ corresponding to $0.08 \times 0.08\lambda_0^2$), and $\theta_0 = 38.2°$ (i.e. $0.04\lambda_0$), where λ_0 is the free space wavelength at 4 GHz. Full-wave simulations performed with ANSYS HFSS 2016 [19] of the benchmark at C-band are depicted in Figure 3.4(b), and these are compared to the responses of a conventional TE_{101} SIW cavity resonator. The resonator wide-band response highlights a spurious free band of more than one octave, while the fundamental mode f_0 (i.e. TEM-mode) has been strongly reduced with respect to the one of a conventional SIW resonator having the same cavity size (i.e. 63% of frequency reduction from 11.3 to 4 GHz).

Figure 3.5(a) highlights the Q_u versus f_0 of the coaxial SIW resonator if the main design parameters are modified one at a time. In particular, the spacing s_p has

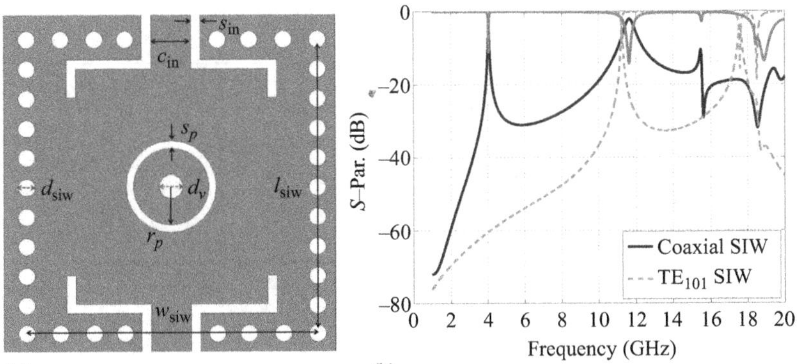

(a) (b)

Figure 3.4 (a) Coaxial SIW resonator. (b) Comparison between simulated responses of the coaxial SIW resonator and its TE_{101}-based equivalent. © *2015 IEEE*

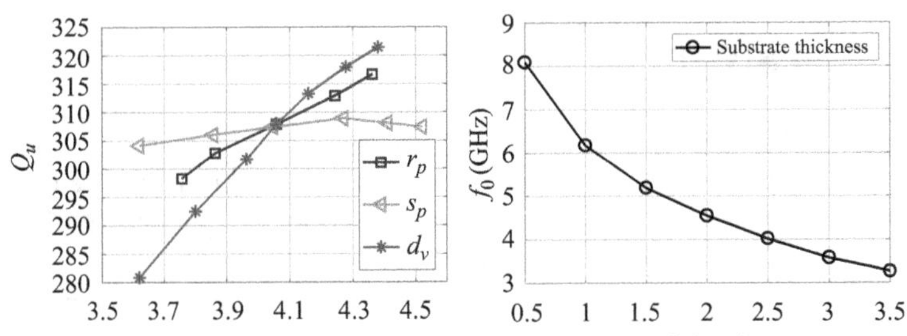

(a) (b)

Figure 3.5 (a) Q_u versus f_0 for the SIW-coaxial resonator changing d_v, s_p and p_p. (b) f_0 versus the substrate thickness h

a dominant effect on f_0, showing a broad tuning with slightly EM performance degradation. In addition, to enhance Q_u is important to reduce the capacitive patch size, and, especially, a wider d_v is needed. Thus, constraints of s_p play a major role in the choice of the fabrication technology. In this context, Figure 3.5(b) depicts the simulated f_0 as function of substrate thickness h. As previously mentioned, f_0 does depend on the coaxial line length that corresponds to h. Since h can be changed by using laminates of different thicknesses, a coarse tuning of f_0 may be achieved. These results set out another prove that the fundamental mode of the proposed topology is the TEM mode, which resonates along the resonator vertical direction. Note that thicker h allows us to increase the resonator Q_u, and to boost the spurious-free band-stop performance of the coaxial resonator, since the the TE_{101} mode does not depend on h.

Finally, Table 3.1 gives the relation between Q_u and the miniaturisation degree for the benchmark resonator, when its slope parameter is increased. Those values

Table 3.1 Q_u versus miniaturisation [13]

SIW topology	Area (mm^2)	ΔArea (%)	Q_u	ΔQ_u(%)
TE_{101}	392	0	294	0
Coaxial $b = 0.0265$	324	−17	308	+5
Coaxial $b = 0.029$	213	−46	274	−7
Coaxial $b = 0.034$	121	−69	238	−19
Coaxial $b = 0.035$	90	−77	230	−22

have been compared to the area and Q_u of a standard TE_{101} SIW resonator centred at 4 GHz. If the building block resonator is designed with a higher b, the SIW cavity size may be strongly reduced to compensate for the higher C_l while maintaining the same f_0. This circuit area reduction also leads to an upward shift of the cavity first spurious mode, widening the stopband bandwidth, so that, an improvement in terms of size and performance can be really achieved at the cost of just moderate Q_u degradation.

3.3 Coaxial SIW BPFs

In this section, the inter-resonator coupling mechanisms that can be used for coupling coaxial SIW resonators are presented and studied. Two different coupling structures will be presented and analysed: a conventional magnetic coupling system based on post-wall irises, and an electric coupling mechanism especially suited for coaxial SIW resonators. Then, some examples of filter designs are presented to prove the potentiality of using coaxial SIW resonators as building blocks of in-line microwave BPFs.

To study the performance of the proposed inter-resonator coupling mechanisms, let us consider a benchmark coaxial SIW resonator having centre frequency 5.5 GHz and using the 1.524 mm-thick Rogers RO4003C dielectric substrate ($\varepsilon_r = 3.55 \pm 0.05$, $\tan\delta = 2.1 \times 10^{-3}$ at 2.5 GHz). Setting $b = 29.4$ mS, the main parameters become: $C_l = 0.825$ pF (i.e. $r_p = 4.2 \times 4.2\,\text{mm}^2$ and $s_p = 0.24$ mm), $Z_0 = 103.5\,\Omega$ (i.e. $d_{via} = 0.6$ mm, $w_{SIW} \times l_{SIW} = 14.6 \times 14.6\,\text{mm}^2$ corresponding to $0.27 \times 0.27\lambda_0^2$ at 5.5 GHz), and $\theta_0 = 18.9°$ (i.e. $0.03\lambda_0$). Weakly-coupled benchmark resonators are considered and simulated in order to estimate $k_{i,i+1}$ in terms of the characteristic frequencies of synchronously tuned coupled resonators, as demonstrated in [18].

3.3.1 Magnetic coupling

The post-wall iris coupling used in SIW presents a structure very similar to the one used in classical metallic waveguide technology. Thus, some via holes have been removed from the post-wall that separates adjacent coaxial SIW resonators. The post-wall iris can be seen as a form of discontinuity, which works by exciting evanescent higher order modes. Its main design parameter is the iris width W which is shown in Figure 3.6(a).

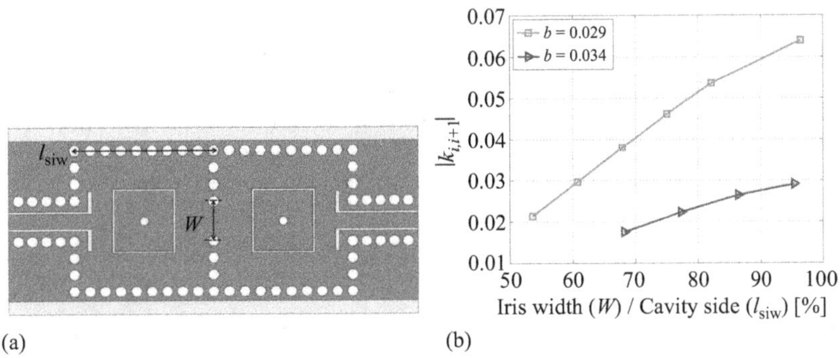

Figure 3.6 (a) Magnetic coupling based on post-wall iris. (b) $k_{i,i+1}$ versus the ratio between W and cavity side l_{SIW} for two values of resonator b.

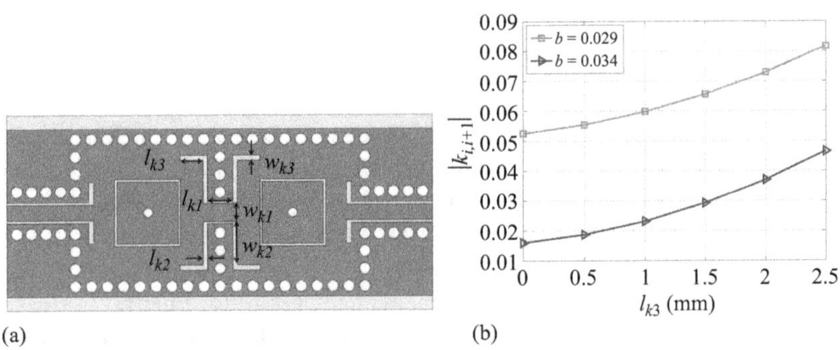

Figure 3.7 (a) Magnetic coupling based on post-wall iris and current CPW probes. (b) $k_{i,i+1}$ variation versus l_{k3}

As Figure 3.6(b) shows, $k_{i,i+1}$ coefficients have been extracted from simulations for two resonator configurations having distinct b, which are 0.029 S and 0.034 S, resulting in very different miniaturisation degrees. Indeed, raising b, the cavity side diminishes from 14.6 mm to 11.5 mm, while C_l increases from 0.825 to 0.95 pF. Magnetic coupling value $k_{i,i+1}$ increases for wider W, nevertheless, the magnitude is strongly reduced for higher b values (see Figure 3.6(b)). This means that only very narrow BPFs can be designed by using such coupling mechanism when a higher miniaturisation is required.

However, to enhance the magnetic coupling between coaxial SIW resonators, coplanar (CPW) probes can be introduced at the top metal layer in a similar manner to the CPW-to-SIW transition, which are used for the external coupling of SIW filters [20]. Thus, current probes have been etched at the centre of the post-wall iris, as it is shown in Figure 3.7(b). The coupling mechanism is mainly controlled by setting w_{k_2} and l_{k_3}. As Figure 3.7(a) shows for $b = (0.029, 0.034)$, $k_{i,i+1}$ increases with longer l_{k_3}. Although this solution allows to achieve higher coupling magnitude, a significant magnitude reduction is again evident for miniaturised coaxial SIW implementations. As previously mentioned, the CPW-to-SIW transition with 90° bend slots created at the ends of the ports, and shown in Figure 3.6(a), are widely used as magnetic external coupling mechanism in standard SIW resonators [20]. This mechanism is also applicable to coaxial SIW resonators, and similarly, external coupling magnitude increases with longer ground plane slots, as for the inter-resonator coupling case.

3.3.2 Electric coupling

The electric coupling scheme between coaxial SIW resonators has been firstly proposed in [21], and it is suitable for implementing external and inter-resonator couplings. This mechanism permits to achieve higher coupling magnitudes compared to conventional magnetic irises for strongly loaded coaxial SIW filter configurations, as proved in [13], while maintaining a simple single-layer fabrication process.

To realise an electric inter-resonator coupling, a capacitive probe based on a high impedance CPW line is introduced between adjacent resonators. The two ends of this CPW line penetrate into the head of the capacitive patches, which means in the region with the highest intensity of EM field (see Figure 3.8(a)). A gap s between the capacitive patch and the CPW probe must be ensured, so that the probe insertion p can control the $k_{i,i+1}$ magnitude (see Figure 3.8(b)). A clear advantage compared to other electric coupling approaches in SIW is the presence of coupling probes only on one metal layer, usually at the top side. Indeed, as it has been demonstrated in [10,22], slots must be etched on the top and bottom layers to generate electric coupling, limiting component integrability. Finally, by inserting input/output ports in the capacitive patchs of the first and last coaxial SIW resonators and ensuring isolating gaps, an electric external coupling can be easily created too. Again, the higher the insertion of the CPW probes is, the lower the Q_e value is. To sum up, the proposed electric coupling is an efficient approach for ensuring higher couplings between coaxial SIW resonators, and this is especially

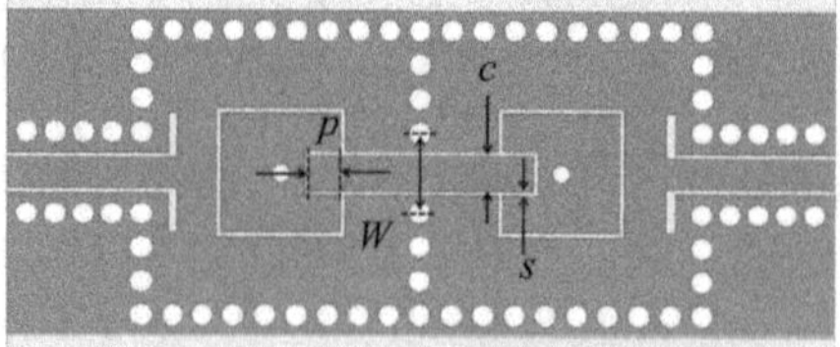

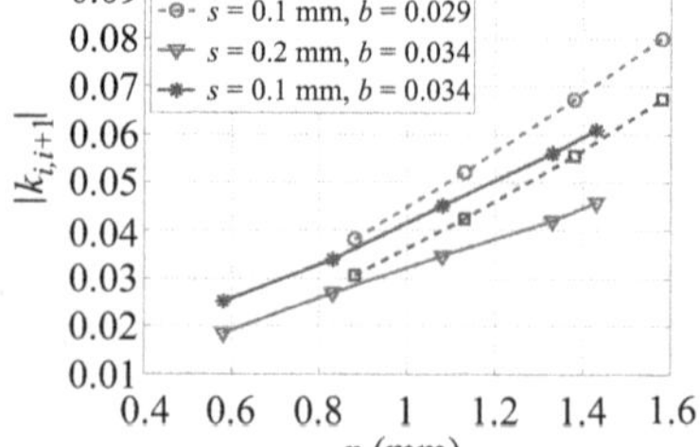

Figure 3.8 *(a) Electric inter-resonant coupling mechanism. (b) $k_{i,i+1}$ versus probe insertion p. © 2015 IEEE*

true when the compactness has to be increased. Hence, this mechanism provides improved design flexibility, which allows us to fulfill stringent filter requirements in terms of bandwidth, size and rejection band.

3.3.3 In-line configuration: design examples

To demonstrate the flexibility of the coaxial SIW topology and its various coupling schemes, this approach is now applied to three in-line BPFs operating at C- and X-bands, which have been manufactured using different fabrication processes. The first example is a three-pole Chebyshev BPF whose synthesised response is centred at 9.8 GHz with fractional bandwidth (FBW) of 5% and 0.05 dB in-band ripple [12]. The filter synthesised response is described by: $J_{01} = J_{34} = 0.0036$, $J_{12} = J_{23} = 0.007$, and input/output $Z_S = 65\ \Omega$. Thus, magnetic coupling schemes have been used for both external and inter-resonator couplings, which are the CPW-to-SIW probes and post-wall iris, respectively.

The filter was fabricated in a 1.524 mm-thick Rogers R 4003C substrate ($\varepsilon_r = 3.55$, $\tan\delta = 2.7 \times 10^{-3}$ at 10 GHz) using standard single-layer PCB processing technology (see Figure 3.9(a)). To reduce the filter size, the resonator b has been set to 14 mS, thus generating $C_l = 0.2$ pF and $Y_0 = 8.22$ mS. The miniaturisation degree is remarkable as the filter size is just $8 \times 17.7\ \text{mm}^2$ (i.e. $0.26 \times 0.58\lambda_0^2$). An equivalent conventional TE_{101}-mode SIW filter would require about $12 \times 36\ \text{mm}^2$ (i.e. $0.4 \times 1.18\lambda_0^2$). Furthermore, the wide-band filter response reveals a spurious-free band of almost one octave, with a rejection level better than 30 dB up to 17 GHz, which represents a major improvement against standard SIW BPFs.

The second in-line filter example is a two-pole Chebyshev BPF that was designed and fabricated in an eight-layer LTCC stack-up [23]. The synthesised filtering response is centred at 10.75 GHz with 1-dB FBW equals to 4.3% and return losses (RL) higher than 15 dB. The use of the LTCC technology allows us to create embedded coaxial SIW resonators, as it is shown in Figure 3.10(a). Thus, a highly loaded configuration having $b = 59.7$ mS is created showing: $C_l = 0.87$ pF, and $Z_0 = 74.9\ \Omega$. Post-wall irises with CPW probes are needed to

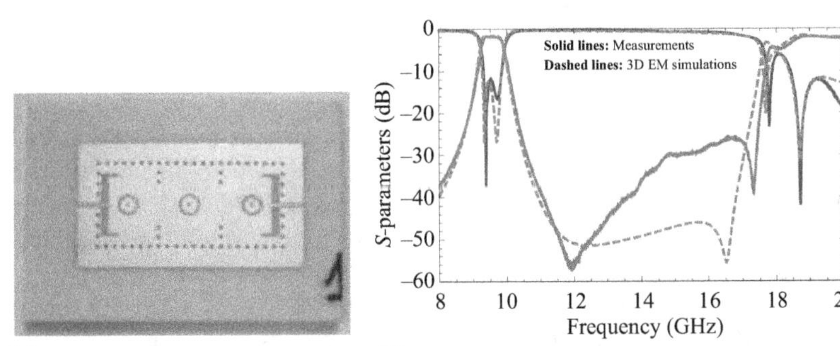

Figure 3.9 Three-pole coaxial SIW filter: (a) photographs, and its wideband (b) measurements (solid) and simulations (dashed). © 2012 IEEE

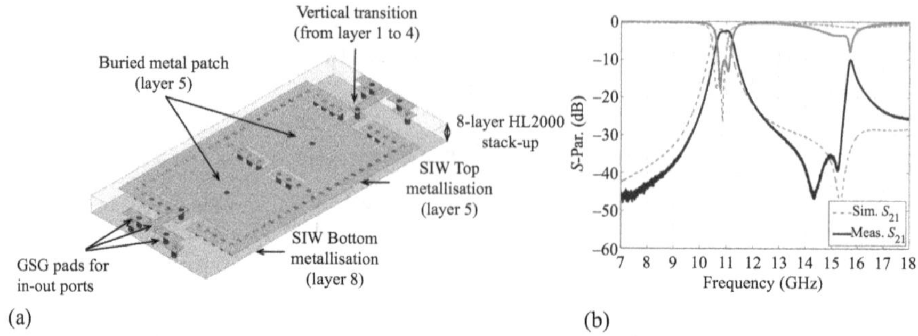

Figure 3.10 Two-pole LTCC filter: (a) 3D filter view and (b) (dashed) simulations, and (solid) measurements

increase the magnetic coupling between coaxial SIW resonators, while standard CPW-to-SIW probes are used for controlling Q_e. In this design, the Heraeus HL2000 LTCC tape ($\varepsilon_r = 7.4 \pm 0.2$, $\tan\delta = 2.6 \times 10^{-3}$ @ 2.5 GHz) was used, whose un-fired and fired thicknesses are 0.134 and 0.091 mm, respectively. This leads to a fired thickness stack-up of $h = 0.73$ mm.

The filter total size including input/output feeding ports is only 5.6×11.8 mm^2 (i.e. $0.2 \times 0.42\lambda_0^2$) showing a prominent compact size. The simulated and measured results of the coaxial SIW BPF are in good agreement, as shown in Figure 3.10(b). Note that in-band RL and insertion loss (IL) are better than 10 and 2.5 dB, respectively.

Finally, as an example of the application of the electric coupling mechanism for filter design, two in-line third-order coaxial SIW filters having centre frequency 5.5 GHz, passband of 250 MHz (FBW = 4.6%) have been designed [13]. The two BPFs of Figure 3.11(a) feature distinct resonator b to demonstrate the improved design characteristics added by using the electric coupling. In Figure 3.11(a,1), the coaxial SIW resonators present b = 0.029 S that gives $C_l = 825$ fF, $Z_0 = 103.5\ \Omega$, resulting in a final filter footprint of 15×45 mm^2 (i.e $0.27\lambda_0 \times 0.82\lambda_0$). Conversely, coaxial SIW resonators in structure of Figure 3.11(a,2) feature an increased $b = 0.034$ S. Hence, C_l and Z_0 become now 950 fF and 89.4 Ω, respectively, while the filter size consequently diminishes down a footprint of just 11.6×34.8 mm^2 (i.e. $0.21\lambda_0 \times 0.63\lambda_0$). This enables a 40% of size reduction compared to the previous filter implementation, and up to a 70% of area reduction if compared to a standard TE_{101}-mode SIW filter centred at 5.5 GHz. Note that this size reduction is incompatible to the design of such filtering response with the use of the magnetic couplings, since a smaller cavity leads to lower $k_{i,i+1}$ values limiting the achievable passband BW.

Both filters have been fabricated in single-layer PCB by using a 1.524 mm-thick Rogers R4003C. A photograph of the filter prototypes is visible in Figure 3.11(a), while the simulated and measured responses are depicted in Figure 3.11(b) and (c), respectively. Measurements have validated the proposed approach for filter miniaturisation, showing good IL (i.e. 1.57 and 1.97 dB at f_0, respectively), RL better than

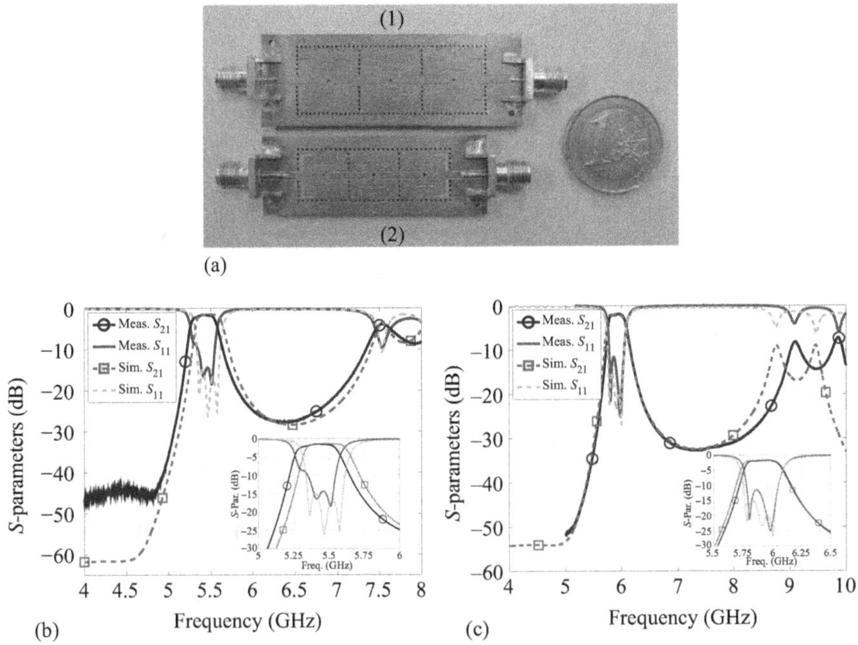

Figure 3.11 *Electric-coupled three-pole BPFs: (a) Photographs: (1) $b = 0.029\,S$, and (2) $b = 0.034\,S$. (b,c) (Dashed) Simulated and (solid) measured filter responses for: (b) $b = 0.029\,S$, and (c) $b = 0.034\,S$. © 2015 IEEE*

11 dB in both cases, and improved spurious free-band (i.e. compact filter with $b = 0.034$ S). However, as Figure 3.11(c) shows, the compact BPF response has been shifted up to 5.9 GHz due to an increase of the air gap width s_p during the PCB fabrication process.

3.3.4 Cross-coupling configuration: design examples

In order to further demonstrate the proposed coupling solution, as well as the miniaturisation degree allowed by the coaxial SIW topology, the design of two quasi-elliptic BPFs implementing advanced filtering responses is now presented. Thus, a four-pole narrow-band BPF [21], and a six-pole BPF [24] have been designed, fabricated and measured in low-cost PCB technology. The first example is a quasi-elliptic BPF featuring a cross coupling between the first and the fourth resonators that has opposite sign with regard to direct inter-resonator couplings. By doing that, the filter selectivity is improved by introducing two TZs located above and below the filter passband. Given that the cross-coupling is based on an electric coupling, input–output and inter-resonator couplings are standard magnetic coupling mechanisms. The filter response is designed to be centred at 5.75 GHz with equi-ripple 1 dB-FBW = 2%. The TZs are set at 5.63 GHz and 5.87 GHz, respectively, corresponding to the centre frequencies of adjacent filters in a multiplexer application with contiguous channels, which leads to the following synthesised response: $k_{12} = k_{34} = 0.019$, $k_{23} = 0.016$, $k_{14} = -0.004$ and $Q_e = 31.4$. $k_{12} = k_{34} = 0.019$, $k_{23} = 0.016$, $k_{14} = -0.004$ and $Q_e = 31.4$.

The filter structure presents a very compact footprint of $21.2 \times 21.2\,\text{mm}^2$ (i.e. $0.86\lambda_g \times 0.86\lambda_g$), which corresponds to 63% of area reduction compared to a four-pole filter based on TE_{101}-mode SIW resonators and implemented in the same dielectric substrate. Indeed, the use of thicker substrate laminate allows us to tightly reduce the coaxial SIW resonator f_0. To do that, a Rogers TMM4 ($\varepsilon_r = 4.5$, $\tan\delta = 2 \cdot 10^{-3}$) having a thickness $h = 3.17$ mm (i.e. $0.13\lambda_g$ at 5.75 GHz) is chosen, and resonators b has been set to 0.016 S (i.e. $\theta_0 = 46.5°$, $C_l = 33.5$ fF and $Z_0 = 78.4\ \Omega$). Furthermore, due to the use of a thick laminate, a surface-mounted device (SMD) self-packaged solution has been developed to enhance filter planar integration, as demonstrated in [21]. Connections are provided to the feeding ports by means of plated half-a-hole vias enabling interconnection to a carrier substrate, which is visible in Figure 3.12(a). A photograph of the filter prototype fabricated by using standard PCB process is shown in Figure 3.12(a), while Figure 3.12(b) shows its measured frequency responses, which are in very good agreement with 3D EM simulations. The upper stop-band rejection is always better than 30 dB up to 12 GHz, which is an important advantage over a standard SIW solution. Finally, the measured IL has worsened, from a simulated value of 2.2 to 4.1 dB, giving an extracted $Q_u = 180$. The degradation is owing to the nickel layer thickness deposited on the copper foil, due to the electroless nickel immersion gold (ENIG) finishing.

The second design to be discussed is a wide-band six-pole BPF, which is designed to have a quasi-elliptic frequency response that satisfies stringent requirements for typical C-band applications ($f_0 = 4$ GHz, equi-ripple BW = 250 MHz). Particularly, the filter must provide rejection better than 35 dB on frequency bands adjacent to the filter passband, flat IL and group delay responses.

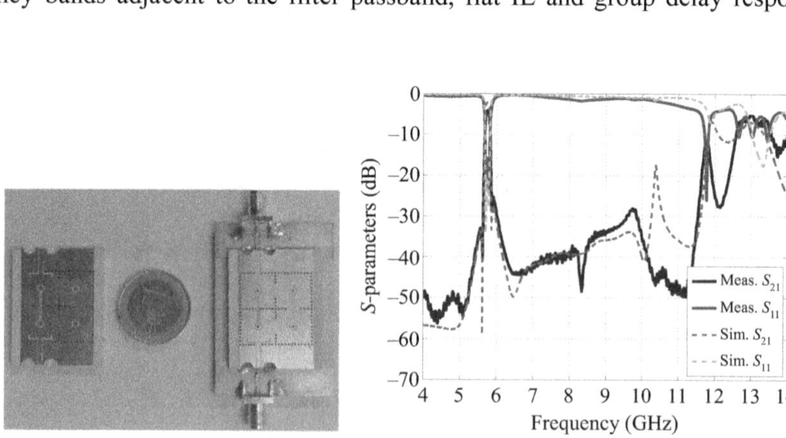

(a) (b)

Figure 3.12 (a) Photography of the filter mounted on a carrier, and (b) (solid) measured and (dashed) simulated wideband BPF response.

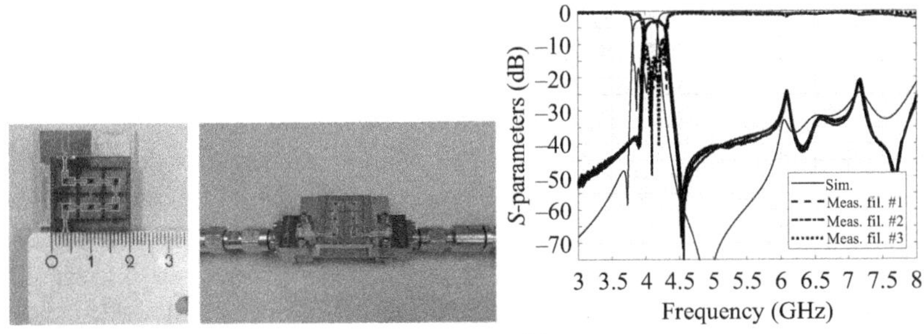

Figure 3.13 6-pole C-band BPF: (a) photographs and (b) measured results of three filter prototypes. © 2017 IEEE

Additional properties that filter features are a very small size, easy integration and hermetically packaged solution.

Thus, electric coupling mechanism is used for all direct and external couplings to achieve a high coupling magnitude. A standard magnetic post-wall iris implements a cross coupling between the second and fifth resonators, generating two TZs located at both sides of the passband. To greatly reduce the resonator size, a 2.54 mm-thick Rogers TMM10i substrate with high permittivity ($\varepsilon_r = 9.8 \pm 0.245$, $\tan \delta = 2 \times 10^{-3}$) has been selected, while resonator *b* is set to 36.5 mS (i.e. $C_l = 1.25$ pF, $Z_0 = 40.8\ \Omega$). By doing that, the filter size is only $20 \times 18.8 \times 2.54$ mm^3 (i.e. $0.27 \times 0.25 \times 0.034\lambda_0^3$), which is a remarkable miniaturisation degree. Again, vertical transitions are created on both filter sides to create an SMD self-packaged solution, thus guaranteeing a very good planar integration. Figure 3.13 shows a photograph of the filter under test, and the measured results for three filter prototypes. The fabricated filter response undergoes a frequency shift that moves the passband at 4.17 GHz (i.e. +4.25% respecting 4 GHz). The in-band IL varies between 2.82 dB value at 4.18 GHz and 4.5 dB at 4.295 GHz. As a result, the extracted Q_u is around 180, which is lower than that extracted from simulations (i.e. $Q_u \approx 250$), again due to the ENIG finishing. Nevertheless, the filter provides high rejection levels, which are better than 35 dB, at just ±250 MHz of the passband centre.

3.3.5 Power-handling capability of coaxial SIW filters

The power-handling capability (PHC) of any device is defined by the thermo-mechanical effects due to the self-heating which can produce, ultimately, the destruction of the device, but also from other physical phenomena such as the air ionisation, also called corona breakdown. The self-heating is linked to the effective applied power to the circuit, therefore, its corresponding power limit is usually named average power-handling capability (APHC). Air ionisation is produced by high voltages in the device, no matter if the effective applied power is low, since signals with low effective power levels can have high peak power levels producing a corona discharge, such as radar signals. Thus, taking into account that the dielectric substrates commonly used in microwave applications present very high limits of dielectric breakdown, mainly in comparison to the air ionisation limits, a corona discharge due to the peak power of the applied signal will probably limit the power handling capability of the circuit. Therefore, air ionisation generally establishes the peak power handling capability PPHC of the device[1].

For the APHC study, the internal heat sources of the SIW component (i.e., conductor and dielectric losses) must be computed. In particular, dielectric loss is treated as a volumetric heat source, whereas conductive loss as a surface heat source. Once they are found, thermal boundary conditions are applied to the circuit and the thermal gradient of temperature in the device is obtained [25]. For the coaxial SIW cavity, its large aspect ratio width/height and the use of outer via holes (along with the central inner conductor) facilitate the heat spreading in the structure, leading to a temperature nearly homogeneous along the cavity, and consequently, an increase of the APHC, since hot spots are avoided [13].

The APHC of a filter depends on the physical structure of the cavity (along with the thermo-mechanical boundary conditions) but also from the kind of filtering function. The filtering function defines the losses per resonator as a function of frequency. The filter cut-off frequencies are usually the limiting frequencies for APHC, since around them, losses are maximum. The APHC of the fourth-order BPF with FBW = 2% centred at 5.75 GHz of Section 3.3.4 was studied in detail and experimentally validated in [13]. It afforded 5 W continuous-wave at one of the cut-off frequencies, where losses were maximum, with no damage and no change in the filtering response. This is a relatively high value for such a narrow-band and miniaturised filter.

For the PPHC analysis, the filtering function is again very important, since it defines the voltage magnification in each resonator as a function of frequency. However, air-ionisation is a phenomenon linked to the electric field rather than to the voltage. So, the geometry of the resonator also plays a fundamental role in defining the PPHC. (Please note that two resonators with the same voltage magnification function may provide different electric field strengths depending on their geometry.) The coaxial SIW cavity presents a region susceptible to a corona discharge, this is the annular gap on the top metal layer. There, the electric field strength can be strong, as gaps in coaxial SIW technology generally present low values for a high degree of miniaturisation. The PPHC of the same filter previously discussed was also experimentally validated in [13]. For the PPHC measurements, pulsed signals with a low duty cycle were applied to the circuit in order to avoid the device self-heating. The critical pressure (i.e., the pressure where power-handling capability is minimum) was found to be 10.4 mbar, at which a corona discharge happened for an applied input power of 1.02 W. At ambient pressure (1,013 mbar), the corona discharge occurred for an input power of 14 W. Figure 3.14 depicts the capture from a video camera at the moment when a corona discharge occurred. As seen, the spark is uniformly originated along the annular gap.

[1]Please note that for space applications, air-ionisation cannot happen. However, an electron avalanche discharge may be still produced, known as multipactor, which may limit the PPHC of the device.

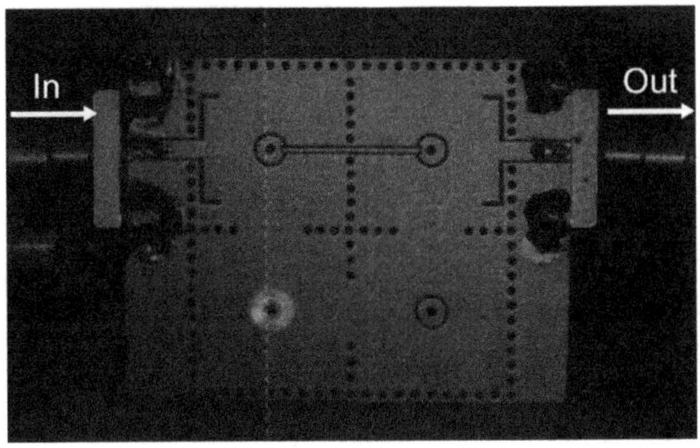

Figure 3.14 *Capture of a corona discharge at the second resonator of the DUT [13]. © 2015 IEEE*

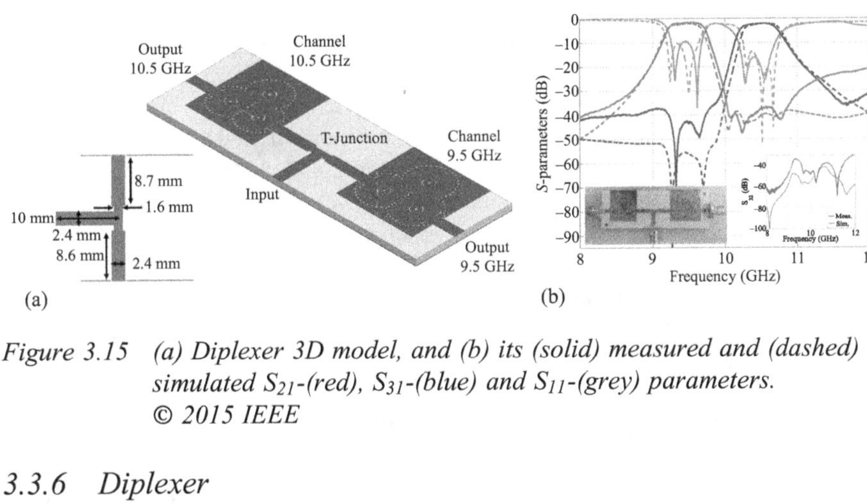

(a) (b)

Figure 3.15 *(a) Diplexer 3D model, and (b) its (solid) measured and (dashed) simulated S_{21}-(red), S_{31}-(blue) and S_{11}-(grey) parameters. © 2015 IEEE*

3.3.6 Diplexer

To conclude this section, an example of a diplexer based on coaxial SIW topology is presented. Both magnetic and electric cross-couplings in circular triplet sections are employed in order to achieve high rejection in compact X-band diplexers with narrow inter-channel separation [26]. Indeed, the coaxial SIW triplet blocks introduce TZs placed below and above the passband depending on the cross-coupling sign.

In this specific design, the objective is to develop a highly compact implementation of an SIW diplexer in a PCB laminate Rogers RO4003C based on the coaxial SIW topology. The diplexer is centred at X-band having channel centre frequencies at 9.5 and 10.5 GHz, respectively, and absolute BWs of 400 MHz. In order to improve rejection between channels, two TZs will be generated at 10.2 and 9.8 GHz, in correspondence with the channel passbands, which means one TZ introduced for each triplet block. Thus, the lower-band channel introduces a TZ at the upper-side by using a magnetic cross-coupling between the first and third resonator. On the contrary, the upper-band filter uses an electric cross-coupling to introduce a TZ at the lower-side. A simple T-junction has been designed for the common port (see Figure 3.15(a)).

Thus, an open-circuit condition has been forced at the junction for the opposite channel filter in order to improve isolation, and minimise channel filter interaction. Additionally, a small step in width on the microstrip T-junction ports has been added for compensating the direct discontinuity. The diplexer has been manufactured using a standard single-side fabrication process. The 9.5 GHz-channel filter size is $18.7 \times 19.5\,\text{mm}^2$ (i.e. $0.59 \times 0.62\lambda_0^2$), while the 10.5 GHz-channel filter size is $16.5 \times 17.2\,\text{mm}^2$ (i.e. $0.58 \times 0.6\lambda_0^2$). The whole device dimensions are compact, being $58.4 \times 18.7\,\text{mm}^2$ that correspond to $1.95 \times 0.62\lambda_0^2$ at 10 GHz. A photo of the fabricated filter is shown in Figure 3.15, together with the simulated and measured results that show very good agreement. As it can be seen, an additional TZ is obtained at the upper-channel filter at 9.34 GHz due to the cancellation of the mixed coupling between non-adjacent resonators. The standard TZs are now located at 9.64 and 10.25 GHz, providing more than 45 dB of rejection, respectively, for the upper and lower channels.

3.4 Advanced topologies for coaxial SIW resonators

The coaxial SIW resonator, as previous demonstrated, presents a big potential for the design of very competitive BPFs with a small footprint and a moderate level of IL. Additionally, it allows for the creation of electric and magnetic couplings in the filter routing paths leading to TZs generation. However, its 2.5D geometry can be still exploited to go one step forward and obtain advanced-feature filtering building blocks. Two examples of that are presented in this section: a coaxial SIW cavity able to support two propagation modes (i.e., a dual-mode cavity) [27,28], and a coaxial SIW single cavity with an inherent direct input-output coupling producing a TZ below or above the passband, as required.

3.4.1 Dual-mode coaxial SIW topology

Figure 3.16(a) shows the layout of the dual-mode coaxial SIW cavity. It is formed by a SIW cavity where two via holes are symmetrically inserted with respect to the cavity centre. Each via hole with diameter d is connected to the circuit ground at

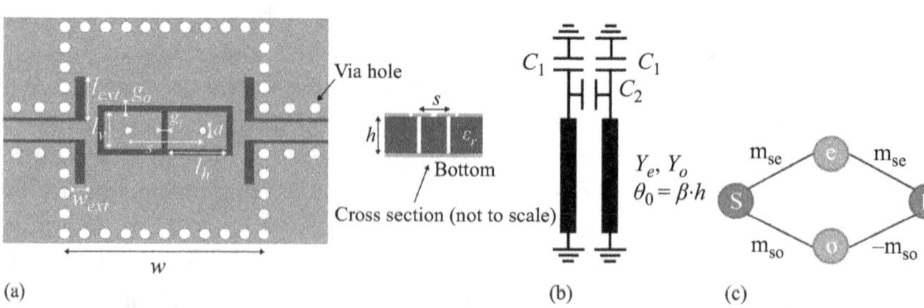

Figure 3.16 (a) Layout of the proposed dual mode coaxial SIW building block; (b) equivalent transmission line model; and (c) coupling routing path. © 2016 IEEE

the bottom and to a capacitive patch at the top. This configuration can be modeled as two TEM-mode combline resonators of length h which are coupled to each other by means of a distributed coupling, which is defined mainly by the separation between via holes s, and by a lumped capacitance associated to the capacitive coupling between metallic patches C_2. Figure 3.16(b) shows the equivalent transmission line model describing the proposed building block. The outer perimeter of each patch $2l_h + l_v$ and the gap g_o determine the capacitance value C_1, whereas C_2 depends on the length l_v and the inside gap g_i. Their values can be computed by means of a quasi-static simulation or by using the approach of [17]. Note that the Cu base thickness t also plays an important role in both capacitance values. This equivalent circuit, due to its symmetry, can be analysed by means of an even–odd mode analysis. The two transcendental equations which provide the two resonant frequencies of the proposed dual-mode cavity are found as:

$$\omega_e C_1 = \frac{Y_e}{\tan\frac{\omega_e}{v}h}; \quad \omega_o(C_1 + 2C_2) = \frac{Y_o}{\tan\frac{\omega_o}{v}h} \tag{3.10}$$

where ω_e and ω_o, and Y_e and Y_o are the angular resonant frequencies and characteristic admittances of the even and odd mode, respectively, and v is the propagation speed. Y_e and Y_o can be computed numerically by using static methods such as conformal transformation or Green's function [29,30]. Their values depend on the ratio d/W (where W is the cavity width), s and the substrate permittivity ε_r. Obviously, as s increases, the ratio Y_o/Y_e approaches to 1 and no distributed coupling appears between lines. For small resonator electrical lengths, as usually happens in coaxial SIW resonators, one can find a closed expression for the ratio between the two resonant frequencies, which facilitates the filter design:

$$\frac{f_o}{f_e} = \sqrt{\frac{C_1}{C_1 + 2C_2}\frac{Y_o}{Y_e}}. \tag{3.11}$$

As deduced from (3.11), it is possible to synthesise the required f_o/f_e by controlling both the coupling between lines (Y_o/Y_e) and C_1 and C_2. The ratio f_o/f_e defines the passband bandwidth of the filter response. Indeed, this topology allows us the design of both narrowband and wideband filters by just choosing properly f_o/f_e. Other characteristics, such as the position of the generated TZ, can also be easily controlled with this topology as will be shown next. It should be noted that f_o/f_e can be set to a value higher or lower than 1, depending on the desired filter performance. Therefore, the proposed building block provides a high degree of flexibility.

For the filter design purpose, the doublet filtering configuration shown in Figure 3.16(c) perfectly models the proposed building block. This configuration provides two poles and one finite TZ [31]. The TZ is placed at the normalised frequency

$$\Omega_{TZ} = \frac{m_{oo}m_{se}^2 - m_{ee}m_{so}^2}{m_{so}^2 - m_{se}^2} \tag{3.12}$$

where m_{ee} and m_{oo} are the normalised even and odd mode frequencies, and m_{se} and m_{so} are the normalised couplings between the input and the even- and odd-modes, respectively. Thus, by properly choosing the design parameters, the TZ can be placed at frequencies lower or higher than the filter passband, as required for the application. The parameters m_{se} and m_{so} are controlled by the penetration of the coplanar feeding line into the cavity through l_{ext} and w_{ext}. For the proposed filtering building block, $m_{se} > m_{so}$ for any geometrical parameter, which means that the input/output port is coupled more strongly to the even mode than to the odd mode.

3.4.1.1 Design examples

In order to validate the proposed topology and to demonstrate its high design flexibility, two filters exhibiting wide- and narrow-band responses are presented, respectively. The filters have been implemented on 1.52 mm-thick RO4003C Rogers substrate, which was previously described. Both filters have been designed at a centre frequency $f_0 = 8$ GHz. For all designs $d = 0.4$ mm and $w = 10.6$ mm. The Q_u of the proposed dual-mode cavity is around 220, as extracted from EM simulations. The first prototype to be discussed is a wide-band BPF with a $\text{FBW}_{3dB} = 20\%$. The TZ is designed to be above the passband. The filter design parameters and dimensions are given in Tables 3.2 and 3.3 (WB Filter), whereas the photograph of the implemented filter is shown in Figure 3.17(a). The filter presents a very compact size of $10.6 \times 10.6\,\text{mm}^2$ (i.e. $0.42 \times 0.42\,\lambda_g^2$, where λ_g is the guided wavelength of the CPW feeding line). A conventional second order SIW filter based on rectangular cavities coupled by means of iris windows would occupy a size around 4 times bigger, without presenting any TZ in the response. Figure 3.17(b) plots the theoretical, simulated (by using ANSYS HFSS 2016 [19]) and measured responses, where a good agreement is observed among all of them. The filter presents a high selectivity in the higher band part due to the generated TZ. The stopband of the filter is spurious free up to 16 GHz, where the SIW cavity TE_{101} mode is excited. The achieved FBW would be very difficult to be obtained

Table 3.2 Filter design parameters

	Y_e (mS)	Y_o (mS)	C_1 (fF)	C_2 (fF)
WB Filter	7.30	13.89	325	74
NB Filter (first stg.)	9.30	11.23	332	52
NB Filter (second stg.)	8.55	11.49	329	42

Table 3.3 Filter dimensions (units in mm)

	l_h	l_v	s	g_i	g_o	l_{ext}	w_{ext}
WB Filter	1.8	2.2	2.95	0.15	0.15	3.5	0.8
NB Filter (first stg.)	2.9	2.0	5.2	0.30	0.30	2.3	0.5
NB Filter (second stg.)	3.23	2.0	4.0	0.30	0.30	2.3	0.5

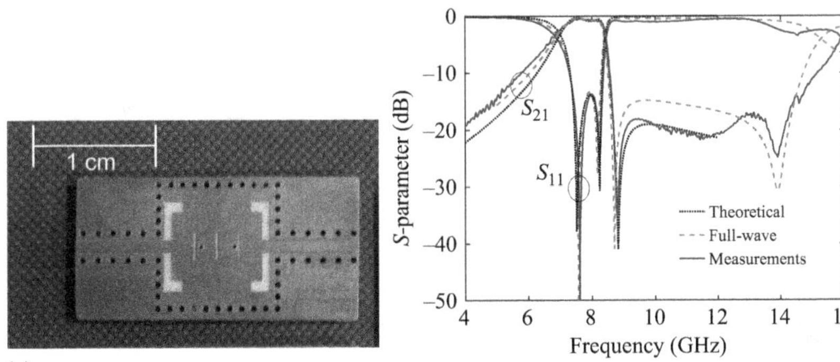

(a) (b)

Figure 3.17 Proposed wideband BPF: (a) prototype photograph and (b) theoretical, full-wave simulated and measured responses of the implemented wideband BPF. For the theoretical response: $m_{se} = 0.62$, $m_{so} = 0.35$, $m_{ee} = 0.70$ and $m_{oo} = -0.38$ with $BW_{3\ dB} = 20\%$. © 2016 IEEE

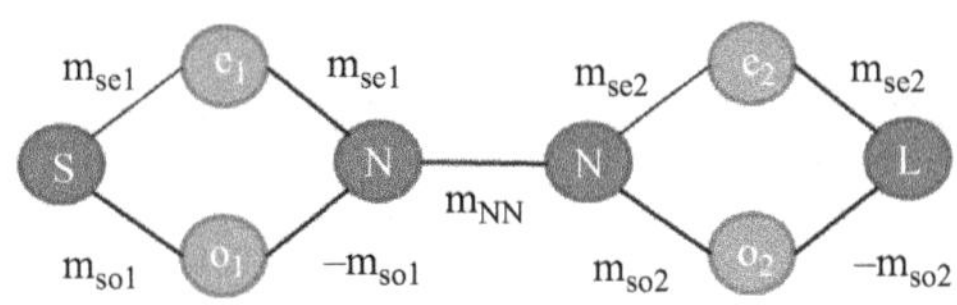

Figure 3.18 Coupling routing scheme of two dual-mode coaxial SIW cavities connected in cascade. © 2016 IEEE

by using conventional SIW filter configurations, or even with the previously discussed coaxial SIW filters of Section 3.3.3.

The second implemented filter is a 4th order narrow-band BPF presenting a quasi-elliptic type response, with $FBW_{3\ dB} = 4\%$. It is formed by cascading two dual-mode coaxial SIW cavities. The coupling routing scheme is depicted in Figure 3.18, whereas the photograph of the filter built is shown in Figure 3.19(a).

In this scheme, both stages can be designed almost independently, the first stage creates the TZ below the passband whereas the second one the TZ placed above, providing in this way a quasi-elliptic filter response with a very high selectivity. Both stages are connected by means of a CPW line, represented in the coupling routing scheme by m_{nn}. The design parameters and sizes are given in Tables 3.2 and 3.3 (NB Filter). Again, the high degree of miniaturisation must be highlighted. For instance, in order to obtain an equivalent response in conventional SIW technology it would be needed four single-mode rectangular cavities in a quadruplet configuration, occupying around four times more than the proposed filtering configuration. Figure 3.19(b) depicts, theoretical, full-wave simulated and measured results. Note that for the theoretical response: $m_{se_1} = 0.83$, $m_{so_1} = 0.56$,

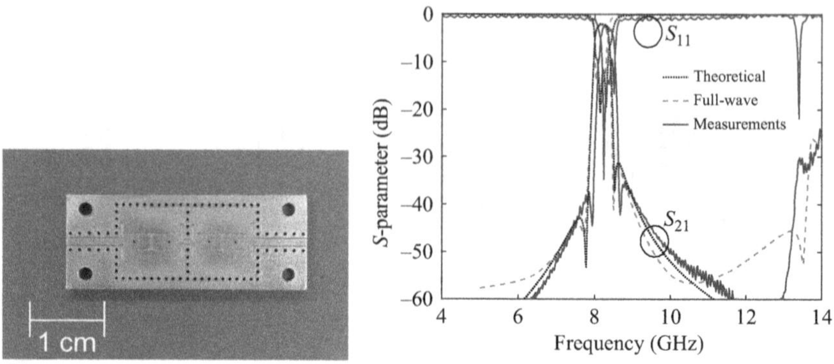

(a) (b)

Figure 3.19 Quasi-elliptic BPF by cascading two dual-mode coaxial SIW cavities: (a) photograph, and (b) theoretical, full-wave simulated and measured responses. © 2016 IEEE

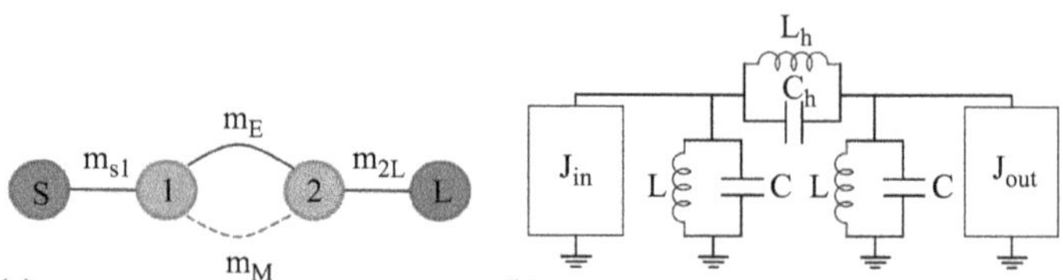

(a) (b)

Figure 3.20 Alternative model for the dual-mode cavity of Figure 3.16: (a) coupling routing path and (b) lumped-element equivalent circuit

$m_{ee_1} = -1.16$, $m_{oo_1} = 0.95$, $m_{se_2} = 0.71$, $m_{so_2} = 0.44$, $m_{ee_2} = 0.67$, $m_{oo_2} = -0.90$ and $m_{NN} = 1.05$, with $FBW_{3\text{ dB}} = 4\%$ and $f_0 = 8.2$ GHz. The measured IL at f_0 is 2.3 dB, and the out-of-band rejection is higher than 30 dB up to around 14 GHz.

3.4.1.2 Alternative model

An alternative circuit can be also used in order to model the filtering physical behavior of the aforementioned dual-mode cavity. Instead of considering a single cavity where two modes are orthogonally excited (as previously modelled), since the electric length of the two via holes is relatively small as compared to the wavelength, this filtering building block can be also seen as two synchronously tuned lumped-element resonators coupled to each other by means of an electric-magnetic hybrid coupling (see Figure 3.20). The magnetic coupling is produced by the coupling between the two via holes, whereas the electric one is due to the capacitance C_2. By using this approach, the TZ generation is produced when the in-line electric and magnetic couplings cancel each other.

Figure 3.20 shows the coupling routing path and its lumped-element equivalent circuit counterpart. In Figure 3.20(a), m_{s1} and m_{2L} indicate the input/output coupling, and m_E and m_M the electric and magnetic coupling produced between resonators. With respect to the lumped-element model of Figure 3.20(b), J_{in} and J_{out} are the input and output inverter parameters, L and C define the even-mode resonance of the coupling configuration, whereas L_h and C_h model the hybrid coupling between resonators and consequently define the TZ frequency location. According to this equivalent circuit, the self-resonant frequency of each lumped-element resonator can be computed as [32]

$$f_0 = \frac{1}{2\pi}\sqrt{\frac{L + L_h}{LL_h(C + C_h)}}, \tag{3.13}$$

whereas the TZ frequency is placed at $f_{TZ} = 1/(2\pi)\sqrt{1/(L_h C_h)}$. This alternative circuit may be more practical than the dual-mode configuration for the design of higher-orders BPFs, since it avoids the use of non-resonant nodes and allows for the design of odd order filters. For the synthesis process, the couplings m_E and m_H of Figure 3.20(a) may be merged and modeled by means of a frequency-dependent inverter [33,34].

3.4.2 Coaxial SIW singlet

The coaxial SIW cavity can be easily modified in order to provide a direct source-to-load coupling path leading to a singlet configuration. Such a novel structure can be used as a building block for implementing advanced filtering responses (such as quasi-elliptic responses) by cascading several of them. Indeed, the main advantage is that each singlet independently introduces and controls its own TZ. Figure 3.21 shows the 3D view, layout and coupling routing path of the coaxial SIW singlet, whose sizes are: $w_{sl} = 1.2$ mm, $g_{sl} = 0.2$ mm, $g_e = 0.23$ mm, $l_e = 1.8$ mm and $g_l = 0.26$ mm. The source/load coupling is implemented by forcing the feeding CPW lines to penetrate the capacitive section of the coaxial SIW resonator, providing a strong coupling even in the case of highly loaded resonators. An additional distributed capacitance is created on the CPW feeding line. This capacitance associated to m_{sl} is controlled by the spacing g_{sl} and width w_{sl}.

On the other hand, the length l_e and the spacing g_e set the value of the capacitive coupling among the feeding CPW line and the coaxial SIW resonator, thus controlling m_{s1}. The proposed coaxial SIW is a first order topology able to produce a TZ either below or above the transmission pole. The singlet configuration has

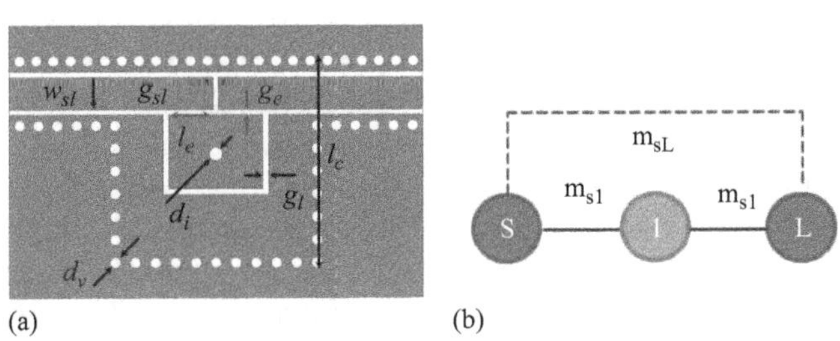

Figure 3.21 Coaxial SIW singlet: (a) top view and (b) coupling routing path

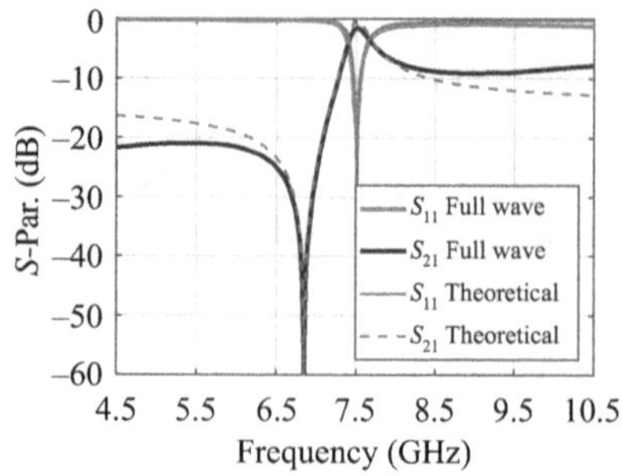

Figure 3.22 Coaxial SIW singlet: (dashed) theoretical and (solid) simulated responses

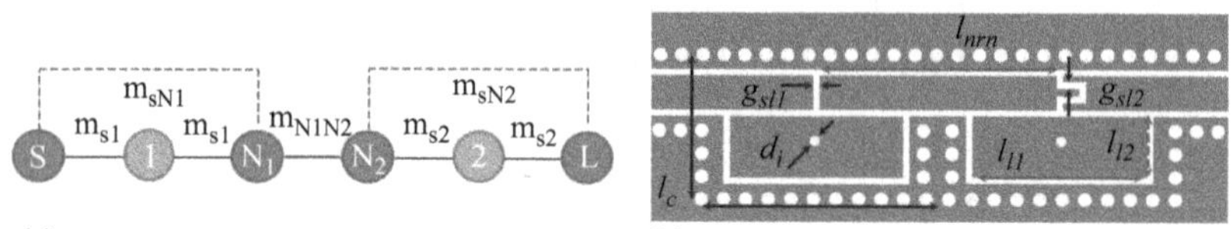

Figure 3.23 Two cascaded coaxial SIW singlets: (a) coupling routing scheme and (b) layout with a quarter-wave CPW NRN

been widely studied and demonstrated in different technologies [35,36]. The TZ position depends on the sign of the source/load coupling m_{sl}. If m_{sl} is electric, the TZ appears below the bandpass and above if it is magnetic. In particular, its position can be analytically computed as $\Omega_{TZ} = m_{s1}^2/m_{sl}$, where Ω_{TZ} is the normalised frequency. Figure 3.22 shows the theoretical and full-wave simulated responses of the particular example of Figure 3.21, which presents a transmission pole at 7.5 GHz with in-band RL = 15 dB and one TZ placed at 6.8 GHz. The associated coupling matrix for this example is: $m_{s1} = -0.69$, $m_{sl} = -0.09$, and $m_{11} = 0$.

3.4.2.1 Design example

To validate the singlet concept, the design of a narrow-band BPF centred at 7 GHz with FBW = 3% implemented on a 1.52 mm-thick Rogers R04003C substrate is herein presented. By cascading two coaxial SIW singlets, a second-order filter with two TZs below the bandpass is generated, since negative cross-coupling is used for each section.

To cascade two adjacent singlets, a non-resonating node (NRN) is used as coupling mechanism. For this purpose, a quarter-wave CPW section created at the top metal layer and embedded into the resonators is used as NRN. The layout and its corresponding coupling routing scheme are shown in Figure 3.23. Varying the NRN length l_{nrn}, the coupling among singlets can be slightly adjusted. The gaps g_{sl_1} and g_{sl_2} independently control, respectively, the position of each TZ. Finally, the filter dimensions are: $w_{sl} = 1.2$ mm, $g_{sl,1} = g_{sl,2} = 0.2$ mm, $l_e = 2.3$ mm, $l_{l1} = 4.9$ mm, $l_{l2} = 1.69$ mm, $g_l = 0.26$ mm, $l_c = 6.7 \times 4.4$ mm^2 and $l_{nrn} = 6.7$ mm.

In order to increase m_{sN2} and bring one TZ closer to the passband, an inter-digital capacitance is created on the top metal layer, as shown in the right-hand side of Figure 3.23. The side wall holes have diameter $d_v = 0.4$ mm, while the inner d_i is set to 0.3 mm. Thus, a more compact coaxial resonator is now obtained, whose parameters are: $b = 27$ mS, $Z_0 = 90\,\Omega$, $\theta_0 = 24°$ and $C_l = 560$ fF. The coupling matrix elements are: $m_{s1} = 1.1$, $m_{s2} = 1.094$, $m_{sN1} = -0.062$, $m_{sN2} = -0.13$, $m_{N1,N2} = 1.05$, and $m_{11} = m_{22} = 0$. Again, the device has been fabricated by using a single-layer PCB process. It is noted that the device presents a compact size of 4.4×13.8 mm^2 (i.e. $0.1 \times 0.3\lambda_0^2$). Simulations, measurements and a photograph of the device are shown in Figure 3.24. The agreement between simulations and measurements is good, mainly in terms of centre frequency and measured IL, the latter being 1.9 dB at 7 GHz. The filter features an improved selectivity below the bandpass, where the rejection is always better than 55 dB.

This kind of implementation for producing a direct source/load coupling in single coaxial SIW cavities may be also applied to the dual-mode coaxial SIW topology presented in Section 3.4.1. By doing this, it can be obtained a doublet configuration able to produce two transmission poles and two TZs. Figure 3.25 shows an example of that. As seen from this example, with just one single cavity a second order quasi-elliptic BPF response has been achieved (see Figure 3.25), where: $g_{sl,d} = 0.2$ mm,

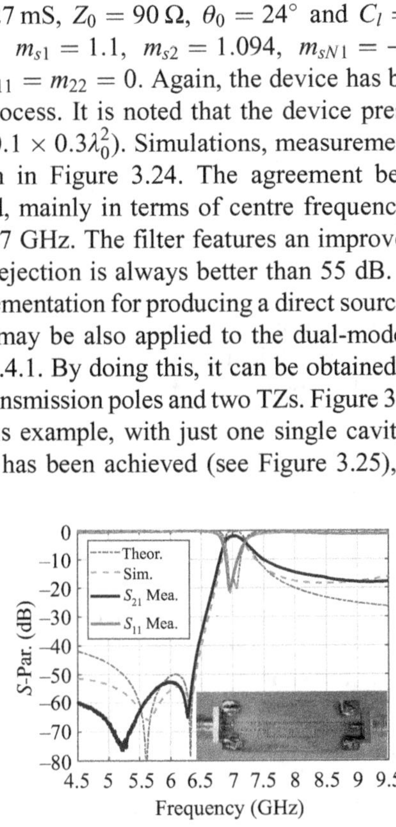

Figure 3.24 Two cascaded coaxial SIW singlets: (dotted) theoretical response, (dashed) simulations, and (solid) measurements, and photograph

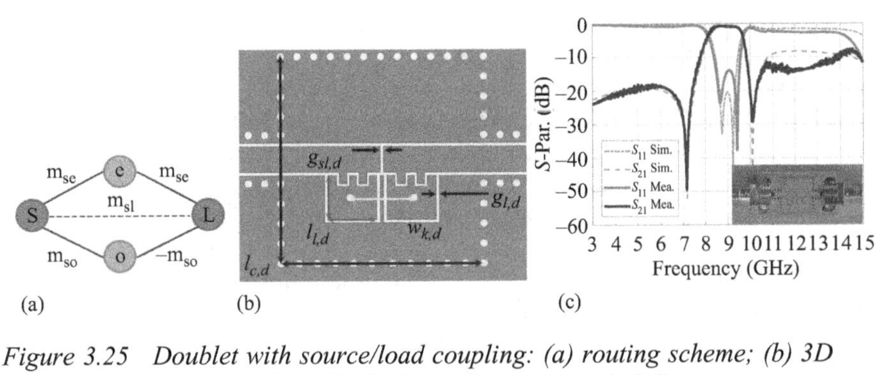

Figure 3.25 Doublet with source/load coupling: (a) routing scheme; (b) 3D layout; and (c) (dashed) simulations and (solid) measurements

$l_{c,d} = 2.2$ mm, $g_{l,d} = 0.21$ mm, $l_{c,d} = 10.3$ mm and $w_{k,d} = 3.25$ mm. The coupling matrix elements for this example are: $m_{se} = 0.629, m_{so} = 0.488, m_{sl} = -0.14$, $m_{ee} = 1.065$ and $m_{oo} = -0.816$.

3.5 Summary

Nowadays, there is a huge interest in developing modern communication systems operating at microwave frequencies, which are expected to improve performance of those presently used in radio and television broadcasting, mobile communication services or satellite payloads. In this context, the SIW technology has gained considerable attention during the last years, and is now fully considered a promising solution for implementing such future microwave (and even millimetre-wave) devices with advantages in terms of low-cost, small size, relatively high power, high Q and high-density integration.

As it has been demonstrated in this chapter, the application of a coaxial topology to the SIW was able to boost considerably the properties of this promising technology, since it enables for an important size reduction preserving an overall good EM performance. The latter has been demonstrated in Section 3.2, where the in-depth analysis of the coaxial SIW performance helped to understand the resonant nature of the structure, resulting in a conclusive equivalent circuit composed by a substrate-integrated coaxial line loaded by a capacitor. The proposed topology showed a wide variety of configurations, which were represented by the variations of the resonator slope parameter, enabling designers to adapt easily the resonator properties for satisfying different requirements.

Additionally, it was demonstrated that the coaxial SIW topology is excellent for the introduction of different coupling schemes. For this aim, several configurations involving magnetic and electric coupling mechanisms have been proposed and studied in detail. Therefore, it is clear that the coaxial SIW topology increased considerably the design flexibility of microwave SIW BPFs, enabling the designer to implement a wide variety of filter topologies and responses.

In this context, the design of coaxial SIW resonator filters with advanced performances has been deeply investigated in Section 3.3. It has been proved that the proposed electric coupling configuration ensures high flexibility in filter design, which allows us to obtain very compact filters by keeping a reasonable level of losses, as well as to increase the maximum achievable bandwidth in coaxial SIW technology. To prove that, a variety of filters have been proposed, where the magnetic and electric coupling schemes have been arranged in order to design a very compact BPF with advanced and very selective responses.

Furthermore, a dual-mode topology based on coaxial SIWs has been analysed in detail, providing a design strategy and an in-depth analysis of the involved coupling mechanisms. Such configuration enables for a further miniaturisation of coaxial SIW filters while providing also more design flexibility. The latter properties have been demonstrated with the design of both narrow and wide-band BPFs, which introduce TZs placed as required for particular applications. Several proof-of-concept filters have been successfully manufactured and tested, whose results are shown in Section 4. Finally, the implementation of singlets, cascaded singlets, and doublets in coaxial SIW technology has been also proposed and proved. This basic singlet topology, which allow the introduction of a TZ below the transmission pole, could be a promising candidate as a building block for implementing advanced filtering responses with improved compactness.

References

[1] Cheng XP, Wu K. Substrate integrated waveguide filters: practical aspects and design considerations. IEEE Microwave Magazine. 2014;15(7):75–83.

[2] Grigoropoulos N, Sanz-Izquierdo B, Young P. Substrate integrated folded waveguides and filters. IEEE Microwave and Wireless Components Letters. 2005;15(12):829–831.

[3] Wang Y, Hong W, Dong Y, *et al.* Half mode substrate integrated waveguide bandpass filter. IEEE Microwave and Wireless Components Letters. 2007;17(4):265–267.

[4] Jin C, Shen Z. Compact triple-mode filter based on quarter-mode substrate integrated waveguide. IEEE Transactions on Microwave Theory and Techniques. 2014;62(1):37–45.

[5] Dong YD, Yang T, Itoh T. Substrate integrated waveguide loaded by complementary split-ring resonators and its applications to miniaturized waveguide filters. IEEE Transactions on Microwave Theory and Techniques. 2009;57(9):2211–2223.

[6] Wu LS, Zhou L, Zhou XL, *et al.* Bandpass filter using substrate integrated waveguide cavity loaded with dielectric rod. IEEE Microwave and Wireless Components Letters. 2009;19(8):491–493.

[7] Levy R. Filters with single transmission zeros at real or imaginary frequencies. IEEE Transactions on Microwave Theory and Techniques. 1976;24(4):172–181.

[8] Potelon B, Favennec J, Quendo C, *et al.* Design of a substrate integrated waveguide filter using a novel topology of coupling. IEEE Microwave Wireless and Components Letters. 2008;18(9):596–598.

[9] Zhu F, Hong W, Chen JX, *et al.* Cross-coupled substrate integrated waveguide filters with improved stopband performance. IEEE Microwave Wireless and Components Letters. 2012;22(12):633–635.

[10] Chen XP, Wu K. Substrate integrated waveguide cross-coupled filter with negative coupling structure. IEEE Transactions on Microwave Theory and Techniques. 2008;56(1):142–149.

[11] Martínez JD, Taroncher M, Boria VE. Capacitively loaded resonator for compact substrate integrated waveguide filters. In: Proceedings of 40-th European Microwave Week (EuMW). Paris, France; 2010. pp. 192–195.

[12] Martínez JD, Sirci S, Taroncher M, *et al.* Compact CPW-fed combline filter in substrate integrated waveguide technology. IEEE Microwave and Wireless Components Letters. 2012;22(1):7–9.

[13] Sirci S, Sánchez-Soriano MA, Martínez JD, *et al.* Design and multiphysics analysis of direct and cross-coupled SIW combline filters using electric and magnetic couplings. IEEE Transactions on Microwave Theory and Techniques. 2015;63(12):4341–4354.

[14] Matthaei GL. Combline band-pass filters of narrow or moderate bandwidth. Microwave Journal. 1963;6(8):82–96.

[15] Rhodes JD. The stepped digital elliptic filter. IEEE Transactions on Microwave Theory and Techniques. 1969;17(4):178–184.

[16] Riblet HJ. An accurate approximation of the impedance of a circular cylinder concentric with an external square tube. IEEE Transactions on Microwave Theory and Techniques. 1983;31(10):841–844.

[17] Lee HS, Eom HJ. Potential distribution through an annular aperture with a floating inner conductor. IEEE Transactions on Microwave Theory and Techniques. 1999;47(3):372–374.

[18] Hong JS, Lancaster MJ. Microstrip filter for RF/microwave applications. 1st ed. New Jersey: John Wiley and Sons, Inc.; 2001.

[19] HFSS, ANSYS Electromagnetics Suite 17.2.0, Version: Release 17.2.0, ANSYS Inc., Canonsburg (PA), USA; 2017. Web site: http://www.ansys.com/Products/Electronics/ANSYS-HFSS.

[20] Deslandes D, Wu K. Integrated transition of coplanar to rectangular waveguides. In: In Proceedings of IEEE MTT-S International Microwave Symposium Digest. 2001;2:619–622.

[21] Sirci S, Gentili F, Martínez JD, *et al.* Quasi-elliptic filter based on SIW combline resonators using a coplanar line cross-coupling. In: In Proceedings of IEEE MTT-S International Microwave Symposium Digest. Phoenix, Arizona (USA), 2015. pp. 1–4.

[22] Gong K, Hong W, Zhang Y, *et al.* Substrate integrated waveguide quasi-elliptic filters with controllable electric and magnetic mixed coupling. IEEE Transactions on Microwave Theory and Techniques. 2012;60(10):3071–3078.

[23] Sirci S, Martínez JD, Boria VE. LTCC packaging of substrate integrated coaxial filters. In: In Proceedings of 6th CNES/ESA International Workshop on Microwave Filters. Toulouse, France; 2015. pp. 1–4.

[24] Martínez JD, Sirci S, Boria VE, *et al.* Practical considerations on the design and optimization of substrate integrated coaxial filters. In: 2017 IEEE MTT-S International Conference on Numerical Electromagnetic and Multiphysics Modeling and Optimization for RF, Microwave, and Terahertz Applications (NEMO). Sevilla, Spain; 2017. pp. 299–301.

[25] Sanchez-Soriano MA, Quere Y, Le Saux V, *et al.* Average power handling capability of microstrip passive circuits considering metal housing and environment conditions. IEEE Transactions on Components, Packaging and Manufacturing Technology. 2014;4(10):1624–1633.

[26] Sirci S, Martínez JD, Vague J, *et al.* Substrate integrated waveguide diplexer based on circular triplet combline filters. IEEE Microwave and Wireless Components Letters. 2015;25(7):430–432.

[27] Sánchez-Soriano MA, Sirci S, Martínez JD, *et al.* Compact dual-mode substrate integrated waveguide coaxial cavity for bandpass filter design. IEEE Microwave and Wireless Components Letters. 2016;26(6):386–388.

[28] Sánchez-Soriano MA, Sirci S, Martínez JD, *et al.* Compact bandpass filters based on a new substrate integrated waveguide coaxial cavity. In: Proceedings of IEEE MTT-S International Microwave Symposium Digest. San Francisco, California (USA); 2016. p. 1–4.

[29] Levy R. Conformal transformations combined with numerical techniques, with applications to coupled-bar problems. IEEE Transactions on Microwave Theory and Techniques. 1980;28(4):369–375.

[30] Borji A, Safavi-Naeini S, Chaudhuri SK. TEM properties of shielded homogeneous multiconductor transmission lines with PEC and PMC walls. In: Proceedings of IEEE MTT-S International Microwave Symposium Digest. vol. 2; 2001. pp. 731–734 vol.2.

[31] Rosenberg U, Amari S. Novel coupling schemes for microwave resonator filters. IEEE Transactions on Microwave Theory and Techniques. 2002; 50(12):2896–2902.

[32] Wang H, Chu QX. An inline coaxial quasi-elliptic filter with controllable mixed electric and magnetic coupling. IEEE Transactions on Microwave Theory and Techniques. 2009;57(3):667–673.

[33] Tamiazzo S, Macchiarella G. Synthesis of cross-coupled filters with frequency-dependent couplings. IEEE Transactions on Microwave Theory and Techniques. 2017;65(3):775–782.

[34] He Y, Wang G, Sun L, *et al.* Direct matrix synthesis for in-line filters with transmission zeros generated by frequency-variant couplings. In: Proceedings of IEEE MTT-S International Microwave Symposium Digest; 2017. pp. 356–359.

[35] Amari S, Rosenberg U, Bornemann J. Singlets, cascaded singlets, and the nonresonating node model for advanced modular design of elliptic filters. IEEE Microwave and Wireless Components Letters. 2004;14(5):237–239.

[36] Bastioli S. Nonresonating mode waveguide filters. IEEE Microwave Magazine. 2011;12(6):77–86.

Chapter 4

Planar lossy filters for satellite transponders

Stéphane Bila[1], *Ahmed Basti*[1], *Aurélien Périgaud*[1], *Serge Verdeyme*[1], *Laetitia Estagerie*[2], *Ludovic Carpentier*[2], *and Hervé Leblond*[3]

4.1 Introduction

Microwave filters are key elements in many communication systems. Regarding the system and the position of the filter within the system, its design has to deal with particular electrical specifications and constraints that concern its weight and footprint. For instance, in satellite transponders, high quality factor (Q) filters based on cavities or dielectric resonators are required for output multiplexers, which have to cope with severe specifications in terms of insertion losses and power handling. For filters in the receiver part, the insertion loss and power-handling performances are less critical, allowing the usage of more compact technologies, which make easier the integration with active circuits also.

For a receive filter, the challenge is to design a compact bandpass filter with a flat response in the passband and a sharp transition in the passband edges. The insertion loss is not crucial, since it does not affect the total noise factor as the filter is placed after the low-noise amplifier and it can be compensated by the amplifier, leaving a room to the design of a lossy filter. Such a filter accepts additional losses, which can be distributed in the network in order to provide a flat transmission in the passband and a sharp selectivity.

This new class of microwave filters has been proposed recently in [1–6], and can be divided into two families mainly. Additional losses are introduced either in individual resonators, forming a network with non-uniform-Q resonators [3,4], or distributed through resistive cross-couplings [4–6].

In this literature, lossy filter prototypes are realized using different technologics and working frequencies making the comparison between different approaches difficult. In this chapter, the two approaches are compared considering the same specifications and the same technology. A classical filter is designed and fabricated first.

[1]XLIM UMR7252, CNRS/Université de Limoges, Limoges, France
[2]CNES, DSO/RF/HNO, Toulouse, France
[3]Thales Alenia Space, Toulouse, France

Afterwards, lossy filters formed on the one hand by a transversal network with non-uniform Q resonators and, on the other hand, by an in-line network with resistive cross couplings are designed and fabricated for comparison. Finally, absorptive lossy filters are designed and fabricated. Such filters integrate the property of attenuating reflected wave that is often a requirement for protecting the receiver subsystem.

4.2 Impact of losses on filter performances

4.2.1 Relation between quality factor and insertion losses

The unloaded quality factor is related to the resonator technology (coaxial, dielectric, planar structure, superconductor ...). Generally, high Q resonators are physically larger and require the use of more expensive technology. For example, dielectric resonator filters can offer a very high quality factor [7], but their volume and cost is much higher than filters in microstrip technology [8]. It is shown in [9], that insertion losses (IL) are inversely proportional to the unloaded quality factor Q_u. For instance, insertion losses in the middle of the bandwidth of a Chebyshev-type bandpass filter can be estimated by

$$IL = 8.686[N - 1.5]\frac{f_0}{\Delta_f Q_u} \tag{4.1}$$

where N is the degree of the filter, f_0 its center frequency, and Δ_f its passband.

Physically, this relationship can be explained by the fact that the group delay of the filter is inversely proportional to its bandwidth. Thus, for a given quality factor, the more the signal remains in the filter, the greater the losses are. In addition, the group delay of a filter always increases near the edges of the bandwidth. Thus, insertion losses near the edges of the bandwidth will be higher than the value in the middle of the bandwidth. Assuming a filter with a uniform distribution of the quality factor between resonators and a cutoff pulsation $\omega = 1$ rad $\cdot$ s^{-1}, the increase in insertion losses can be estimated as [10]:

$$\Delta IL(\omega) \approx 8.686\left(\frac{T_g(\omega)}{Q_u}\right) \tag{4.2}$$

Figure 4.1 shows the variation of insertion losses as a function of different Q values. One can observe a degradation of the flatness in the passband, which results in a rounding of passband edges. This variation, associated with less distinct transmission zeros, results in a degradation of the out-of-band rejections.

Figure 4.2 shows the displacement of the poles of the transfer function in the complex plane in the ideal case without losses (infinite Q) and the real case with losses (finite Q). Considering losses in the transfer function causes the poles to shift to the left in the complex plane. This shift is inversely proportional to the quality factor of the resonators:

$$a = \frac{Q_p - Q_0}{Q_p} \tag{4.3}$$

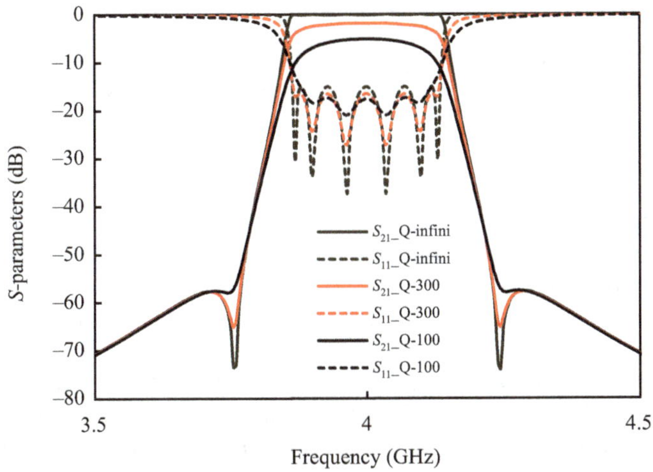

Figure 4.1 Filter responses for different values of the unloaded quality factor

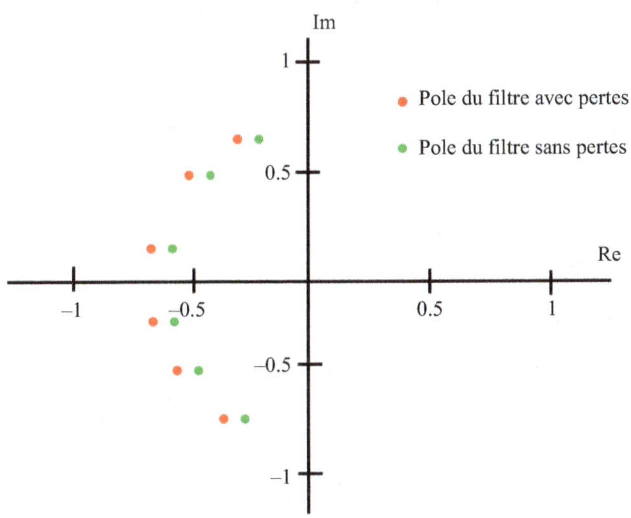

Figure 4.2 Typical displacement of the poles in the complex plane, considering finite and infinite quality factors (example of a 6 pole transfer function)

where Q_0 is the quality factor of the initial filtering function and Q_p is that of the modified function.

If all the resonators have the same finite quality factor, it is therefore possible to evaluate the offset α and compensate it upstream to maintain the same transfer function, excepted a fix amount losses.

This predistortion technique has been the subject of several studies [11–14].

4.2.2 Compensation by predistortion

The predistortion technique, which was developed in 1940 [13], is a method of total or partial compensation. The principle is to shift the poles (p) of the transfer function to the right of the complex plane:

$$p \rightarrow p - \alpha \tag{4.4}$$

Thus, the transfer function of the filter in the presence of losses corresponds to a conventional lossless transfer function, with a constant additional attenuation, whatever the frequency.

There are some disadvantages associated with the predistortion technique, which suffers from poor bandwidth adaptation. Indeed, the polynomial reflection function is evaluated by applying the relation of conservativity, and the displacement of poles and zeros of reflection destroys the matching condition. It should be noted that this approach leads to several solutions [13] for the realization of the filter since several choices are possible to determine the reflection zeros.

In order to limit the previous disadvantages, an adaptive or partial predistortion can be applied to compensate the rounding of the bandwidth edges while keeping a correct matching. Details of these techniques are presented, respectively, in [12] and [14].

Relatively high return loss levels limit the use of these filters to systems or subsystems where the power level is low. In the payload of a satellite, these filters are used for input multiplexers, placing a circulator that directs the reflected signal to a suitable load. Since the input signal of the input multiplexer is relatively weak, the power dissipated by the load remains low. The subsystem introduces more losses but makes it possible to limit the variation in amplitude in the bandwidth with a smaller device.

4.2.3 Synthesis of lossy filters

The synthesis of lossy filters is intended to achieve a filter having an improved flatness in the bandwidth, playing on the distribution of losses. Thus, compared to a filter resulting from a conventional synthesis, this type of filter will have for a number of resonators and a given maximum quality factor, a higher relative loss level but a flatness equivalent to a much higher quality factor. In comparison with the predistortion technique, this approach degrades the level of adaptation much less. Figure 4.3 shows the expected filter response for different Q values in the case of lossy synthesis. One can see a decline of insertion loss, but still a very flat and very selective response. This result is obtained by using resonators with different quality factors and/or introducing lossy couplings. In the following paragraphs, different approaches proposed in the literature for lossy filter synthesis are presented.

4.2.3.1 Lossy resonators and resistive cross-couplings for optimizing transfer function

The first approach, proposed in [15], consists in multiplying the polynomial lossless transfer function by a constant attenuation factor $K < 1$. This transfer function

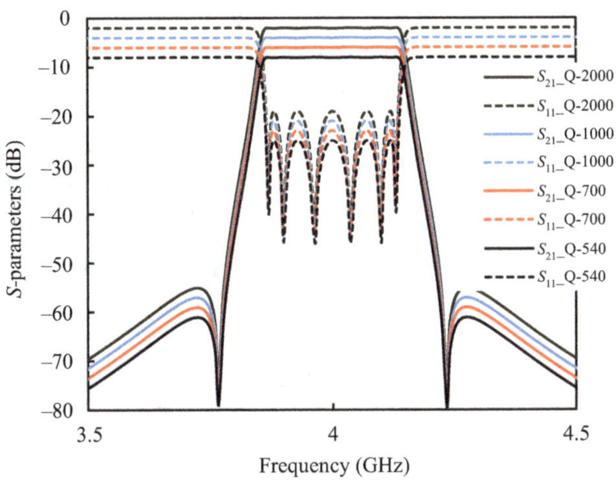

Figure 4.3 Lossy filter response for different Q values

includes finite losses; consequently, the condition of unit conservativity connecting scattering parameters S_{21} and S_{11} does not apply. Thus, it is difficult to understand how to form the reflection function and then the input impedance, in order to synthesize the network exactly. This problem is solved using a new pre-distortion technique, which is applicable only in the case of a symmetric quadripole, where an approximation can be made to compute S parameters, based on the theory of odd and even modes. The network comprises coupled resonators with different Q factors at its input and output.

An improvement of this technique is presented in [16]. The technique consists in distributing the losses created by attenuation factor K. Losses are distributed through the network by adding a finite pole-zero pair to the loss transfer function. The additional pole corresponds to the addition of a resonator within the network, while the additional zero creates a parallel signal transmission path. These additional elements make possible to improve electrical performance, including flatness in the passband, matching and selectivity. The parallel path is made by simply placing a resistor between two non-adjacent resonators of the network to create a non-resonant resistive coupling.

4.2.3.2 Resistive couplings and non-resonating nodes for attenuating transfer and reflection functions

The previous approach is extended in [4] and [5] considering different attenuation factors $K_{ij} < 1$ for transfer and reflection functions. In this general case, several methods for direct synthesis of the lossy coupling matrix have been developed in the literature [4–6,17,18].

A special case is to consider a constant attenuation ($K_{11} = K_{21} = K_{22} = K < 1$). From a network point of view, this particular case is equivalent to placing two identical attenuators at the input and output ports of a lossless filter. In the general case, the synthesis technique makes possible to have filtering functions with different attenuation levels for the input and output reflections [4].

It is then generally preferable to distribute the losses in the network using resistive couplings and non-resonant nodes [5] to simplify the manufacturing process. The distribution of losses within the network is achieved by applying a series of hyperbolic rotation on the complex coupling matrix.

4.2.3.3 Filter networks with heterogeneous Q resonators

Initially, the preferred way to implement lossy filters was to introduce resistive couplings. In some cases, the introduction of resonators with different quality factors can further optimize performance. On the other hand, several authors proposed the implementation of lossy filters based solely on heterogeneous quality factor resonators.

Reference [19] shows that the flatness of the filter with resonators having non-uniform quality factors can be optimized by placing the more dissipative elements at the ends of the network. Apart from an optimization of quality factors and couplings, the generalization of this approach has not been made.

Arranging the lossy filter into a transversal network allows optimizing the flatness in the bandwidth more easily, by controlling the distribution of losses through the parallel paths. In this case, each resonator or each path can be optimized to find the desired performance. A theoretical approach is presented in [3] to explain the influence of losses in each path of the transversal network.

4.2.3.4 Lossy redundant structures

In order to be relatively comprehensive in our description, we have to report on a different approach developed as in [20]. In this approach, the circuit is split into two subnetworks, one symmetrical and the other antisymmetrical, and each of these subnets is redundantly duplicated.

A compromise between redundancy (thus circuit size) and degrees of freedom in the design of the filter is necessary. These degrees of freedom can be exploited to play with the return loss and the selectivity or reduce the quality factors of resonators.

The technique consists in determining the even and odd sub-networks. This step is generally followed by a simplification in order to reduce the network toward N resonators (non-redundant network).

4.3 Reference design: hairpin microstrip filter

The objective of this work is to design a compact microstrip filter centered at 3.8 GHz with an 800-MHz bandwidth, considering the specifications detailed in Table 4.1. The initial design is achieved by a hairpin microstrip filter. The resonators are fabricated on an alumina substrate having a height of 0.254 mm. The substrate permittivity is $\varepsilon_r = 9.9$ and its loss tangent is $\tan\delta = 0.0002$. The metallization is 5 μm thick. Using this technology, the unloaded quality factor of each hairpin resonator is around 95. In order to fulfill the previous requirements, a six-pole Chebyshev

Table 4.1 Specifications for this work. Reprinted with permission from Reference [1]; © 2014 IEEE

Parameter	Value	Unit
Center frequency (f_0)	3,800	MHz
Bandwidth	800	MHz
Insertion loss (IL)	<5	dB
Flatness (IL variation)	<0.7	dB
Return loss (RL)	>18	dB
Attenuation at $f_0 \pm$ 1,000 MHz	35	dBc
Attenuation at $f_0 \pm$ 2,000 MHz	20	dBc

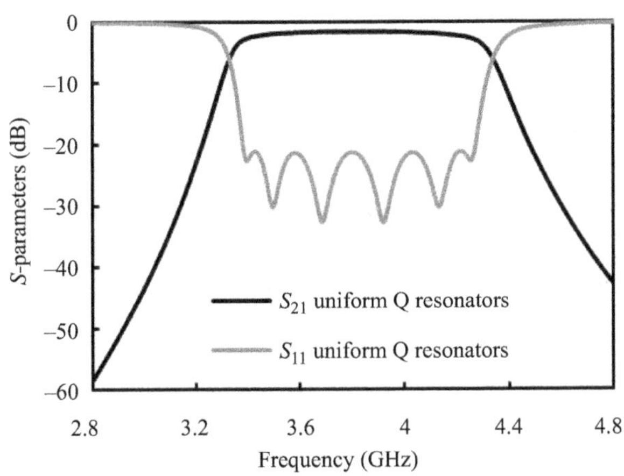

Figure 4.4 Synthesized (Chebyshev) response for the classical hairpin filter (reference design with uniform-Q resonators). Reprinted with permission from Reference [1]; © 2014 IEEE

transfer function is synthesized [19]. The synthesized response is given in Figure 4.4. The response fulfills the specifications, except for the flatness, which attains 1 dB.

The filter can be structured either as an in-line network or as a transversal network, as shown in Figures 4.5 and 4.6. One can note that the transversal network is a transversal pair network, since this particular topology is a practical form for implementing lossy filters. The corresponding coupling matrices are given together with the two coupling networks. Diagonal terms ($j0.05$) in the coupling matrices are placed here for modeling the finite Q value of resonators.

The reference filter is implemented as an in-line network using hairpin resonators as shown in Figure 4.7. The structure has a surface of 210 mm^2. The simulations have been performed using *Momentum*™ included in Agilent Advanced Design System™ (ADS) and dimensions have been optimized by parameter extraction [20]. A prototype has been fabricated and measured. The comparison

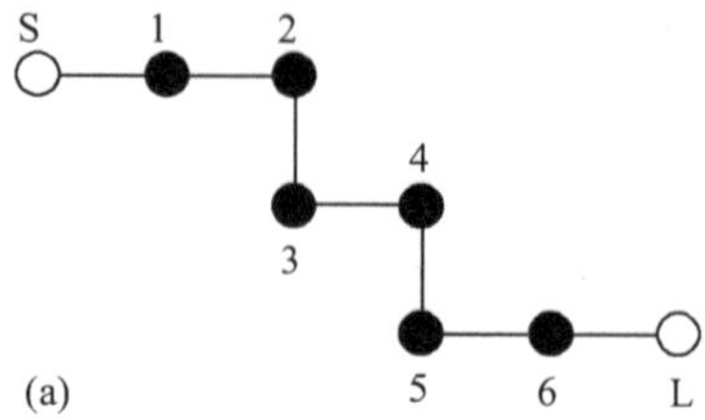

(a)

	S	1	2	3	4	5	6	L
S	0	1.002	0	0	0	0	0	0
1	1.002	j0.05	0.843	0	0	0	0	0
2	0	0.843	j0.05	0.611	0	0	0	0
3	0	0	0.611	j0.05	0.583	0	0	0
4	0	0	0	0.583	j0.05	0.611	0	0
5	0	0	0		0.611	j0.05	0.843	0
6	0	0	0	0	0	0.843	j0.05	1.002
L	0	0	0	0	0	0	1.002	0

(b)

Figure 4.5 Coupling diagram (a) and coupling matrix (b) for realizing the initial filter as an in-line network with uniform-Q resonators (Q = 95). Reprinted with permission from Reference [1]; © 2014 IEEE

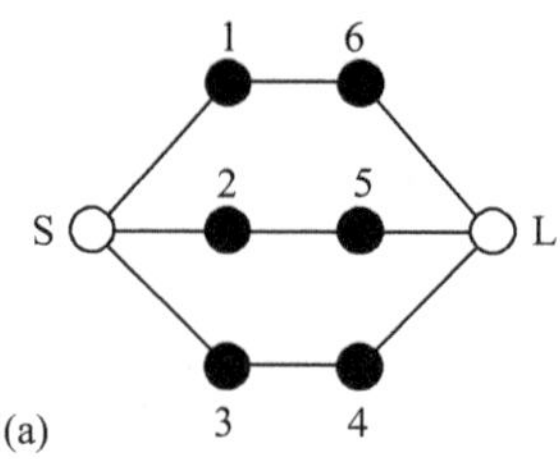

(a)

	S	1	2	3	4	5	6	L
S	0	0.638	0.629	0.447	0	0	0	0
1	0.638	j0.05	0	0	0	0	0.356	0
2	0.629	0	j0.05	0	0	–0.971	0	0
3	0.447	0	0	j0.05	1.199	0	0	0
4	0	0	0	1.199	j0.05	0	0	0.447
5	0	0	–0.971		0	j0.05	0	0.629
6	0	0.356	0	0	0	0	j0.05	0.638
L	0	0	0	0	0.447	0.629	0.638	0

(b)

Figure 4.6 Coupling diagram (a) and coupling matrix (b) for realizing the initial filter as a transversal network with uniform-Q resonators (Q = 95). Reprinted with permission from Reference [1]; © 2014 IEEE

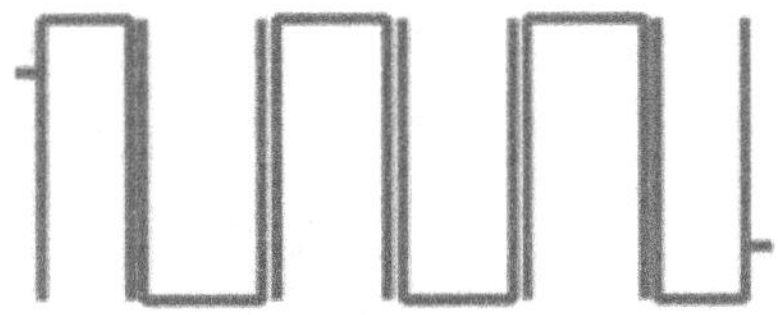

Figure 4.7 Layout of the classical Chebyshev filter implemented as aligned hairpin coupled resonators. Reprinted with permission from Reference [1]; © 2014 IEEE

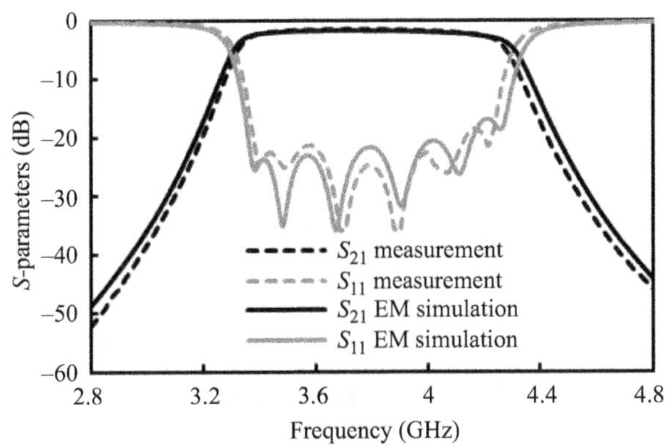

Figure 4.8 Classical Chebyshev filter: measurements and EM simulation. Reprinted with permission from Reference [1]; © 2014 IEEE

between measured and simulated responses is presented in Figure 4.8. A slight reduction in the bandwidth is observed compared to EM simulations. The return loss in the bandwidth is still better than 20 dB and the insertion loss is comparable to the simulated one, around 1.9 dB. The measured flatness is 1.2 dB, i.e., slightly degraded compared to the theory.

In order to explain the difference between measurements and full wave simulations, a sensitivity analysis has been performed considering typical manufacturing tolerances. An 8-μm tolerance for transmission line dimensions and a 1% variation in the permittivity value have been considered for the Monte-Carlo analysis performed with Agilent ADS and presented in Figure 4.9. The permittivity variation mainly affects the frequency, while geometrical variations also affect the return loss. Excluding flatness constraints, more than 70% of realizations meet the specifications.

The objective of this work is now to improve the flatness by designing lossy filters, considering margins available on the insertion losses. Several approaches are compared. The electrical performances of this conventional filter serve as reference for the next designs.

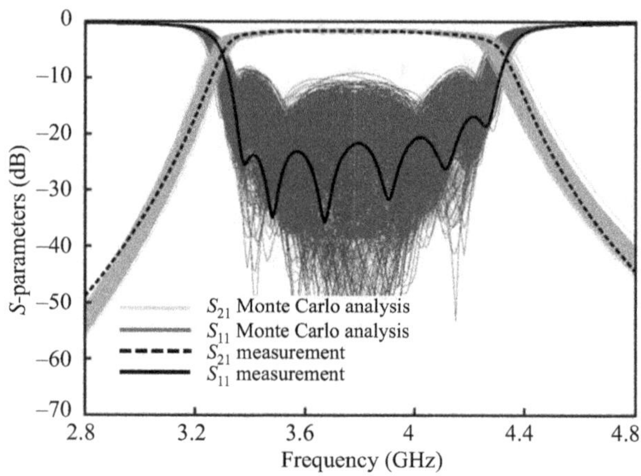

Figure 4.9 Classical Chebyshev filter: Monte-Carlo analysis considering typical manufacturing tolerances (8 μm for transmission line dimensions and 1% for relative permittivity). Reprinted with permission from Reference [1]; © 2014 IEEE

4.4 Design of lossy filters for improved flatness

The principle of lossy filters consists to add and to distribute losses into the coupled-resonator network for retrieving a lossless-like transfer function, *i.e.* having the same flatness and the same selectivity, at a sacrifice on the absolute level of losses. Two ways are possible for realizing a lossy filter. The first way consists in introducing losses in the network through resistive cross-couplings [18] and the second in introducing additional losses directly within some resonators [3]. These two approaches may be mixed also [16]. In this work, lossy filters are designed, on one hand as in-line networks with resistive cross-couplings and on the other hand as transversal networks with heterogeneous-Q resonators.

4.4.1 Inline network with resistive cross couplings

The introduction of resistive cross-couplings (RCCs) in the network allows adjusting the flatness in the passband [22,23]. The number of RCCs in the network is an important parameter, which impacts the performance of the filter directly, in particular its insertion loss and flatness.

For comparison, two lossy filters have been synthesized, with 2 and 4 resistive cross-couplings, respectively, using the approach and software developed in [21] and [22] respectively. The coupling diagrams and coupling matrices can be found in Figures 4.10 and 4.11. The lossy filter responses are compared together with the classical filter response in Figure 4.12. One can observe a slight reduction in the selectivity for the lossy filters.

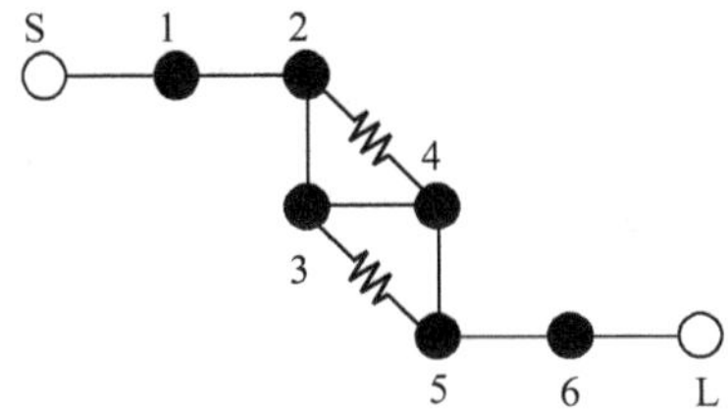

(a)

	S	1	2	3	4	5	6	L
S	0	1.304	0	0	0	0	0	0
1	1.304	j0.059	1.191	0	0	0	0	0
2	0	1.191	j0.091	0.696	j0.031	0	0	0
3	0	0	0.696	j0.091	0.679	j0.031	0	0
4	0	0	j0.031	0.679	j0.091	0.696	0	0
5	0	0	0	j0.031	0.696	j0.091	1.191	0
6	0	0	0	0	0	1.191	j0.059	1.304
L	0	0	0	0	0	0	1.304	0

(b)

Figure 4.10 Coupling diagram: (a) and coupling matrix (b) of the in-line network with 2 RCCs. Reprinted with permission from Reference [1]; © 2014 IEEE

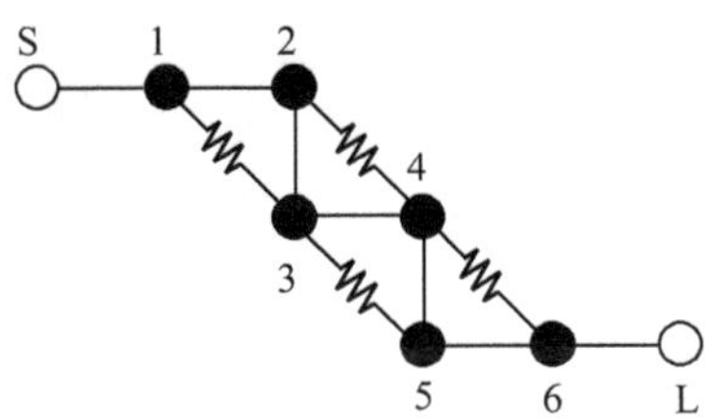

(a)

	S	1	2	3	4	5	6	L
S	0	1.304	0	0	0	0	0	0
1	1.304	j0.064	1.168	j0.004	0	0	0	0
2	0	1.168	j0.115	0.713	j0.055	0	0	0
3	0	j0.004	0.713	j0.119	0.675	j0.055	0	0
4	0	0	j0.055	0.675	j0.119	0.713	j0.004	0
5	0	0	0	j0.055	0.713	j0.115	1.168	0
6	0	0	0	0	j0.004	1.168	j0.064	1.304
L	0	0	0	0	0	0	1.304	0

(b)

Figure 4.11 Coupling diagram: (a) and coupling matrix (b) of the in-line network with 4 RCCs. Reprinted with permission from Reference [1]; © 2014 IEEE

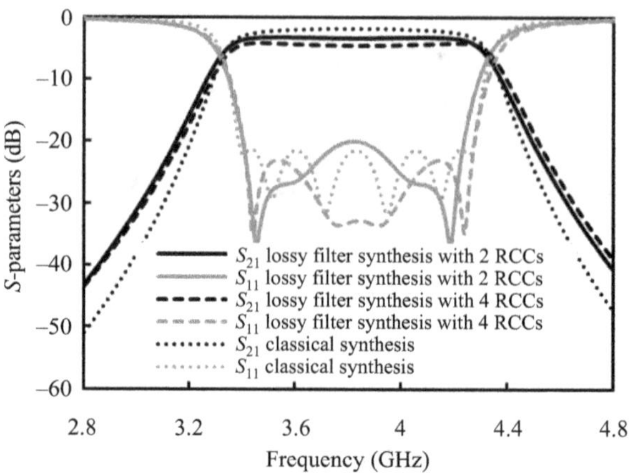

Figure 4.12 Synthesized responses for the lossy filters with 2 and 4 RCCs. Reprinted with permission from Reference [1]; © 2014 IEEE

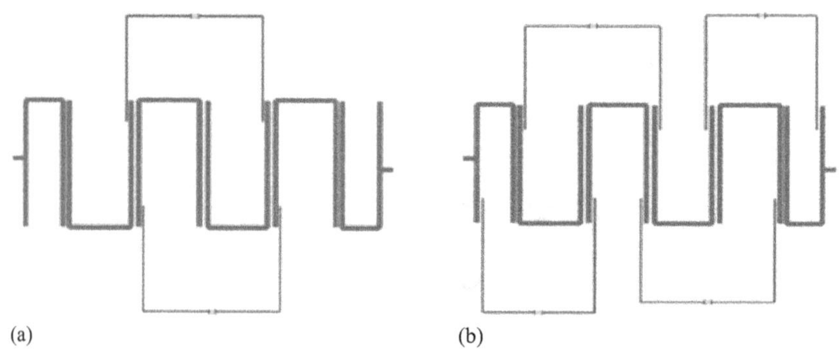

Figure 4.13 Layout of lossy filters implemented as in-line coupled hairpin resonators with (a) two RCCs and (b) four RCCs. Reprinted with permission from Reference [1]; © 2014 IEEE

The flatness is around 0.4 dB for the structure with 2 RCCs and 0.3 dB for the structure with 4 RCCs. At the same time, the insertion loss is found to be 3.4 dB for the first filter (2 RCCs) and 4.2 dB for the second one (4 RCCs). The RCCs are implemented using engraved Ta_2N resistors ($R_\square = 50\ \Omega \cdot$ sq) placed between quarter-wave transmission lines, as shown in Figure 4.13.

The filters were designed and optimized by parameter extraction using *Momentum*^{TM} for full wave simulations. Figure 4.14 shows the simulated scattering parameters compared to theoretical ones for both lossy filters. The lossy filter with two RCCs presents a flatness of 0.4 dB, with a maximum insertion loss of 3.1 dB.

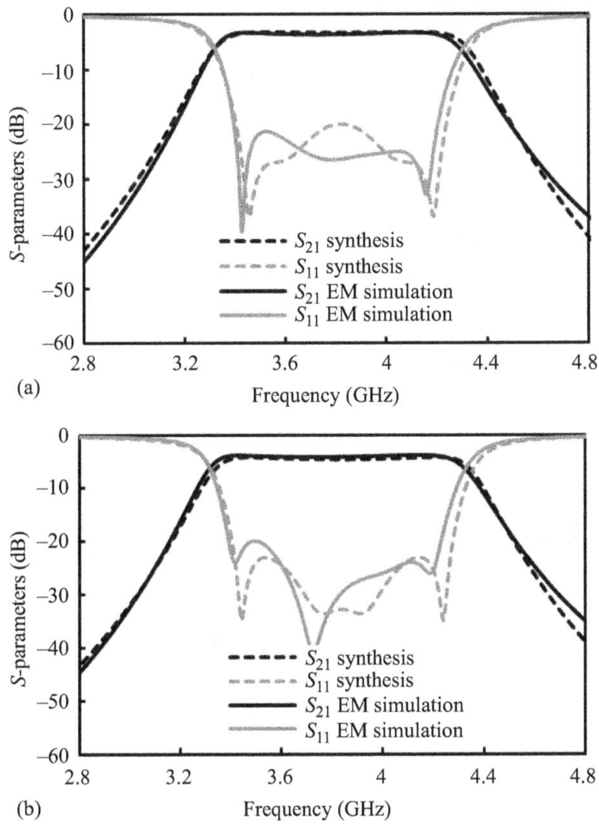

Figure 4.14 Simulated response of the lossy filters with (a) two RCCs and (b) four RCCs. Reprinted with permission from Reference [1]; © 2014 IEEE

The lossy filter with four RCCs presents a flatness of 0.3 dB, and a maximum insertion loss of 4.2 dB. The filters were fabricated using the previous substrate. A photograph is displayed in Figure 4.15. The lossy filter with 2 RCCs has a size of 290 mm^2 and the lossy filter with 4 RCCs a size of 300 mm^2.

Measurements are presented in Figure 4.16. The measured performances remain comparable to the simulated ones for the two fabricated filters except a slight reduction in the bandwidth and a slight degradation of the flatness (+0.1 dB).

A Monte-Carlo analysis has been performed considering the same manufacturing tolerances. A 2% tolerance is also considered for engraved resistors. Figure 4.17 compares the measurements together with the Monte-Carlo analysis. About 65% of realizations met the specifications for the filter with 2 RCCs, and 60% considering the filter with 4 RCCs.

Performances at higher frequency are displayed in Figure 4.18. One can observe that lossy filters provide better attenuation of the second harmonic.

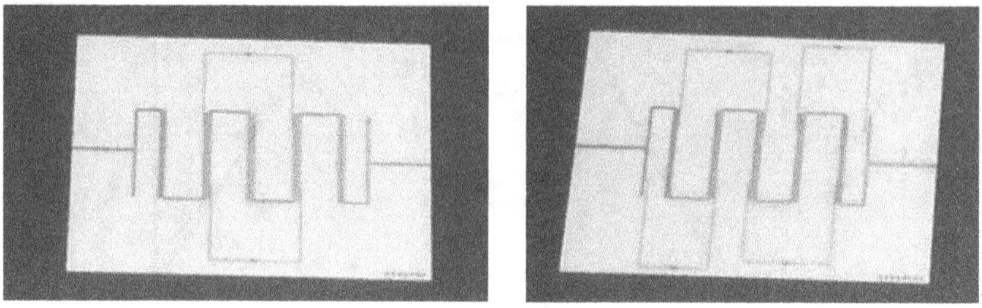

(a) (b)

Figure 4.15 Fabricated lossy filters with (a) two RCCs and (b) four RCCs Reprinted with permission from Reference [1]; © 2014 IEEE

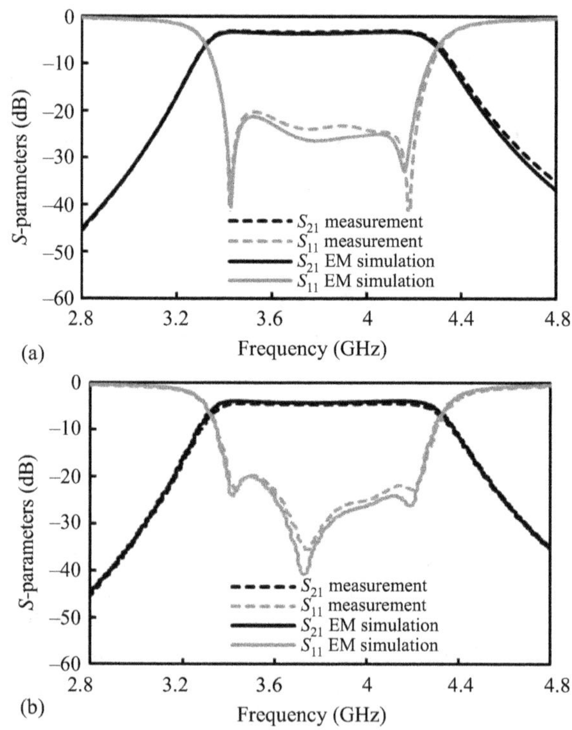

Figure 4.16 Lossy filters with (a) two RCCs and (b) four RCCs: measurements and EM simulations. Reprinted with permission from Reference [1]; © 2014 IEEE

The lossy filter with two RCCs satisfies all the specifications. Adding two additional resistive cross-couplings provides a very small improvement in terms of flatness, while the degradation of insertion loss becomes significant. The equivalent quality factor (Q) of the lossy filter with two RCCs is about 300, *i.e.* three times higher compared to the real one.

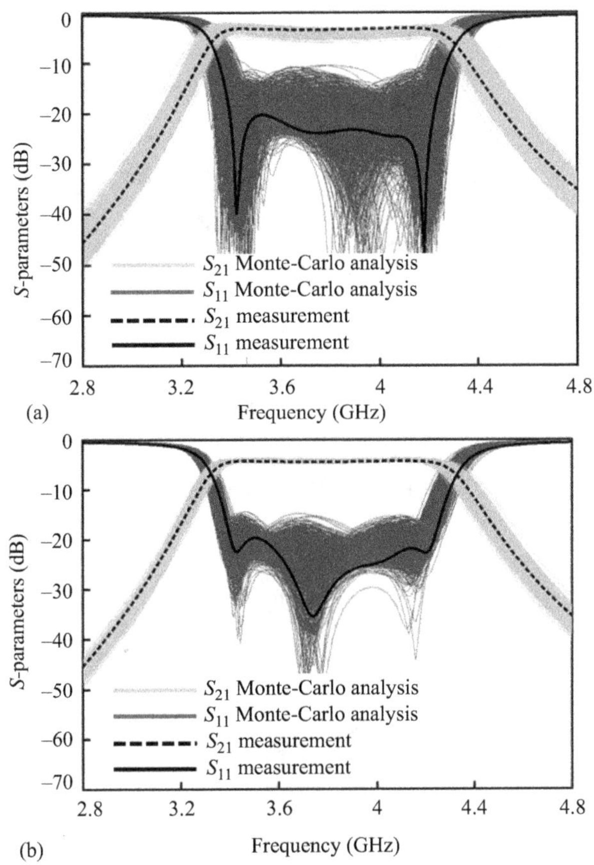

Figure 4.17 Lossy filters with (a) two RCCs and (b) four RCCs: Monte-Carlo analysis considering typical manufacturing tolerances (8 μm for transmission line dimensions, 1% for relative permittivity, and 2% for engraved resistors). Reprinted with permission from Reference [1]; © 2014 IEEE

4.4.2 Transversal network with non-uniform Q resonators

In the case of a transversal network, the signal follows multiple paths. Each path contributes in the filter response almost independently. Using a transversal pair network, the path comprising the stronger coupled resonators mainly contributes to insertion loss in the passband edges. For these resonators, the higher the Q is, the higher the selectivity is. Moreover, the insertion loss is also maximized in the passband edges; consequently these resonators need to be high-Q. Other paths, i.e., with weakly coupled resonators, contribute to insertion loss in the middle of the passband. Consequently, the quality factor of such resonators can be degraded for increasing the insertion loss in the middle of the passband, reducing by this way the insertion loss variation in the bandpass filter response.

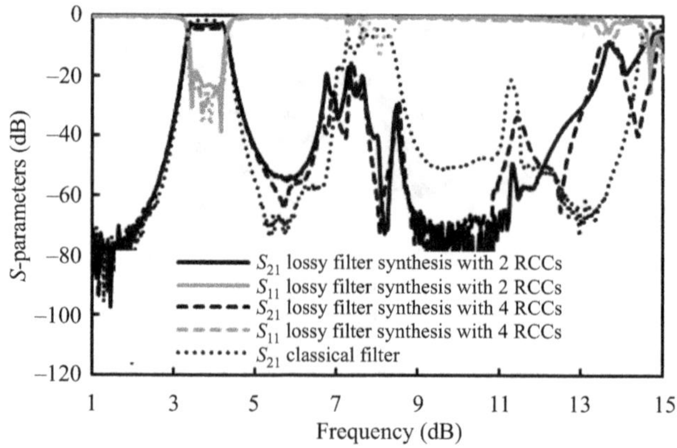

Figure 4.18 Measured out-of-band performances for lossy filters with resistive cross-couplings. Reprinted with permission from Reference [1]; © 2014 IEEE

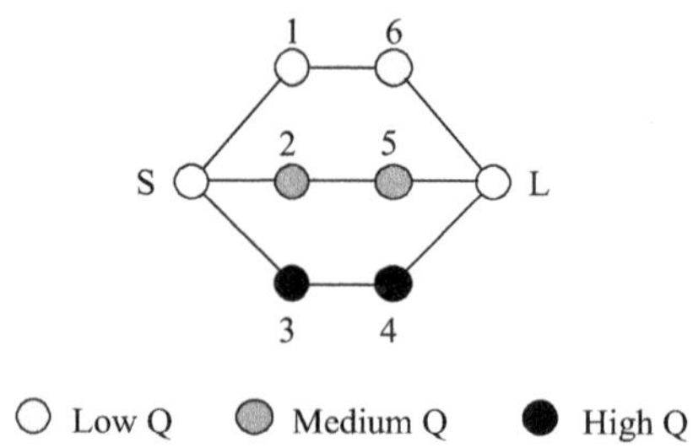

Figure 4.19 Six-pole transversal filter network with non-uniform Q resonators. Reprinted with permission from Reference [1]; © 2014 IEEE

The transversal pair network presented in Figure 4.19 can be synthesized as a lossy filter network by optimizing a coupling matrix obtained from a general Chebyshev filtering function, as introduced in [19]. Initially, a coupling matrix with uniform Q resonators is considered. The synthesis consists in optimizing the couplings and the quality factors of each pair for achieving the desired performances. It is clear that the quality factors can only be degraded in this case.

In order to fit with the specifications in Table 4.1, the coupling matrix synthesized from the Chebyshev filtering function (Figure 4.6) is optimized considering now non-uniform Q resonators using a lumped element equivalent circuit defined with Agilent ADS. The optimized coupling matrix is given in Figure 4.20. The imaginary parts of diagonal terms in the coupling matrix reflect the quality factors of individual resonators. One can see that Q factors are degraded from 95 ($r_i = 0.05$), respectively, to 57 ($r_i = 0.08$) and 35 ($r_i = 0.13$) in the two other paths.

The initial and optimized responses are compared in Figure 4.21. One can observe that the transversal network with non-uniform Q has improved performances in terms

	S	1	2	3	4	5	6	L
S	0	0.653	0.645	0.462	0	0	0	0
1	0.653	j0.13	0	0	0	0	0.374	0
2	0.645	0	j0.08	0	0	−1.026	0	0
3	0.462	0	0	j0.05	1.249	0	0	0
4	0	0	0	1.249	j0.05	0	0	0.462
5	0	0	−1.026		0	j0.08	0	0.645
6	0	0.374	0	0	0	0	j0.13	0.653
L	0	0	0	0	0.462	0.645	0.653	0

Figure 4.20 Coupling matrix of transversal network with non-uniform Q resonators. Reprinted with permission from Reference [1]; © 2014 IEEE

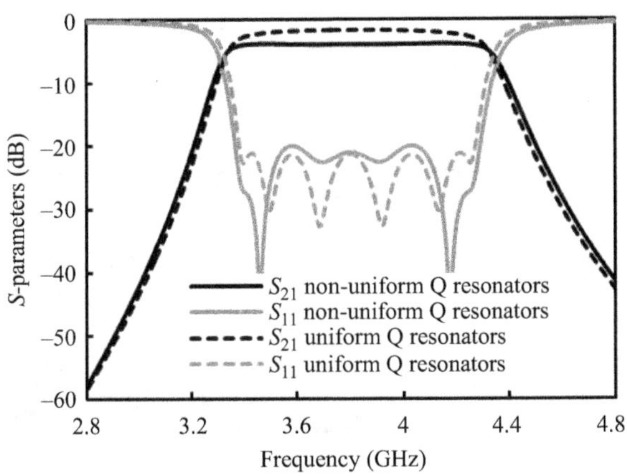

Figure 4.21 Synthesized response of the transversal network with non-uniform Q resonators. Reprinted with permission from Reference [1]; © 2014 IEEE

of flatness while maintaining a similar selectivity and return loss level. The insertion loss variation reaches now 0.1 dB in the passband (compared to 1 dB for the transversal network with uniform Q resonators) and the minimum insertion loss is around 3.3 dB (+1.7 dB compared to the transversal network with uniform Q resonators).

The filter is designed with *Momentum*™ using microstrip hairpin resonators. With 240-μm-wide lines, the achieved Q factor is equal to 95. By reducing the width at 70 μm, the Q factor is diminished to 57. In order to achieve a lower Q factor ($Q = 35$), engraved resistors are introduced in the resonators. Figure 4.22 shows the layout of the six-pole transversal filter with non-uniform Q resonators. Its size covers 480 mm^2 and the scattering parameters obtained from EM simulations are shown in Figure 4.23.

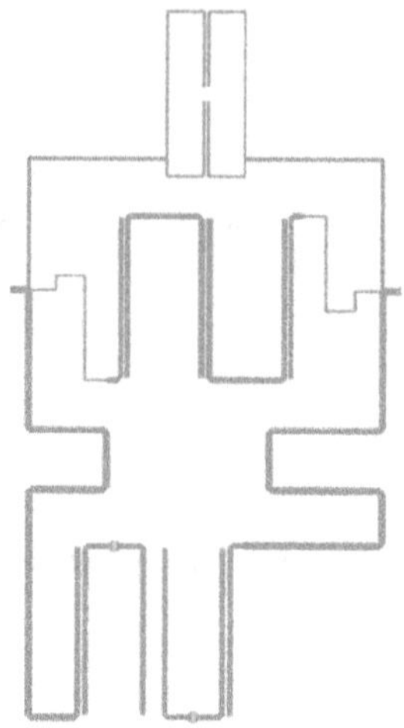

Figure 4.22 Layout of the transversal filter with non-uniform Q resonators. Reprinted with permission from Reference [1]; © 2014 IEEE

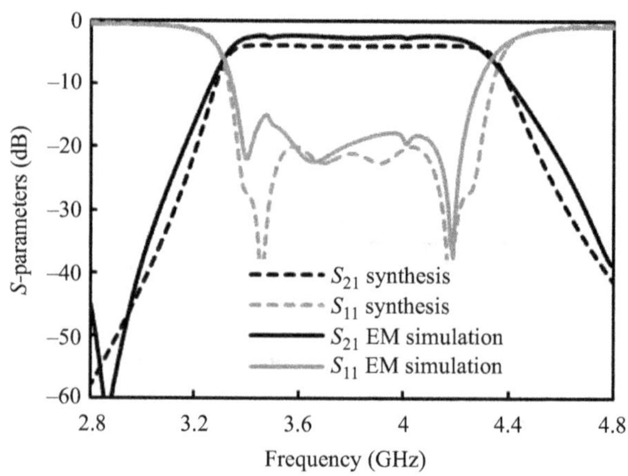

Figure 4.23 Simulated response of the transversal filter with non-uniform Q resonators. Reprinted with permission from Reference [1]; © 2014 IEEE

The filter has been fabricated (Figure 4.24) and characterized. A good agreement is found between measurements and EM simulations as shown in Figure 4.25. Only a slight reduction in the bandwidth is observed. The fabricated filter presents an insertion loss variation of 0.4 dB, the minimum insertion loss is about 2.2 dB and the return loss is better than 18 dB.

The sensitivity analysis presented in Figure 4.26 has been performed considering the previous tolerances. With this transversal design, only 37% of the

Figure 4.24 Fabricated transversal filter with non-uniform Q resonators: Reprinted with permission from Reference [1]; © 2014 IEEE

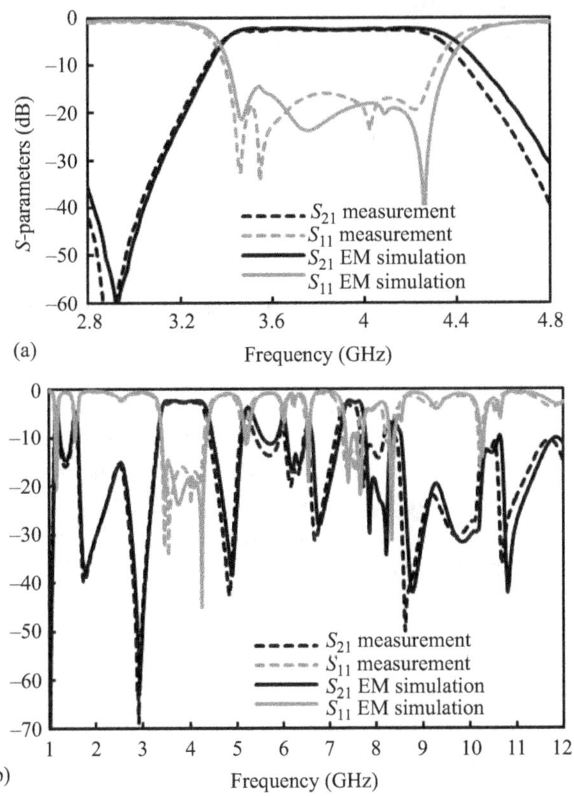

Figure 4.25 Measurements of the transversal filter with non-uniform Q resonators: (a) around the passband and (b) in a broad band. Reprinted with permission from Reference [1]; © 2014 IEEE

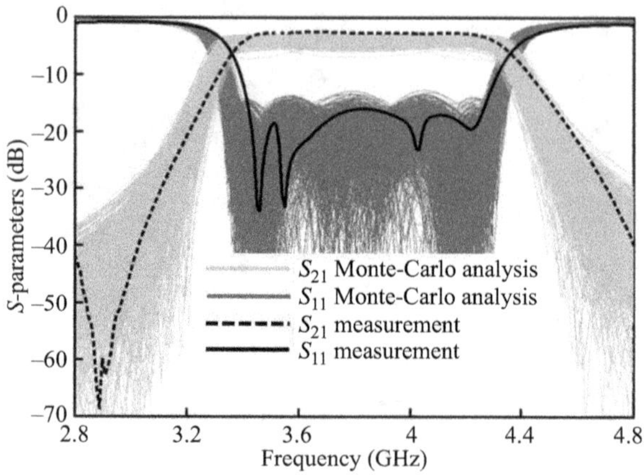

Figure 4.26 Transversal lossy filter: Monte-Carlo analysis considering typical manufacturing tolerances (8 μm for transmission line dimensions, 1% for relative permittivity, and 2% for engraved resistors). Reprinted with permission from Reference [1]; © 2014 IEEE

realizations met the specifications. Moreover, evaluating the out-of-band performance, spurious transmissions are found close to the passband, violating the specifications in terms of rejection in the stopband. For this reason, the lossy filter with 2 RCCs is preferred for our application. However, the spurious transmission could be attenuated by introducing lowpass cells in the structure.

4.5 Design of absorptive lossy filters for attenuation of reflected waves

A frequent need for the design of receivers is to attenuate reflected waves in the stopband in order to protect circuits placed before. The protection is generally achieved by introducing attenuators before the filter, which affect both the reflected and transmitted signals. Using lossy filters, the transmitted signal is already attenuated in order to improve the flatness. Attenuation of the reflected signal is then achieved by modifying slightly the architecture, without additional insertion loss in the passband.

Following the procedure detailed in [4,16,22], the attenuation is distributed together with the losses among the network, using resistive couplings and non-resonating nodes (NRN). The previous in-line network has been modified as presented in Figure 4.27, in order to design two lossy filters with symmetric and asymmetric levels of attenuation.

4.5.1 Symmetric absorptive lossy filter

In this configuration, both the transmission and reflection functions will be attenuated with the same factor ($K = 0.63$). Figure 4.28 shows the ideal response of this

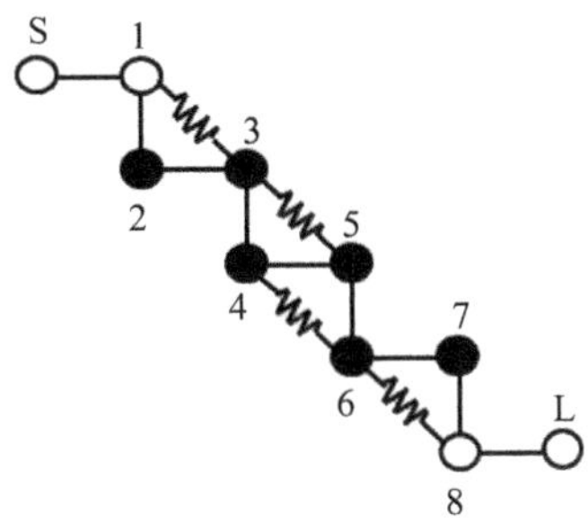

Figure 4.27 Coupling network for absortive lossy filters (symmetric and asymmetric configurations). Reprinted with permission from Reference [1]; © 2014 IEEE

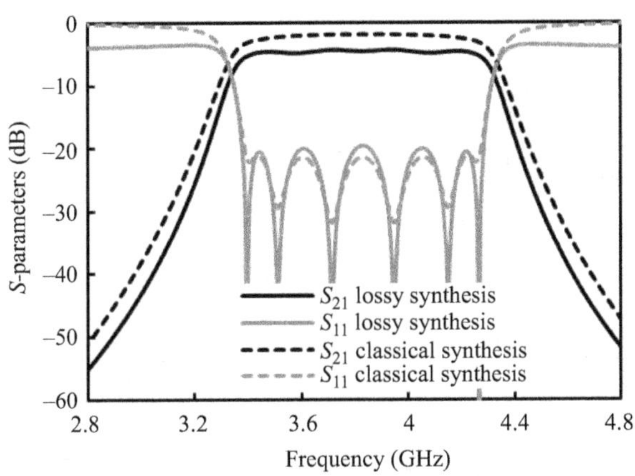

Figure 4.28 Synthesized response of the symmetric absorptive lossy network. Reprinted with permission from Reference [1]; © 2014 IEEE

filter, and the coupling matrix synthesized following the procedure detailed in [21] is given in Figure 4.29. Again, the filter is implemented using microstrip hairpin resonators as shown in Figure 4.30. The filter is dimensioned by parameter extraction from full wave simulations with *Momentum*. The simulated scattering parameters are presented in Figure 4.31. As expected, the attenuation in the stop-band is around 4 dB ($K = 0.63$). The insertion loss is slightly higher (4.3 dB) and the insertion loss variation attains 0.6 dB. The filter was fabricated using the same substrate (Figure 4.32). Measurements in Figure 4.33 are found in good accordance with the simulation, except a slight reduction in the bandwidth and a slight degradation of the flatness (+0.05 dB).

The sensitivity analysis performed with the previous tolerances is presented in Figure 4.34. For this design with non-resonating nodes, 59% of realizations met the specifications.

	S	1	2	3	4	5	6	7	8	L
S	0	0.3662	0	0	0	0	0	0	0	0
1	0.3662	j0.0315	–0.3173	j0.0051	0	0	0	0	0	0
2	0	-0.3173	j0.0529	0.8326	0	0	0	0	0	0
3	0	j0.0051	0.8326	j0.0753	0.5969	j0.0175	0	0	0	0
4	0	0	0	0.5969	j0.0977	0.5384	j0.0449	0	0	0
5	0	0	0	j0.0175	0.5384	j0.0701	0.5969	0	0	0
6	0	0	0	0	j0.0449	0.5969	j0.2330	0.8326	j0.1352	0
7	0	0	0	0	0	0	0.8326	j0.0527	–0.3173	0
8	0	0	0	0	0	0	j0.1352	–0.3173	j0.0670	0.3662
L	0	0	0	0	0	0	0	0	0.3662	0

Figure 4.29 *Coupling matrix of the symmetric absorptive lossy network. Reprinted with permission from Reference [1]; © 2014 IEEE*

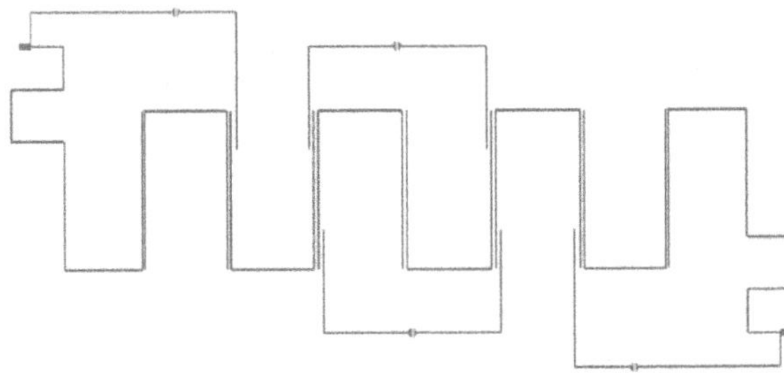

Figure 4.30 Layout of the symmetric absorptive lossy filter. Reprinted with permission from Reference [1]; © 2014 IEEE

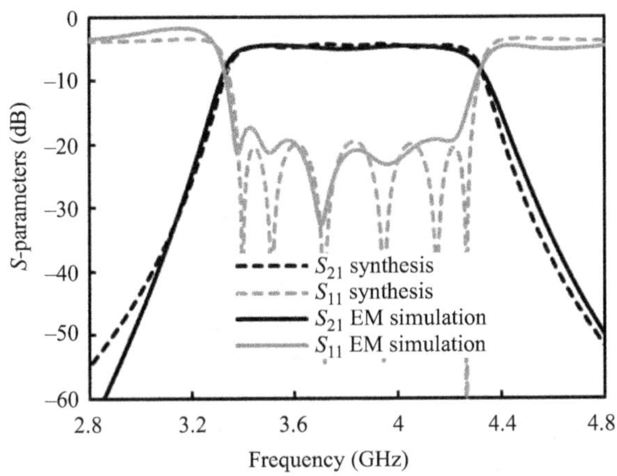

Figure 4.31 Simulated response of the symmetric absorptive lossy filter. Reprinted with permission from Reference [1]; © 2014 IEEE

Figure 4.32 Fabricated absorptive lossy filters (symmetric and asymmetric attenuations). Reprinted with permission from Reference [1]; © 2014 IEEE

4.5.2 Asymmetric absorptive lossy filters

An asymmetric configuration has been designed applying the same procedure, with attenuation levels set to 6 and 1 dB, respectively, at port 1 and port 2 (Figure 4.36). The network remains identical, but the synthesis leads to another coupling matrix,

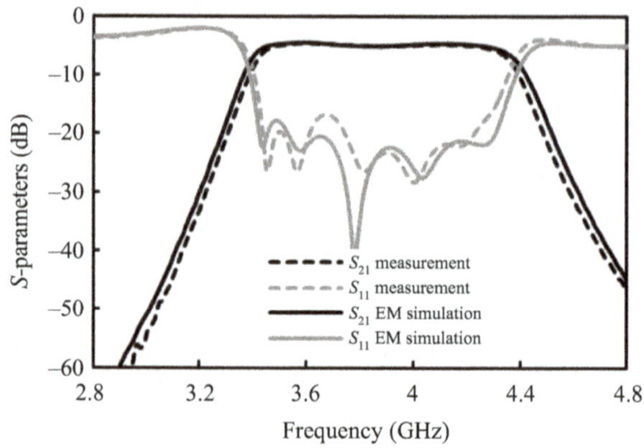

Figure 4.33 Mesured response of the symmetric absorptive lossy filter. Reprinted with permission from Reference [1]; © 2014 IEEE

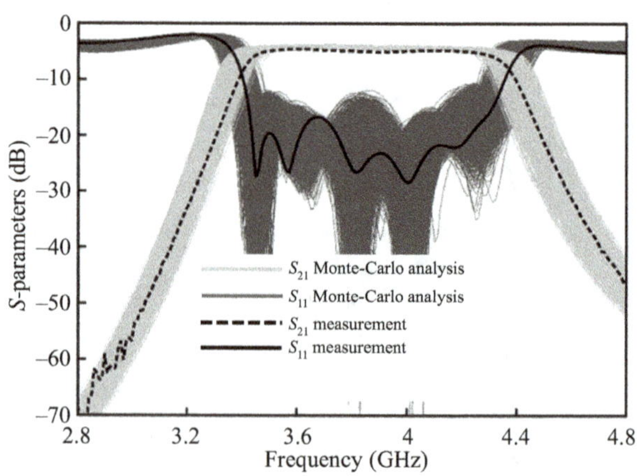

Figure 4.34 Symmetric absorptive lossy filter: Monte-Carlo analysis considering typical manufacturing tolerances (8 μm for transmission line dimensions, 1% for relative permittivity, and 2% for engraved resistors). Reprinted with permission from Reference [1]; © 2014 IEEE

as shown in Figure 4.35. The layout is presented in Figure 4.37 and the simulated response in Figure 4.38.

The insertion loss and the flatness are 4 and 0.6 dB, respectively. The reflection parameter is 18 dB in the stopband but presents a rebound around 3.3 GHz. As shown in Figure 4.39, measurements agree well with simulations except a slight reduction in the bandwidth and a degradation of the flatness (+0.3 dB). The

	S	1	2	3	4	5	6	7	8	L
S	0	0.3940	0	0	0	0	0	0	0	0
1	0.3940	j0.0113	−0.3974	j0.0167	0	0	0	0	0	0
2	0	−0.3974	j0.0523	0.8543	0	0	0	0	0	0
3	0	j0.0167	0.8543	j0.0792	0.6093	j0.0111	0	0	0	0
4	0	0	0	0.6093	j0.0647	0.5728	j0.0128	0	0	0
5	0	0	0	j0.0111	0.5728	j0.0634	0.5917	0	0	0
6	0	0	0	0	j0.0128	0.5917	j0.1171	0.7908	j0.0517	0
7	0	0	0	0	0	0	0.7908	j0.0527	−0.3660	0
8	0	0	0	0	0	0	j0.0517	−0.3660	j0.0430	0.3875
L	0	0	0	0	0	0	0	0	0.3875	0

Figure 4.35 Synthesized response of the asymmetric absorptive lossy network. Reprinted with permission from Reference [1]; © 2014 IEEE

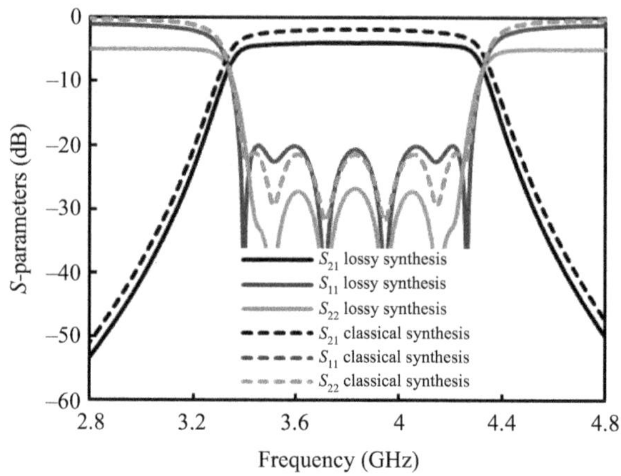

Figure 4.36 Coupling matrix of the asymmetric absorptive lossy network

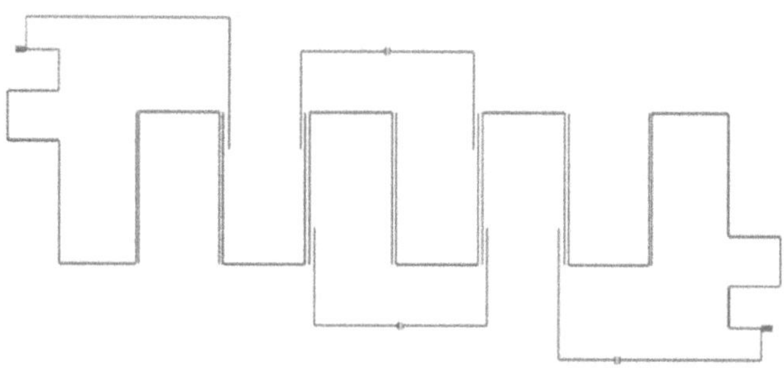

Figure 4.37 Layout of the asymmetric absorptive lossy filter. Reprinted with permission from Reference [1]; © 2014 IEEE

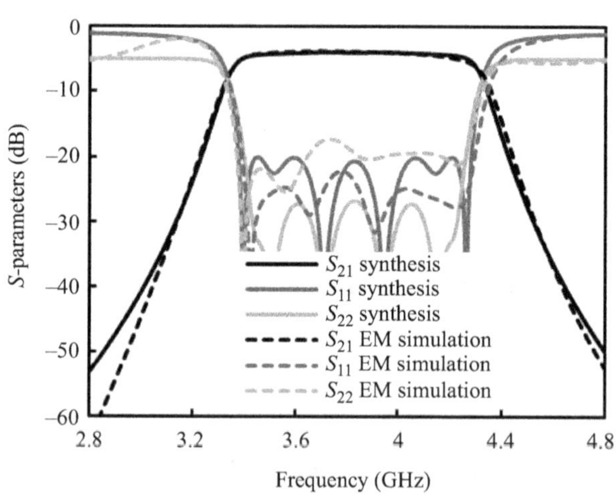

Figure 4.38 Simulated response of the asymmetric absorptive lossy filter. Reprinted with permission from Reference [1]; © 2014 IEEE

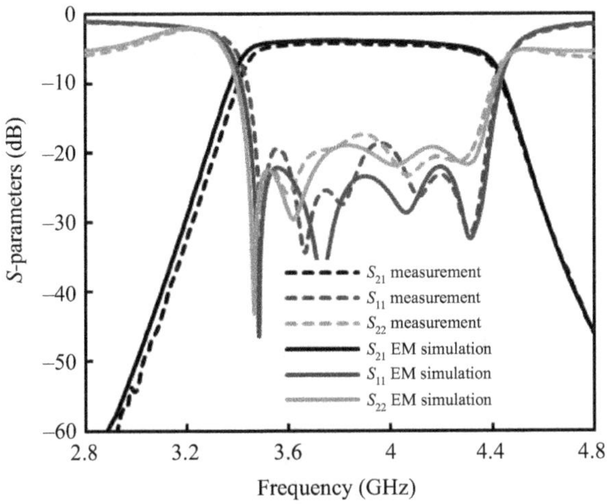

Figure 4.39 Measured response of the asymmetric absorptive lossy filter. Reprinted with permission from Reference [1]; © 2014 IEEE

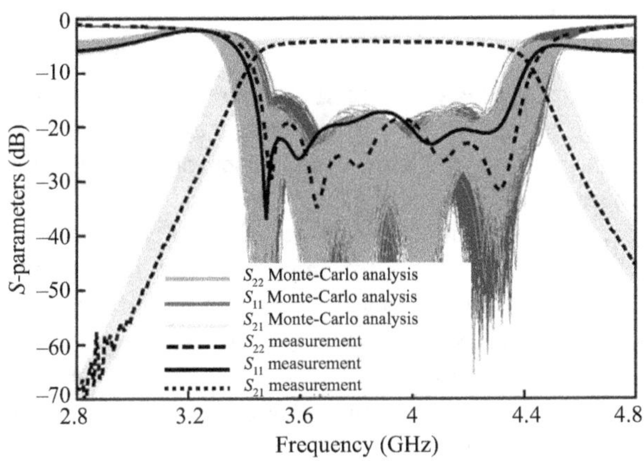

Figure 4.40 Asymmetric absorptive lossy filter: Monte-Carlo analysis considering typical manufacturing tolerances (8 μm for transmission line dimensions, 1% for relative permittivity, and 2% for engraved resistors). Reprinted with permission from Reference [1]; © 2014 IEEE

sensitivity analysis presented in Figure 4.40, shows that 62% of the realizations met the specifications.

Finally, the out-of-band performance (shown in Figure 4.41) is satisfying for both symmetric and asymmetric absorptive lossy filters, considering the specifications given in Table 4.1.

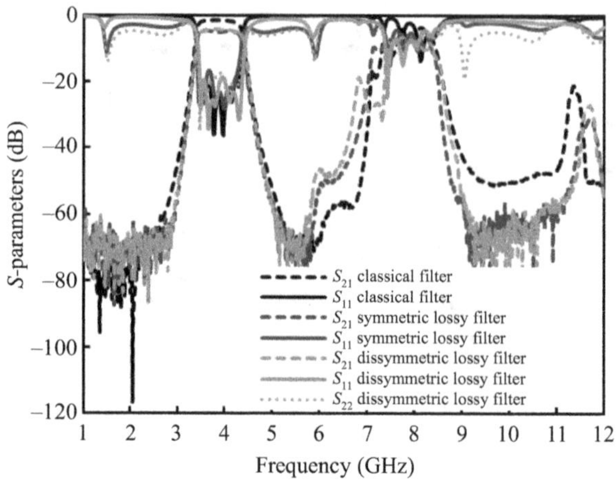

Figure 4.41 Absorptive lossy filters: measured out-of-band performances. Reprinted with permission from Reference [1]; © 2014 IEEE

4.6 Summary

Several designs of lossy filters for receivers in satellite transponders have been investigated in order to improve the performance in terms of flatness compared to a classical hairpin filter design. Flatness and insertion loss performances for all solutions are summarized in Figure 4.42.

The reference design is a microstrip filter made of coupled hairpin resonators. Two approaches have investigated for improving the flatness of the reference design. Lossy filter designs have been compared to the reference, considering the same specifications and the same technology. An in-line network with resistive cross-couplings and a transversal network with heterogeneous Q resonators have been designed and fabricated first. Theoretically, the transversal network leads to better performances in terms of flatness, but its implementation is generally difficult, especially considering the input/output junctions between multiple paths, which are naturally dispersive, causing spurious transmissions in the stopband. Moreover, a Monte Carlo analysis has been performed, showing higher sensitivity of this later solution with respect to manufacturing tolerances. Consequently, considering measured performances reported in Table 4.2, in-line networks with resistive cross-couplings appear to be the best solution for implementing our receiver filter. In particular, the version with two RCCs appears as a good compromise between flatness and insertion loss.

Finally, the in-line network has been transformed introducing non-resonant nodes and additional resistive cross-couplings in order to design absorptive lossy filters. The resulting filters allow attenuating the reflected wave substantially with a reduced impact on the absolute level of losses.

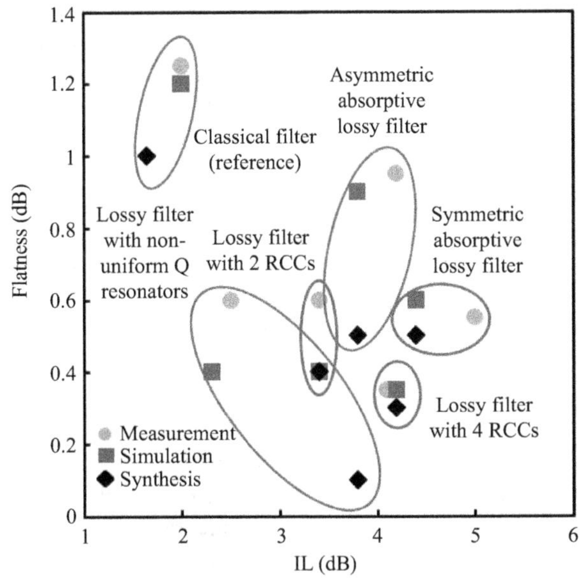

Figure 4.42 Flatness versus IL for all filters designed in this work. Reprinted with permission from Reference [1]; © 2014 IEEE

Table 4.2 Measured performances for all filters designed in this work. Reprinted with permission from Reference [1]; © 2014 IEEE

	IL (dB)	Flatness (dB)	RL (dB)	20-dB rejection bandwidth (GHz)	Surface (mm^2)
Reference filter	1.9	1.2	20	2.5	210
Transversal lossy filter with non-uniform *Q* resonators	2.5	0.5	18	0.5	480
In-line lossy filter with two RCCs	3.4	0.45	20	2.5	290
In-line lossy filter with four RCCs	4.2	0.35	20	2.5	300
Symmetric absorptive lossy filter (four RCCs and two NRNs)	5	0.55	19	2.5	610
Asymmetric absorptive lossy filter (four RCCs and two NRNs)	4.2	0.9	19	2.5	610

References

[1] A. Basti, A. Perigaud, S. Bila, S. Verdeyme, L. Estagerie, and H. Leblond "Design of microstrip lossy filters for receivers in satellite transponders," *IEEE Trans. Microw. Theory Techn.*, vol. 62, no. 9, pp. 2014–2024, 2014.

[2] A. Basti, S. Bila, S. Verdeyme, A. Perigaud, L. Estagerie, and H. Leblond "Comparison of two approaches for the design of microstrip lossy filters," in *Proc. of 33rd Eur. Microw. Conf. (EuMC), Nuremberg*, Oct. 2013, pp. 21–24.

[3] M. Meng and I. C. Hunter, "The design of parallel connected filter networks with non-uniform Q resonators," in *2012 IEEE MTT-S Int. Microw. Symp. Dig.*, Montreal (QC), Jun. 2012, pp. 1–3.

[4] V. Miraftab and M. Yu, "Advanced coupling matrix and admittance function synthesis techniques for dissipative microwave filters," *IEEE Trans. Microw. Theory Techn.*, vol. 57, no. 10, pp. 2429–2438, 2009.

[5] V. Miraftab and M. Yu, "Generalized lossy microwave filter coupling matrix synthesis and design using mixed technologies," *IEEE Trans. Microw. Theory Techn.*, vol. 56, no. 12, pp. 3016–3027, 2008.

[6] L. Szydlowski, A. Lamecki, and M. Mrozowski, "Design of microwave lossy filter based on substrate integrated waveguide (SIW)," *IEEE Microw. Wireless Compon. Lett.*, vol. 21, no. 5, pp. 249–251, 2011.

[7] I. C. Hunter, J. D. Rhodes and V. Dassonville "Dual mode filters with conductor loaded dielectric resonators", *IEEE Trans. Microw. Theory Techn.*, vol. 47, no. 12, pp. 2304–2311, 1999.

[8] K. Konno, "Small size combline microstrip narrow BPF", *IEEE Microwave Theory and Techniques Symposium,* Albuquerque, NM, USA, juin 1992.

[9] I. C. Hunter, R. Ranson, A.C. Guyette, and A. Abunjaileh, "Microwave filter design from a systems perspective", *IEEE Microwave Magazine*, October 2007.

[10] G. Matthaei, L. Young, and E. M. T. Jones, *Microwave filters, impedance-matching networks, and coupling structures*. Norwood, MA: Artech House, 1980, p. 152.

[11] A. E. Williams, W. G. Bush, R. R. Bonetti, "Predistortion techniques for multicoupled resonator filters", *IEEE Trans. Micr. Theory Tech*, vol. 33, no. 5, 1985.

[12] S. Bila, P. Lenoir, D. Baillargeat, and S. Verdeyme, "Multiple solutions for the synthesis of microwave filters with predistorted transfer functions", *Int. J. RF Micr. Computer-Aided Eng.*, vol. 17, no. 1, 2007.

[13] S. Darlington, "Synthesis of a reactance-four pole with prescribed insertion loss characteristics," *J. Math. Phys.*, vol. 18, no. 1939, pp. 257–353.

[14] M. Yu, W-C Tang, A. Malarky, V. Dokas, R. Cameron, and Y. Wang, "Predistortion technique for cross-coupled filters and its application to satellite communication systems", *IEEE Trans. Micr. Theory Tech*, vol. 51, no. 12, 2003.

[15] B. S. Senior, I. C. Hunter, and J. D. Rhodes, "Synthesis of lossy filters," in *Proc. of 32nd Eur. Microw. Conf.*, 2002, pp. 1–4.

[16] A. Guyette, I. Hunter, and R. Pollard, "The design of microwave bandpass filters using resonators with nonuniform Q," *IEEE Trans. Micr. Theory Tech*, vol. 54, no. 11, pp. 3914–3922, 2006.

[17] J. Mateu, A. Padilla, C. Collado, *et al.*, "Synthesis of 4th order lossy filters with uniform Q distribution," in *2010 IEEE MTT-S Int. Microw. Symp. Dig.*, Anaheim (CA), May 2010, pp. 568–571.

[18] L. Szydlowski, A. Lamecki, and M. Mrozowski, "Synthesis of coupled-lossy resonator filters," *IEEE Microw. Wireless Compon. Lett.*, vol. 20, no. 7, pp. 366–368, 2010.

[19] R. J. Cameron, "General coupling matrix synthesis methods for Chebyshev filtering functions," *IEEE Trans. Microw. Theory Techn.*, vol. 47, no. 4, pp. 433–442, 1999.
[20] S. Bila, D. Baillargeat, M. Aubourg, *et al.*, "Finite element modelling for the design optimization of microwave filters," *IEEE Trans. Magn.*, vol. 40, no. 2, pp. 1472–1475, 2004.
[21] A. Padilla, J. Mateu, C. Collado, *et al.*, "Comparison of lossy filters and predistorted filters using novel software," in *2010 IEEE MTT-S Int. Microw. Symp. Dig.*, Anaheim (CA), May 2010, pp. 1720–1723.
[22] http://www.tsc.upc.edu/lossyfilters.
[23] M. Yu and V. Miraftab, "Shrinking microwave filters," *IEEE Microw. Mag.*, vol. 9, no. 5, pp. 40–54, 2008.

Chapter 5

Microstrip extracted-pole lossy filters

Zhou Zhou[1] *and Jiasheng Hong*[1]

5.1 Introduction

There have been increasing demands for advanced RF/microwave filters other than conventional lossless synthesis filters in order to meet stringent requirements from RF/microwave systems, especially from modern wireless communications systems. Conventional microwave lossless filter synthesis has been focused for years since it is the fundamental theory to design filters, which can be regarded as 'matured' to some extent with any filter transfer function being synthesized to the desired filter topology using exact synthesis procedures [1–3]. However, these microwave lossless synthesis filters are designed in the limitation of microwave filter itself without considering from the perspective of the whole system, which can reduce the system compatibility.

The rapid development of microwave wireless applications has also stimulated increasing requirements for compact and high-performance microstrip bandpass filter [4]. Normally, for microstrip line filters, the quality factor can achieve no more than 250 using conventional conductors and dielectric substrates, which means that taking the limited resonators quality factor (Q) into consideration in filter design is desired. This is because that higher Q often suffers from bigger resonator sizes to make the overall size larger, while lower Q usually leads to degraded insertion loss performance in the passband and rounding passband edges, especially in narrowband filters [5]. Accepting some additional insertion loss and knowing the fact that the loss can be compensated for by the already-existing microwave amplifiers in the system [5–7], the sharp response of the filter transmission can be restored. The lossy circuit extraction techniques can be used to improve the return loss by using non-uniform dissipation and modified topologies with additional loss.

In this chapter, a type of microstrip extracted-pole lossy filter is presented [8,9], which is realized by introducing resistive cross couplings into a microstrip extracted-pole filter to achieve a flat passband. The high selectivity is achieved by introducing two transmission zeros using two extracted poles, which can be

[1]Institute of Sensors, Signals and Systems, School of Engineering and Physical Sciences, Heriot-Watt University, UK

adjusted independently. In the first part of the chapter, microwave filter design from a system perspective is briefly introduced. Microwave lossy filter and the extracted-pole filter techniques are then described. After that, the design and the microstrip implementation of extracted-pole lossy filters are presented in detail. Simulation tools from [10] and [11] are used for theory circuit model and full-wave electromagnetic simulations.

5.2 Filter design from system perspective

Consider microwave filters being used in the receive side cascaded with the low-noise amplifier (LNA) for interference filtering and signal channelization [12]. For a conventional filter and amplifier cascade structure shown in Figure 5.1(a), a low-loss filter (Filter 1) is often placed before the LNA. The absolute insertion loss of the filter 1, also known as the preselect filter, can result in more negative effect on the overall noise figure of the receiver. Although higher Q resonators such as cavity or coaxial resonators may be used in the preselect filter design to achieve a lower insertion loss and sharper band edge rolloff, mass and volume are also very critical considerations from a physical perspective. The absolute insertion loss in the passband is often a secondary parameter that can be traded for size reduction. Alternatively, to reduce the insertion loss of the preselect filter, the other requirements of the first filter can be relaxed by moving, for instance, some selectivity and out-of-band rejection requirements to the second filter (Filter 2). In this case, the absolute insertion loss value of the filter 2 is no longer critical since the LNA can easily make up the loss of the filter by setting the gain to a higher value, while the most critical electrical parameters are in band performance such as the insertion loss variation and the sharper band edge rolloff. Another approach has emerged to split the amplifier into two stages with a filter in between as shown in Figure 5.1(b), which makes it also possible to enhance the

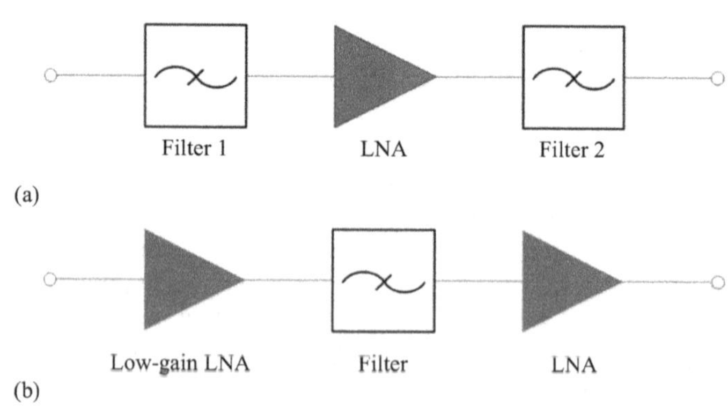

Figure 5.1 Distributed filter with (a) two filter and (b) two LNAs. Adapted with permission from [12]; © 2008 IEEE

performance or at least maintain similar performance by using lower Q resonators for the filter design.

Adaptive pre-distortion and lossy circuit techniques are the most promising approaches for trading the absolute insertion loss to improve the passband performance. The adaptive pre-distortion technique [13–16] assumes no change to the filter topology resulted in degraded return loss performance. As a result to compensate for the return loss performance, in practice, pre-distorted filters are usually connected with nonreciprocal devices such as isolators and circulators. An alternative approach to improve the response is using lossy circuit techniques [17–29]. The main aim of lossy technology is to accept additional insertion loss without changing the passband shape of the filter function. Compared with the pre-distortion, the lossy synthesis is based on the absorption rather than the reflection of power, which can improve return loss.

5.3 Microwave lossy filter technique

The principle of microwave lossy synthesis consists of adding and distributing losses into the coupled resonators network. There are two normal ways to realizing microwave lossy filter: one is using resistive cross-couplings (RCCs) to introduce losses in inline network and the other is using non-uniform Q resonators in transversal network. These two methods can be also mixed in some complex designs. The synthesis of lossy filters may be composed of different steps starting from the design requirements to a possible circuit design. Figure 5.2 shows a stepwise approach for a generalized synthesis technique [5], where two types of synthesis are involved: polynomial synthesis and circuit synthesis.

The theoretical progress for the direct synthesis of lossy polynomials in the literature [5,6] has been so far limited to multiplying both reflection and

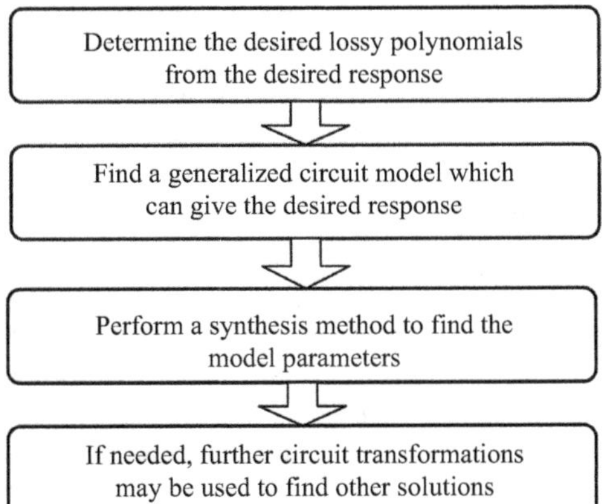

Figure 5.2 Generalized steps for direct synthesis of lossy filters. Adapted with permission from [5]; © 2008 IEEE

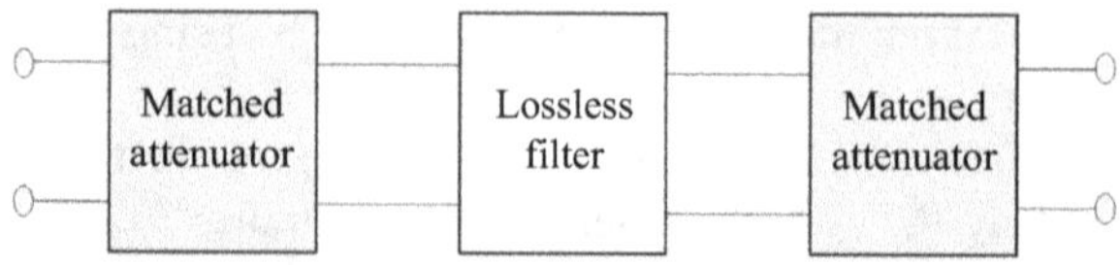

Figure 5.3 Equivalent lossy filter network with same attenuation. Adapted with permission from [5]; © 2008 IEEE

transmission polynomials of a lossless function by a constant attenuation factor of $K < 1$ as

$$\begin{aligned} S_{11\_lossy} &= KS_{11\_lossless} \\ S_{21\_lossy} &= KS_{21\_lossless} \end{aligned} \tag{5.1}$$

From the network perspective, this special case is equivalent to placing two identical matched attenuators at input and output ports of a lossless filter with attenuation factors of $\sqrt{K}$, as in Figure 5.3. The two identical matched attenuators can then be made using different kinds of resistive networks. This method is also suitable for an initial lossy coupling matrix synthesis for cases, where both reflection and transmission coefficients are multiplied by K. This case, however, is not the most general form of lossy polynomials. On the other hand, the direct circuit synthesis appears to be more flexible for handling different cases and the circuit model may result from a coupling matrix with complex entries for the lossy filter.

Numerous microwave lossy filters have been proposed in [5–9], [17–29], which can be realized and divided into two families. Additional losses are introduced either in individual resonators, forming a network with non-uniform Q resonators or distributed through RCCs. In the following section, the two approaches are described with theory design examples.

5.3.1 RCC in inline network

The introduction of RCCs in the network allows adjusting the flatness in the passband. For example, a four pole lossy filter with RCCs centred at 1 GHz with a FBW of 10% is demonstrated. The coupling diagram is given in Figure 5.4(a). Node 1, 2, 3 and 4 are resonators in the inline network with two RCCs between nodes 1 and 3, 2 and 4. The equivalent circuit model is shown in Figure 5.4(b) with all units for each parameters. All resonators are supposed to be synchronously tuned resonating at the same frequency and having the same unloaded quality factor Q. The couplings between resonators are represented by the ideal admittance- or J-inverters, i.e. $J12$, $J23$ and $J34$. The input/output (I/O) or external couplings are represented by the J-inverters of JE. The RCC is represented by R_{lossy}. Table 5.1 lists a set of circuit element values for a lossy filter of this type with uniform Q of 100 and 10% fractional bandwidth (FBW) centred at f_0 of 1.0 GHz.

The comparison of frequency response with/without RCC is presented in Figure 5.5. It can clearly been seen that without RCC the passband response is upward rounded while the introduction of RCC makes the passband flatter with a

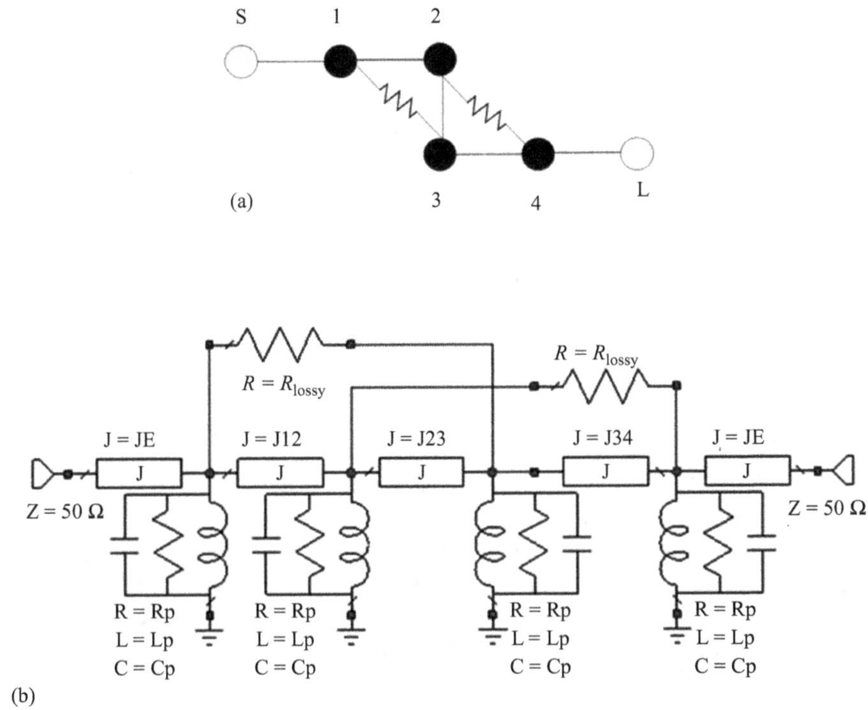

Figure 5.4 Four-pole lossy filter with RCCs: (a) coupling diagram and (b) equivalent circuit model

Table 5.1 Element values for four-pole lossy filter circuit with uniform Q of 100 and FBW = 10% centred at f_0 = 1.0 GHz

JE	0.0207 S	Cp	31.83 pF	R_{lossy}	1,000 Ω
J12 = J34	0.01822 S	Lp	0.7958 nH		
J23	0.0140 S	Rp	500 Ω		

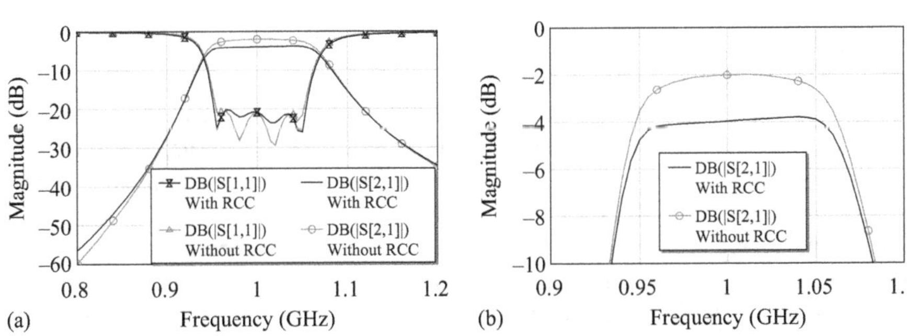

Figure 5.5 (a) Frequency responses with/without RCC and (b) zoom in passband

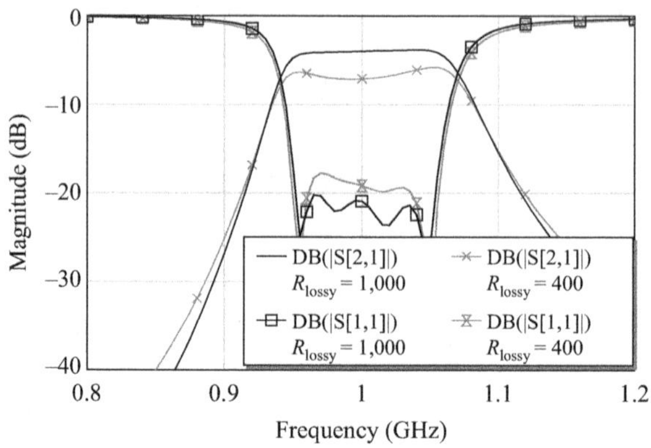

Figure 5.6 Frequency response with different values of RCC

smaller insertion loss variation across the passband. Varying the RCC, namely, the resistor value of R_{lossy} can adjust the flatness in the passband. As displayed in Figure 5.6, a fairly flat passband is attained when the resistor value is 1,000 Ω. When the resistor value is 400 Ω, the insertion loss response in the passband is downward rounded, indicating too much loss being introduced to the mid-band by the RCC.

5.3.2 Non-uniform *Q* resonators for transversal network

In the case of a transversal filter network, the signal follows multiple paths so that each path contributes in the filter response almost independently. Using a transversal pair network, the path comprising the stronger coupled resonators mainly contributes to insertion loss in the passband edges. For these resonators, the higher the *Q*, the higher the selectivity. Moreover, the insertion loss is also maximized in the passband edges that consequently these resonators need to be high *Q*. Other paths, i.e., with weakly coupled resonators, contribute to insertion loss in the middle of the passband. As a result, the quality factor of such resonators can be degraded for increasing the insertion loss in the middle of the passband, reducing by this way the insertion loss variation in the bandpass filter response.

The transversal pair network presented in Figure 5.7(a) has two transversal paths between the source and the load: one has two low *Q*-coupled resonators (path 1) and the other path has two high *Q*-coupled resonators (path 2). Therefore, the path 1 contributes the insertion loss in the middle of the passband, which should have bigger external couplings and smaller couplings between resonators to dominate the performance of mid passband; while path 2 affects the passband edges performance, which should have smaller external couplings and bigger couplings between resonators to dominate the actual FBW. With the compensation of these two paths, the passband insertion loss can be adjusted to be flat while maintaining good matching in the passband. Figure 5.7(b) shows the equivalent circuit model of this topology defined with microwave office AWR [10], where L_{pm}, $C_{pm}(m = 1 \text{ and } 2)$ are the

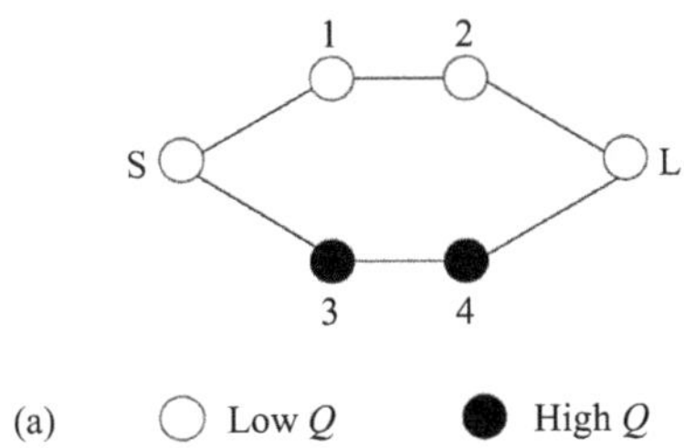

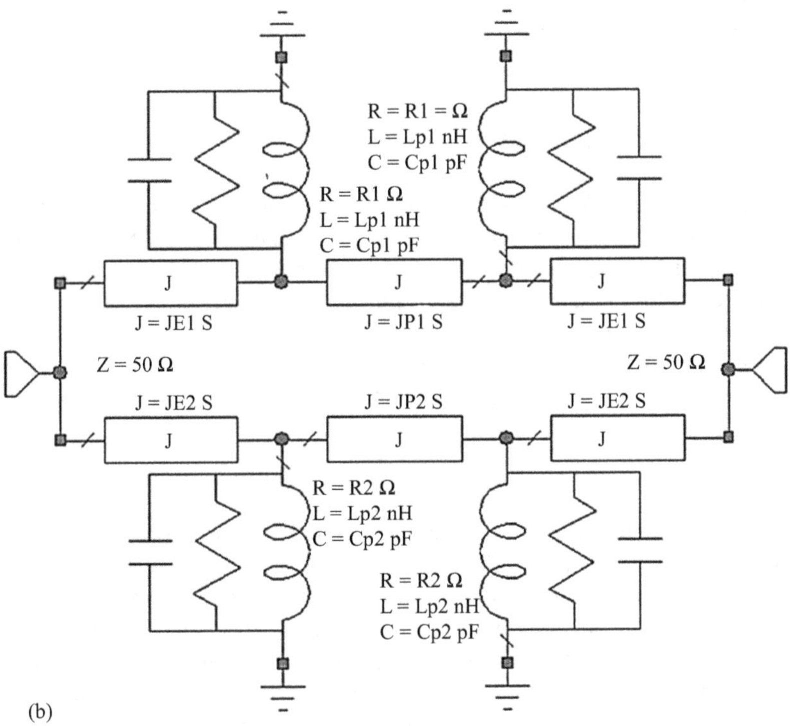

Figure 5.7 (a) Four-pole transversal filter network with non-uniform Q resonators and (b) its equivalent circuit model

circuit parameters of resonators with R_m representing the resonator loss. The ideal (*frequency-invariant*) J-inverters are used to represent the couplings. Table 5.2 lists a set of element values of this circuit model with non-uniform Q and FBW = 10% centred at $f_0 = 1.0$ GHz.

The circuit responses of this four-pole transversal lossy filter with uniform/non-uniform Q resonators are compared in Figure 5.8(a) and (b). As can be seen, the transversal network with non-uniform Q, i.e. low-Q of 70 and high-Q of 350, has improved performance in terms of flatness and selectivity while maintaining

Table 5.2 Element values for four-pole transversal lossy filter with non-uniform Q and FBW = 10% centred at $f_0 = 1.0$ GHz

JE_1	0.01758 S	Lp_1	0.7958 nH	Q_1	70
JE_2	0.0105 S	Cp_1	31.83 pF	R_1	350 Ω
JP_1	0.01381 S	Lp_2	0.7958 nH	Q_2	350
JP_2	−0.02556 S	Cp_2	31.83 pF	R_2	1,750 Ω

good passband matching. Figure 5.8(c) shows that other combinations of non-uniform Q can also improve the passband flatness, thought with different absolute passband insertion losses. Note that only Q_1 and Q_2 are changed in the circuit model for plotting these responses.

5.4 Extracted-pole filter technique

Extracted pole filters [30–32] are well known to generate the transmission zeros (TZs) by dangling resonators rather than cross couplings [33] in a specified topology. Benefiting from this feature, filter design with extracted pole techniques can have a very flexible layout. It is useful for many applications, especially when a cross-coupled topology cannot be realized because of the restricted mechanical configuration. Furthermore, every TZ is uniquely linked to the resonant frequency of the dangling resonator, so it can be controlled independently. Because of this, a less sensitive network can be obtained.

5.4.1 Synthesis procedure

The basic idea of extracted pole synthesis theory was first developed in [30] and the complete description of the synthesis procedure can be found in [31]. Additional corrections are proposed in [34] for fully canonical cases. References [35–38] developed the non-resonating node (NRN) synthesis technique, which is indicated that extracted pole and NRN synthesis techniques are mathematically equivalent by [31]. Recently, [39] presented an approach to synthesize the cross-coupled network with both resonator nodes and NRN nodes. The NRN synthesis is an extension of extracted pole and coupling matrix theory by introducing new node elements in the coupling matrix. Although this technique has been used, there is no explicit design procedure because it is hard to find or model a physical structure to accurately link with the extracted pole elements in the synthesized prototype.

The extracted-pole synthesis procedure starts from the lowpass quasi-elliptic transfer function. Strictly speaking, it can only be performed on a complex conjugate symmetrical network. Therefore, it will only work on an even-order transfer function with a fourth-order minimum. Two important types of extraction cycles are included in the synthesis procedure. The initial cycles of the extracted-pole synthesis involve extracting a unity impedance phase shifter from both ends of the network with complex conjugate symmetry, shown in Figure 5.9. A phase shifter θ is extracted at both sides of the network N in Figure 5.9(a) with a remaining

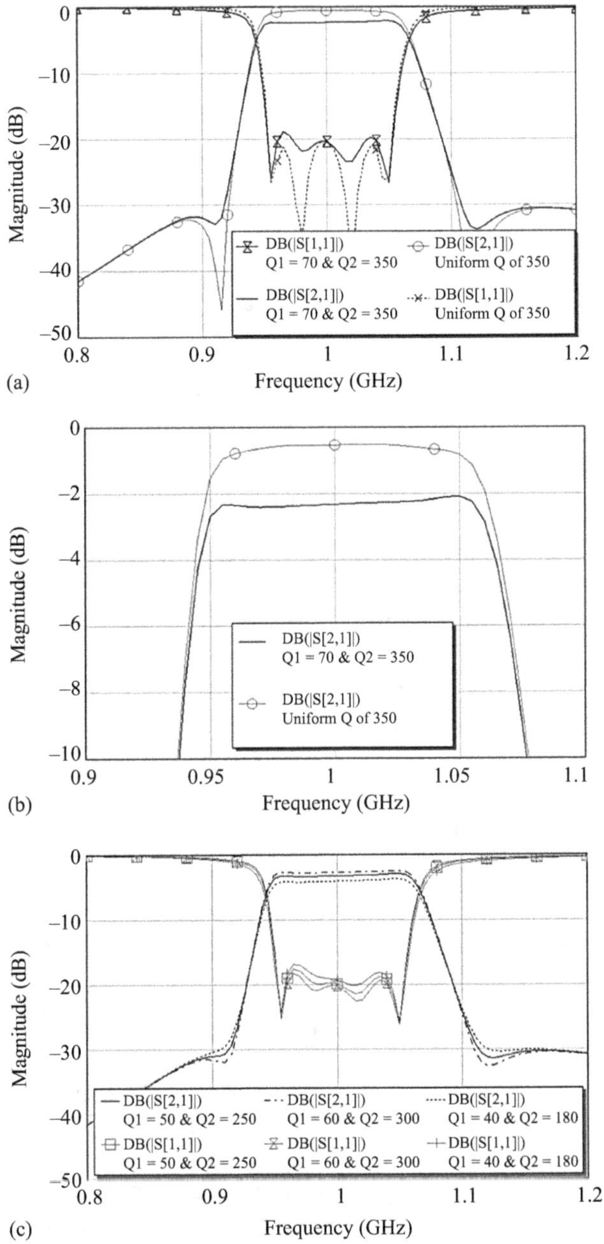

Figure 5.8 (a) Circuit response of four-pole transversal filter network with uniform/non-uniform Q resonators; (b) zoom in passband and (c) responses for other combinations of non-uniform Q

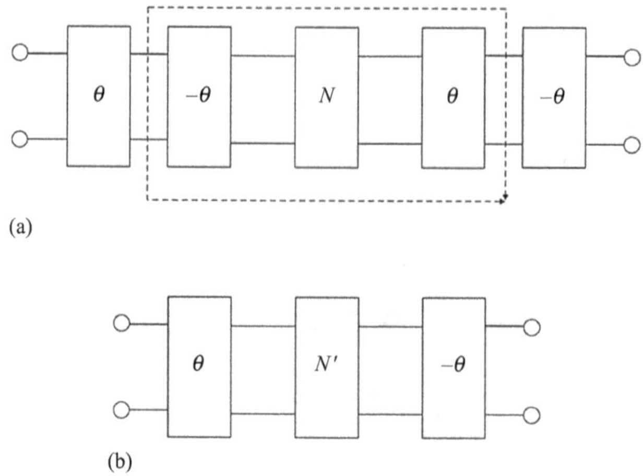

Figure 5.9 Equivalent networks for extracting phase shifters

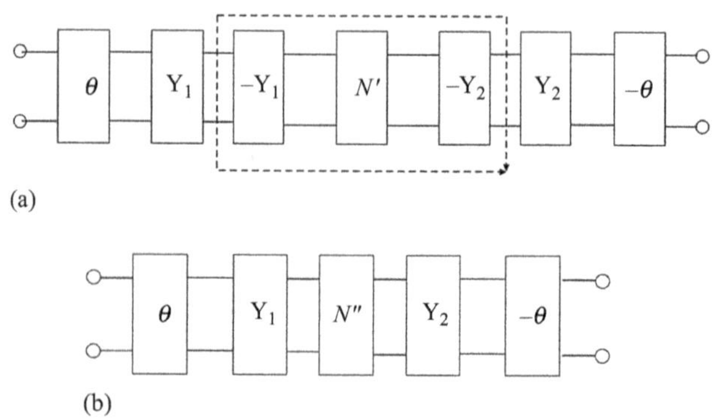

Figure 5.10 Equivalent networks for extracting shunt admittances

equivalent network N' in Figure 5.9(b). The next cycle is to extract a complementary pair of the complex axis TZs (real-frequency zeros) from both ends of the remaining network, displayed in Figure 5.10. After extracting a phase shifter θ, a pair of TZs represented by shunt admittances of Y_1 and Y_2 is extracted from the network N' in Figure 5.10(a) with a remaining equivalent network N'' in Figure 5.10(b). This process will be repeated until all complex zeros are extracted from the transfer function. The remaining network can then be extracted using a cascade synthesis. A detailed example can be found in [4].

5.4.2 Design example

A five-pole filter centred at 3 GHz with 10% FBW and two extracted poles, which have a single pair of TZs at the lowpass prototype frequency $p = \pm j\Omega_Z = \pm j2.0$,

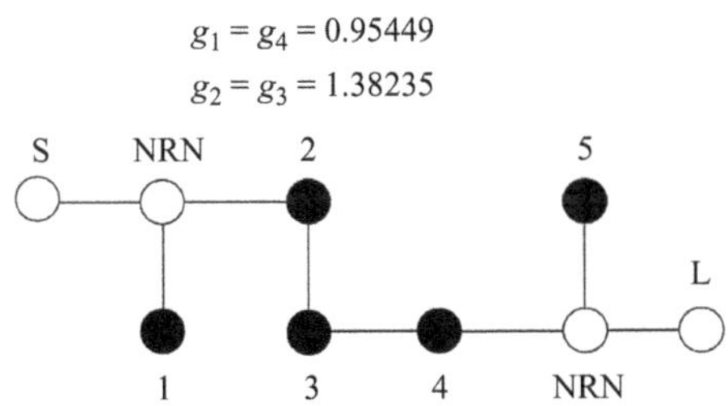

Figure 5.11 Coupling topology of lossless five-pole extracted-pole filter

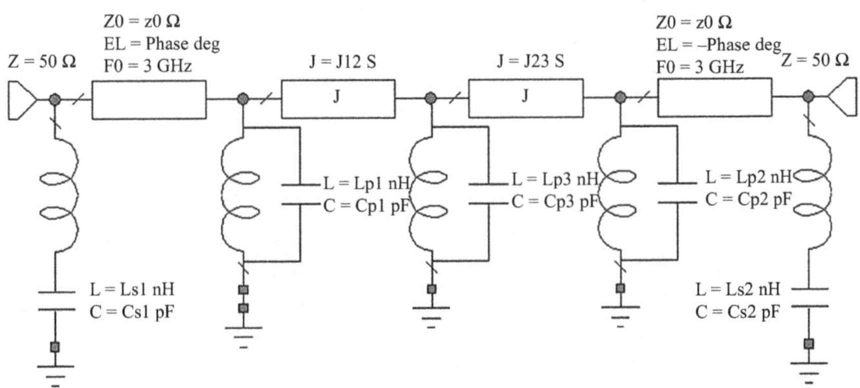

Figure 5.12 The equivalent circuit model of lossless five-pole extracted-pole filter

is designed, where Ω_Z is the position of Tz in lowpass prototype. The lossless coupling topology is shown in Figure 5.11. Node 1 and 5 are two extracted-poles connected to the NRN nodes, while node 2, 3, 4 are resonators in the inline network.

Firstly, a lowpass quasi-elliptic function prototype is designed to convert to an equivalent bandpass circuit of Figure 5.12. The five poles are odd numbers, which is not a standard even symmetric model for the synthesis described in the previous section. Nevertheless, we could start to extract from a standard four poles symmetric model and then extend it to the five poles model by using optimization. The element g-values of the symmetrical four-pole cross-coupled lowpass prototype filter are

$$g_1 = g_4 = 0.95449$$
$$g_2 = g_3 = 1.38235$$

The detailed synthesis of this four-pole example can be found in [4] with the extracted lowpass prototype parameters: $\theta = 33.96^\circ$, $b_1 = 2.58$, $B_1 = 1.20$, $C_1 = 4.15$, which together with an assumption of $B_2 = 0$ and $C_2 = C_1$ are used as the initial values to calculate the bandpass parameters of

$$
\begin{aligned}
C_{s1} &= \frac{b_1}{\omega_0}\left(\frac{1}{\text{FBW}} + \frac{\Omega_Z}{2}\right)^{-1} & L_{s1} &= \frac{1}{\omega_0 b_1}\left(\frac{1}{\text{FBW}} - \frac{\Omega_Z}{2}\right) \\
C_{s2} &= \frac{b_1}{\omega_0}\left(\frac{1}{\text{FBW}} - \frac{\Omega_Z}{2}\right)^{-1} & L_{s2} &= \frac{1}{\omega_0 b_1}\left(\frac{1}{\text{FBW}} + \frac{\Omega_Z}{2}\right) \\
C_{p1} &= \frac{1}{\omega_0}\left(\frac{C_1}{\text{FBW}} + \frac{B_1}{2}\right) & L_{p1} &= \frac{1}{\omega_0}\left(\frac{C_1}{\text{FBW}} - \frac{B_1}{2}\right)^{-1} \\
C_{p2} &= \frac{1}{\omega_0}\left(\frac{C_1}{\text{FBW}} - \frac{B_1}{2}\right) & L_{p2} &= \frac{1}{\omega_0}\left(\frac{C_1}{\text{FBW}} + \frac{B_1}{2}\right)^{-1} \\
C_{p3} &= \frac{1}{\omega_0}\left(\frac{C_2}{\text{FBW}} + \frac{B_2}{2}\right) & L_{p3} &= \frac{1}{\omega_0}\left(\frac{C_2}{\text{FBW}} - \frac{B_2}{2}\right)^{-1}
\end{aligned}
\tag{5.2}
$$

where ω_0 is the centre angular frequency, Ω_Z is the position of Tz in lowpass prototype and FBW is the fractional bandwidth. The symbol b_1 is a residue constant of the extracted pole, B_1, B_2 and C_1, C_2 are the frequency invariant susceptance and capacitance, respectively [4].

The above formulas are normalized by the source impedance $g_0 = 1$. The impedance scaling is required to remove the $g_0 = 1$ normalization and adjusts the filter to work for any value of the source impedance denoted by Z_0. In principle, applying the impedance scaling upon a filter network in such a way that

$$
\begin{aligned}
L &\to Z_0 L \\
C &\to C/Z_0
\end{aligned}
\tag{5.3}
$$

has no effect on the response shape.

The initial coupling coefficients between adjacent coupled resonators can be estimated as

$$
\text{M}_{1,2} = \frac{\text{FBW}}{\sqrt{g_1 g_2}} \quad \text{and} \quad \text{M}_{2,3} = \frac{\text{FBW}}{\sqrt{g_2 g_3}} \tag{5.4}
$$

In the circuit model, the ideal J-inverters are used to represent the coupling coefficients, which can be obtained by

$$
J_{1,2} = \text{M}_{1,2} \times \sqrt{\omega_{p_1} C_{p_1} \omega_{p_3} C_{p_3}} \text{ and } J_{2,3} = \text{M}_{2,3} \times \sqrt{\omega_{p_2} C_{p_2} \omega_{p_3} C_{p_3}} \tag{5.5}
$$

Note that the phase shifters associated the extracted poles at the input and output have opposite signs and the phase element used in the equivalent circuit is an ideal, frequency-independent and reciprocal model described by the S-parameters of $S_{11} = S_{22} = 0$ and $S_{21} = S_{12} = e^{-j\theta}$.

Table 5.3 Element values for lossless five-pole extracted-pole filter with terminal impedance $z_0 = 50\ \Omega$

FBW	10%	Cs_1	0.2526 pF	Ls_1	9.115 nH
f_0	3.0 GHz	Cs_2	0.3088 pF	Ls_2	11.14 nH
Phase	36.4°	Cp_1	43.85 pF	Lp_1	0.0666 nH
J12	0.05312S	Cp_2	42.27 pF	Lp_2	0.0642 nH
J23	0.05372 S	Cp_3	46.98 pF	Lp_3	0.0602 nH

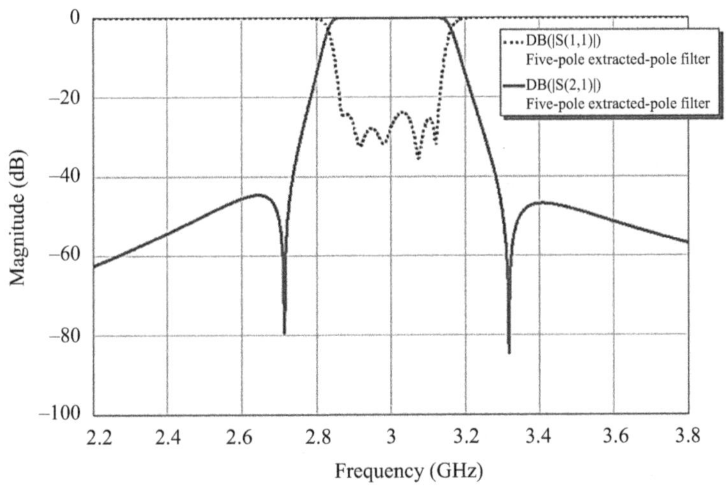

Figure 5.13 Responses of lossless five-pole extracted-pole filter

An optimization can then be operated for the desired response obtaining the final element values given in Table 5.3, in which the optimized lowpass parameters are $\theta = 36.4°$, $b_1 = 2.619$, $B_1 = 1.488$, $C_1 = 4.058$, $B_2 = 0.19$ and $C_2 = 4.418$.

The circuit modelling responses of the lossless five-pole extracted-pole filter are presented in Figure 5.13. It can be seen that the two extracted poles not only introduce two transmission zeros in the S_{21} response for the selectivity enhancement but also contribute to the passband, indicated by the five transmission poles seen in the passband return loss (S_{11}) response.

5.5 Microstrip extracted-pole lossy filter

5.5.1 Design of five-pole extracted-pole lossy filter

In the previous section, a lossless circuit model of five-pole extracted pole filter is described in detail. In this section, we convert it to a lossy model by introducing additional loss to make the passband flat with the same selectivity. One easy way is to introduce one RCC between resonator 2 and 4 as shown in Figure 5.14.

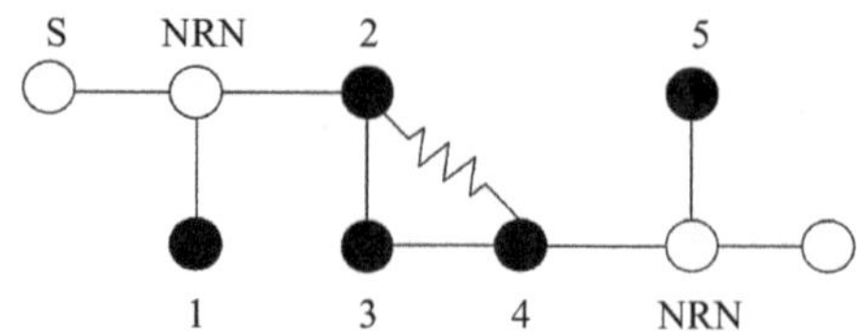

Figure 5.14 Coupling topology of five-pole extracted-pole lossy filter

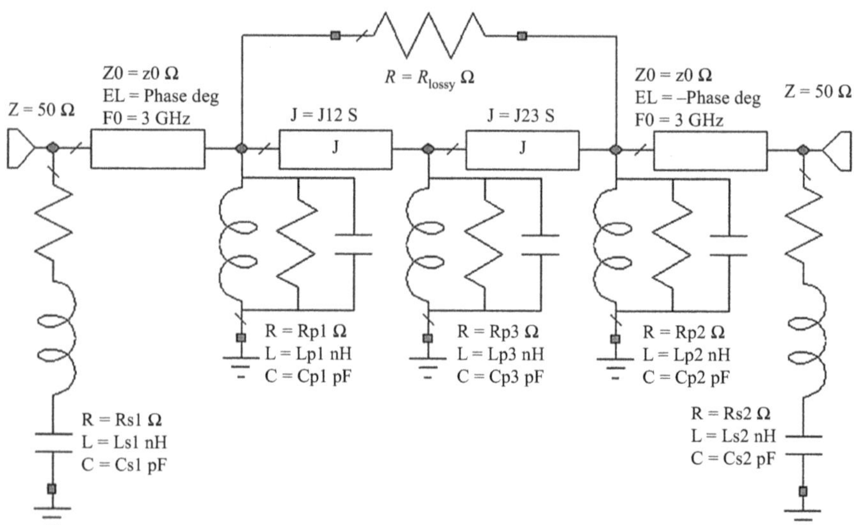

Figure 5.15 Equivalent circuit model of five-pole extracted-pole lossy filter

Table 5.4 Element values for five-pole extracted-pole lossy filter with terminal impedance $z_0 = 50\ \Omega$

FBW	10%	Cp_1	54.22 pF	Lp_3	0.054 nH
f_0	3.0 GHz	Cp_2	52.69 pF	Rs_1	1.22 Ω
Phase	36.4°	Cp_3	52.29 pF	Rs_2	1.22 Ω
J12	0.0527 S	Ls_1	9.36 nH	Rp_1	158.8 Ω
J23	0.05412 S	Ls_2	11.45 nH	Rp_2	158.8 Ω
Cs_1	0.246 pF	Lp_1	0.053 nH	Rp_3	162.3 Ω
Cs_2	0.3 pF	Lp_2	0.052 nH	R_{lossy}	220 Ω

Following the procedure discussed above, the equivalent circuit model can be set up with AWR [10] as given in Figure 5.15. The element values are displayed in Table 5.4 with the uniform resonators $Q = 160$. The response comparison between lossless and lossy models is shown in Figure 5.16(a) and the group delay performance is illustrated in Figure 5.16(b).

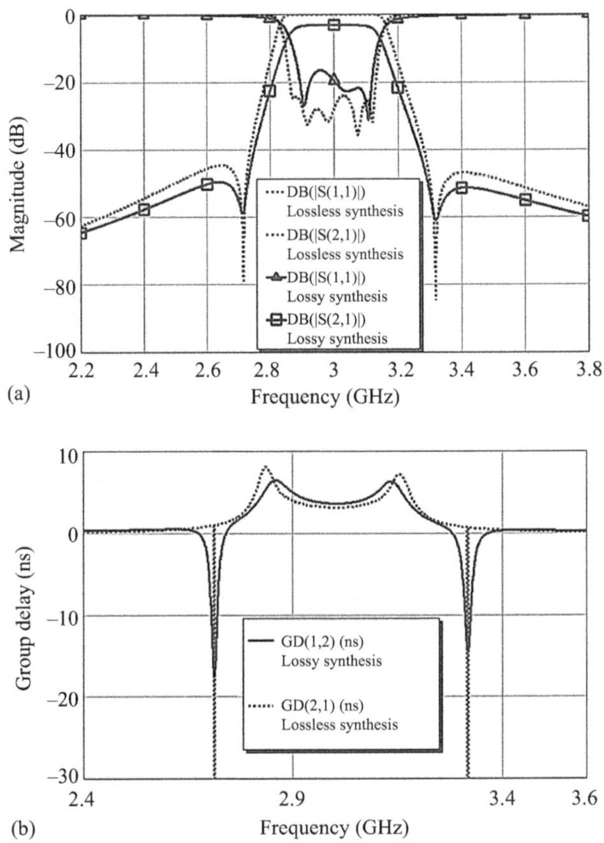

Figure 5.16 Response comparison of lossy and lossless synthesis of five-pole extracted-pole filter: (a) S-parameter and (b) group delay

5.5.2 Design of six-pole extracted-pole lossy filter

5.5.2.1 Theoretical circuit model

The lossless and lossy coupling topologies of proposed filter are shown in Figure 5.17. Node 1 and 6 are two extracted-poles connected to the NRNs, while node 2, 3, 4 and 5 are resonators in the inline network. In the lossy coupling topology, two RCCs are added between node 2 and node 4, node 3 and node 5, respectively.

For computer-aided synthesis and design, the equivalent circuit models of Figure 5.18(a) and (b) are deployed. L_{sm}, C_{sm} and R_{sm}, ($m = 1$ and 2) are the circuit elements of the two extracted-pole resonators with R_{sm} representing the resonator losses in the lossy case. L_{pm}, C_{pm} and R_{pm} ($m = 1$ to 4) are the circuit elements of the four parallel resonators with R_{pm} representing the resonator losses in the lossy case. The ideal (frequency-invariant) J inverters are used to represent the couplings. R_{lossy} represents the resistive coupling introduced for the lossy

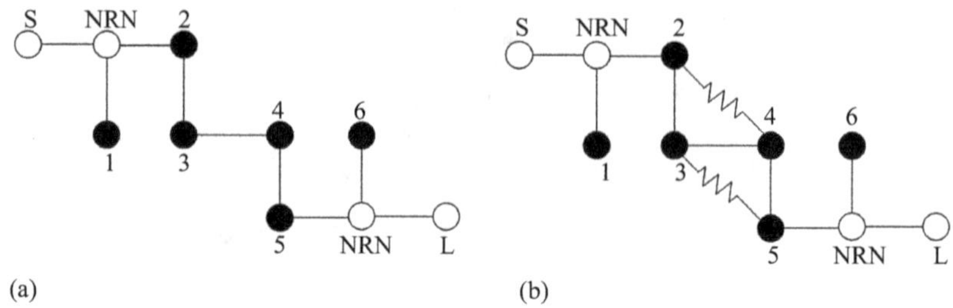

Figure 5.17 (a) Lossless coupling topology and (b) lossy coupling topology. Reprinted with permission from [9]; © 2018 by John Wiley & Sons, Inc

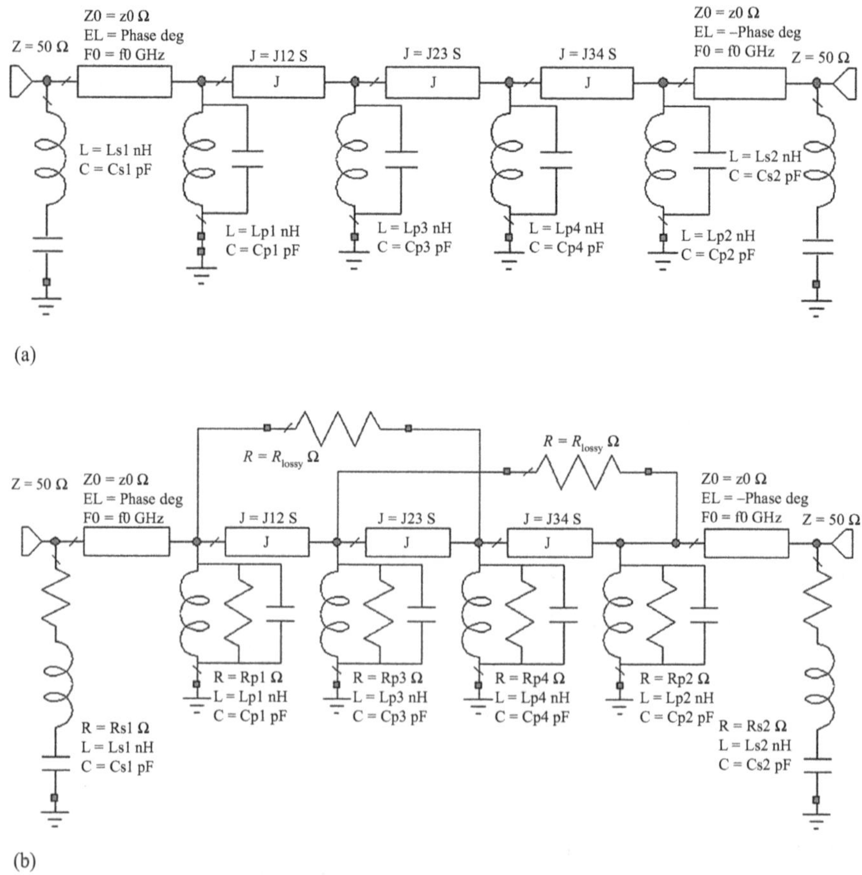

Figure 5.18 (a) Lossless circuit model and (b) circuit model of lossy synthesis. Adapted with permission from [9]; © 2018 by John Wiley & Sons, Inc

technique. The phase shifter elements associated with the extracted-poles use an ideal model having the *S*-parameters of $S_{11} = S_{22} = 0$ and $S_{21} = S_{12} = e^{-j\theta}$.

The initial circuit element values can first be calculated based on the method described in [4] for the lossless circuit model of Figure 5.18(a). By applying the frequency transformation and equating slope reactances as well as slope susceptances, we can obtain:

$$
\begin{aligned}
C_{s1} &= \frac{b_1}{\omega_0}\left(\frac{1}{\text{FBW}} + \frac{\Omega_z}{2}\right)^{-1} & L_{s1} &= \frac{1}{\omega_0 b_1}\left(\frac{1}{\text{FBW}} - \frac{\Omega_z}{2}\right) \\
C_{s2} &= \frac{b_1}{\omega_0}\left(\frac{1}{\text{FBW}} - \frac{\Omega_z}{2}\right)^{-1} & L_{s2} &= \frac{1}{\omega_0 b_1}\left(\frac{1}{\text{FBW}} + \frac{\Omega_z}{2}\right) \\
C_{p1} &= \frac{1}{\omega_0}\left(\frac{C_1}{\text{FBW}} + \frac{B_1}{2}\right) & L_{p1} &= \frac{1}{\omega_0}\left(\frac{C_1}{\text{FBW}} - \frac{B_1}{2}\right)^{-1} \\
C_{p2} &= \frac{1}{\omega_0}\left(\frac{C_1}{\text{FBW}} - \frac{B_1}{2}\right) & L_{p2} &= \frac{1}{\omega_0}\left(\frac{C_1}{\text{FBW}} + \frac{B_1}{2}\right)^{-1} \\
C_{p3} &= \frac{1}{\omega_0}\left(\frac{C_2}{\text{FBW}} + \frac{B_2}{2}\right) & L_{p3} &= \frac{1}{\omega_0}\left(\frac{C_2}{\text{FBW}} - \frac{B_2}{2}\right)^{-1} \\
C_{p4} &= \frac{1}{\omega_0}\left(\frac{C_2}{\text{FBW}} - \frac{B_2}{2}\right) & L_{p4} &= \frac{1}{\omega_0}\left(\frac{C_2}{\text{FBW}} + \frac{B_2}{2}\right)^{-1}
\end{aligned}
\tag{5.6}
$$

where ω_0 is the midband angular frequency, Ω_Z is the position of TZ in lowpass prototype and FBW is the fractional bandwidth. The symbol b_1 is a residue constant of the extracted pole, B_1, B_2 and C_1, C_2 are the frequency invariant susceptance and capacitance, respectively. As an example, given $b_1 = 2.52, B_1 = 1.464, B_2 = 0.15$, $C_1 = 4.12$ and $C_2 = 4.31$, Table 5.5 shows the element values of Figure 5.18(a) for a lossless extracted-pole filter with a single pair of TZs at the lowpass prototype frequency $p = \pm j\Omega_Z = \pm j2.0$, and a 6% FBW centred at $f_0 = 2$ GHz.

Table 5.5 Element values for six-pole extracted-pole lossless filter with terminal impedance $z_0 = 50\ \Omega$

FBW	6%	Cp_2	108.1 pF
f_0	2 GHz	Cp_3	114.4 pF
Phase	35.2°	Cp_4	114.2 pF
J12	0.0561 S	Ls_1	24.74 nH
J23	0.0521 S	Ls_2	27.89 nH
J34	0.0572 S	Lp_1	0.05857 nH
Cs_1	0.227 pF	Lp_2	0.05733 nH
Cs_2	0.256 pF	Lp_3	0.05545 nH
Cp_1	110.5 pF	Lp_4	0.05533 nH

Table 5.6 Element values for six-pole extracted-pole lossy filter with terminal impedance $z_0 = 50\ \Omega$

FBW	6%	Cp_2	107.4 pF	Rs_1	2.132 Ω
f_0	2.0 GHz	Cp_3	133.4 pF	Rs_2	2.132 Ω
Phase	37.2°	Cp_4	133.3 pF	Rp_1	117.4 Ω
J12	0.04892 S	Ls_1	25.56 nH	Rp_2	117.4 Ω
J23	0.05492 S	Ls_2	28.82 nH	Rp_3	95.47 Ω
J34	0.04972 S	Lp_1	0.05894 nH	Rp_4	95.47 Ω
Cs_1	0.2197 pF	Lp_2	0.05786 nH	R_{lossy}	223 Ω
Cs_2	0.2478 pF	Lp_3	0.04749 nH		
Cp_1	109.4 pF	Lp_4	0.04747 nH		

To extend to the lossy synthesis, two RCCs with R_{lossy} are added as shown in Figure 5.18(b). The loss in the resonators themselves is also needed to consider, which are related to the Q factor of the resonators given below, for series and parallel resonators, respectively:

$$R_{s_m} = \frac{\omega_{s_m} L_{s_m}}{Q_{s_m}} \quad m = 1, 2 \tag{5.7}$$

$$R_{p_m} = \frac{Q_{p_m}}{\omega_{p_m} C_{p_m}} \quad m = 1, 2, 3, 4 \tag{5.8}$$

where ω_{s_m} and ω_{p_m} are the angular resonant frequencies of the series and parallel resonators, respectively. Assuming an equal Q of 160 for the microstrip resonators to be used, the parameters values in Table 5.6 are extracted from this procedure with a slight optimization to adjust the better response for the circuit model of Figure 5.18(b) for an extracted-pole lossy filter with a 6% FBW centred at $f_0 = 2$ GHz. Note that the corresponding lowpass prototype parameters are $b_1 = 2.439, B_1 = 1.256, B_2 = 0.05, C_1 = 4.088$ and $C_2 = 5.028$.

The responses of lossless and lossy synthesizes are also compared in Figure 5.19. It can be seen that the two extracted poles introduce two transmission zeros for the selectivity enhancement. Even with the low-Q resonators in the lossy synthesis, the passband insertion loss is also very flat indicating that the two RCCs work well. To see the effect of the RCCs clearly, Figure 5.20 shows the passband performance of 6% FBW with different values of resistors in the RCCs. For this example, the best value of resistor in RCCs is 223 Ω.

The lossy synthesis performance with different FBWs is shown in Figure 5.21. As we can see, the circuit model responses are good with flat passband and high selectivity in different FBWs from narrowband to wideband. Table 5.7 lists the lossy circuit element values at the same centre frequency of 2 GHz and with the same equal Q of 160 for all resonators, which are calculated using (5.6) with the same lowpass prototype parameters as those associated with Table 5.6. The positions of two transmission zeros in different FBWs' response are then kept the same in the lowpass prototype synthesis, i.e. $p = \pm j\Omega_Z = \pm j2.0$. However, the slope of the passband edges is much sharp as the FBW decreases. It is also more obvious to achieve a flat passband when RCCs are applied in narrowband.

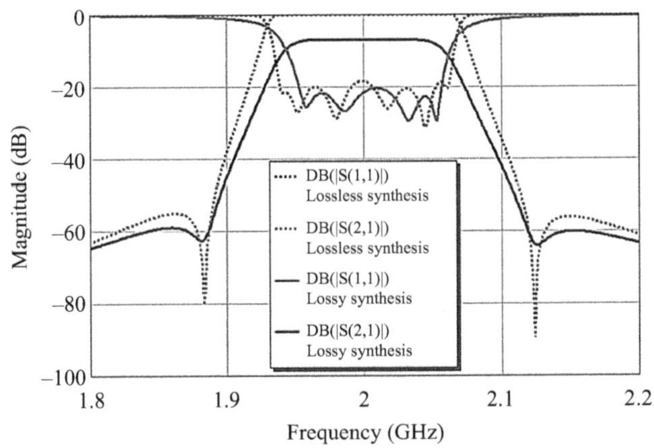

Figure 5.19 Responses of lossless and lossy synthesis

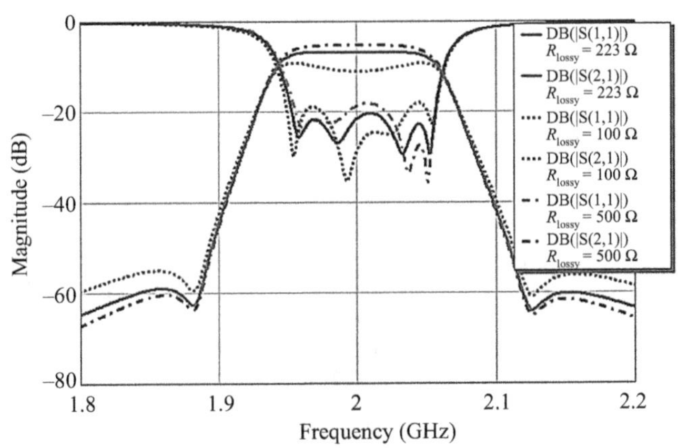

Figure 5.20 Lossy synthesis performance with different values of RCCs. Adapted with permission from [9]; © 2018 by John Wiley & Sons, Inc

To find the circuit element values for a centre frequency f_0 (GHz) other than 2 GHz from Table 5.7, only the *LC* element values need to be modified. To do so, define a frequency scale factor

$$sf = \frac{2}{f_0} \tag{5.9}$$

The new *LC* element values can then be calculated by

$$\begin{aligned} L &\to sf \times L \\ C &\to sf \times C \end{aligned} \tag{5.10}$$

Figure 5.22 illustrates the responses of 10% FBW at different centre frequencies, which are modelled by the circuit model of Figure 5.18(b) applying the *LC* conversion of (5.10).

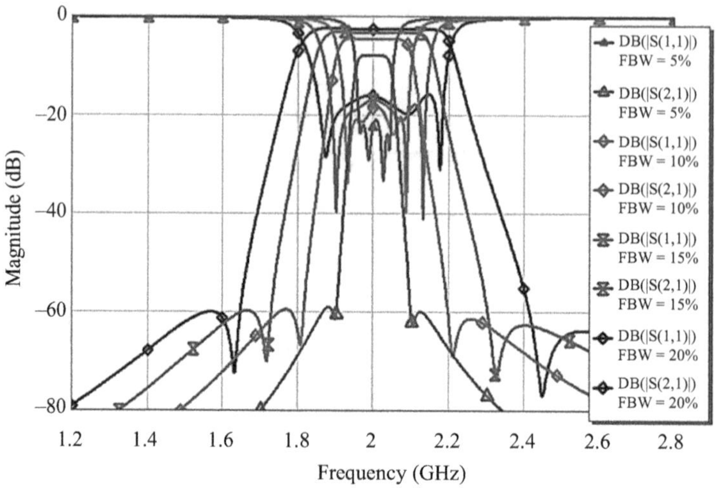

Figure 5.21 Lossy synthesis performance with different FBWs

Table 5.7 Element values for six-pole extracted-pole lossy filter with different FBWs

FBW	Cs1 (pF)	Ls1 (nH)	Rs1 (Ω)	Cs2 (pF)	Ls2 (nH)	Rs2 (Ω)
5%	0.1848	31.0	2.559	0.2043	34.26	2.559
10%	0.3529	14.68	1.275	0.4313	17.94	1.275
15%	0.5063	9.244	0.8445	0.685	12.51	0.8445
20%	0.647	6.525	0.6277	0.9704	9.788	0.6277
FBW	Cp1 (pF)	Lp1 (nH)	Rp1 (Ω)	Cp2 (pF)	Lp2 (nH)	Rp2 (Ω)
5%	131.1	0.04904	97.85	129.1	0.04829	97.85
10%	66.06	0.09885	195.7	64.06	0.09586	195.7
15%	44.37	0.1494	293.6	42.38	0.1427	293.6
20%	33.53	0.2008	391.6	31.53	0.1889	391.6
FBW	Cp3 (pF)	Lp3 (nH)	Rp3 (Ω)	Cp4 (pF)	Lp4 (nH)	Rp4 (Ω)
5%	160.1	0.03958	79.55	160	0.03956	79.55
10%	80.06	0.07917	159.1	79.98	0.07909	159.1
15%	53.39	0.1188	238.7	53.31	0.1186	238.7
20%	40.05	0.1584	318.2	39.97	0.1581	318.2
FBW	R_{lossy} (Ω)	Phase (Degree)	J12 (S)	J23 (S)	J34 (S)	$f_0 = 2$ GHz
5%	220	37.2	0.04892	0.05492	0.04972	$Q = 160$
10%	314					$Z_0 = 50$ Ω
15%	435					
20%	570					

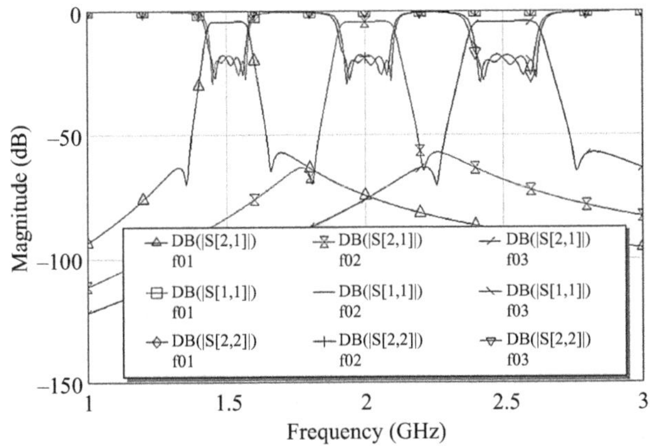

Figure 5.22 Lossy synthesis responses with the same FBW at different f_0

To find the circuit element values for a uniform resonator Q other than the assumed Q of 160 from Table 5.7, only the R element values need to be modified. To do so, define a scale factor of Q

$$qs = \frac{Q}{160}$$

The new R element values can then be calculated by

$$\begin{aligned} R_p &\to qs \times R_p \\ R_s &\to R_s/qs \end{aligned} \tag{5.11}$$

Figure 5.23 demonstrates the responses of 10% FBW for different uniform-Q, which are simulated with the circuit model of Figure 5.18(b) applying the R conversion of (5.11).

5.5.2.2 Filter implementation

Referring to the circuit model of Figure 5.18(b), for the implementation of six-pole microstrip extracted-pole filter, a new set of design parameters, i.e. the external Q-factor $Q_{ext,m}$ and the coupling coefficients $M_{m,m+1}$, is derived as follows

$$\begin{aligned} Q_{\text{ext,sm}} &= \frac{\omega_{\text{sm}} L_{s_m}}{Z_0}, m = 1, 2 \\ Q_{\text{ext,pm}} &= \omega_{pm} C_{pm} Z_0, m = 1, 2 \\ M_{1,2} &= \frac{J_{1,2}}{\sqrt{\omega_{p_1} C_{p_1} \omega_{p_3} C_{p_3}}} \\ M_{2,3} &= \frac{J_{2,3}}{\sqrt{\omega_{p_3} C_{p_3} \omega_{p_4} C_{p_4}}} \\ M_{3,4} &= \frac{J_{3,4}}{\sqrt{\omega_{p_4} C_{p_4} \omega_{p_2} C_{p_2}}} \end{aligned} \tag{5.12}$$

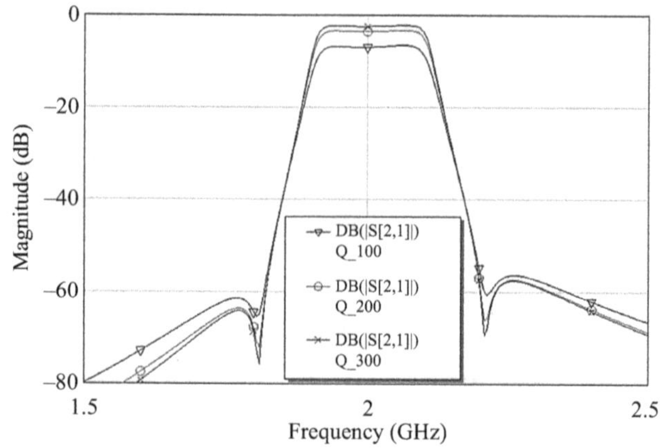

Figure 5.23 Lossy synthesis responses with different uniform resonator Q

where ω denotes the angular resonance frequency and $Z_0 = 50\,\Omega$ is the terminal impedance.

For example, the external Q-factors and coupling coefficients based on the lossy synthesis of Table 5.6 can then be determined by (5.12), and their values are given as:

$$
\begin{aligned}
Q_{ext,s1} &= Q_{ext,s2} = 6.821 \\
Q_{ext,p1} &= Q_{ext,p2} = 68.13 \\
M_{1,2} &= 0.03252 \\
M_{2,3} &= 0.03307 \\
M_{3,4} &= 0.03308
\end{aligned}
$$

For microstrip realization, two open-loop resonators are used to implement the extracted-pole resonators while another four stepped-impedance open-loop resonators are deployed to realize the four shunt resonators in the circuit model with a compact size. To realize the RCCs in the microstrip line, two cross couplings with surface mount resistors are introduced between resonator 2 and resonator 4, resonator 3 and resonator 5, respectively. Full-wave EM simulations with Sonnet ***em*** [11] are then carried out to extract the external Q and coupling coefficient against the physical dimensions based on the theory in [4]. Figure 5.24 shows the designed filter layout on a 1.27-mm-thick dielectric substrate with a relative dielectric constant of 10.2.

The designed filter is then fabricated, as displayed in Figure 5.25(a), and tested with a microwave network analyser. The measured passband performance is plotted in Figure 5.25(b). The measured filter has high selectivity with two TZs: the lower side is located at 1.75 GHz while 1.97 GHz at high side. The passband is quite flat with a variation only around 0.3 dB, though the absolute insertion loss of the

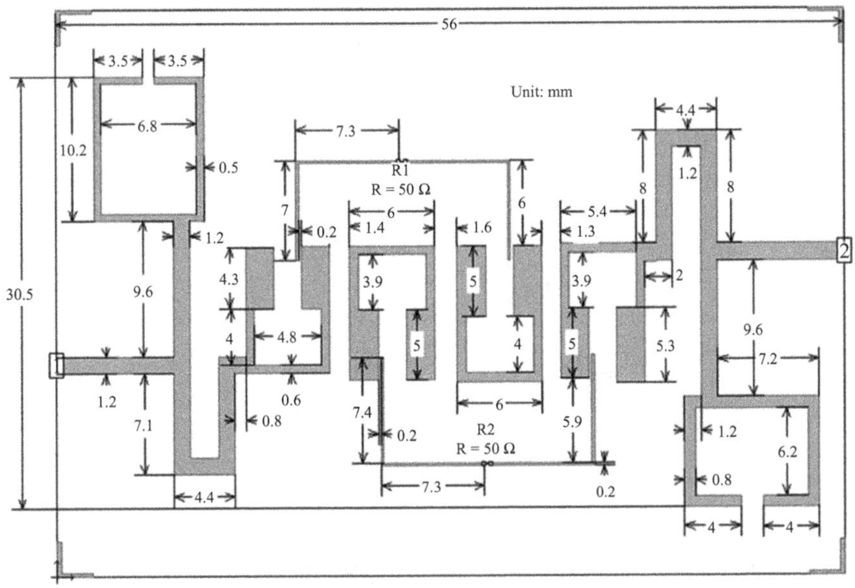

Figure 5.24 Layout of a designed microstrip extracted-pole lossy filter on a 1.27-mm-thick dielectric substrate with a relative dielectric constant of 10.2. Adapted with permission from [9]; © 2018 by John Wiley & Sons, Inc

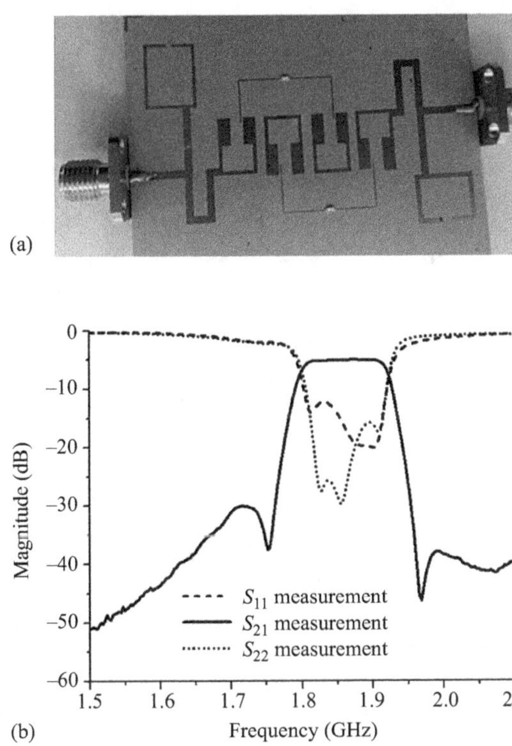

Figure 5.25 (a) Photo of fabricated microstrip extracted-pole lossy filter and (b) measured passband performance. Adapted with permission from [9]; © 2018 by John Wiley & Sons, Inc

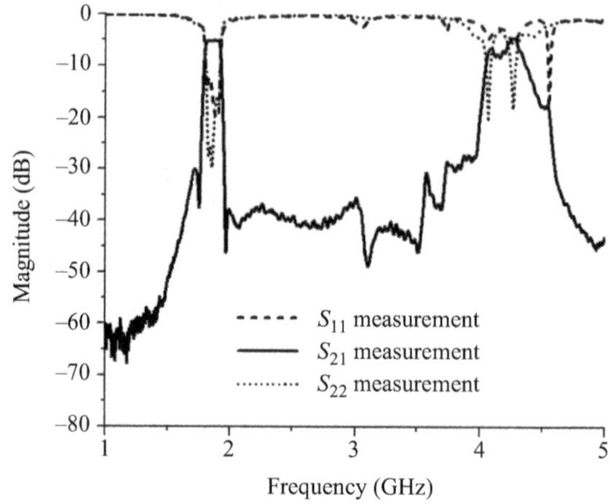

Figure 5.26 Measured wideband response of the microstrip extracted-pole lossy filter of Figure 5.25(a). Adapted with permission from [9]; © 2018 by John Wiley & Sons, Inc

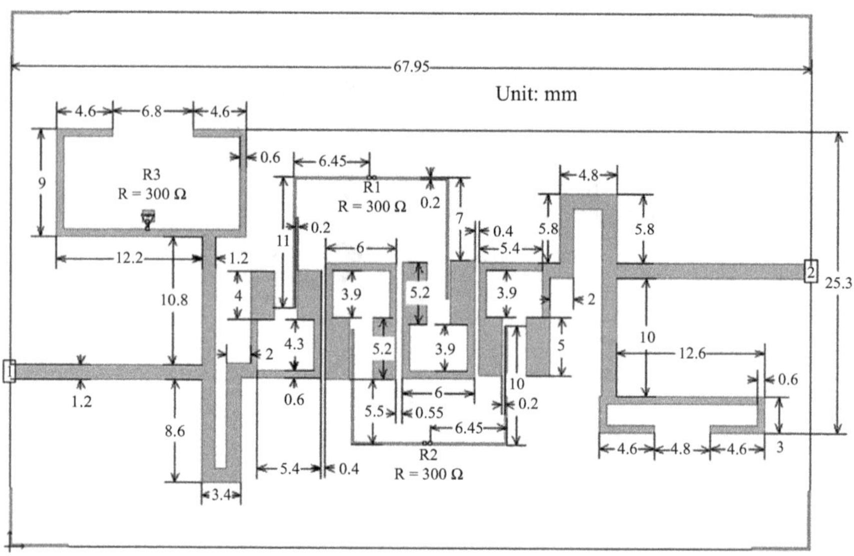

Figure 5.27 Layout of a designed 20% FBW microstrip extracted-pole lossy filter on a 1.27-mm-thick dielectric substrate with a relative dielectric constant of 10.2. Adapted with permission from [9]; © 2018 by John Wiley & Sons, Inc

passband is 6.5 dB. Such a passband flatness and selectivity would be equivalent to that of a conventional 6-pole filter with a uniform Q of 600, which means this type of lossy filter can increase the equivalent Q to achieve a higher performance in terms of passband flatness and selectivity.

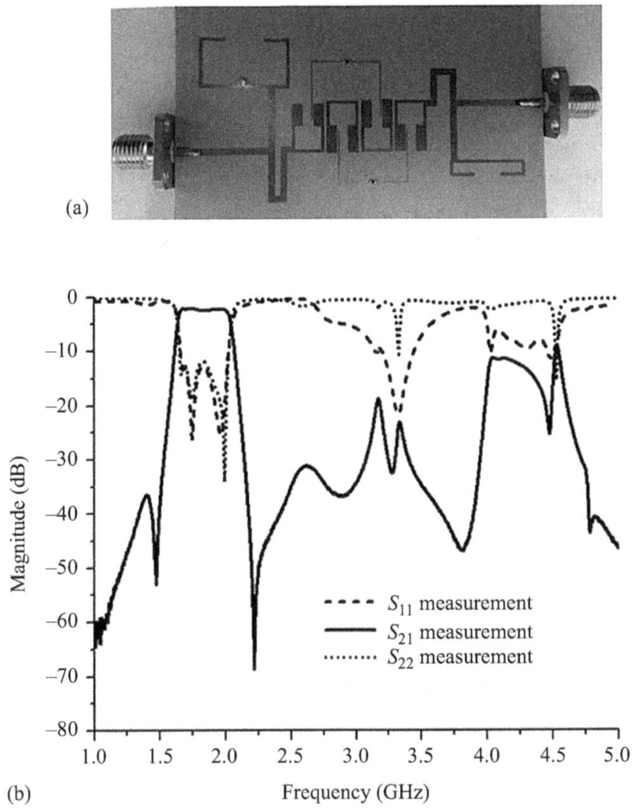

Figure 5.28 (a) Photo of fabricated 20% FBW microstrip extracted-pole lossy filter and (b) measured wideband response. Adapted with permission from [9]; © 2018 by John Wiley & Sons, Inc

The measured wideband response of the filter is illustrated in Figure 5.26. The first spurious occurs at 2.25 times the centre frequency resulting from the implemented stepped-impedance open-loop resonators. This is a wider upper stopband as compared with a filter using conventional half-wavelength hairpin resonators.

Another microstrip extract-pole lossy filter with a wider FBW of 20% has been implemented following the same procedure as described above. Figure 5.27 illustrates the designed layout of the microstrip lossy filter. An additional resistor loading is applied in the middle of open-loop resonator for the low side extracted-pole to improve the upper stopband performance. The fabricated filter and the measured wideband response of the filter are shown in Figure 5.28. A flat passband is obtained with the insertion loss variation measured within 0.3 dB. The measured filter has also two TZs located at 1.48 and 2.2 GHz, respectively, to achieve a good selectivity. A 20-dB upper stopband is extended to 4 GHz benefited by deploying the stepped-impedance half-wavelength resonators as well as the additional loading resistor.

5.6 Summary

In this chapter, the design of microstrip lossy extracted-pole filters has been described. This type of filter can have both of high selectivity and flat passband. The high selectivity is contributed by using extracted-pole technique and the flat passband is achieved by introducing RCCs in the inline network. The independently controllable extracted-pole transmission zeros result in an inline asymmetrical filter topology with asynchronously tuned resonators. The direct circuit synthesis of this type of filter has been described with numerous examples providing a more direct insight into the theory. The proposed equivalent circuit and tabulated elements are ready to hand with most commercially available circuit modelling tools. Two six-pole microstrip filters of this type have been designed and demonstrated experimentally showing promising filtering characteristics. The improvement of passband flatness is more significant and obvious in the narrowband than that of the wideband.

References

[1] A. E. Atia and A. E. Williams. 'Narrow-bandpass waveguide filters'. *IEEE Trans. Microw. Theory Tech.* 1972, vol. MTT-20, pp. 258–265.

[2] A. E. Atia, A. E. Williams, and R. W. Newcomb. 'Narrow-band multiple-coupled cavity synthesis'. *IEEE Trans. Circuits Syst.* 1974, vol. CAS-21(9), pp. 649–655.

[3] R. J. Cameron, C. Kudsia, and R. Mansour. *Microwave Filters for Communication Systems*, 1st edn. Hoboken, NJ: Wiley; 2007.

[4] J. Hong, *Microstrip Filters for RF/Microwave Applications*. 2nd Edition, John Wiley & Sons, Inc., New Jersey, 2011.

[5] V. Miraftab, M. Yu. 'Generalized lossy microwave filter coupling matrix synthesis and design using mixed technologies'. *IEEE Trans. Microw. Theory Tech.* 2008, vol. 56(12), pp. 3016–3027.

[6] A. Guyette, I. Hunter, and R. Pollard. 'The design of microwave bandpass filters using resonators with non-uniform Q'. *IEEE Trans. Microw. Theory Tech.* 2006, vol. 54(11), pp. 3914–3922.

[7] A. Guyette, I. Hunter, and R. Pollard. 'Exact synthesis of microwave filters with non-uniform dissipation', *Proceedings of the IEEE MTT-S Int. Microw. Symp. Dig*; Hawaii, HI, Jun 2007. pp. 537–540.

[8] Z. Zhou and J.-S. Hong. 'Novel lossy microstrip filter with extracted-pole technique'. *Proceedings of the 46th Eur. Microw. Conf*; London, Oct 2016. pp. 120–123.

[9] Z. Zhou and J.-S. Hong. 'Computer-aided design of microstrip extracted-pole lossy filter'. *Int. J. RF and Microw. Comput-Aided Eng*. 2018, vol. 10 (1002).

[10] AWR Microwave Office ver.11.

[11] Sonnet ***em*** ver.13.

[12] M. Yu and V. Miraftab. 'Shrinking microwave filters'. *IEEE Microw. Mag.* 2008, vol. 9(5), pp. 40–54.

[13] R. M. Livingston. 'Predistorted waveguide filters'. *Proceedings of the IEEE MTT-S Int. Microwave Symp. Dig.*, 1969. pp. 291–297.

[14] A. E. Williams, W. G. Bush, and R. R. Bonetti. 'Predistortion technique for multicoupled resonator filters'. *IEEE Trans. Microwave Theory Tech.* 1985, vol. MTT-33, pp. 402–407.

[15] M. Yu, V. Dokas, W.-C. Tang, and R. Cameron. 'Novel adaptive predistortion technique for cross coupled filters'. *Proceedings of the IEEE MTT-S Int. Microwave Symp. Dig*; Philadelphia, PA, June 8–13, 2003. pp. 929–932.

[16] M. Yu, W.-C. Tang, A. Malarky, V. Dokas, R. Cameron, and Y. Wang. 'Novel adaptive predistortion technique for cross coupled filters and its application to satellite communication systems'. *IEEE Trans. Microwave Theory Tech.* 2003, vol. 51(12), pp. 2505–2514.

[17] M. Meng, I. C. Hunter, and J. D. Rhodes. 'The design of parallel connected filter networks with non-uniform Q resonators'. *IEEE Trans. Microw. Theory Tech.* 2013, vol. 61(1), pp. 372–381.

[18] V. Miraftab and M. Yu. 'Advanced coupling matrix and admittance function synthesis techniques for dissipative microwave filters'. *IEEE Trans. Microw. Theory Tech.* 2009, vol. 57(10), pp. 2429–2438.

[19] L. Szydlowski, A. Lamecki, and M. Mrozowski. 'Design of microwave lossy filter based on substrate integrated waveguide (SIW)'. *IEEE Microw. Wireless Compon. Lett.* 2011, vol. 21(5), pp. 249–251.

[20] I. C. Hunter, A. Guyette, and R. Pollard. 'Passive microwave receive filter networks using low-resonators'. *IEEE Microw. Mag.* 2005, vol. 6(3), pp. 1959–1962.

[21] C.-K. Liao, C.-Y. Chang, and J. Lin. 'A vector-fitting formulation for parameter extraction of lossy microwave filters'. *IEEE Microw. Wireless Compon. Lett.* 2007, vol. 17(4), pp. 277–279.

[22] L. Szydlowski, A. Lamecki, and M. Mrozowski. 'Synthesis of coupled lossy resonator filters'. *IEEE Microw. Wireless Compon. Lett.* 2010, vol. 20(7), pp. 366–368.

[23] A. Padilla, J. Mateu, C. Collado, *et al.* 'Comparison of lossy filters and predistorted filters using novel software'. *Proceedings of the IEEE MTT-S Int. Microw. Symp. Dig*; Anaheim, CA, USA, May 2010. pp. 1720–1723.

[24] M. Oldoni, G. Macchiarella, G. Gentili, and C. Ernst. 'A new approach to the synthesis of microwave lossy filters'. *IEEE Trans. Microw. Theory Tech.* 2010, vol. 58(5), pp. 1222–1229.

[25] Jia Ni, W. Xing Tang, J.-S. Hong, and R.H. Geschke. 'Design of microstrip lossy filter using an extended doublet topology'. *Microwave and Wireless Components Letters, IEEE.* 2014, vol. 24(5), pp. 318–320.

[26] A. Basti, A. perigaud, S. Bila, S. Verdeyme, L. Estagerie, and H. Leblond. 'Design of microstrip lossy filters for receivers in satellite transponders'. *IEEE Trans. Microw, Theory Tech.* 2014, vol. 62(9), pp. 0018–9480.

[27] M. Meng and K.-L. Wu. 'An analytical approach to computer-aided diagnosis and tuning of lossy microwave coupled resonator filters'. *IEEE Trans. Microw. Theory Tech.* 2009, vol. 57(12), pp. 3188–3195.

[28] J. Mateu, A. Padilla, C. Collado, *et al.* 'Synthesis of 4th order lossy filters with uniform Q distribution'. *Proceedings of the 2010 IEEE MTT-S Int. Microw. Symp. Dig*; Anaheim (CA), May 2010. pp. 568–571.

[29] L. F. Qiu, L. S. Wu, W. Y. Yin, and J. F. Mao. 'A flat-passband microstrip filter with nonuniform-Q dual-mode resonators'. *IEEE Microw. Wireless Compon. Lett.* 2016, vol. 26(3), pp.183–185.

[30] R. J. Cameron, C. Kudsia, and R. Mansour. *Microwave Filters for Communication Systems*, 1st ed. Hoboken, NJ: Wiley; 2007.

[31] J. D. Rhodes and R. J. Cameron. 'General extracted pole synthesis technique with applications to low-loss TE011 mode filters'. *IEEE Trans. Microw. Theory Tech.* vol. MTT-28, 1980, pp. 1018–1028.

[32] R. J. Cameron, H. Gregg, C. J. Radcliffe, and J. D. Rhodes. 'Extracted pole filter manifold multiplexing'. *IEEE Trans. Microw. Theory Tech.* vol. MTT-30, 1982, pp. 1041–1050.

[33] R. J. Cameron. 'General coupling matrix synthesis methods for Chebyshev filtering functions'. *IEEE Trans. Microw. Theory Tech.* 1999, vol. 47(4), pp. 433–442.

[34] J. R. Montejo-Garai, J. A. Ruiz-Cruz, J. M. Rebollar, M. J. Padilla-Cruz, A. Oñoro-Navarro, and I. Hidalgo-Carpintero. 'Synthesis and design of in-line N-order filters with N real transmission zeros by means of extracted poles implemented in low-cost rectangular H-plane waveguide'. *IEEE Trans. Microw. Theory Tech.* 2005, vol. 53(5), pp. 1636–1642.

[35] S. Amari and U. Rosenberg. 'New in-line dual- and triple-mode cavity filters with nonresonating nodes'. *IEEE Trans. Microw. Theory Tech.* 2005, vol. 53(4), pp. 1272–1279.

[36] S. Amari and G. Macchiarella. 'Synthesis of inline filters with arbitrarily placed attenuation poles by using nonresonating nodes'. *IEEE Trans. Microw. Theory Tech.* 2005, vol. 53(10), pp. 3075–3081.

[37] A. E. Atia and A. E. Williams. 'General TE011-mode waveguide bandpass filters'. *IEEE Trans. Microw. Theory Tech.* 1976, vol. MTT-24, pp. 640–648.

[38] S. Amari and U. Rosenberg. 'New building blocks for modular design of elliptic and self-equalized filters'. *IEEE Trans. Microw. Theory Tech.* 2004, vol. 52(2), pp. 721–736.

[39] S. Tamiazzo and G. Macchiarella. 'Synthesis of cross-coupled prototype filters including resonant and non-resonant nodes'. *IEEE Trans. Microw. Theory Tech.* 2015, vol. 63(10), pp. 3408–3415.

Chapter 6
Planar reflectionless filters

Matthew A. Morgan[1]

6.1 Introduction

It is found in many circumstances that the reactive termination presented by conventional filters in their stop bands is harmful to overall system performance, as a consequence of the standing waves that develop between the filter and neighboring components. Even out-of-band, these interactions can have a negative impact on dynamic range, stability, power capacity, and a host of other subtle performance parameters, especially when filters are combined with non-linear devices such as mixers, multipliers, or compressed-mode amplifiers. In such situations, it is advantageous to develop a filter that blocks stop-band energy by absorption rather than by reflection.

Absorptive filters have a long history, comprising mostly empirical experimentation, though some satisfactory theoretical work had been done as early as the 1920s on what came to be called constant-resistance networks [1]: low-order sections which could be cascaded to reproduce transfer functions of arbitrary complexity, save for some unavoidable flat insertion loss (even when using ideal components). Other researchers have focused on susceptance-annulling techniques, whereby an otherwise conventional filter would be 'corrected' with a complementary driving-point admittance in parallel at one of the ports [2]—a technique closely related to contiguous diplexer design and capable of producing filters that are absorptive only from one side.

It was not until fairly recently that networks had been found that are capable of yielding high-order responses with perfect real-valued impedance matching from both ports and at all frequencies from DC to infinity (in theory), and without the excess insertion loss of the older constant-resistance chains. These are the defining characteristics of filters that we call *reflectionless* [3].

6.2 Reflectionless network topology

By far the most common lumped-element topology for conventional filters is the ladder network, shown for a low-pass filter prototype in Figure 6.1(a). The labels,

[1]National Radio Astronomy Observatory, Central Development Lab, USA

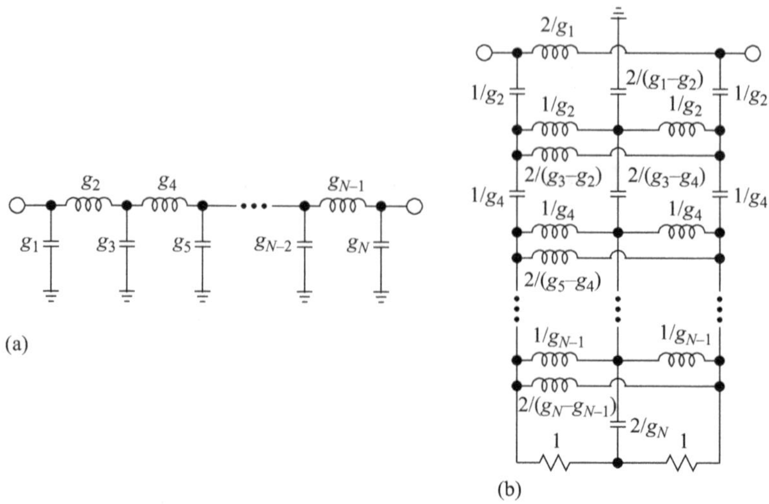

Figure 6.1 (a) Ladder topology for conventional filters and (b) a reflectionless filter topology

g_k, may be thought of as normalized values, corresponding to a system characteristic impedance of 1 Ω, and a cutoff frequency of 1 rad/s. To scale the elements for any other characteristic impedance, $Z_0 = 1/Y_0$, and for any cutoff frequency, ω_c, we may apply the usual scaling laws,

$$L_k = \frac{g_k Z_0}{\omega_c} \tag{6.1a}$$

$$C_k = \frac{g_k Y_0}{\omega_c} \tag{6.1b}$$

Microwave textbooks abound with tables of the prototype element values, g_k, required in the ladder topology to yield the most common, optimal filter transfer functions, such as Butterworth or Chebyshev. Other transfer functions—Legendre-Papoulis, Achieser–Zolotarev, and limitless others without names—are infrequently tabulated, but just as easily realized with an appropriate selection of prototype values. Since none of the elements are resistive, however, no assignment of values is sufficient to make this circuit reflectionless. That requires a different topology entirely.

Though very many such topologies are possible [3–6], the first that we will be concerned with is shown in Figure 6.1(b). It is labeled with the same prototype element parameters as the ladder topology, and indeed the same scaling laws and the exact same tables found in countless filter textbooks may be used (with some caveats, discussed in Section 6.2.4). Both of the filters in Figure 6.1 are, of course, the low-pass prototypes for these topologies. The straightforward transformation techniques to realize high-pass, band-pass, or band-stop varieties based on these prototypes are well-known and described in the academic literature.

To understand why the latter topology is reflectionless, we employ a technique known as *even- and odd-mode analysis*.

6.2.1 Even- and odd-mode analysis

The topology in Figure 6.1(b) is symmetric about its two ports. This means that its full two-port behavior may be analyzed in terms of its response to two particular stimuli: the *even-mode*, wherein the two ports are excited with identical signals in-phase, and the *odd-mode*, wherein the ports are excited with identical signals 180 degrees out-of-phase.

In the even-mode case, since both ports are always held at the same potential, there can be no current passing laterally from one side of the symmetry line to the other—for its direction would be indeterminate. It is permissible, then, to analyze the circuit under this excitation by considering only half the elements, with the symmetry line comprising an open-circuit boundary. This is analogous to the magnetic-wall boundary condition applied in certain symmetric electromagnetic simulations to speed up computation time. The *even-mode equivalent circuit* for the reflectionless filter is thus shown in Figure 6.2(a).

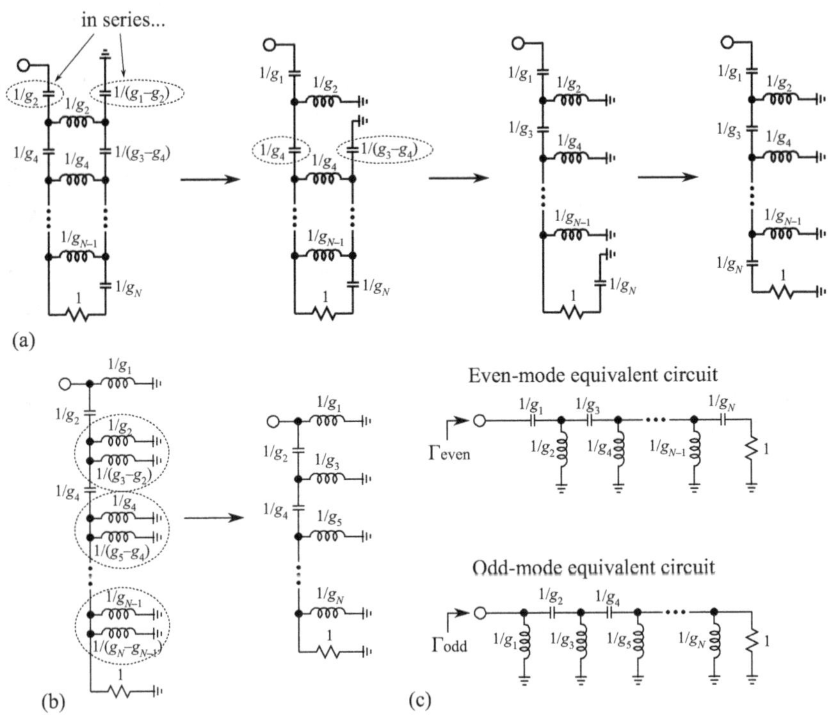

Figure 6.2 (a) Simplification of the even-mode equivalent circuit; (b) simplification of the odd-mode equivalent circuit; and (c) final even- and odd-mode equivalent circuits for the reflectionless filter

The immediate form of the even-mode equivalent circuit in this case is that of a balanced ladder. It may be simplified by recognizing that the two capacitors, $1/g_2$ and $1/(g_1 - g_2)$, nearest the top of the figure at the far left are in series with one another, and may be combined. The resulting network thus has a capacitor with normalized value $1/g_1$ followed by a $1/g_2$ inductor and then the remainder of the balanced ladder elements. The next two capacitors are similarly combined in series, and so on. Finally, the termination resistor and the final capacitor with which it is in series may exchange places. This leaves an unbalanced ladder, just like that in Figure 6.1(a), only this time it is the high-pass form. The termination resistor has unit normalized value, so that the impedance looking into the even-mode network is that of a high-pass filter with a matched termination at its output—reflective at low frequency, but matched at high frequency.

In the odd-mode excitation, where the two ports are driven with identical signals 180 degrees out-of-phase, the two ports are constrained to have opposite polarity at all times. This ensures that any nodes lying on the line of symmetry have no voltage potential with respect to ground, for again the polarity of that potential would be indeterminate. The odd-mode equivalent circuit is thus shown in Figure 6.2(b). Simplification of this circuit is trivial, as the circled pairs of inductors are obviously in parallel. Like the even-mode equivalent circuit, the final result in this case is an unbalanced ladder comprising a high-pass filter with matched output termination.

The final even- and odd-mode circuits are thus redrawn in Figure 6.2(c). Although quite similar, these two sub-circuits are not identical. In fact, they are *dual forms* of the general high-pass filter ladder, where the inductors and capacitors have been exchanged, and series elements replaced with shunt elements. Each is a terminated high-pass filter in its own right, but as duals, they have the property that their reflection coefficients are additive inverses of one another. That is, $\Gamma_{\text{even}} = -\Gamma_{\text{odd}}$.

The significance of this result lies in how the even- and odd-mode equivalent responses combine to create the scattering parameters, s_{ij}, of the original two-port network. Consider that the even- and odd-mode reflection coefficients may each be understood as a superposition of the individual-port scattering parameters. In the even-mode case, the reflection coefficient is simply the reflection from port one added to the transmission from port two (or the reflection from port two added to the transmission from port one) when both ports are stimulated. Therefore,

$$\Gamma_{\text{even}} = s_{11} + s_{12} = s_{22} + s_{21} \tag{6.2}$$

On the other hand, the odd-mode reflection coefficient is the superposition of the reflection from port one and the transmission from port two with the equivalent negative input signal, or

$$\Gamma_{\text{odd}} = s_{11} - s_{12} = s_{22} - s_{21} \tag{6.3}$$

Since the network is symmetric and reciprocal, we can also say that $s_{11} = s_{22}$, and $s_{12} = s_{21}$. Thus, we have for any symmetric network

$$s_{11} = s_{22} = \frac{1}{2}(\Gamma_{\text{even}} + \Gamma_{\text{odd}}) \tag{6.4a}$$

$$s_{21} = s_{12} = \frac{1}{2}(\Gamma_{\text{even}} - \Gamma_{\text{odd}}) \tag{6.4b}$$

In the present case, since the even- and odd-mode equivalent circuits are duals ($\Gamma_{\text{even}} = -\Gamma_{\text{odd}}$), we are left with

$$s_{11} = s_{22} = 0 \tag{6.5a}$$

$$s_{21} = s_{12} = \Gamma_{\text{even}} = -\Gamma_{\text{odd}} \tag{6.5b}$$

Equation (6.5a) is, of course, the desired reflectionless condition.

6.2.2 *Transfer function of the reflectionless filter*

Section 6.1.1 showed that the transmission response, s_{21}, of the reflectionless filter in Figure 6.1(b) is given by the reflection coefficient of the even-mode equivalent circuit and the negative of the reflection coefficient of the odd-mode equivalent circuit, both of which have the topology of high-pass filters—but what exactly is the reflection coefficient of a terminated high-pass filter? Like the transfer function, of course, it depends on the prototype parameters chosen. For the present example, let us assume the prototype parameters have been taken from the standard tables for a Chebyshev filter. Although most textbooks don't say it explicitly, these parameters when applied to the ladder topology more specifically yield a Chebyshev *type I* filter, having ripples in its pass band while being monotonic in the stop band. That is, the magnitude of the transfer function (in both cases, even- and odd-mode) is given by

$$|H(\omega)| = \frac{1}{\sqrt{1 + \varepsilon^2 T_N^2(\omega^{-1})}} \tag{6.6}$$

where ε is called the ripple factor, $T_N(\cdot)$ is the Nth-order Chebyshev polynomial of the first kind, and the inverse frequency, ω^{-1}, has been used since these equivalent circuits are of the high-pass form.

By conservation of energy, the magnitude of the reflection coefficients are given by

$$|\Gamma_{\text{even}}|^2 = |\Gamma_{\text{odd}}|^2 = 1 - |H(\omega)|^2 \tag{6.7a}$$

$$= 1 - \frac{1}{1 + \varepsilon^2 T_N^2(\omega^{-1})} = \frac{\varepsilon^2 T_N^2(\omega^{-1})}{1 + \varepsilon^2 T_N^2(\omega^{-1})} = \frac{1}{1 + \varepsilon^{-2} T_N^{-2}(\omega^{-1})} \tag{6.7b}$$

$$\therefore |\Gamma_{\text{even}}| = |\Gamma_{\text{odd}}| = \frac{1}{\sqrt{1 + \varepsilon^{-2} T_N^{-2}(\omega^{-1})}} \tag{6.7c}$$

which corresponds precisely to the transfer function curve of a low-pass Chebyshev *type II* filter, having ripples in its stop band while being monotonic in the pass band.

As an example, the reflection coefficients of the even- and odd-mode equivalent circuits for our reflectionless filter, given Chebyshev prototype parameters with ripple factor $\varepsilon = 0.2187$ and assuming the order $N = 7$, are plotted in Figure 6.3. As promised by the even- and odd-mode analyses, these are additive inverses of one

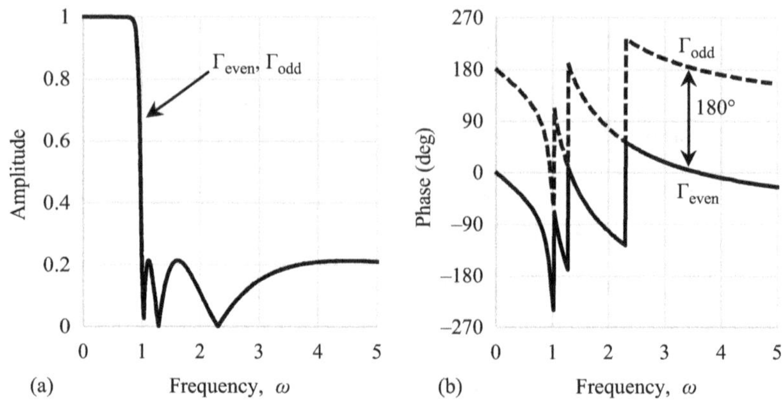

Figure 6.3 (a) Amplitude and (b) phase of the even- and odd-mode equivalent circuit reflection coefficients. For this example, $\varepsilon = 0.2187$ and $N = 7$

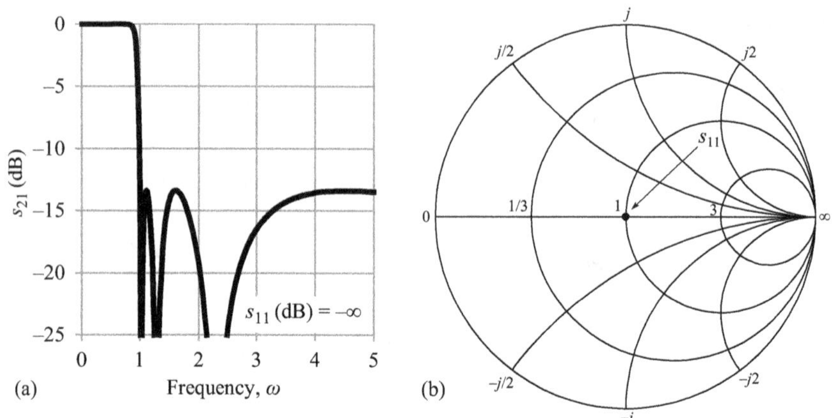

Figure 6.4 Frequency response of the seventh-order reflectionless filter with ripple factor $\varepsilon = 0.2187$: (a) transmission coefficient (S_{21}) on a dB scale and (b) reflection coefficient (S_{11}) on a Smith chart

another. The s_{11} and s_{22} therefore vanish, while s_{21} and s_{12} assume the transfer function of a Chebyshev type II low-pass filter.

The final low-pass transfer function and reflection coefficient are plotted on a dB scale and a Smith chart, respectively, in Figure 6.4.

6.3 Other approaches and limitations

The most obvious disadvantage of the reflectionless filter topology we have presented is the number of elements that is required—approximately 2.5 times as many as the simple (reflective) ladder topology in Figure 6.1(a). This is not atypical

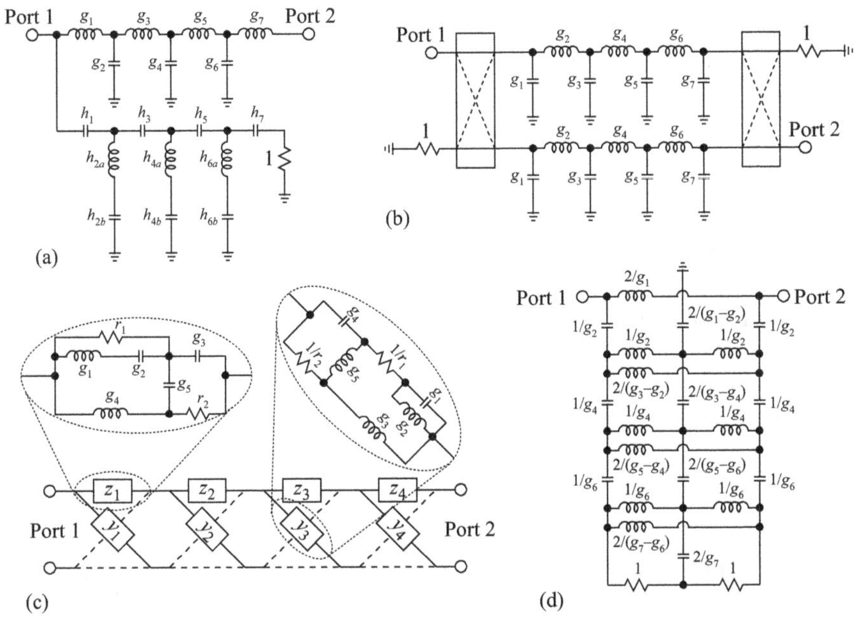

Figure 6.5 Several possible topologies for absorptive filters of seventh order: (a) complementary-susceptance diplexer; (b) hybrid-coupled balanced filter; (c) constant-resistance lattice chain; and (d) the reflectionless filter described in this chapter

for absorptive filter topologies, however. A number of different approaches are shown in Figure 6.5, each requiring at least twice as many elements as the standard non-absorptive filter, and having other pros and cons which will be discussed briefly in the following sections.

6.3.1 Diplexers or complementary-susceptance networks

Illustrated in Figure 6.5(a) is a diplexer comprising a low-pass filter in the primary signal path, and a terminated, complementary, high-pass filter attached in parallel with the input port. Given an appropriate choice of element values, such a structure can be made to become essentially reflectionless from the input side [7]. It is not strictly necessary that the complementary susceptance network take the form of a high-pass filter with a single resistive termination, but a theorem by Sidney Darlington guarantees that such a solution will always exist [8], in which case the termination may be replaced with a second output port. If so configured, and the primary signal path realizes a Chebyshev type I filter, then the secondary signal path will implement a Chebyshev type II filter (or vice-versa). In any case, while the common input port may be reflectionless, the separate output ports will each present to the external network the same reflection coefficient as a conventional filter in their stop-bands.

6.3.2 Hybrid-coupled balanced filter

The structure in Figure 6.5(b) comprises two identical low-pass filters coupled with ideal quadrature hybrids at input and output. Like a balanced amplifier, any reflection from the sub-filter inputs, which must be the same, will add constructively at the termination resistors while canceling at the ports due to the phasing properties of the hybrids.

The problem with this structure is that it doesn't really exist! There is no such thing as an ideal quadrature hybrid, and practical hybrids are inherently band-limited in their behavior. That may not seem important if the filter itself is designed to select a narrow frequency range, but keep in mind that it is the out-of-band effective impedance that gives a structure such as this its purpose. Reflectionless filters, by the definition given at the beginning of this chapter, are obligated to present a well-controlled impedance at all frequencies, making them inherently broadband circuits, regardless of how narrow their pass-bands may be.

6.3.3 Constant-resistance lattice chains

A cascade chain of constant-resistance lattice sections such as those proposed by Otto Zobel [1] is shown in Figure 6.5(c). Notable in this list for being intrinsically a differential-mode solution, it is also the only one shown that carries the penalty of excess loss beyond that introduced by the imperfect quality of real elements. That situation arises because the methodology is built upon the idea that an arbitrarily complex transfer function may be factored into first- and second-degree terms. Each term is then realized as an individual lattice section comprising dual impedances in the through- and crossed-branches, respectively. (Higher than second-order lattices are possible, but become increasingly complex and difficult to synthesize.) The catch is that while a seventh-order transfer function may be passive and realizable, there is no guarantee that the first- and second-degree factors it is composed of are individually passive and realizable themselves. This necessitates the synthetic introduction of losses in each lattice stage to make them passive.

The exact amount of excess loss that is required depends on the transfer function chosen and the order of the filter, but can be quite sizable if the filter is highly selective [3]. For Chebyshev type II filters of this order, it is typically a couple of dB. For type I filters, it is quite often many tens of dB!

Another great disadvantage of this approach is the number of elements. While all of the topologies shown require at least twice as many elements as a conventional reflective filter, this one is especially greedy. A seventh-order filter, as shown, requires three second-order lattices, and one first-order lattice. Each lattice is made of four branches (more specifically, two pairs of dual branches). Each branch in the second-order lattices will generally require seven elements to realize the necessary impedance function, for a total of 28 elements per second-order lattice. Three of those, plus a first-order lattice comprised of eight elements yields an extraordinary total of 92 elements to build this filter.

6.3.4 Limiting ripple factor of the reflectionless topology

While discussing the disadvantages of various absorptive-filter topologies, it is only fair to mention one of the key limitations of the new reflectionless filter topology being promoted here. Close examination of Figure 6.5(d) reveals that some of the elements are constrained to be inversely proportional to the difference in consecutive prototype parameters, e.g. $2/(g_k - g_{k\pm1})$, where k is always odd. While the parameters, g_k, may be chosen to some extent arbitrarily, if any odd-numbered parameter in the list was less than either of its neighbors, the topology as shown would somewhere require a negatively valued reactive element.

This fact limits the transfer functions that such a topology can realize in practice. The Butterworth response is strictly forbidden, and the Chebyshev responses are constrained to a certain range of ripple factors. In the case of a type II response, as indicated here, this means that the peak rejection per stage is limited. The minimum ripple factor that can be achieved with this topology is given by [3]

$$\varepsilon_{\min}(N) = \sqrt{e^{4\tanh^{-1}(e^{-\beta})} - 1} \tag{6.8}$$

where

$$\beta = 2N \sinh^{-1}\left(\sqrt{\frac{1}{2}\sin\left(\frac{\pi}{N}\right)\tan\left(\frac{\pi}{N}\right)}\right) \tag{6.9}$$

and where N is the order of the filter. The maximum stop-band rejection for a type II filter is then given by

$$R_{\max}(dB) = 10\log_{10}\left(1 + \varepsilon_{\min}{}^{-2}\right) \tag{6.10}$$

This limit is only very weakly dependent upon order; for $N > 3$ it always lies within 13.43 dB $\pm$ 0.06 dB. As this is only a modest rejection level, it is most common to configure reflectionless filters according to this limit, in which case the top capacitor, $2/(g_1 - g_2)$, and the bottom inductor, $2/(g_7 - g_6)$, both vanish—they become infinite in value, or equivalently a short circuit in the case of the capacitor, and an open circuit in the case of the inductor. Deeper rejection is then conventionally achieved by cascading—something which reflectionless filters are uniquely qualified to do, given their good, broadband impedance match [9].

Although cascading stages to achieve deep rejection inevitably requires a large number of elements, it is particularly suited to planar, monolithic implementations since the extra stages are identical and require only a copy and paste operation during layout. (Contrast this with the cascaded, constant-resistance, lattice networks of Figure 6.5(c) where the 92 elements are all potentially unique.)

Alternatively, one may consider distributing filter stages throughout the signal path to better optimize dynamic range. It is often more optimal to filter out unwanted spectrum a little at a time, say in between gain stages and other components, rather than all at the beginning or all at the end. This helps to keep the spurious signal levels low at all points in the system rather than allowing them to build up to peak intensity before finally filtering them out in one step. Doing this with conventional filters would be risky, as it maximizes the opportunity for an unexpected out-of-band interaction to wreak havoc with the system performance. With reflectionless filters, however, spreading them around the signal path is not only safe, but can actually mitigate the harmful interactions between other components. Just as small attenuators, or pads, are often liberally distributed throughout complex electronic systems to suppress standing waves, reflectionless filters can be used as frequency-selective pads to perform the same function out-of-band without compromising in-band gain.

6.4 Scaled prototype parameter tables

Conventionally, at least in academic circles, the cutoff frequency of Chebyshev type II filters is specified according to the onset of the stop-band, when the rejection has first reached the level of the ripple peaks (the point labeled ω_{stop} in Figure 6.6). This is the standard that results from utilizing the Chebyshev prototype parameter tables found in most textbooks. Table 6.1 gives these parameters for the special case

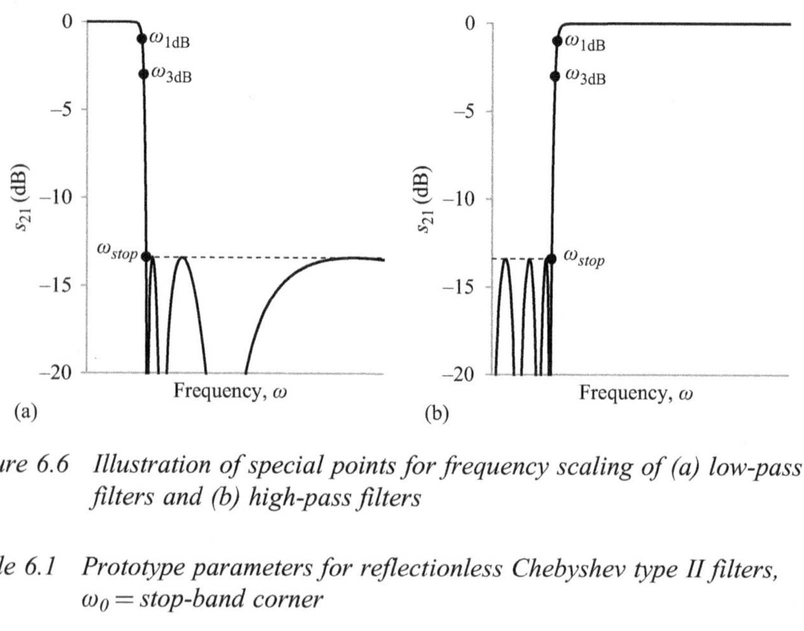

Figure 6.6 Illustration of special points for frequency scaling of (a) low-pass filters and (b) high-pass filters

Table 6.1 Prototype parameters for reflectionless Chebyshev type II filters, ω_0 = stop-band corner

N	ε	g_1	g_2	g_3	g_4	g_5	g_6	g_7	g_8
3	0.1925	1.155	1.155						
5	0.2164	1.337	1.337	2.164					
7	0.2187	1.377	1.377	2.280	1.498				
9	0.2192	1.392	1.392	2.315	1.531	2.378			
11	0.2194	1.400	1.400	2.330	1.544	2.407	1.561		
13	0.2194	1.404	1.404	2.338	1.550	2.420	1.572	2.433	
15	0.2195	1.406	1.406	2.343	1.554	2.427	1.578	2.445	1.583

where the ripple factor, ε, corresponds to the minimum value achievable in a single-stage reflectionless filter. Since it is known that the list of parameters is symmetric about the center, only the first half of the values in each row is shown.

In practice, however, filter requirements for electronic systems are more often stated in terms of the 1 dB or 3 dB cutoff corner ($\omega_{1\mathrm{dB}}$ and $\omega_{3\mathrm{dB}}$ in Figure 6.6). This requires scaling the prototype parameters according to the transfer function and the order of the filter, N. Specifically, we must locate the frequency, ω_{IL}, at which the Chebyshev type II prototype response in (6.7c) achieves the desired insertion loss, IL, in dB,

$$IL = 10\log|s_{21}|^{-2} = 10\log|\Gamma_{\text{even}}|^{-2} = 10\log\left(1+\varepsilon^{-2}T_N^{-2}\left(\omega_{IL}^{-1}\right)\right) \quad (6.11\text{a})$$

$$\therefore T_N\left(\omega_{IL}^{-1}\right) = \cosh\left(N\cosh^{-1}\left(\omega_{IL}^{-1}\right)\right) = \frac{1}{\varepsilon\sqrt{10^{IL/10}-1}} \quad (6.11\text{b})$$

$$\omega_{IL}^{-1} = \cosh\left(\frac{1}{N}\cosh^{-1}\left(\frac{1}{\varepsilon\sqrt{10^{IL/10}-1}}\right)\right) \quad (6.11\text{c})$$

We then scale the prototype parameters such that the desired cutoff corner now appears at unit frequency,

$$g_k' = \frac{g_k}{\omega_{IL}} = g_k\cosh\left(\frac{1}{N}\cosh^{-1}\left(\frac{1}{\varepsilon\sqrt{10^{IL/10}-1}}\right)\right) \quad (6.12)$$

For convenience, in this section we have carried out these calculations and presented prototype tables for Chebyshev type II reflectionless filters that can be applied directly to the topology in Figure 6.1(b), scaled according to these cutoff standards (Table 6.2 for 1 dB corner specification, and Table 6.3 for 3 dB corner specification). In all cases, the maximum stop-band rejection (minimum ripple factor) has been selected.

Table 6.2 Prototype parameters for reflectionless Chebyshev type II filters, $\omega_0 = 1$ dB corner

N	ε	g_1	g_2	g_3	g_4	g_5	g_6	g_7	g_8
3	0.1925	1.788	1.788						
5	0.2164	1.568	1.568	2.537					
7	0.2187	1.496	1.496	2.477	1.627				
9	0.2192	1.464	1.464	2.435	1.611	2.501			
11	0.2194	1.448	1.448	2.411	1.597	2.490	1.615		
13	0.2194	1.438	1.438	2.396	1.588	2.480	1.611	2.493	
15	0.2195	1.432	1.432	2.387	1.582	2.472	1.607	2.490	1.612

Table 6.3 Prototype parameters for reflectionless Chebyshev type II filters, $\omega_0 = 3$ dB corner

N	ε	g_1	g_2	g_3	g_4	g_5	g_6	g_7	g_8
3	0.1925	1.521	1.521						
5	0.2164	1.470	1.470	2.379					
7	0.2187	1.446	1.446	2.394	1.573				
9	0.2192	1.434	1.434	2.384	1.577	2.450			
11	0.2194	1.428	1.428	2.377	1.575	2.455	1.593		
13	0.2194	1.424	1.424	2.372	1.572	2.455	1.595	2.468	
15	0.2195	1.421	1.421	2.369	1.570	2.453	1.595	2.471	1.600

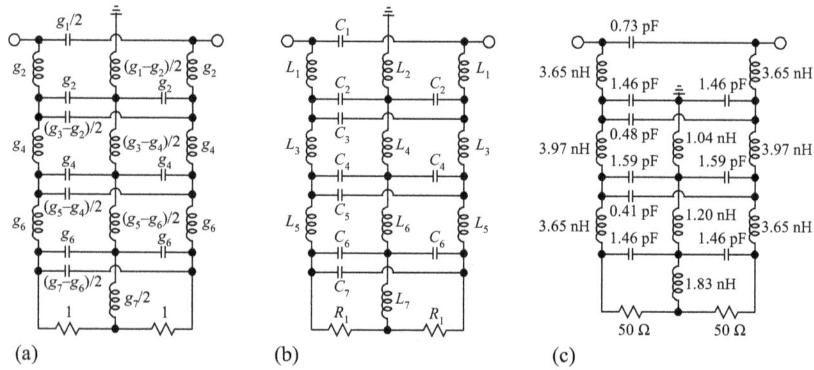

Figure 6.7 (a) High-pass transformed, seventh-order, reflectionless filter topology; (b) scaled element labeling; and (c) final element values for 1 dB corner at 3 GHz

6.5 Planar lumped-element implementations

Although the reflectionless topology, like other absorptive filters, requires a relatively large number of components, the symmetry and consequent repetition of element values makes it especially suited to monolithic implementation. Copied and mirrored components on a layout are essentially free, except for the extra real estate they occupy. The fine lithography afforded by planar, printed-circuit processes also permit filters of very compact size to be made, with tight tolerances suitable for high-frequency operation, even into the millimeter-wave band [10].

As an example, consider a high-pass filter designed for a stop-band corner of 3 GHz. We begin by taking a high-pass transformation of the topology in Figure 6.5(d) that is, inductors and capacitors are exchanged, and the normalized element values are inverted, as shown in Figure 6.7(a). To realize the filter, we must first calculate the values of the physical elements, scaled according to the cutoff frequency and the characteristic impedance we wish to match to. We use the prototype parameters from Table 6.1, row $N = 7$, and the element labels in Figure 6.7(b). The values are determined as follows,

$$C_1 = \frac{g_1 Y_0}{2\omega_0} = \frac{(1.377)(50\,\Omega)^{-1}}{2(2\pi \cdot 3\text{ GHz})} = 0.73\text{ pF} \tag{6.13a}$$

$$C_2 = \frac{g_2 Y_0}{\omega_0} = \frac{(1.377)(50\,\Omega)^{-1}}{(2\pi \cdot 3\text{ GHz})} = 1.46\text{ pF} \tag{6.13b}$$

$$C_3 = \frac{(g_3 - g_2) Y_0}{2\omega_0} = \frac{(2.280 - 1.377)(50\,\Omega)^{-1}}{2(2\pi \cdot 3\text{ GHz})} = 0.48\text{ pF} \tag{6.13c}$$

$$C_4 = \frac{g_4 Y_0}{\omega_0} = \frac{(1.498)(50\,\Omega)^{-1}}{(2\pi \cdot 3\text{ GHz})} = 1.59\text{ pF} \tag{6.13d}$$

$$C_5 = \frac{(g_5 - g_4) Y_0}{2\omega_0} = \frac{(2.280 - 1.498)(50\,\Omega)^{-1}}{2(2\pi \cdot 3\text{ GHz})} = 0.41\text{ pF} \tag{6.13e}$$

$$C_6 = \frac{g_6 Y_0}{\omega_0} = \frac{(1.377)(50\,\Omega)^{-1}}{(2\pi \cdot 3\text{ GHz})} = 1.46\text{ pF} \tag{6.13f}$$

$$C_7 = \frac{(g_7 - g_6) Y_0}{2\omega_0} = \frac{(1.377 - 1.377)(50\,\Omega)^{-1}}{2(2\pi \cdot 3\text{ GHz})} = 0 \tag{6.13g}$$

$$L_1 = \frac{g_2 Z_0}{\omega_0} = \frac{(1.377)(50\,\Omega)}{(2\pi \cdot 3\text{ GHz})} = 3.65\text{ nH} \tag{6.13h}$$

$$L_2 = \frac{(g_1 - g_2) Z_0}{2\omega_0} = \frac{(1.377 - 1.377)(50\,\Omega)}{2(2\pi \cdot 3\text{ GHz})} = 0 \tag{6.13i}$$

$$L_3 = \frac{g_4 Z_0}{\omega_0} = \frac{(1.498)(50\,\Omega)}{(2\pi \cdot 3\text{ GHz})} = 3.97\text{ nH} \tag{6.13j}$$

$$L_4 = \frac{(g_3 - g_4) Z_0}{2\omega_0} = \frac{(2.280 - 1.498)(50\,\Omega)}{2(2\pi \cdot 3\text{ GHz})} = 1.04\text{ nH} \tag{6.13k}$$

$$L_5 = \frac{g_6 Z_0}{\omega_0} = \frac{(1.377)(50\,\Omega)}{(2\pi \cdot 3\text{ GHz})} = 3.65\text{ nH} \tag{6.13l}$$

$$L_6 = \frac{(g_5 - g_6) Z_0}{2\omega_0} = \frac{(2.280 - 1.377)(50\,\Omega)}{2(2\pi \cdot 3\text{ GHz})} = 1.20\text{ nH} \tag{6.13m}$$

$$L_7 = \frac{g_7 Z_0}{2\omega_0} = \frac{(1.377)(50\,\Omega)}{2(2\pi \cdot 3\text{ GHz})} = 1.83\text{ nH} \tag{6.13n}$$

$$R_1 = Z_0 = 50\,\Omega \tag{6.13o}$$

and are shown in Figure 6.7(c).

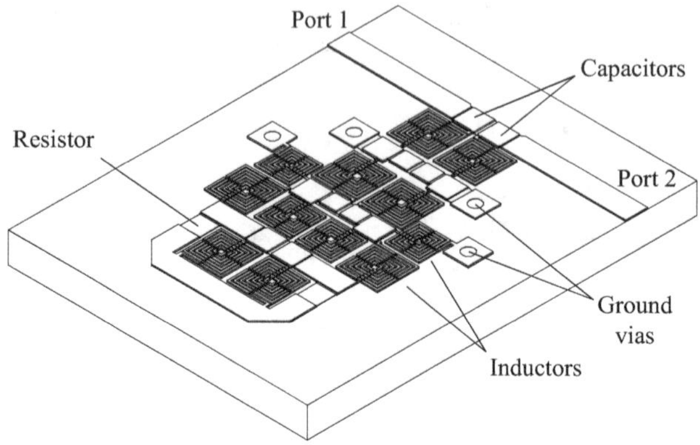

Figure 6.8 Planar implementation of a seventh-order, high-pass reflectionless filter

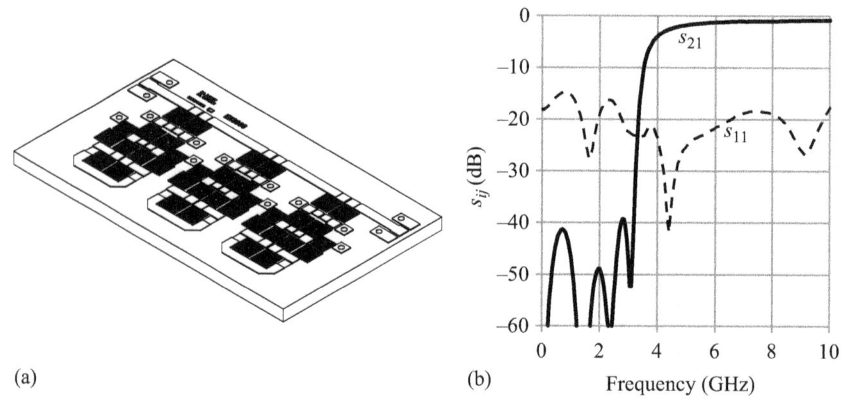

Figure 6.9 (a) Three-stage, high-pass reflectionless filter die and (b) measured performance

Having determined the nominal element values, it remains simply to realize them in the layout. Typically, the capacitors will be interdigital or Metal-Insulator-Metal (MIM) stackups, and the inductors may be realized using planar spirals. It may be helpful to split some of the elements into series or parallel pairs for convenience and symmetry, as shown in Figure 6.8.

This particular example was borrowed from a commercially available part [11] shown in Figure 6.9(a). To increase the rejection, three identical stages were cascaded on a single die. The measured die performance is plotted in Figure 6.9(b). Non-ideality and parasitic coupling between the reactive elements shift the effective cutoff frequency slightly and make the reflection coefficient imperfect, however the return loss is still close to 20 dB over the pass band, stop band, and transition band.

6.6 Delta-wye transformation for reduced impedance

Returning to the low-pass prototype of Figure 6.1(b), it is evident that the topology has a repeating pattern of inductors in groups of three. In fact, each set of inductors is arranged in a ring, or *delta*, configuration. This allows one to apply the well-known delta-wye transformations to these inductor groups, resulting in a modified topology.

Consider the first trio of inductors, for example, having normalized values $1/g_2$ and $2/(g_3{-}g_2)$, as indicated in Figure 6.10(a). The delta-wye transformation formula gives

$$z_1 = \frac{z_b z_c}{z_a + z_b + z_c} = \frac{\left(g_2^{-1}\right)\left(g_2^{-1}\right)}{2(g_3 - g_2)^{-1} + g_2^{-1} + g_2^{-1}} = \frac{g_3 - g_2}{2g_2 g_3} = \frac{1}{2}\left(\frac{1}{g_2} - \frac{1}{g_3}\right) \tag{6.14a}$$

$$z_2 = z_3 = \frac{z_a z_c}{z_a + z_b + z_c} = \frac{2(g_3 - g_2)^{-1} g_2^{-1}}{2(g_3 - g_2)^{-1} + g_2^{-1} + g_2^{-1}} = \frac{1}{g_3} \tag{6.14b}$$

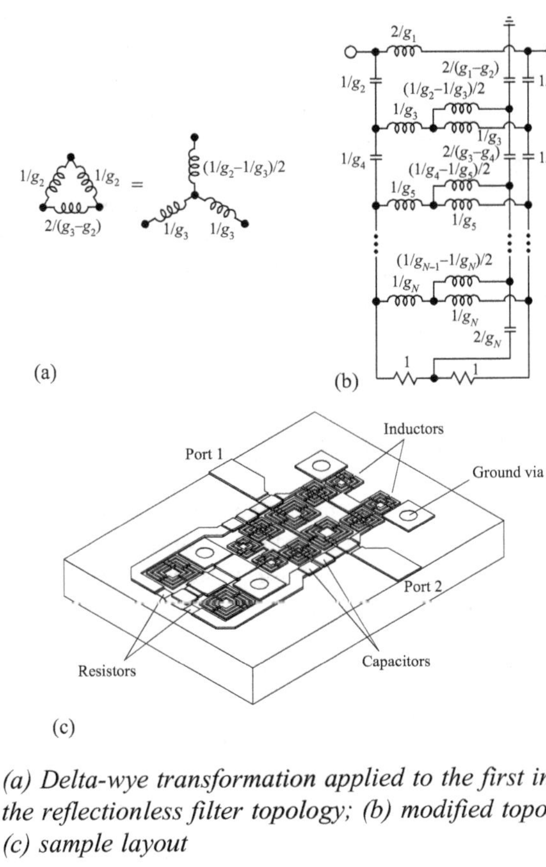

Figure 6.10 (a) Delta-wye transformation applied to the first inductor group in the reflectionless filter topology; (b) modified topology; and (c) sample layout

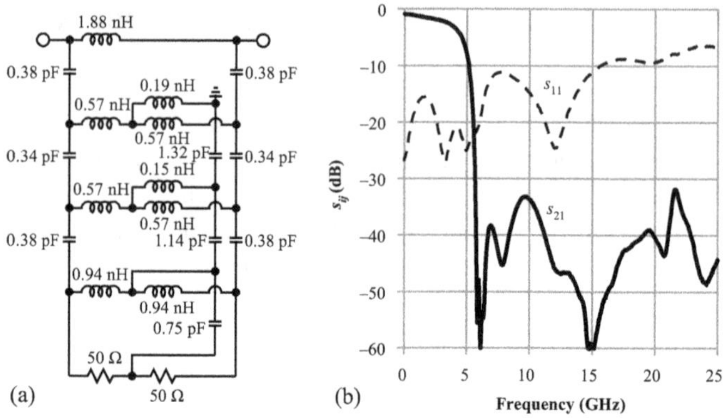

Figure 6.11 *(a) Schematic of a single-stage, seventh-order reflectionless low-pass filter; and (b) measured performance of a three-stage die*

Proceeding similarly for the rest of the inductor groups, we arrive at the modified topology in Figure 6.10(b).

The significance is that, in all cases, the transformation results in smaller inductors overall than in the original topology. In planar implementations, this helps to minimize real estate, reduce layout parasitics, and can be helpful in maximizing the Q of the inductors. The same transformation applied to a high-pass filter would result in larger capacitors, and to a band-pass filter would yield series resonators with lower impedance. Importantly, none of the data in Tables 6.1–6.3 (nor the applicable tables in any other textbook) need changing.

An example of a low-pass filter layout utilizing the delta-wye transformation is shown in Figure 6.10(c). As before, some elements in the core schematic have been split for symmetry reasons and ease of layout. A single-stage schematic and measured die performance for a three-stage commercial part based on this design [12] are shown in Figure 6.11.

6.7 Non-canonical filter responses

It must be re-iterated that the general reflectionless topology presented at the beginning of this chapter, Figure 6.1(b), is but one of many, and in itself is capable of realizing many different kinds of responses besides Chebyshev. Several of these alternatives will be explored briefly in this section.

6.7.1 Achieser–Zolotarev reflectionless filters

The topology introduced in Section 6.1 is as flexible as the more widely known ladder topology it was compared to. The prototype parameters, g_k, may take on any values whatsoever, and the filter that results will still be reflectionless. The only limitation in this topology is a practical one—that the resultant reactive element values must remain positive.

That constraint, when applied to the Chebyshev optimization criteria, led to a limitation in the achievable rejection level of the filter in one stage. In some applications, however, the rejection need not be constant over the entire range from the stop-band corner to infinity. Rejection close-in to the band edge may be most critical. For these situations, the Achieser–Zolotarev (or simply Zolotarev) type II filter response may be more appropriate, and with it, much deeper rejection over a limited stop-band range may be realized.

Two examples of Zolotarev type II filters realizable within the reflectionless topology for $N = 9$ are given in Figure 6.12. Whereas the Chebyshev response was limited to a rejection level of about 13.4 dB, these have nearly 20 dB and 30 dB over bandwidths of 1.8:1 and 2.5:1, respectively. The defining characteristic of the Zolotarev response is that it has one large ripple (the outermost stop-band ripple in the type II case) while the rest of the ripples are smaller and equal in height [13].

The mathematics of the Zolotarev polynomials on which the filter response is based are beyond the scope of this chapter, however it is generally parameterized in terms of the order, N, the ripple factor for the smaller ripples, ε, and a third free parameter, λ^{-1}, which is the bandwidth ratio of the deeper stop-band rejection (for example, if $\lambda^{-1} = 2$, the smaller ripples occupy a stop-band having an octave bandwidth). Whereas the Chebyshev type II response is realizable in the reflectionless configuration when ε is not too small, the Zolotarev type II response is constrained such that ε and λ aren't both simultaneously too small (λ^{-1} too large), however one can be traded off for the other. As with the Chebyshev response, the critical feature of the limiting case is when $g_1 = g_2$ (and consequently $g_{N-1} = g_N$). Tables of Zolotarev prototype parameters that meet this criterion and that are suitable for use with this reflectionless topology are given in Table 6.4. Since the list of parameters is symmetric about the center, only the first half are shown in each row.

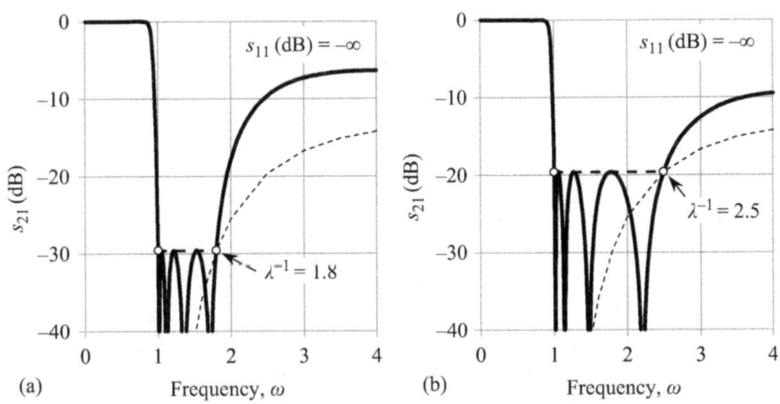

Figure 6.12 Achieser–Zolotarev reflectionless filters of order $N = 9$: (a) $\lambda^{-1} = 1.8$ and (b) $\lambda^{-1} = 2.5$. The thin dashed line represents the limiting stop-band rejection as a function of stop-band bandwidth (λ^{-1})

Table 6.4 Prototype parameters for reflectionless Zolotarev type II filters

N	ε	λ^{-1}	Rejection (dB)	g_1	g_2	g_3	g_4	g_5
5	0.2164	*	13.49	1.3375	1.3375	2.1641		
5	0.2150	3	13.55	1.3363	1.3363	2.1649		
5	0.1993	2.5	14.18	1.3225	1.3225	2.1751		
5	0.1543	2	16.33	1.2808	1.2808	2.1998		
5	0.1250	1.8	18.13	1.2510	1.2510	2.2107		
5	0.08955	1.6	21.00	1.2108	1.2108	2.2155		
5	0.05073	1.4	25.91	1.1576	1.1576	2.2010		
5	0.01608	1.2	35.88	1.0886	1.0886	2.1424		
7	0.2187	*	13.41	1.3769	1.3769	2.2801	1.4983	
7	0.2162	4	13.50	1.3746	1.3746	2.2839	1.4939	
7	0.2067	3.5	13.87	1.3659	1.3659	2.2989	1.4770	
7	0.1858	3	14.77	1.3462	1.3462	2.3321	1.4397	
7	0.1483	2.5	16.67	1.3084	1.3084	2.3926	1.3704	
7	0.09090	2	20.86	1.2415	1.2415	2.4847	1.2555	
7	0.06399	1.8	23.90	1.2031	1.2031	2.5249	1.1943	
7	0.03767	1.6	28.49	1.1559	1.1559	2.5565	1.1241	
7	0.01573	1.4	36.07	1.0984	1.0984	2.5620	1.0461	
7	0.00277	1.2	51.15	1.0286	1.0286	2.5091	0.9622	
9	0.2192	*	13.39	1.3921	1.3921	2.3150	1.5313	2.3783
9	0.2189	5.5	13.40	1.3921	1.3921	2.3156	1.5307	2.3791
9	0.2162	5	13.50	1.3897	1.3897	2.3208	1.5247	2.3881
9	0.2098	4.5	13.75	1.3836	1.3836	2.3336	1.5100	2.4102
9	0.1979	4	14.24	1.3723	1.3723	2.3578	1.4824	2.4527
9	0.1782	3.5	15.12	1.3529	1.3529	2.3989	1.4360	2.5277
9	0.1480	3	16.69	1.3214	1.3214	2.4658	1.3621	2.6554
9	0.1053	2.5	19.60	1.2717	1.2717	2.5689	1.2499	2.8719
9	0.05329	2	25.48	1.1946	1.1946	2.7127	1.0889	3.2355
9	0.03314	1.8	29.60	1.1533	1.1533	2.7750	1.0101	3.4349
9	0.01628	1.6	35.77	1.1044	1.1044	2.8278	0.9246	3.6589
9	0.00508	1.4	45.88	1.0465	1.0465	2.8502	0.8358	3.8741
9	$5.03 \cdot 10^{-4}$	1.2	65.97	0.9779	0.9779	2.8039	0.7494	3.9890

*Largest λ^{-1} for each order corresponds to a Chebyshev prototype.

6.7.2 Filters with uniform element values

In some cases, convenience and ease of manufacturability may outweigh the desire to have a mathematically optimal frequency response. There are variants of the reflectionless topology presented at the beginning of this chapter, for example, in which all of the elements of like-kind have equal value. One such variant for $N = 7$ is shown in Figure 6.13(a). (In fact, in the sequence of the development of modern reflectionless filters, it was these topologies that were discovered first [14,15].)

The uniformity of element values makes tuning of such filters especially straightforward, as there are but two free parameters—the inductance value and the capacitance value (the resistors being fixed to the characteristic impedance of the system). The ratio of the reactive elements sets the characteristic impedance, while

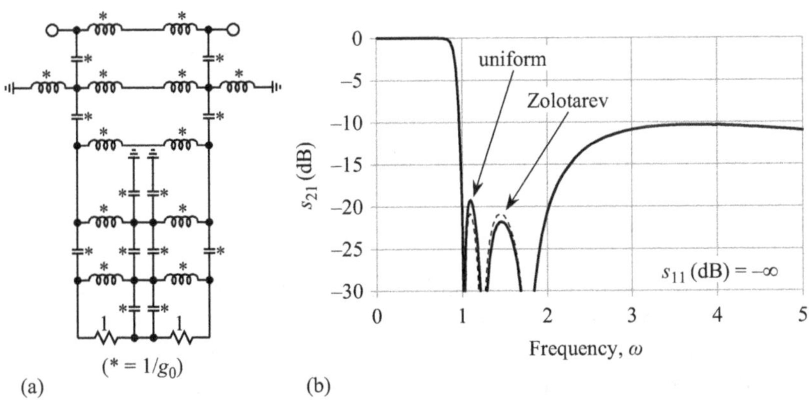

Figure 6.13 (a) Reflectionless filter with all equal element values and (b) frequency response where $g_0 = 1.25$ and that of a Zolotarev filter ($\lambda^{-1} = 2$) for comparison

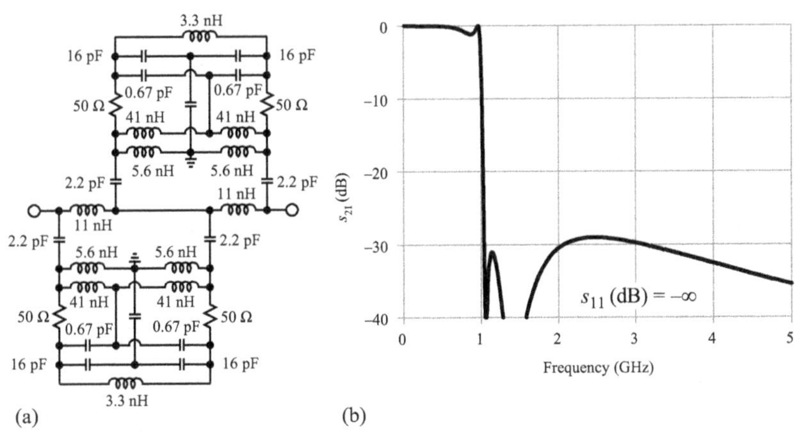

Figure 6.14 (a) Pseudo-elliptic reflectionless filter topology and (b) frequency response

their product determines the cutoff frequency. In planar implementations, where the inductors typically require the most work, one must optimize the layout of but a single inductor and then fill in the rest by copy and paste, regardless of the order of the filter. The response of this circuit is shown in Figure 6.13(b), along with a Zolotarev reflectionless filter for comparison.

6.7.3 Pseudo-elliptic and other topologies

Finally, it is worth illustrating that the techniques described in this chapter for creating reflectionless filter topologies are capable of a much wider variety of responses than those shown here. As one example, Figure 6.14 shows a two-stage quasi-elliptic filter involving a topology quite unlike any of the others previously demonstrated. A more complete explanation of the methods for deriving and improving reflectionless topologies is given in [3].

6.8 Transmission-line reflectionless filters

Thus far in this chapter, we have focused on lumped-element designs, but the symmetric reflectionless filter concept extends to the transmission-line regime as well [3,16]. Consider the transmission-line network in Figure 6.15(a). We assume it is a commensurate-line network—in which all transmission lines are equal in length, in this case a quarter wavelength at the center frequency of the pass band—and that all lines are labeled with their normalized characteristic impedance. The characteristic impedances of the coupled-line sections are given as follows,

$$z_e = \frac{2\rho}{\rho+1} \tag{6.15a}$$

$$z_o = \frac{2}{\rho+1} \tag{6.15b}$$

$$\rho = \frac{z_e}{z_o} = 1 + 2\left(x - x^{-1}\right) + 2\sqrt{\left(x - x^{-1}\right)\left(1 + x - x^{-1}\right)} \tag{6.15c}$$

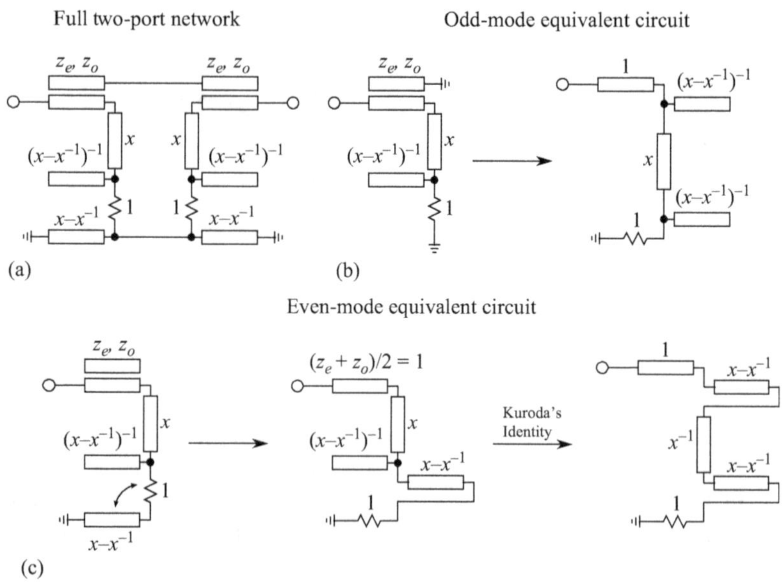

Figure 6.15 (a) Transmission-line reflectionless filter; (b) simplification of odd-mode equivalent circuit; and (c) simplification of even-mode equivalent circuit. All transmission-lines are a quarter-wavelength long at center frequency, and are labeled with their normalized characteristic impedance. Constraints on z_e and z_o are given in the text

The odd-mode equivalent circuit, Figure 6.15(b), is found, as in the lumped-element case, by assigning a virtual short to each node on the line of symmetry. The resulting equivalent circuit may then be simplified by applying a transmission-line identity to the coupled lines, resulting in an additional open-circuited stub near the input port with normalized impedance

$$z_{stub} = 2z_e z_o \frac{z_e + z_o}{(z_e - z_o)^2} = 2\left(\frac{2\rho}{\rho+1}\right)\left(\frac{2}{\rho+1}\right)\frac{\left(\frac{2\rho}{\rho+1}\right) + \left(\frac{2}{\rho+1}\right)}{\left(\left(\frac{2\rho}{\rho+1}\right) - \left(\frac{2}{\rho+1}\right)\right)^2} = \frac{4\rho}{(\rho-1)^2} \tag{6.16a}$$

$$= \frac{4\left(1 + 2(x - x^{-1}) + 2\sqrt{(x - x^{-1})(1 + x - x^{-1})}\right)}{\left(2(x - x^{-1}) + 2\sqrt{(x - x^{-1})(1 + x - x^{-1})}\right)^2} = \frac{1}{x - x^{-1}} \tag{6.16b}$$

There is also a residual transmission line at the port with normalized characteristic impedance given by the average of the coupled line characteristic impedances,

$$z_{line} = \frac{1}{2}(z_e + z_o) = \frac{1}{2}\left(\frac{2\rho}{\rho+1} + \frac{2}{\rho+1}\right) = 1 \tag{6.17}$$

as shown in the figure. Clearly, this line represents a port extension which alters the phase of the frequency response without affecting its amplitude.

The even-mode equivalent circuit in Figure 6.15(c), on the other hand, is found by leaving open each node that appears on the line of symmetry. Simplification is done in two steps. In the first step, the coupled-lines are replaced with a simple transmission-line having normalized characteristic impedance again equal to the average of the coupled-line impedances (unity), in accordance with a known transmission-line identity for this configuration. At the same time, the resistor and the short-circuited stub with which it is in series are exchanged. This leaves a series-connected, short-circuited stub.

In step two, the remaining open-circuited stub is replaced with a series-connected, short-circuited stub using Kuroda's identity. Note that due to the relative values of the line and the stub—x and $(x–x^{-1})^{-1}$, respectively—the application of Kuroda's identity in this case results in the inversion of their values. That is,

$$z'_{stub} = \frac{x^2}{x + (x - x^{-1})^{-1}} = x - x^{-1} = (z_{stub})^{-1} \tag{6.18a}$$

$$z'_{line} = \frac{x(x - x^{-1})^{-1}}{x + (x - x^{-1})^{-1}} = x^{-1} = (z_{line})^{-1} \tag{6.18b}$$

The topology and impedance parameters of the original structure were thus carefully chosen such that we would once again arrive at even- and odd-mode equivalent circuits which are duals of one another. Parallel-connected, open-circuited

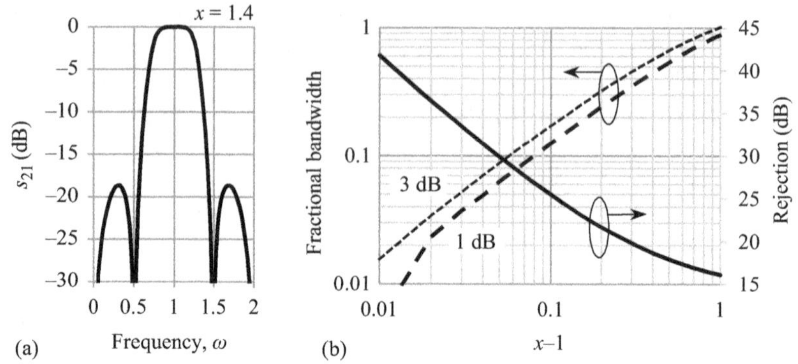

Figure 6.16 (a) Frequency response of the transmission-line reflectionless filter with open-circuited stubs, where x = 1.4 and (b) fractional bandwidth (at 1 and 3 dB corners) and peak stop-band rejection versus the design parameter, x

stubs are the duals of series-connected, short-circuited stubs, and cascade transmission lines are the duals of other cascade lines having the inverse characteristic impedance. This guarantees that the reflection coefficients of the even- and odd-mode equivalent circuits are once again additive inverses, $\Gamma_{\text{even}} = -\Gamma_{\text{odd}}$. This condition, as we saw before in Section 6.1.1, ensures that the original two-port filter is reflectionless.

A typical frequency response for this structure is shown in Figure 6.16(a). The variable x, which must be greater than one (lest some of the transmission lines take on negative impedance), is a free parameter that affects the shape of the response curve. Smaller values of x yield narrower pass bands with somewhat greater peak rejection in the stop-bands, while larger values permit wider pass bandwidths. A plot of these features versus $x-1$ is shown in Figure 6.16(b). Typically, x must fall somewhere between about 1.2 and 1.8 to realize practical line impedances.

There is further an alternate topology which replaces the open-circuited stub with a quarter-wave cascade resonator. This topology is shown in Figure 6.17(a). The coupled-line parameters are the same as before, given in (6.15). Simplification of the even- and odd-mode equivalent circuits to illustrate their duality proceeds much as before, except that Kuroda's identity is applied in the even-mode circuit twice, Figure 6.17(c). Note that the line and stub once again have the special relationship described earlier for which Kuroda's identity simply inverts their values while converting between a shunt stub and a series stub. As the identity is applied here twice, the resulting stub is of the same configuration and has the same impedance as the original, only displaced by two quarter-wave units, with the intervening transmission lines in reverse order. For an extensive list of transmission line identities such as this, which may be useful in the development of reflectionless filters, the reader is referred to [3].

The response of this filter is shown in Figure 6.18. Not as flat in the pass-band as the preceding topology, it does have the advantage of greater peak rejection in the stop band for given line impedances and bandwidth.

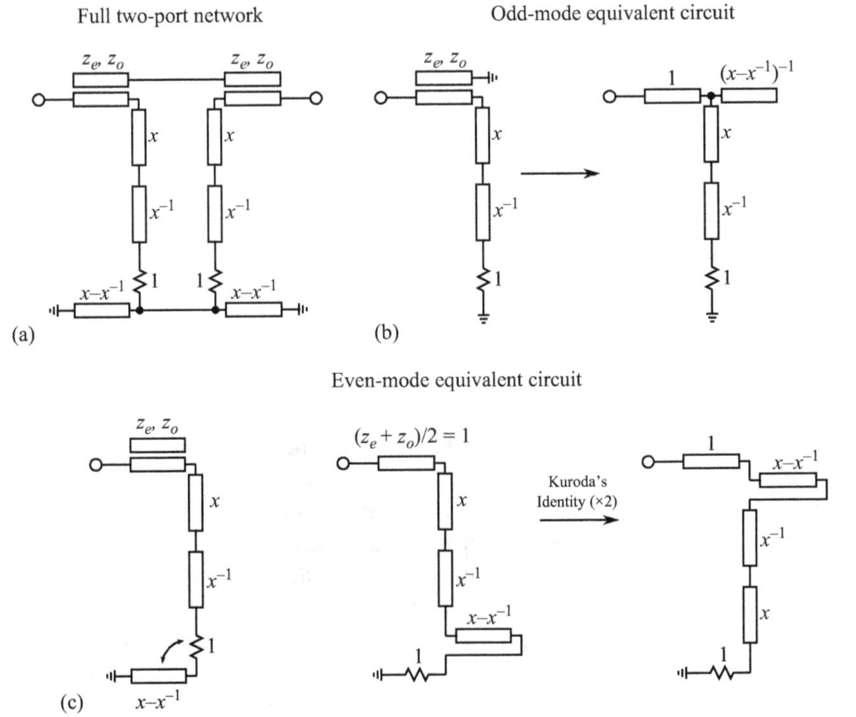

Figure 6.17 (a) Alternate transmission-line reflectionless filter; (b) simplification of odd-mode equivalent circuit; and (c) simplification of even-mode equivalent circuit. All transmission-lines are a quarter-wavelength long at center frequency, and are labeled with their normalized characteristic impedance. Constraints on z_e and z_o are given in the text

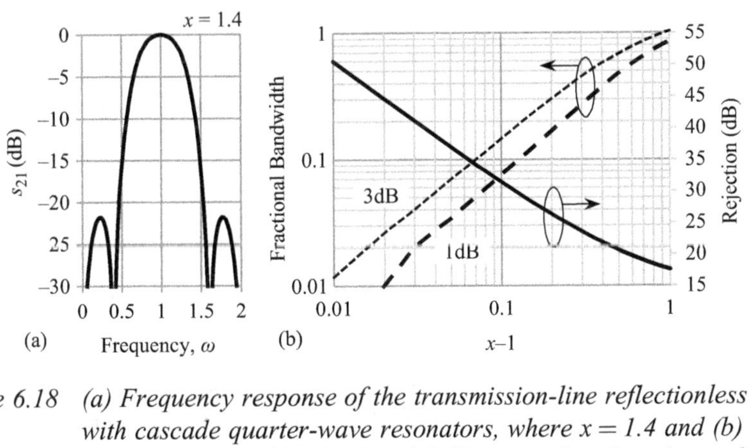

Figure 6.18 (a) Frequency response of the transmission-line reflectionless filter with cascade quarter-wave resonators, where x = 1.4 and (b) fractional bandwidth (at 1 and 3 dB corners) and peak stop-band rejection versus the design parameter, x

Higher-order topologies than these do exist [3], generally comprising additional cascaded quarter-wavelength or stub resonators, but these tend to have higher stop-band peaks. The third-order designs presented here are deemed to be the most practical.

6.9 Chebyshev and Zolotarev type I filters

Recall from Section 6.1.2 that the stop-band spectrum of a Chebyshev type I filter (having ripples in the pass-band) corresponds to a type II filter (having ripples in the stop-band) with the opposite frequency response, e.g. low-pass instead of high-pass, and vice-versa. The same can be said of the Zolotarev filters. Thus the coupling between the incident signal of a type II reflectionless filter and the associated termination resistor is equivalent to the transfer function of a type I filter. This suggests that a type I reflectionless filter, whether it be Chebyshev, or Zolotarev, or something else, may be obtained by simply terminating one of the standard outputs and extracting the output instead from the terminals of the opposite resistor. Since the resistor is not grounded, it would require a broadband balun to isolate the differential signal at these terminals without loading the rest of the circuit to ground.

Take, for example, the high-pass reflectionless filter from Figure 6.7(a). It has been redrawn in Figure 6.19(a) with the usual output terminated and a balun in place of the input-side termination. The corresponding frequency response is a Chebyshev type I low-pass filter. Moreover, it is still reflectionless from both input and output (in as much as the balun has good performance over a wide bandwidth). Similarly, the transmission-line based type II band-pass filters of Section 6.7 could be converted to reflectionless band-stop filters with type I response curves by similarly employing a balun at the termination ports, as in Figure 6.19(b).

In either case, if all four potential outputs are utilized, the structure becomes a dual-directional contiguous diplexer which is totally reflectionless from all ports.

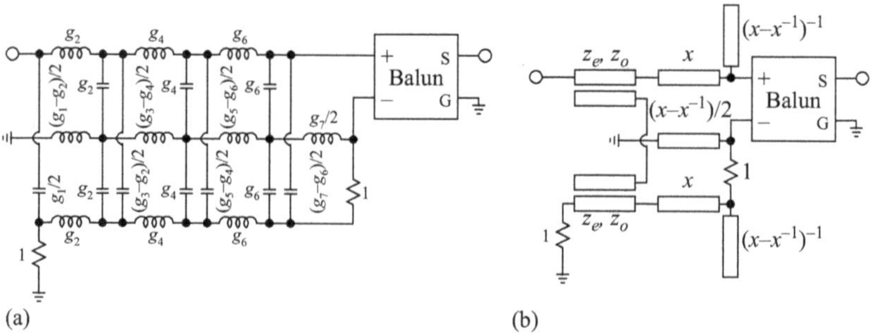

Figure 6.19 Chebyshev type I reflectionless filters using established topologies with a broadband balun: (a) low-pass filter using lumped-elements and (b) band-stop filter using transmission lines

6.10 Summary

The harmful effects that standing waves may have on an electronic system are well known—gain ripples, unexpected resonances, degraded stability and dynamic range, sensitivity to manufacturing tolerance and environmental factors, etc. The symptoms of standing waves are less direct when they occur out-of-band, but the negative impact they have on system performance is no less real or important. Countless designers have had the experience of selecting a part that was thoroughly characterized in a 50 Ω laboratory environment, only to be surprised when it falls short of the expected behavior in a real assembly embedded with other components. Problems that arise from stop-band mismatches are in some ways even more troublesome than those in-band because they are too often not fully appreciated until after the first prototype is built, by which time some critical design and packaging decisions have already been made. Reflectionless filters provide a comforting safeguard against these concerns.

Among the most critical problems that arises uniquely from out-of-band reflections is the proliferation of spurious mixing products. Mixers are notorious for producing a rich spectrum of intermodulation tones in response to CW inputs. In most applications, the vast majority of these is immediately filtered out. This can actually make things worse, however, as the spurs that are blocked by a conventional filter have nowhere to go but back into the mixer, where they can mix again and produce new products at different frequencies, some of them in-band. At best, the energy that was originally put into these spurs is eventually dissipated by the parasitic losses of the mixer, filter, or interconnects. At worst, it may radiate into the circuit housing, or find its way back into the signal path at some other point not protected by the filter. In any case, the mixer-filter interaction is bound to have an effect on the linearity and spurious response of the application [17].

If a reflectionless filter is used instead, these spurs are absorbed immediately upon creation, allowing them no chance to re-multiply inside the mixer, or build-up in a standing wave that radiates into the housing. Indeed the mixer now sees at its output port the same thing it saw when it was being characterized on the laboratory bench—a broadband 50 Ω impedance match.

Switching transients emitted by ADC's at their analog inputs can have similar disastrous effects if not properly terminated. The same anti-aliasing filter that was intended to mute the out-of-band spectrum incident upon the ADC may inadvertently reflect these transients back into it, directly impacting the spur-free dynamic range. A reflectionless filter here would ensure that this does not occur.

Multipliers and broadband amplifiers too are not immune to the reactive impedances presented to them at their ports outside of the intended operating band. The phase of the nearly unit-magnitude reflection coefficient presented by conventional filters in their stop-bands often varies quite rapidly with frequency. If coupled to an amplifier or multiplier at a critical harmonic or idler band, this can easily lead to an assembly whose dynamic range and/or efficiency is surprisingly sensitive to manufacturing tolerances, ambient temperature, or aging.

The critical lesson in this discussion is that the spectral components which are blocked by a conventional filter are not lost. Indeed, fundamentally an ordinary filter does nothing to remove these components, it merely redirects them back toward the source in hopes that they will somehow dissipate elsewhere before degrading performance. Experienced designers know that this is often not the case—more likely, the errant signals will end up creating side-effects that one at least may hope are less harmful to the system than those that would have occurred had the filter not been present. Rarely, however, are such side-effects fully anticipated in advance.

For applications where the risks associated with out-of-band mismatch is too great to accept, reflectionless filters provide an effective countermeasure.

References

[1] Zobel O. 'Distortion correction in electrical circuits with constant resistance recurrent networks'. *The Bell System Technical Journal*. 1928;7(3):438–534.

[2] Khalaj-Amirhosseini M. and Khalaj-Amirhosseini A. 'Reflectionless filters with arbitrary transfer functions'. *Journal of Telecommunications*. 2016;**34**(2):1–3.

[3] Morgan M. *Reflectionless Filters*. Norwood, MA: Artech House; 2017.

[4] Morgan M. *Reflectionless filters*. U.S. Patent No. 8,392,495, 5 March 2013. People's Republic of China Patent No. 201080014266.1, 30 July 2014.

[5] Morgan M. *Sub-network enhanced reflectionless filter topology*. U.S. Patent No. 9,705,467, 11 July 2017. Taiwan Patent No. I581494, 1 May 2017.

[6] Morgan M. *Optimal response reflectionless filters*. U.S. Patent Application No. 15/298,459, 20 October 2016.

[7] Khalaj-Amirhosseini M. and Taskhiri M. 'Two-fold reflectionless filters of inverse-Chebyshev response with arbitrary attenuation'. *IEEE Transactions on Microwave Theory and Techniques*. 2017;**65**(11):4616–4620.

[8] Darlington S. 'Synthesis of reactance 4-poles which produce prescribed insertion loss characteristics'. *Journal of Mathematical Physics*. 1939;**18**(4):257–355.

[9] 'Advantages of cascading reflectionless filters'. *Mini-Circuits Application Note*. 2015;AN-75-008, available at http://www.minicircuits.com/app/AN75-008.pdf

[10] 'Reflectionless filters improve linearity and dynamic range'. *Microwave Journal*. 2015;**58**(8):42–50.

[11] XHF-53H+,Mini-Circuits Datasheet, available at https://www.minicircuits.com/pdfs/XHF-53H+.pdf

[12] XLF-252H+,Mini-Circuits datasheet, available at https://www.minicircuits.com/pdfs/XLF-252H+.pdf

[13] Levy R. 'Characteristics and element values of equally terminated Achieser–Zolotarev quasi-low-pass filters'. *IEEE Transactions on Circuit Theory*. 1971;**18**(5):538–544.

[14] Morgan M. and T. Boyd. 'Theoretical and experimental study of a new class of reflectionless filters'. *IEEE Transactions on Microwave Theory and Techniques*. 2011;**59**(5):1214–1221.

[15] Morgan M. and T. Boyd. 'Reflectionless filter structures'. *IEEE Transactions on Microwave Theory and Techniques*. 2015;**63**(4):1263–1271.

[16] Morgan M. *Transmission-line reflectionless filters*. U.S. Patent Application No. 14/927,881, 30 October 2015.

[17] 'Pairing mixers with reflectionless filters to improve system performance'. *Mini-Circuits Application* Note. 2015; AN-75-007, available at http://www.minicircuits.com/AN75-007.pdf

Chapter 7

Absorptive planar bandstop filters

Yo-Shen Lin[1] and Jhih-Ying Shao[1]

7.1 Introduction

Bandstop filters (BSFs) are widely used for rejecting unwanted blocking and interference signals in RF/microwave systems. Most of the conventional BSF designs are reflective ones. This means that the unwanted signals are not consumed by the BSF but instead are reflected from the filter's input port. However, there are cases that the reflected power will leads to performance degradation of adjacent circuit components. A reflectionless or absorptive bandstop filter (ABSF) that can absorb the unwanted signals within the stopband is thus required. Several ABSF designs have been proposed in recent years [1–18]. In [1,2], the lumped-element bridged-T notch filter technique [3,4] is extended to form first-order absorptive microwave notch filters, in which the signal cancellation between multiple signal paths is utilized to achieve BSFs with an absorptive stopband as well as an infinite attenuation. Built upon the perfect notch concept in [1], a class of second-order networks which simultaneously exhibits an ideal bandstop resonance with infinite stopband attenuation and perfect impedance matching at all frequencies is proposed in [5]. In [6], a new class of reflectionless filter is proposed, and lumped-element filter prototypes with theoretically perfect input/output matching at all frequencies are demonstrated. They are then converted to transmission line-based ones in [7] so as to be better suited for high-frequency designs. To achieve more compact circuit size, a lumped-element realization of the ABSF design concept in [2] is presented in [8]. In addition, ABSFs with tunable stopband are demonstrated in [2,9–13], which can be used to protect sensitive receivers of reconfigurable RF front-ends. On the other hand, multi-harmonic absorption filters [14,15] are realized by the combination of a multi-band rejection filter and a terminated broadband band-pass filter, which is targeted for harmonic absorption of power amplifier so as to avoid the instability issue generated by the reflected harmonics.

In this chapter, a simple and very effective way of achieving ABSF designs is presented. This is based on the introduction of one lossy resonator to a conventional BSF [16–18]. The proposed design method is capable of achieving both narrowband

[1]Department of Electrical Engineering, National Central University, Taiwan

and wideband ABSF designs. In addition, ABSF of high filter order can be easily achieved to enhance the frequency selectivity.

7.2 Filter prototypes and design concept

Shown in Figures 7.1 and 7.2 are the two filter prototypes of proposed ABSF designs. They are based on introducing one lossy resonator to a conventional reflective BSF. As illustrated in Figure 7.1(a), a resistor R is introduced to the first resonator near Port 1. At the stopband center frequency f_0, the filter prototype can be simplified as Figure 7.1(b). If R is set as equal to the port reference impedance Z_0, the input power at Port 1 will then be totally absorbed by R. This leads to an absorptive stopband at f_0. Similarly, for the filter prototype based on series LC resonators in Figure 7.2(a), at the stopband center frequency f_0 the input power at Port 1 can also be absorbed if $R = Z_0$ and this leads to no reflection at Port 1. Although multiple lossy resonators can also be used, it will be shown later by practical design examples that a single lossy

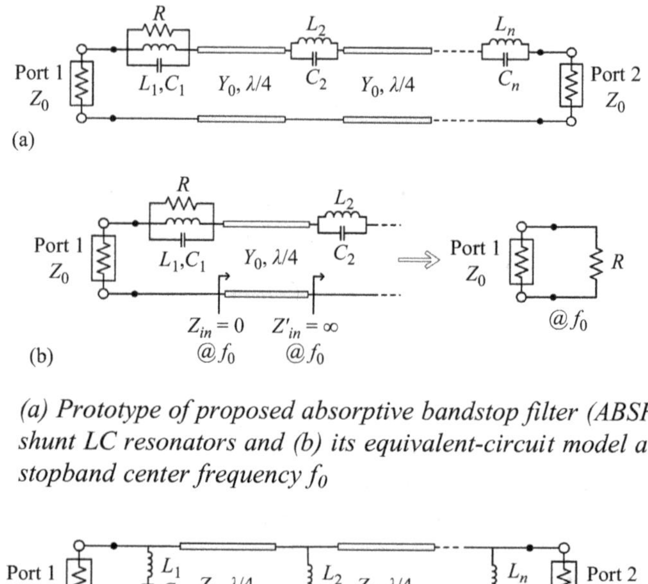

Figure 7.1 *(a) Prototype of proposed absorptive bandstop filter (ABSF) based on shunt LC resonators and (b) its equivalent-circuit model at the stopband center frequency f_0*

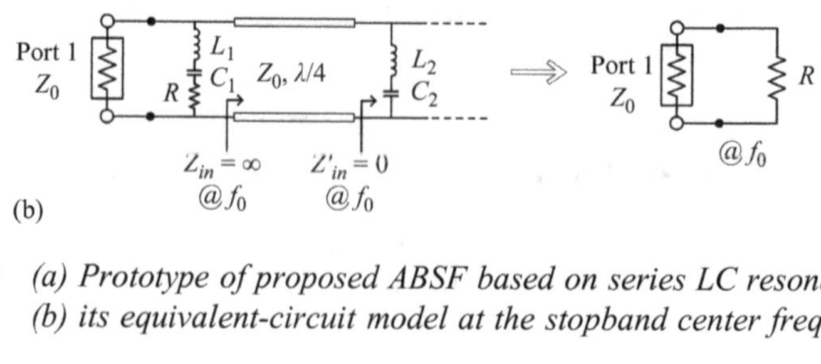

Figure 7.2 *(a) Prototype of proposed ABSF based on series LC resonators and (b) its equivalent-circuit model at the stopband center frequency f_0*

resonator at Port 1 is sufficient to achieve the required absorption while maintaining minimum degradation on the passband insertion loss and stopband selectivity.

7.3 Narrowband absorptive bandstop filters

7.3.1 Filter design

In this section, we start with a design example of narrowband ABSF [17] based on the design concept in Section 7.2. The filter circuit structure is shown in Figure 7.3, which is extended from a conventional narrowband BSF. To achieve compact circuit size and wide upper passband bandwidth, quarter-wavelength ($\lambda/4$) resonators are adopted and the coupled-line structure is used to achieve the coupling between the main signal path and the $\lambda/4$ resonators. The filter design starts from that of the conventional reflective BSF, i.e., $R = \infty$ in Figure 7.3, and the resistor R is then introduced to realize an absorptive stopband. Here a third-order ABSF is used as an example to demonstrate the design procedure, and the circuit model with $R = \infty$ is shown in Figure 7.4(a). Under TEM assumption and by adopting the equivalent transmission-line model of coupled-line [19], each coupled-line section in Figure 7.4(a) can be first equivalent to the combination of a series short stub of characteristic impedance Z_C' and a unit element of characteristic impedance Z_U' as shown in the middle of Figure 7.4(b). Since the electrical lengths of the series short stub and the unit element are equal, the Kuroda's identity [20] can then be applied and the coupled-line section is further equivalent to the combination of a unit element of characteristic impedance Z_U and a shunt open stub of characteristic impedance Z_C in Figure 7.4(b). In this way, the equivalent transmission-line model of this third-order reflective BSF can be obtained as in Figure 7.4(c). Next, by setting Z_U in Figure 7.4(b) equals to the port reference impedance Z_0, the leftmost $\lambda/4$ line section near Port 1 can be neglected and the resulted circuit structure becomes a typical third-order stub BSF with $\lambda/4$ open stubs. By considering the $\lambda/4$ open stubs as shunt resonators and the $\lambda/4$ connection lines as immittance inverters, it can be easily synthesized according to given specifications using the insertion loss method [19]. The associated design equations can then be obtained as listed below:

$$Z_{C1} = \frac{4Z_0}{\pi FBW g_0 g_1} \tag{7.1}$$

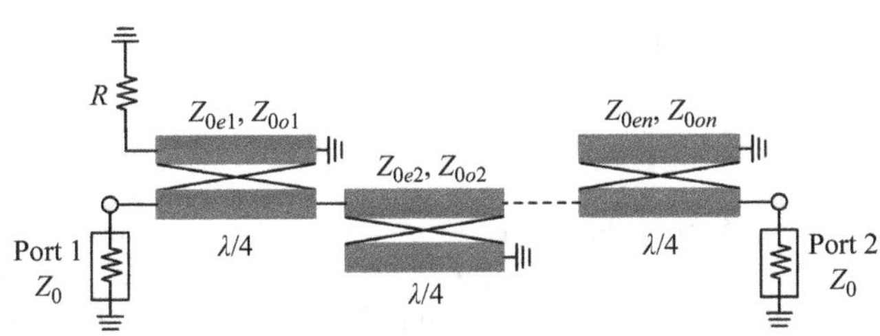

Figure 7.3 Circuit model of the proposed narrowband ABSF, which is achieved by introducing one lossy resonator to a conventional reflective bandstop filter

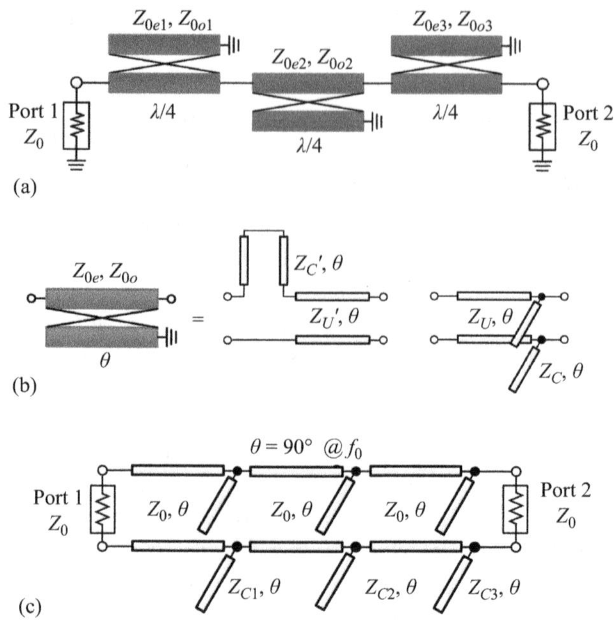

Figure 7.4 (a) Conventional third-order narrowband reflective BSF based on quarter-wavelength resonator; (b) development of equivalent transmission-line models for the coupled-line section; and (c) equivalent transmission-line model of the third-order narrowband reflective BSF in Figure 7.4(a)

$$Z_{C2} = \frac{4Z_0 g_0}{\pi FBW g_2} \tag{7.2}$$

$$Z_{C3} = \frac{4Z_0}{\pi FBW g_3 g_4} = Z_{C1} \tag{7.3}$$

$$Z_{0ei} = Z_0\left(1 + \frac{1}{\sqrt{1 + Z_{Ci}/Z_0}}\right) \tag{7.4}$$

$$Z_{0oi} = Z_0\left(1 - \frac{1}{\sqrt{1 + Z_{Ci}/Z_0}}\right) \tag{7.5}$$

where *FBW* is the stopband fractional bandwidth, and g_i's are the element values of the lowpass prototype.

After the design of the reflective BSF in Figure 7.4 is completed, a resistor *R* is attached to the open end of the first coupled-line section as shown in Figure 7.5(a) to realize a narrowband ABSF. Based on the transmission-line model in Figure 7.4(c), the input impedance Z_{in} looking into the second coupled-line section in Figure 7.5(a) is infinite at the stopband center frequency f_0. Therefore, the filter circuit model at f_0 can be further simplified as Figure 7.5(b). To let the resistor *R* absorb all the input

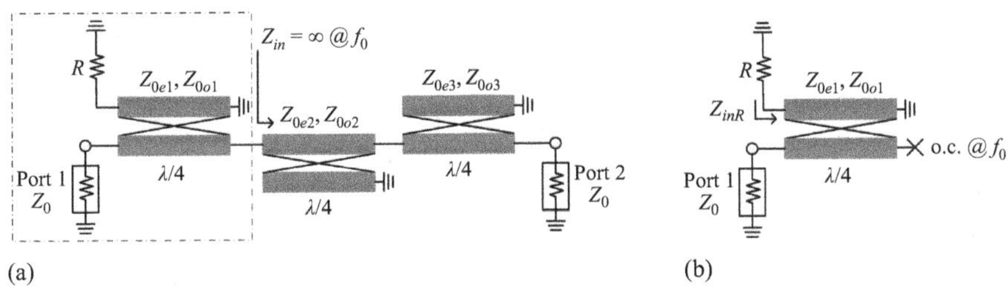

Figure 7.5 (a) Circuit model of a narrowband ABSF of third-order and (b) its equivalent-circuit model at the stopband center frequency f_0

power at Port 1, its value is set equal to the input impedance Z_{inR} looking into the open end of the first coupled-line section. As a result, the input power at f_0 will not propagate to Port 2 of the ABSF in Figure 7.5(a) due to the infinite Z_{in}, but it will be totally consumed by the resistor R. For a resistive termination Z_0 at Port 1 and under TEM assumption, the required resistance R can be obtained as

$$R = Z_{inR} = Z_0 \left(\frac{Z_{0e1} + Z_{0o1}}{Z_{0e1} - Z_{0o1}} \right)^2 \tag{7.6}$$

The design procedure of this narrowband ABSF can then be summarized as follows. First, the resistor is neglected (i.e., $R = \infty$) and the coupled-line BSF is designed as a reflective one using (7.1) to (7.5). Second, the resistor R is introduced and the required resistance is calculated by (7.6). To examine the achieved performance, a third-order narrowband ABSF is designed with a stopband center frequency $f_0 = 2$ GHz, a FBW = 5%, and a third-order Chebyshev response with 0.01 dB ripple. The reference impedances of Port 1 and 2 are set as $Z_0 = 50\ \Omega$. From (7.1) to (7.3), the Z_{Ci}'s can be first obtained as $Z_{C1} = Z_{C3} = 2{,}023.61\ \Omega$ and $Z_{C2} = 1{,}312.22\ \Omega$. The corresponding even- and odd-mode impedances of coupled-line can then be obtained using (7.4) and (7.5) as: $Z_{0e1} = Z_{0e3} = 57.76\ \Omega$, $Z_{0o1} = Z_{0o3} = 42.24\ \Omega$, $Z_{0e2} = 59.58\ \Omega$, and $Z_{0o2} = 40.42\ \Omega$. Finally, the resistance R is calculated using (7.6) to give $R = 2{,}073.61\ \Omega$.

Shown in Figure 7.6 are the circuit simulated results of this third-order narrowband ABSF, and those for the original reflective BSF (i.e., $R = \infty$) is also included for comparison. As observed in Figure 7.6(a), the narrowband ABSF exhibits a bandstop response centered at $f_0 = 2$ GHz. The insertion loss response is very similar to that of the reflective BSF except for the degradation at the two stopband edges, which somewhat reduces the filter selectivity. On the other hand, the input reflection coefficient $|S_{11}|$ is equal to zero or $-\infty$ dB at f_0 while that for the reflective BSF is equal to one or 0 dB. As a result, in Figure 7.6(b) the power dissipated by the ABSF is 100% at f_0, while the power dissipation is zero at all frequencies for the reflective BSF. This verifies that the narrowband ABSF in Figure 7.5(a) can successfully absorb the input power at f_0 using the additional resistor R. As the frequency moves away from f_0, $|S_{11}|$ increases until a maximal value is reached. This is because Z_{inR} in Figure 7.5(b) is purely resistive and Z_{in} in Figure 7.5(a) is infinite only at f_0. From the simulated frequency response of Z_{inR} in Figure 7.7, it can be observed that $|Z_{inR}|$ becomes larger

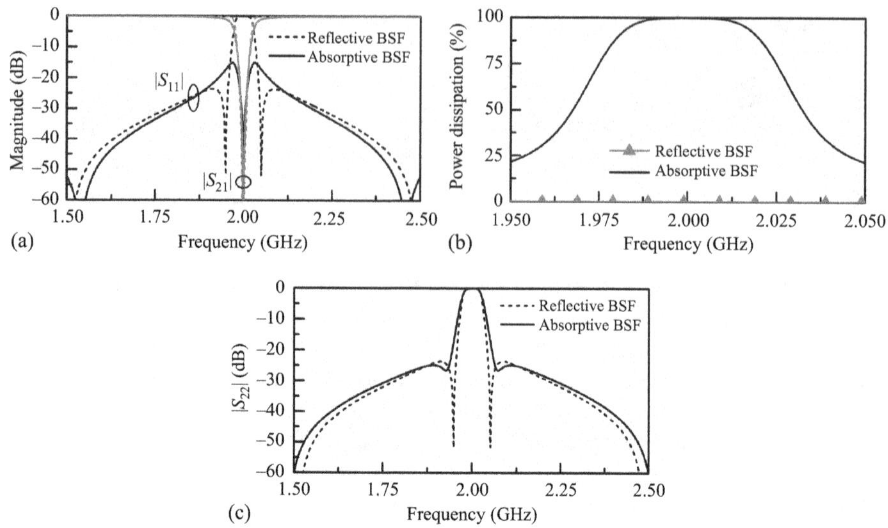

Figure 7.6 *(a) Circuit simulated frequency responses of the third-order narrowband ABSF and the conventional reflective BSF with the input at Port 1; (b) power dissipated in the filter with the input at Port 1, which is calculated by $1 - |S_{11}|^2 - |S_{21}|^2$; and (c) output reflection coefficient at Port 2*

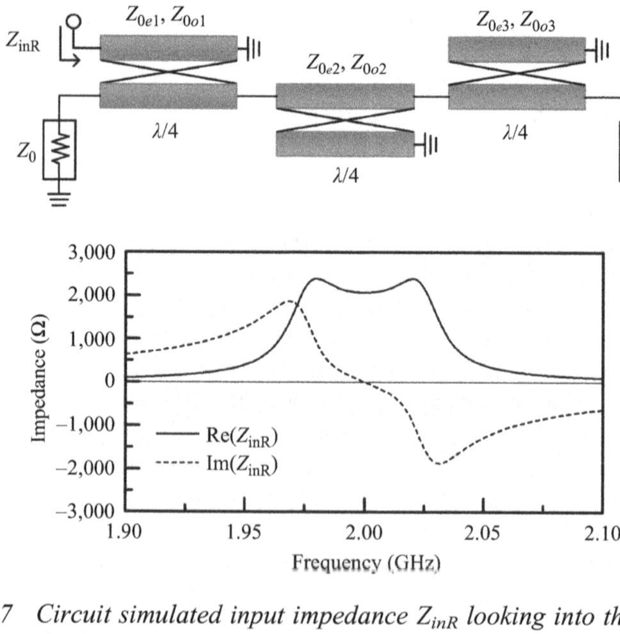

Figure 7.7 *Circuit simulated input impedance Z_{inR} looking into the open end of the first coupled-line section of the narrowband ABSF with $f_0 = 2$ GHz in Figure 7.5(a)*

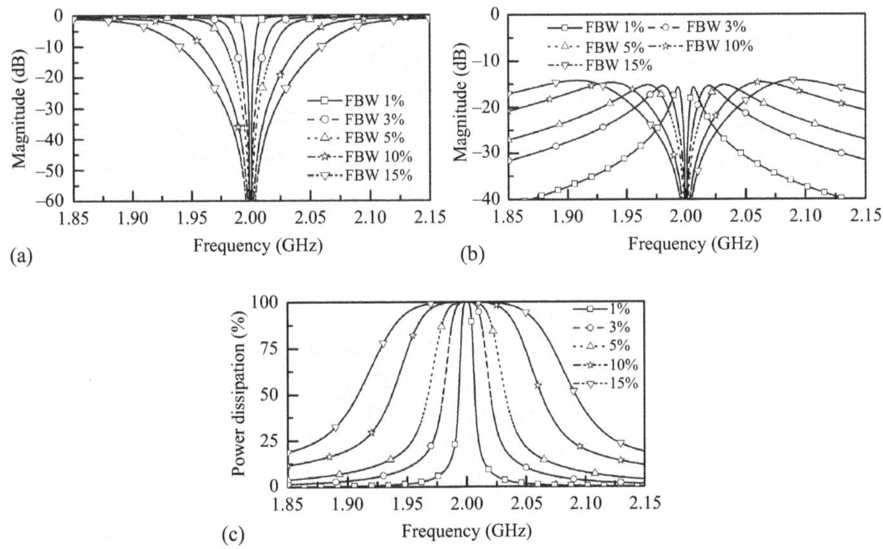

Figure 7.8 Effect of the fractional bandwidth (FBW) on the bandstop performance of the third-order narrowband ABSF designed with 0.01-dB ripple Chebyshev responses. Circuit simulated (a) $|S_{21}|$; (b) $|S_{11}|$ frequency responses; and (c) power dissipated in the filter with the input at Port 1

than its value at f_0, i.e., 2,073.61 Ω at the two stopband edges. As a result, the fixed resistor $R = 2{,}073.61$ Ω in Figure 7.5(a) can no longer achieve good impedance matching there, which give rises to the larger $|S_{11}|$ found at the stopband edges. As for the output reflection coefficient $|S_{22}|$ shown in Figure 7.6(c), the narrowband ABSF exhibits similar response to that of the conventional BSF because the lossy resonator is not used at the output port.

Although the design equation (7.6) for the resistor R is derived from the impedance matching condition at the stopband center frequency only, the resulted ABSF can handle a ripple bandwidth up to about 15% for Chebyshev bandstop responses [17]. Figure 7.8 compares the frequency response of several third-order ABSF designs with FBW ranging from 1% to 15%. They are all designed with a Chebyshev response of 0.01 dB ripple. The circuit parameters are obtained using (7.1) to (7.6) and they are listed in Table 7.1. All these designs feature $S_{11} = 0$ and 100% power dissipation at f_0 as expected. However, the maximal input reflection coefficient gradually increases as FBW increases as shown in Figure 7.8(b). Specifically, the maximal reflection coefficient increases from −15.73 dB for FBW = 1% to −14.21 dB for FBW = 15%. This somewhat limits the maximal achievable FBW of proposed narrowband ABSF. Anyway, the higher reflection coefficient for ABSFs with larger FBW can be easily tackled by trimming the resistor R, which will be discussed later.

Figure 7.9 compares the circuit simulated results of proposed third-order ABSF with the same FBW of 5% but different passband ripples. The corresponding circuit

Table 7.1 Circuit parameters of the narrowband ABSF designs in Figure 7.8 with different FBWs

FBW	***R*** **(Ω)**	Z_{0e1}, Z_{0o1} **(Ω)**	Z_{0e2}, Z_{0o2} **(Ω)**
1%	10,168.04	53.51, 46.49	54.34, 45.65
3%	3,422.68	56.04, 43.96	57.48, 42.52
5%	2,073.61	57.76, 42.24	59.58, 40.42
10%	1,061.80	60.85, 39.15	63.31, 36.69
15%	724.54	63.13, 36.87	66.01, 33.99

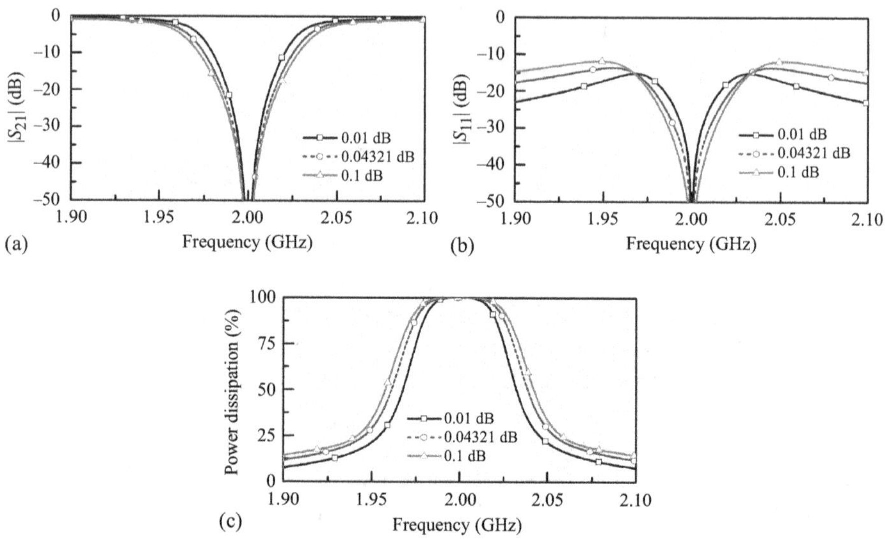

Figure 7.9 Effect of the passband ripple level on the bandstop performance of the third-order narrowband ABSF of 5% FBW. Circuit simulated (a) $|S_{21}|$ and (b) $|S_{11}|$ frequency responses; and (c) power dissipated in the filter with the input at Port 1

Table 7.2 Circuit parameters of the narrowband ABSF designs in Figure 7.9 with different passband ripples

Impedance	**0.01-dB ripple**	**0.04321-dB ripple**	**0.1-dB ripple**
R (Ω)	2,073.61	1,545.14	1,284.26
Z_{0e1}, Z_{0o1} (Ω)	57.76, 42.24	58.99, 41.01	59.87, 40.13
Z_{0e2}, Z_{0o2} (Ω)	59.58, 40.42	60.19, 38.81	60.38, 39.62
Z_{0e3}, Z_{0o3} (Ω)	57.76, 42.24	58.99, 41.01	59.87, 40.13

parameters are listed in Table 7.2. Again, all three designs feature $S_{11} = 0$ and 100% power dissipation at f_0. As the ripple increases from 0.01 dB to 0.1 dB, better stopband selectivity is achieved at the expense of larger maximal reflection coefficient. Anyway, the design with 0.1-dB ripple that has a maximal input reflection

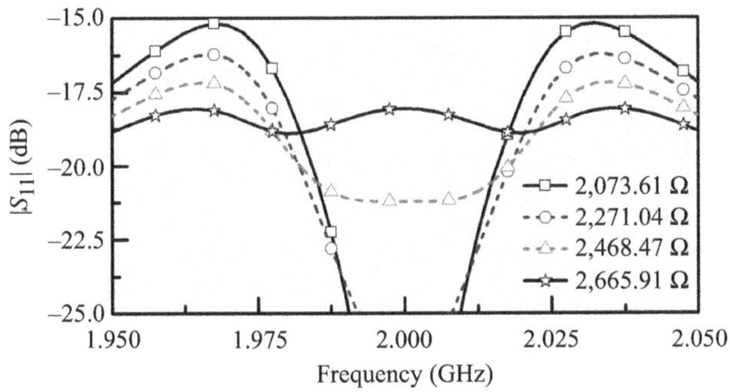

Figure 7.10 Effect of the resistance of R on the $|S_{11}|$ frequency response of the third-order narrowband ABSF

coefficient of −11.86 dB still exhibits high in-band power dissipation as shown in Figure 7.9(c). In addition, as will be shown next the higher input reflection coefficient can be improved by trimming the resistor R.

Shown in Figure 7.10 is a sensitivity analysis of the resistor R on the $|S_{11}|$ response of the third-order ABSF design in Figure 7.6. As R increases from the designed value of 2,073.61 Ω, it is found that the $|S_{11}|$ at f_0 increases due to impedance mismatch. However, both the maximal reflection coefficients above and below f_0 decrease. This is because the input impedance Z_{inR} at the stopband edges is higher than its value at f_0 according to Figure 7.7. Therefore, a better impedance matching at the stopband edges can be achieved by a larger resistance. In addition, when R is set equal to 2,655.91 Ω, the maximal reflection coefficient reduces to a level that is about equal to the reflection coefficient at f_0, which becomes a nearly equal-ripple $|S_{11}|$ response. This implies that the $|S_{11}|$ at the stopband edges can be simply improved by increasing the value of R at the expense of some increase in $|S_{11}|$ at f_0. On the other hand, the S_{21} response does not change much with the varying resistance. This can be examined in detail by referring to the comparison between the two extreme cases with $R = 2{,}073.61$ and 2,665.91 Ω in Figure 7.11. By increasing R from 2,073.61 to 2,665.91 Ω, the maximal reflection coefficient reduced from −15.18 to −18.06 dB in Figure 7.11(b), while the difference in $|S_{21}|$ between the two cases shown in Figure 7.11(a) is negligible. According to Figure 7.11(c), the power dissipation reduced from 100 to 98.44% after R increases from 2,073.61 to 2,665.91 Ω. Anyway, such a 1.56% difference is usually acceptable in practical applications. Therefore, for the ABSF designs with a larger FBW in Figure 7.8 and the ones with a larger passband ripple in Figure 7.9, one may increase the value of R to reduce the maximal reflection coefficient at the expense of little degradation in power dissipation at f_0.

To achieve an even better frequency selectivity, higher-order narrowband ABSF can also be developed. Based on the equivalent circuit model development procedure illustrated in Figure 7.4, the equivalent transmission-line model of a nth order reflective BSF is first obtained as shown in Figure 7.12. By setting $Z_{Ui}|_{i=2 \text{ to } n} = Z_0$, the associated design equations can be obtained in the same way as those for the third-order one, and the results are given in (7.7) to (7.12). The nth order reflective BSF can then be synthesized according to the desired specifications. For odd-order

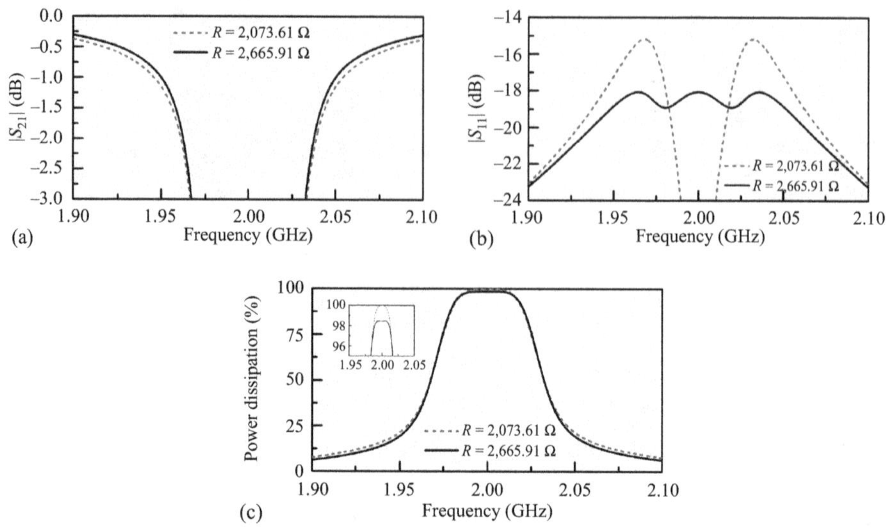

Figure 7.11 Effect of the resistance R on the bandstop performance of the third-order narrowband ABSF for R = 2,073.61 and 2,665.91 Ω. Circuit simulated (a) $|S_{21}|$ and (b) $|S_{11}|$ frequency responses and (c) power dissipated in the filter with the input at Port 1

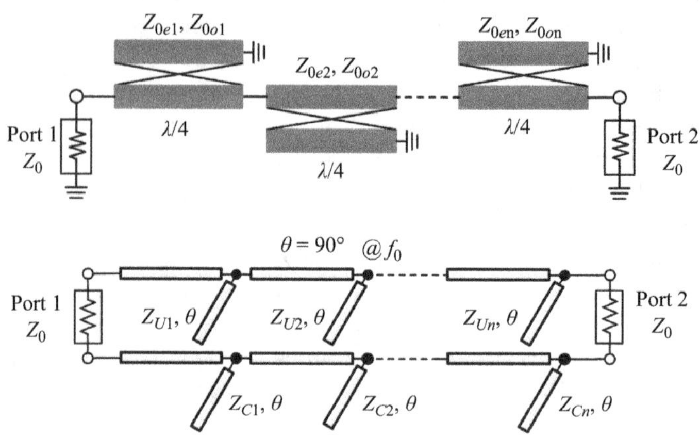

Figure 7.12 Circuit model and equivalent transmission-line model of an nth-order reflective BSF

Chebyshev responses, it can be found that $Z_{U1} = Z_0 = Z_{Ui}|_{i=2 \text{ to } n}$ and $Z_{Ci} = Z_{C(n-i+1)}|_{i=1 \text{ to } (n-1)/2}$, and the resulted filter structure will be symmetrical. For even-order Chebyshev responses, due to $g_{n+1} \neq 1$ and the asymmetry in element values g_i's, the resulted filter structure will not be symmetrical. Therefore, the odd-order Chebyshev response is preferred for circuit implementation. Once the design of the *n*th-order reflective BSF is complete, the resistor *R* is introduced to the open

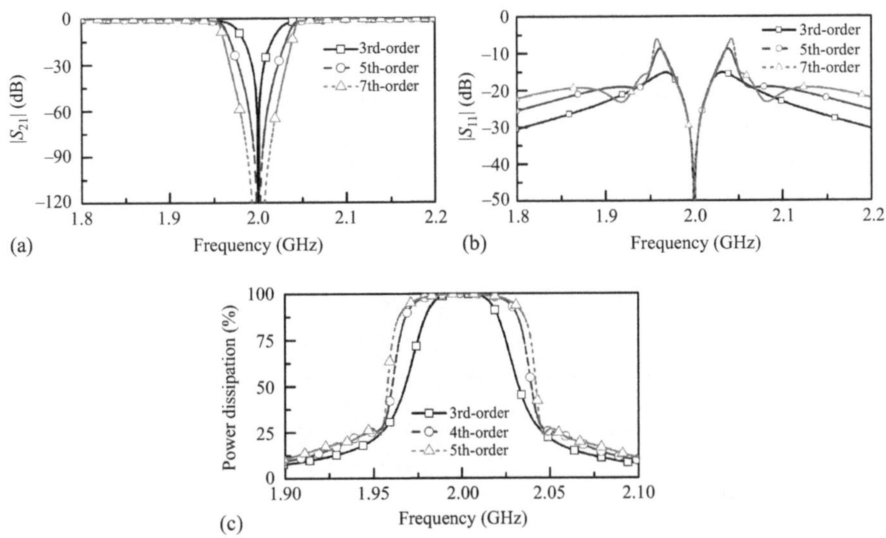

Figure 7.13 Effect of filter order on the bandstop performance of the narrowband ABSF. Circuit simulated (a) $|S_{21}|$ and (b) $|S_{11}|$ frequency responses; and (c) power dissipated in the filter with the input at Port 1

end of the first coupled-line section near Port 1 to achieve a nth-order ABSF as illustrated in Figure 7.3. Again, the resistance R can be determined using (7.6).

$$Z_{Cn} = \frac{4Z_0}{\pi FBW g_n g_{n+1}} \tag{7.7}$$

$$Z_{Ci} = \left.\frac{16Z_0^2}{(\pi FBW)^2 Z_{C(i+1)} g_i g_{i+1}}\right|_{i=n-1 \text{ to } 1} \tag{7.8}$$

$$\begin{cases} Z_{U1} = Z_0\sqrt{g_0 g_{n+1}} & \text{for even } n \\ Z_{U1} = Z_0\sqrt{g_0 / g_{n+1}} & \text{for odd } n \end{cases} \tag{7.9}$$

$$Z_{Ui} = Z_0\big|_{i=2 \text{ to } n-1} \tag{7.10}$$

$$Z_{0ei} = Z_{Ui}\left(1 + \frac{1}{\sqrt{1 + Z_{Ci}/Z_0}}\right) \tag{7.11}$$

$$Z_{0oi} = Z_{Ui}\left(1 - \frac{1}{\sqrt{1 + Z_{Ci}/Z_0}}\right) \tag{7.12}$$

Figure 7.13 compares the bandstop frequency response of three narrowband ABSFs of third-, fifth-, and seventh-order. They are all designed with a stopband center frequency $f_0 = 2$ GHz, a FBW = 5%, a $Z_0 = 50\ \Omega$, and a 0.01-dB Chebyshev response. The corresponding circuit parameters can be obtained as in Table 7.3. As the filter order increases, the stopband selectivity gets better as shown in

Table 7.3 Circuit parameters of the narrowband ABSF designs in Figure 7.13 with different filter orders

Impedance	Third-order	Fifth-order	Seventh-order
R (Ω)	2,073.61	1,733.41	1,647.63
Z_{0e1}, Z_{0o1} (Ω)	57.76, 42.24	58.49, 41.51	58.71, 41.29
Z_{0e2}, Z_{0o2} (Ω)	59.58, 40.42	61.04, 38.96	61.38, 38.62
Z_{0e3}, Z_{0o3} (Ω)	57.76, 42.24	62.08, 37.92	62.67, 37.33
Z_{0e4}, Z_{0o4} (Ω)	–	61.04, 38.96	62.27, 37.73
Z_{0e5}, Z_{0o5} (Ω)	–	58.49, 41.51	62.67, 37.33
Z_{0e6}, Z_{0o6} (Ω)	–	–	61.38, 38.62
Z_{0e7}, Z_{0o7} (Ω)	–	–	58.71, 41.29

Figure 7.13(a). On the other hand, the maximal reflection coefficient also increases with the filter order. They are −15.18, −8.74, and −6.12 dB for the third-, fifth-, and seventh-order designs, respectively. Anyway, according to Figure 7.13(c), the higher-order designs still feature better power dissipation as compared with the lower-order ones due to the lower transmission coefficient within the stopband. As a result, the bandwidth for better than 90% power dissipation also increases with filter order, which are 1.9, 3.1, and 3.4% for the third-, fifth-, and seventh-order designs, respectively. In addition, in case that a smaller maximal reflection coefficient is desired, one can increase the resistance R as demonstrated in Figure 7.10.

7.3.2 Filter implementation

According to the filter synthesis method in Section 7.3.1, narrowband BSFs with an absorptive stopband can be easily designed for a given specification. As an example, a third-order ABSF is implemented in microstrip using a RO4350B substrate ($\varepsilon_r = 3.48$, $\tan\delta = 0.004$, and thickness $h = 1.524$ mm). The specifications are: $f_0 = 2$ GHz, FBW = 5%, $Z_0 = 50\ \Omega$, and 0.01-dB Chebyshev response. The required Z_{0ei}, Z_{0oi}, and R have been given in the first column of Table 7.3. By using computer-aided design tools, e.g. ADS LineCalc, the corresponding line width, gap width, and line length of each microstrip coupled-line section can be easily obtained and they are given in Figure 7.14(a). The circuit simulated frequency response of the microstrip ABSF is shown in Figure 7.14(b), in which the conductor and dielectric losses are neglected first. Although $S_{21} = 0$ at $f_0 = 2$ GHz is obtained as expected, the frequency at which $|S_{11}|$ reaches its minimum shifts to a lower frequency (i.e., 1.995 GHz) as compared with that of the ideal ABSF response given in the same figure. In addition, the $|S_{11}|$ response is no longer symmetrical around f_0.

The deviation between ideal and microstrip ABSFs in Figure 7.14 is due to the non-equal even- and odd-mode effective dielectric constants of the inhomogeneous microstrip coupled-line. As illustrated in Figure 7.15(a), the input reflection coefficient Γ_{in} for the ideal coupled-line reaches its maximum and becomes equal to 1 at the stopband center frequency f_0 of 2 GHz. On the other hand, for the microstrip coupled-line, Γ_{in} reaches its maximum at f_0 as well but the maximum now becomes $\Gamma_{max} = 1\angle\theta_\Gamma = 1\angle{-46.5^\circ}$. By employing the ABCD parameters of the coupled-line

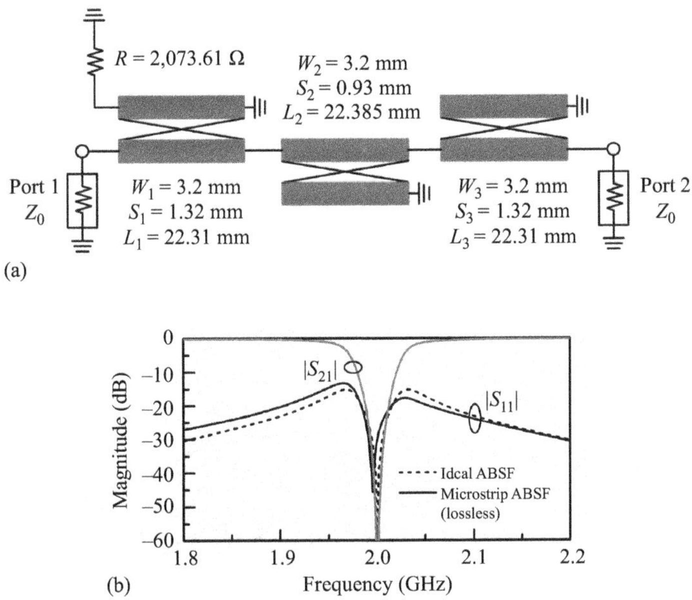

Figure 7.14 (a) Microstrip circuit model and (b) circuit-simulated frequency responses of a microstrip narrowband ABSF of third-order

network in an inhomogeneous medium [21], it is found that the value of θ_Γ depends on the ratio between even- and odd-mode effective dielectric constants of the microstrip coupled-line. The nonzero θ_Γ makes the assumption that the input impedance of the second coupled-line section is infinite at f_0 (i.e., $Z_{in} = \infty$ at f_0 in Figure 7.5) no longer valid, and the nonzero reactive part of Z_{in} leads to the shift in frequency response observed in Figure 7.14(b).

To solve this issue, the modified microstrip coupled-line structure in Figure 7.15(b) can be adopted. It is achieved by decoupling part of the microstrip coupled-line section near the short end. Specifically, the length L_2 of the microstrip coupled-line is reduced and a microstrip short stub of length L_{S2} is introduced to shift Γ_{max} back to infinity while maintaining the original length of the $\lambda/4$ resonator, i.e. $L_2 + L_{S2} \approx \lambda/4$ at f_0 and thus $S_{21} = 0$ at f_0. As for the line width W_{S2} of the microstrip short stub, it is set equal to the line width W_2 of the coupled-line to keep the $\lambda/4$ resonator a uniform one. After that, a microstrip line of length L_{t2} at Port 2 is added so as to achieve 90° phase change from Port 1 to 2 that is required for realizing the immittance inverter, i.e. $L_2 + L_{t2} \approx \lambda/4$ at f_0 and thus $\angle S_{21} = 90^\circ$ at f_0. For the line width W_{t2} of this additional microstrip line, it is also set equal to the coupled-line width W_2 to avoid additional discontinuity effects. In addition, due to the reduction of coupled-line length L_2, the gap width S_2 is reduced so as to maintain the same amount of coupling between the main line and the $\lambda/4$ resonator. As shown in Figure 7.15(b), Γ_{in} of the modified microstrip coupled-line structure is now almost coincided with that of the ideal coupled-line and $\Gamma_{max} = 1$ at $f_0 =$ 2 GHz. Although fine-tuning of the lengths L_2, L_{S_2}, L_{t2} and the gap width S_2 are

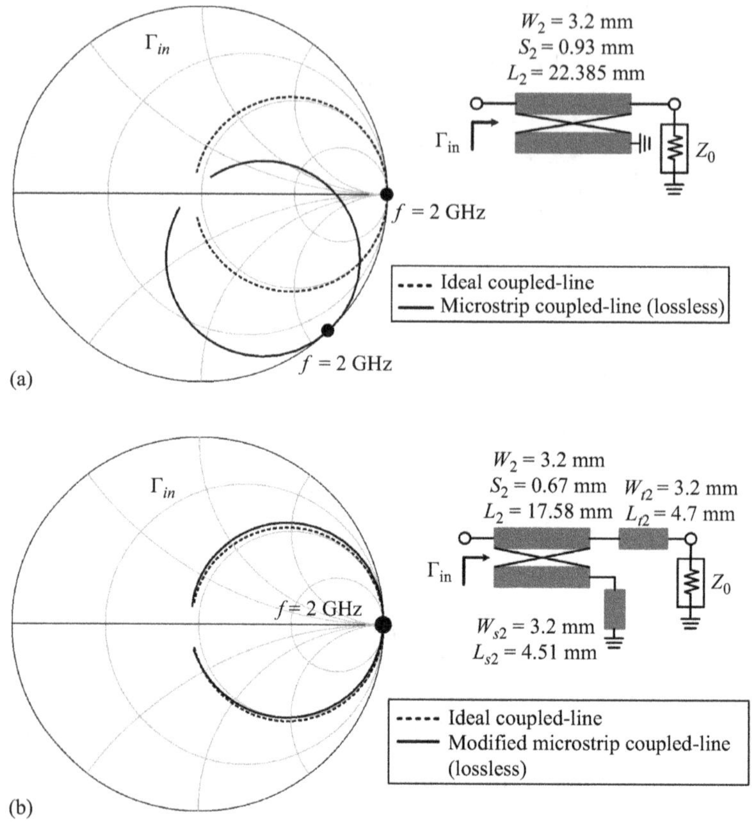

Figure 7.15 (a) Comparison between the input reflection coefficients of the second coupled-line section in Figure 7.14(a) using ideal and microstrip coupled-line models and (b) comparison between the input reflection coefficients of the ideal and the modified microstrip coupled-lines

required, they can be done very efficiently in a circuit simulator according to the design guidelines given above.

By incorporating the modified microstrip coupled-line structure in Figure 7.15(b), the microstrip circuit structure of the third-order narrowband ABSF becomes the one shown in Figure 7.16(a). Its circuit simulated frequency response with no conductor and dielectric losses is shown in Figure 7.16(b), and it is compared with that of the ideal ABSF based on ideal coupled-lines. The frequency of minimum $|S_{11}|$ for the modified microstrip ABSF now shifted back to the desired value of 2 GHz, and good agreement with the frequency response of the ideal ABSF is observed. The circuit simulated results of the modified microstrip ABSF with the presence of conductor and dielectric losses is shown in Figure 7.16(c). Some deviation from the ideal ABSF response is observed and the stopband rejection at $f_0 = 2$ GHz reduces to 36.96 dB due

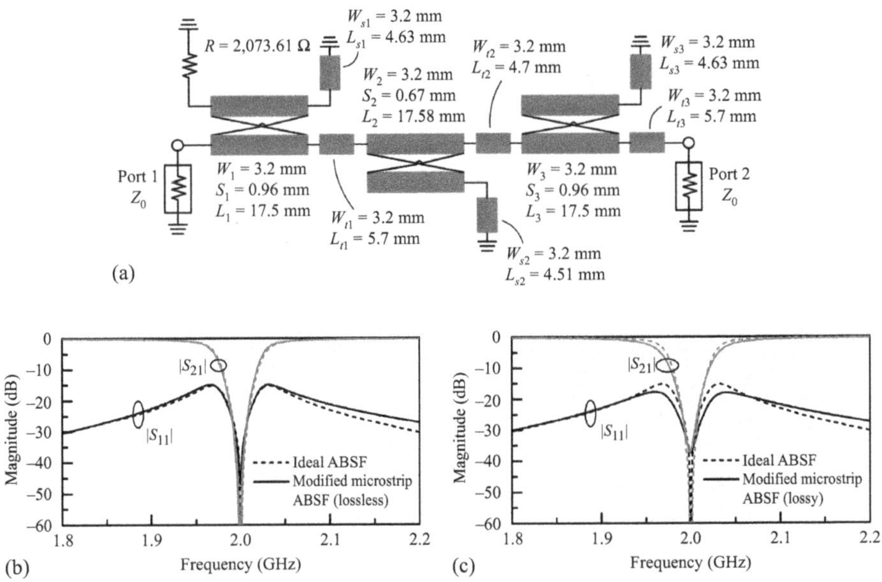

Figure 7.16 (a) Microstrip circuit model of the third-order narrowband ABSF employing the modified microstrip coupled-line structure; (b) circuit-simulated frequency responses neglecting conductor and dielectric losses; and (c) circuit-simulated frequency responses with the presence of conductor and dielectric losses

to the finite quality factor. Anyway, a nearly symmetrical $|S_{11}|$ response about f_0 is maintained and the input reflection coefficient at 2 GHz is only −37.36 dB. This implies that the 99.96% of the input power at f_0 is absorbed, which is only a 0.04% drop compared with that of the ideal ABSF.

To validate the above analysis and the effectiveness of the design modification, the layout of the modified microstrip ABSF in Figure 7.16(a) is generated as shown in Figure 7.17(a) and its frequency response is examined using a full-wave EM simulator, e.g. Ansys HFSS. Based on the initial values listed in Figure 7.16(a), the geometrical parameters are fine-tuned in the full-wave EM simulator to take the discontinuity effects of the vias and the microstrip bends as well as the high-frequency parasitic and proximity effects into account. Shown in Figure 7.17(b) are the circuit and EM simulated results. Good agreement between them is observed. The EM simulated stopband rejection at f_0 further reduces to 31.59 dB, which is mainly due to the radiation loss that is not considered in the circuit simulation.

The microstrip ABSF design is then realized and the photograph of the fabricated microstrip ABSF is shown in Figure 7.18. The measured and EM simulated results are shown in Figure 7.19. Since the required resistor of 2,073.61 Ω is not available in practice, a 2,200 Ω chip resistor of size 0402 is used in the fabricated filter instead. According to the analysis in Figure 7.10, this would increase the reflection coefficient at the stopband center frequency. As a result, the measured

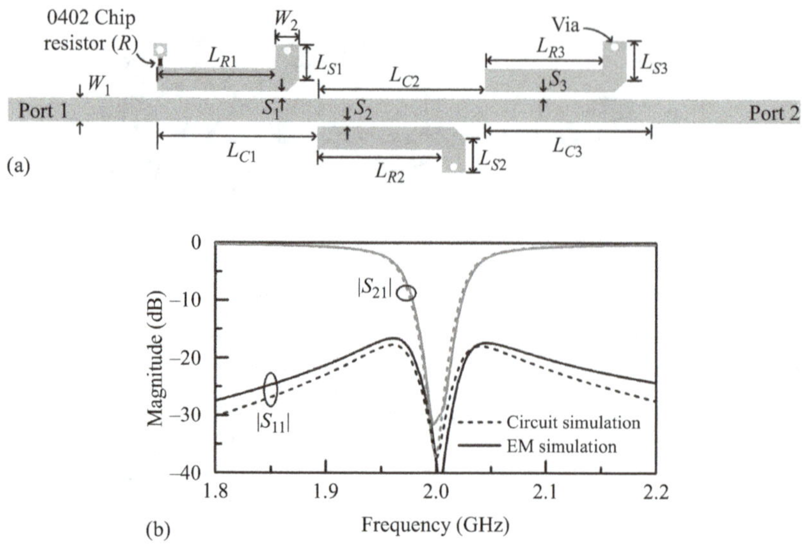

Figure 7.17 *(a) Layout of the modified microstrip narrowband ABSF of third order. Geometrical parameters: $W_1 = 3.3$, $W_2 = 3.2$, $S_1 = 1.0$, $S_2 = 0.73$, $S_3 = 1.0$, $L_{C1} = 21.6$, $L_{C2} = 22.5$, $L_{C3} = 22.5$, $L_{R1} = 15.9$, $L_{R2} = 16.7$, $L_{R3} = 15.9$, $L_{S1} = 4.91$, $L_{S2} = 4.8$, and $L_{S3} = 5.55$. Units in mm. $R = 2.2$ kΩ; and (b) circuit and EM simulated results. Reprinted with permission from Reference [17]; © 2015 IEEE*

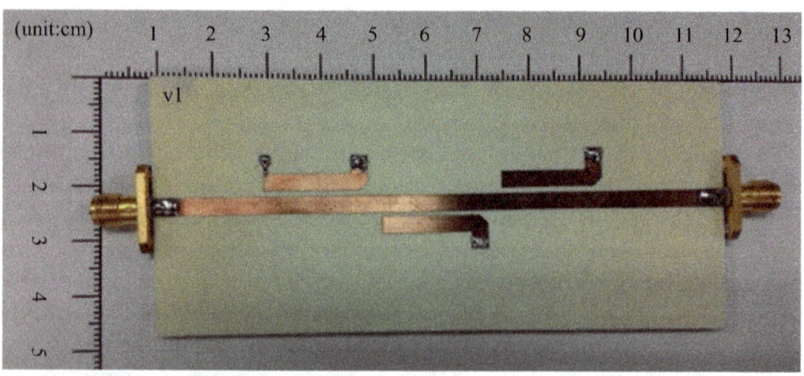

Figure 7.18 *Circuit photograph of the modified microstrip narrowband ABSF of third order. Reprinted with permission from Reference [17]; © 2015 IEEE*

reflection coefficient at 2 GHz increases by 9.49 dB to −28.66 dB as compared with the EM simulated one. Still, low input reflection coefficient around f_0 is observed, which validates the nonreflective property of this filter within the stopband. For the measured $|S_{21}|$ response, some shift in the stopband center frequency

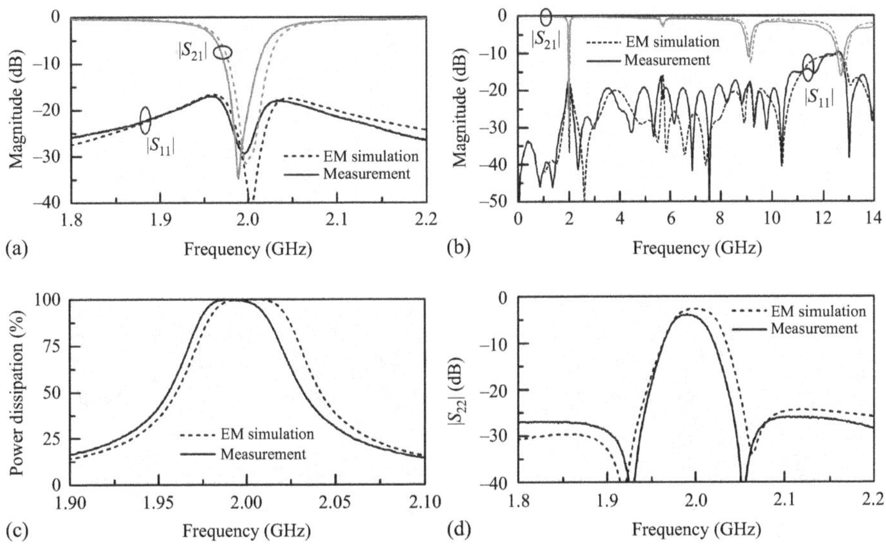

Figure 7.19 Measured and EM simulated results of the modified microstrip narrowband ABSF of third order: (a) narrowband and (b) wideband $|S_{11}|$ *and* $|S_{21}|$ *frequency responses and (c) power dissipated in the filter with the input at Port 1. (d)* $|S_{22}|$ *frequency responses*

is observed and the maximal stopband rejection is 34.83 dB at 1.989 GHz. This is mainly caused by the fabrication error of the filter structure. Anyway, the overall agreement between measured and EM simulated results is good. The measured 20-dB rejection bandwidth is from 1.982 to 2 GHz, which corresponds to a 0.9% fractional bandwidth. Good input matching at Port 1 is achieved over a wide frequency range and the measured input return loss is better than 16 dB from dc up to 10.72 GHz. This is because the resistor R can also absorb the input power at harmonic resonances. As shown in Figure 7.19(c), the measured in-band power dissipation with the input at Port 1 has a maximum of 99.8% at 1.989 GHz. It is larger than 90% from 1.974 to 2.012 GHz, and the corresponding fractional bandwidth is 1.9%. In addition to the high-level of stopband power dissipation, the microstrip ABSF also exhibits low passband insertion loss with good passband impedance matching. According to Figure 7.19(b), the measured $|S_{21}|$ is within 1 dB from dc to 1.913 GHz and from 2.075 to 5.567 GHz. The measured and EM simulated output reflection coefficients of the modified microstrip ABSF are shown in Figure 7.19(d). The input power at Port 2 is mostly reflected within the stopband as expected for the resistor R is only introduced to the first resonator near Port 1.

According to the analysis in Figure 7.10, the maximal input reflection coefficient can be further improved by increasing the resistance R. Shown in Figure 7.20 are the measured results for the same microstrip ABSF but with a 0402 chip resistor R of 3,010 Ω instead. This resistor is chosen to achieve a nearly equal-ripple $|S_{11}|$ response. As observed in Figure 7.20(a), the $|S_{21}|$ response is almost identical to that

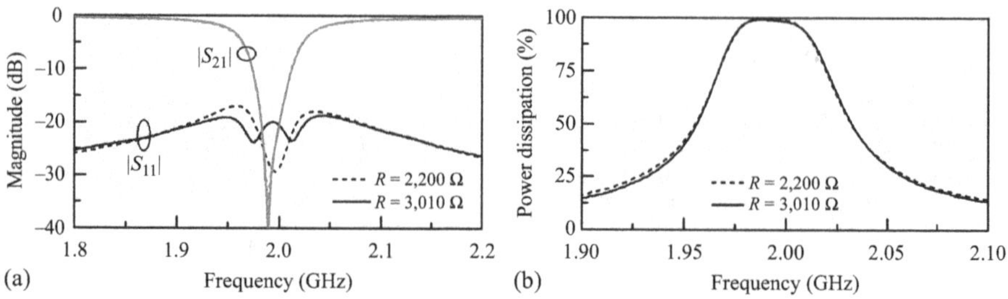

Figure 7.20 Measured and EM simulated results of the modified microstrip narrowband ABSF of third order with the resistor R increased to reduce the input reflection coefficient: (a) narrowband frequency responses and (b) power dissipated in the filter with the input at Port 1

of the $R = 2{,}200\ \Omega$ case, but the maximal input reflection coefficient improved from 16.95 to 18.71 dB. Regarding the input power dissipation shown in Figure 7.20(b), the maximal power dissipation reduces to 99.1% at 1.987 GHz but the bandwidth for >90% power dissipation is also from 1.974 to 2.012 GHz, which corresponds to a 1.9% fractional bandwidth.

7.3.3 Filter implementation with half-wavelength resonators

Based on the design concept illustrated in Figure 7.1 and 7.2, the implementation of a narrowband ABSF is not limited to the ones with $\lambda/4$ resonators as in Figure 7.3. Shown in Figure 7.21(a) is a narrowband ABSF design based on $\lambda/2$ resonators, which is obtained by realizing each series *LC* resonator in Figure 7.2(a) with an L-shape $\lambda/2$ resonator that is coupled to the main signal path through a $\lambda/4$ coupled-line. The shunt resistor *R* at the open end of the first $\lambda/2$ resonator is to absorb the input signal at Port 1 within the stopband. As shown in Figure 7.21(a), since the input impedance Z_{in} looking into the second coupled-line section is infinite at the stopband center frequency f_0, the filter's equivalent-circuit model at f_0 can be simplified as that in Figure 7.21(b). By setting $Z_{inL} = Z_{inR}$ in Figure 7.21(b), the required *R* to absorb all the input power at f_0 can be obtained as

$$R = \frac{4Z_0^3}{\left(Z_{0e1} - Z_{0o1}\right)^2} \tag{7.13}$$

The design procedure of the narrowband ABSF with $\lambda/2$ resonators can then be demonstrated by the following example. Here, a microstrip ABSF is realized with a stopband center frequency $f_0 = 2$ GHz, a ripple bandwidth of 5%, and a fifth-order Chebyshev response with 0.01-dB ripple. The RO4350B ($\varepsilon_r = 3.48$, $\tan\delta = 0.004$, and thickness $h = 1.524$ mm) microwave laminate is employed for circuit implementation. Similarly, the design starts from that of a reflective BSF (i.e., $R = \infty$ in Figure 7.21) with the same specifications, which can be done following the design procedure given in [20]. Specifically, by setting the line impedance equal to the

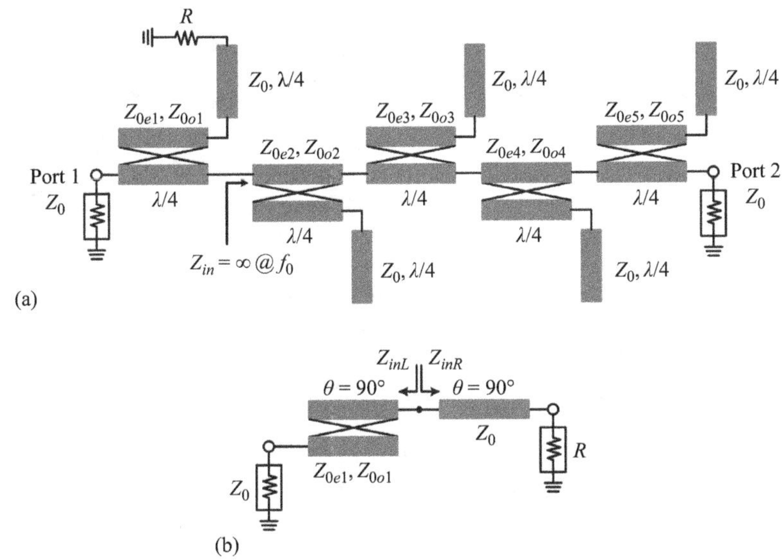

Figure 7.21 (a) Circuit model of proposed narrowband ABSF based on half-wavelength resonators and (b) its equivalent-circuit model at the stopband center frequency f_0

port reference impedance Z_0 for the main signal path between Port 1 and 2, the normalized slope parameter required for each resonator is given by:

$$\frac{x_i}{Z_0} = \frac{b_i}{Y_0} = \frac{g_0}{g_i FBW} \qquad (i = 1 \text{ to } n) \tag{7.14}$$

After that, the resistor R is introduced to achieve an absorptive stopband, and its initial value can be obtained using (7.13). Based on the circuit model in Figure 7.21(a), the microstrip filter layout is first obtained as in Figure 7.22(a). Here, all the line impedances are set as 50 Ω, which corresponds to a line width of 3.3 mm. According to (7.14), the required normalized slope parameters of the five resonators are obtained as: $x_1/Z_0 = x_5/Z_0 = 26.4429$, $x_2/Z_0 = x_3/Z_0 = 15.3265$, and $x_3/Z_0 = 12.6797$. The gap width S_i of each coupled-line section is then determined according to the required x_i/Z_0 with the aid of the EM simulation-based design chart in Figure 7.23. Since the variation of gap width S_i also leads to some shifts in resonant frequency, the line length of each L-shape resonator is properly tuned so as to make it resonant at the desired center frequency of 2 GHz. The resulted geometrical parameters are shown in the caption of Figure 7.22. Next, a shunt resistor R is introduced at the open end of the first L-shape resonator. According to the line/gap width of the first coupled-line section, the initial value of R is calculated as 1,247 Ω using (7.13). It is then increased to 1.8 kΩ so as to achieve a better overall input return loss around f_0 as suggested by the analysis in Figure 7.10. In addition, due to the discontinuity effect of the ground via, the line length L_{S1} has been slightly reduced after the introduction

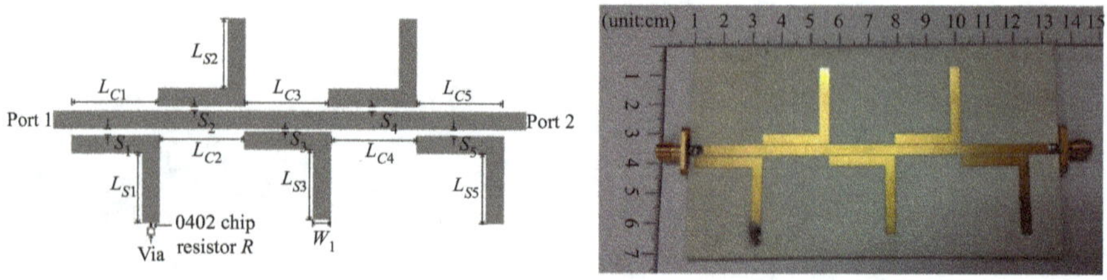

Figure 7.22 *(a) Layout and (b) circuit photograph of proposed fifth-order microstrip narrowband ABSF using based on L-shape half-wavelength resonators. (Geometrical parameters:* $W_1 = 3.3$*,* $L_{C1} = L_{C5} = 22.3$*,* $L_{S1} = 22.1$ *,* $S_1 = S_5 = 0.7$*,* $L_{C2} = L_{C4} = 22.35$*,* $L_{S2} = L_{S4} = 22.75$*,* $S_2 = S_4 = 0.35$*,* $L_{C3} = 22.25$*,* $L_{S3} = 22.65$*,* $S_3 = 0.4$*, and* $L_{S5} = 22.6$*. All units in mm.* $R = 1.8$ *kΩ)*

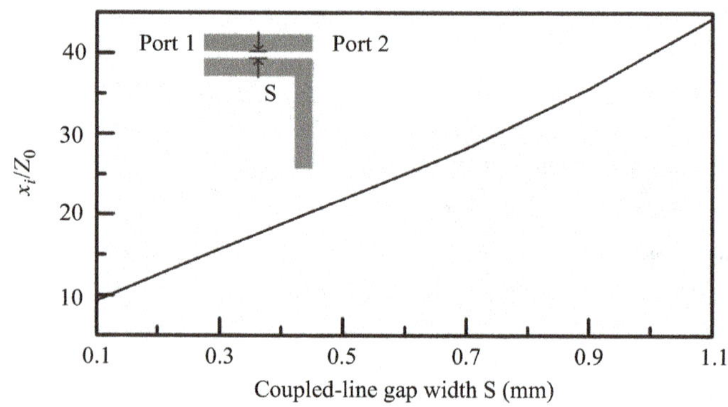

Figure 7.23 *Design chart for determining the coupled-line gap width in Figure 7.22 according to the required normalized reactance slope parameter*

of *R*. The resistor *R* is implemented using a 0402 chip resistor, and a photograph of the fabricated filter is shown in Figure 7.22(b).

Shown in Figure 7.24 are the measured and EM simulated results of this microstrip ABSF with L-shape $\lambda/2$ resonators. When the input is at Port 1, the measured result demonstrates an absorptive stopband around 2 GHz with good frequency selectivity. The maximal measured attenuation is 43.3 dB at 1.999 GHz, and the measured stopband rejection is better than 30 dB from 1.989 to 2.017 GHz. The measured $|S_{11}|$ is better than −20.8 dB in the same frequency range. The corresponding power dissipation is shown in Figure 7.24(b), and the measured power dissipation is better than 99% from 1.984 to 2.019 GHz. Notably, a good passband performance is also achieved. According to Figure 7.24(c), the measured passband insertion loss is within 1 dB and the measured passband return loss is better than 20 dB from dc to 1.91 GHz and from 2.11 to 3.63 GHz.

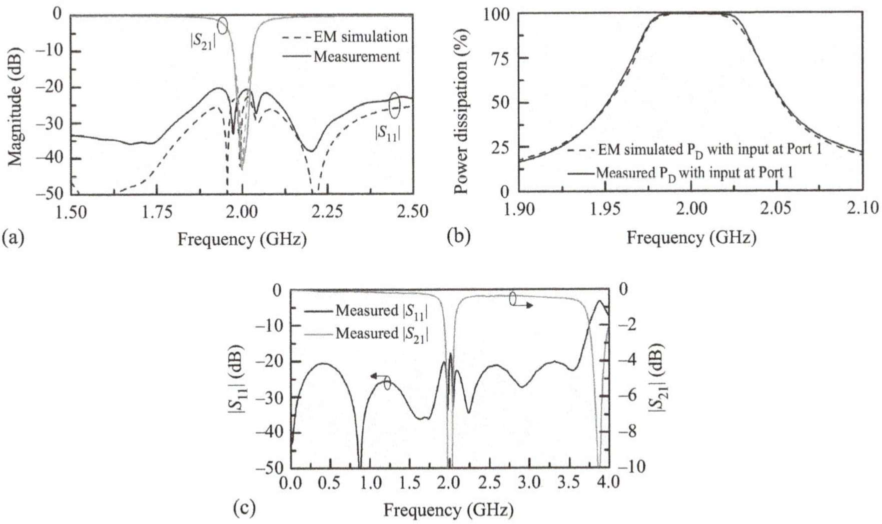

Figure 7.24 Measured and EM simulated results of the fifth-order microstrip narrowband ABSF using based on L-shape half-wavelength resonators: (a) narrowband frequency response; (b) power dissipated by the ABSF when the input is at Port 1, which is defined by $P_D = 1 - |S_{11}|^2 - |S_{21}|^2$*; and (c) measured wideband frequency responses*

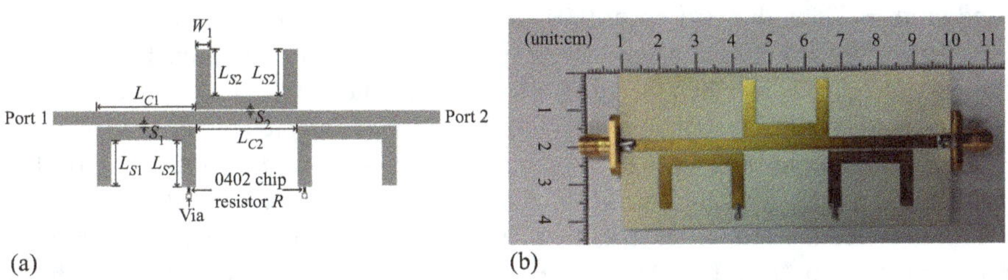

Figure 7.25 (a) Layout and (b) circuit photograph of proposed third-order microstrip narrowband ABSF using two lossy resonators. (Geometrical parameters: $W_1 = 3.3$*,* $L_{C1} = 22.9$*,* $L_{S1} = 11.5$*,* $L_{S2} = 11.8$*,* $S_1 = 0.6$*,* $L_{C2} = 23.4$*, and* $S_2 = 0.35$*. All units in mm.* $R = 1.24\ k\Omega$*)*

The two microstrip ABSF designs presented so far are evolved from the prototypes in Figures 7.1 and 7.2, and they are reflectionless at Port 1 only for the resistor R is only introduced to the first resonator near Port 1. For applications that requires reflectionless properties at both the input and output ports of the BSF, lossy resonators can be introduced to both the first and last resonators to achieve this. Shown in Figure 7.25(a) is the layout of a 3rd-order microstrip narrowband

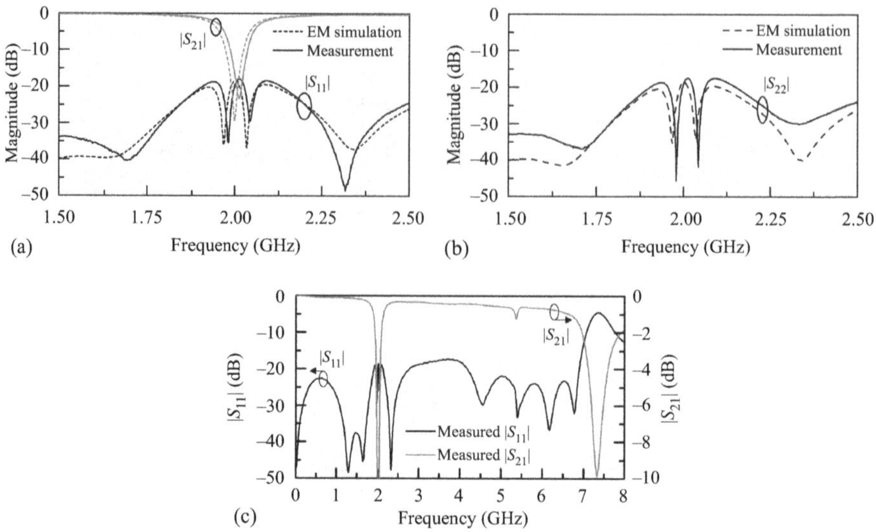

Figure 7.26 Measured and EM simulated results of proposed third-order microstrip narrowband ABSF using two lossy resonators: (a) $|S_{11}|$ and $|S_{21}|$; (b) $|S_{22}|$; and (c) measured wideband frequency responses

ABSF with two lossy resonators. In order to achieve a symmetrical circuit structure, U-shape $\lambda/2$ resonators are adopted and they are coupled to the main signal path through $\lambda/4$ coupled-line sections. The specifications are $f_0 = 2$ GHz, ripple bandwidth = 5% with a 3rd-order 0.01-dB Chebyshev response. It is first designed as a reflective BSF through the conventional BSF synthesis approach [20]. Two resistors of 1.24 kΩ are then introduced to the open ends of the 1st and 3rd U-shape resonators to absorb the input signals. The filter is implemented using also the same RO4350B substrate, and the circuit photograph is shown in Figure 7.25(b). The measured and EM simulated results are shown in Figure 7.26. A bandstop response with non-reflective properties at both ports is observed as expected. The measured stopband rejection is better than 20 dB from 2.005 to 2.020 GHz, while both the measured input and output return losses are better than 17 dB in the same frequency range. The measured power dissipation is better than 97% from 2.002 to 2.023 GHz. As for the passband performance, the measured passband insertion loss is within 1 dB and the measured passband return loss is greater than 17 dB from dc to 1.891 GHz and from 2.142 to 5.340 GHz.

7.4 Wideband absorptive bandstop filters

7.4.1 Basic filter structure

Although the design concept illustrated in Figures 7.1 and 7.2 are based on the filter's equivalent circuit model at the stopband center frequency f_0 only, in addition to the narrowband ABSFs in Section 7.3, it can also be used to realize ABSFs with a

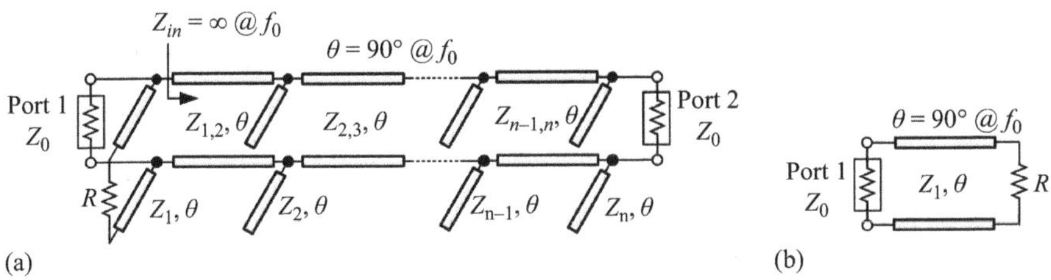

Figure 7.27 (a) Basic circuit model of the proposed wideband ABSF, which is achieved by introducing one lossy resonator to a conventional stub BSF and (b) its equivalent-circuit model at the stopband center frequency f_0

wide absorptive stopband bandwidth up to about 40%. Shown in Figure 7.27(a) is the basic circuit model of proposed wideband ABSF [18]. It is based on introducing one lossy resonator to a conventional wideband stub BSF consisting of $\lambda/4$ open stubs and $\lambda/4$ connecting lines. To achieve a more compact circuit size and more uniform stub impedances, the optimal microwave BSF [22–24] is the preferred choice for implementing the wideband stub BSF. It is optimal because the $\lambda/4$ connecting lines, i.e., the unit elements between stubs in Figure 7.27(a) are incorporated in the design such that significantly steeper attenuation characteristics can be obtained for the same number of stubs than is possible for stub BSFs designed with redundant unit elements [20,24]. Specifically, an optimal microwave BSF with n stubs features a $(2n - 1)$th order bandstop response.

The operation principle of the proposed wideband ABSF is explained as follows. Since all transmission line sections in Figure 7.27(a) are $\lambda/4$ long, the input impedance Z_{in} looking into the right-hand side of the first shunt open stub is infinite at the stopband center frequency f_0. Therefore, the filter circuit model at f_0 can be further simplified as Figure 7.27(b). To have the resistor R absorb all the input power at Port 1, its value is set according to

$$R = Z_1^2/Z_0 \tag{7.15}$$

by using the well-known $\lambda/4$ transformer design formula. As a result, the input power at f_0 will not propagate to Port 2 of this wideband ABSF due to the infinite Z_{in}, but it will be totally consumed by the resistor R instead. This is the same as the design concept illustrated in Figure 7.1 and 7.2.

The design procedure of the wideband ABSF can be summarized as follows. First, the resistor is neglected (i.e., $R = \infty$) and the stub BSF is designed as a reflective one using the design tables available in [20]. Specifically, the line impedances are determined by

$$\begin{aligned} Z_i &= Z_0/g_i \\ Z_{i,i+1} &= Z_0/J_{i,i+1}, \end{aligned} \tag{7.16}$$

where Z_0 is the port reference impedance while g_i and $J_{i,i+1}$ are the element values of optimal microwave BSF. Next, the resistor R is introduced and its initial value is given by (7.15). To illustrate the achievable performance, a wideband ABSF with four stubs (i.e., $n = 4$) is designed with a stopband center frequency $f_0 = 2$ GHz, a FBW of 0.5, and a Chebyshev response with a ripple constant $\varepsilon = 0.1005$. The reference impedances of Port 1 and 2 are both set as $Z_0 = 50\ \Omega$. By referring to the element value table in [20], the line impedances can be determined by (7.16) as $Z_1 = Z_4 = 123.86\ \Omega$, $Z_2 = Z_3 = 69.33\ \Omega$, $Z_{1,2} = Z_{3,4} = 60.84\ \Omega$, and $Z_{2,3} = 63.21\ \Omega$. The initial value of R can then be calculated using (7.15) to give $R = 306.85\ \Omega$.

Shown in Figure 7.28 are the circuit simulated S-parameters of this wideband ABSF, and those for its reflective BSF counterpart (i.e., $R = \infty$) is also included for comparison. As observed in Figure 7.28(a), the wideband ABSF exhibits a bandstop response centered at $f_0 = 2$ GHz. The insertion loss response is very similar to that of the reflective BSF except for some degradation at the two stopband edges. Still, good selectivity is achieved for the proposed wideband ABSF since the reflective BSF with four stubs features a seventh-order Chebyshev response. The 30-dB rejection bandwidth of the wideband ABSF is from 1.75 to 2.25 GHz or 25% in terms of fractional bandwidth, and it is almost identical to that of the reflective BSF. On the other hand, the ABSF's input reflection coefficient $|S_{11}|$ in Figure 7.28(b) is equal to zero or $-\infty$ dB at f_0 while that of the reflective BSF is equal to one or 0 dB. As a result, in Figure 7.28(c) the power dissipated by the wideband ABSF is 100% at f_0, while it is zero at all frequencies for the conventional reflective BSF. These verify that this wideband ABSF can successfully absorb the input power at f_0

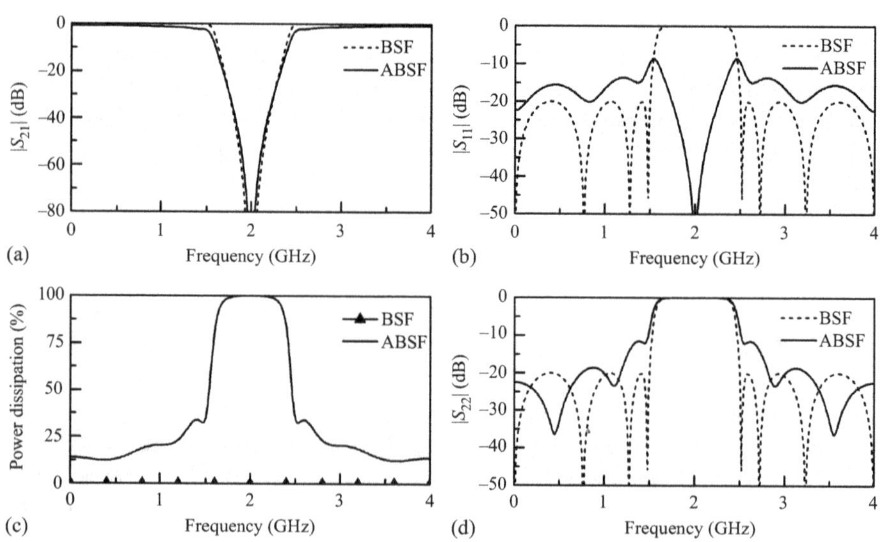

Figure 7.28 Circuit simulated (a) $|S_{21}|$ and (b) $|S_{11}|$ frequency responses of a proposed wideband ABSF design example ($f_0 = 2$ GHz, FBW = 0.5 and n = 4) and its reflective BSF counterpart; (c) power dissipated in the ABSF with the input at Port 1, which is calculated by $1 - |S_{11}|^2 - |S_{21}|^2$; and (d) output reflection coefficient $|S_{22}|$ at Port 2

by introducing the additional resistor R. As the frequency moves away from f_0, the $|S_{11}|$ of the proposed wideband ABSF increases until a maximal value is reached. This is because Z_{in} in Figure 7.27(a) is infinite only at f_0 and $R = Z_1^2/Z_0$ can only achieve perfect impedance matching at f_0 in Figure 7.27(b). As a result, the power dissipation also reduces as the frequency moves away from f_0. As for the output reflection coefficient $|S_{22}|$ shown in Figure 7.28(d), the wideband ABSF exhibits a similar response to that of the conventional reflective BSF because no lossy resonator is used at the output port.

According to Figure 7.28(b), the $|S_{11}|$ of this wideband ABSF exceeds -10 dB at the two stopband edges, which may not be acceptable for some applications. Similar to the analysis in Figure 7.10, this can be improved by modifying the resistance of R. Shown in Figure 7.29 is the input impedance Z_{inR} looking into the open end of the first open stub where R is connected to. It can be observed that Z_{inR} is purely resistive at $f_0 = 2$ GHz with a value of 306.85 Ω. As the frequency moves away from f_0, Z_{inR} becomes a complex value and $|Z_{inR}|$ becomes higher than its value at f_0. This is the cause of the larger $|S_{11}|$ at the two stopband edges for the resistor $R = Z_1^2/Z_0$ no longer matched to Z_{inR} there. It also implies that the $|S_{11}|$ at the stopband edges can be simply improved by increasing the value of R at the expense of some increase in $|S_{11}|$ at f_0.

Shown in Figure 7.30 is the circuit simulated input reflection coefficients of this wideband ABSF design under different values of R. As R increases from the initial value of 306.85 Ω, it can be found that the input reflection coefficient $|S_{11}|$ at f_0 increases due to impedance mismatch. However, the maximal input reflection

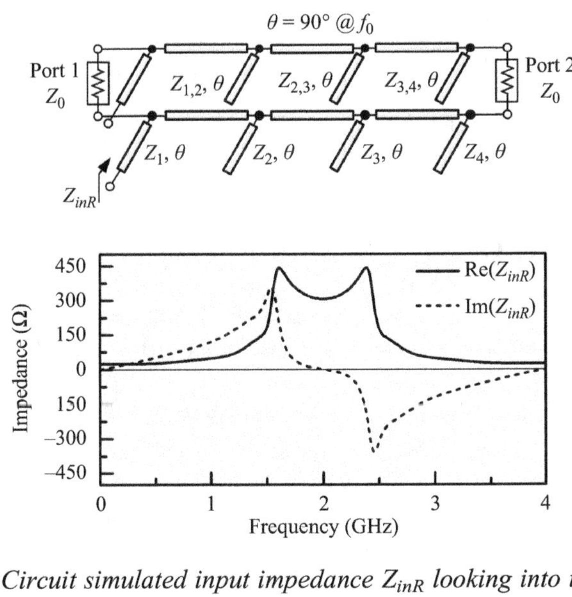

Figure 7.29 Circuit simulated input impedance Z_{inR} looking into the open end of the first open stub of the wideband ABSF with $f_0 = 2$ GHz, $FBW = 0.5$ and $n = 4$

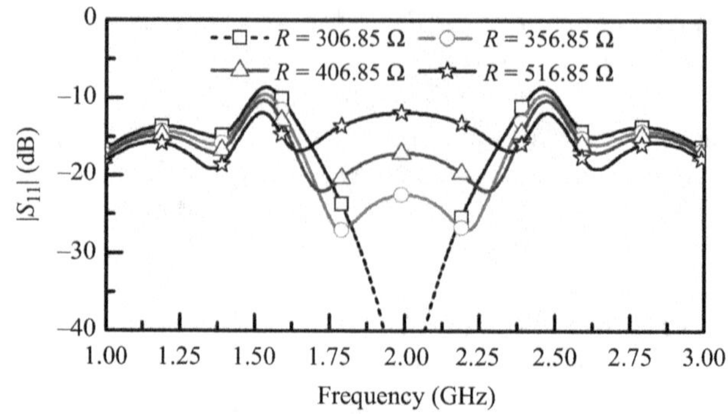

Figure 7.30 Effect of R on the circuit simulated $|S_{11}|$ frequency response of the wideband ABSF with $f_0 = 2$ GHz, FBW = 0.5 and n = 4

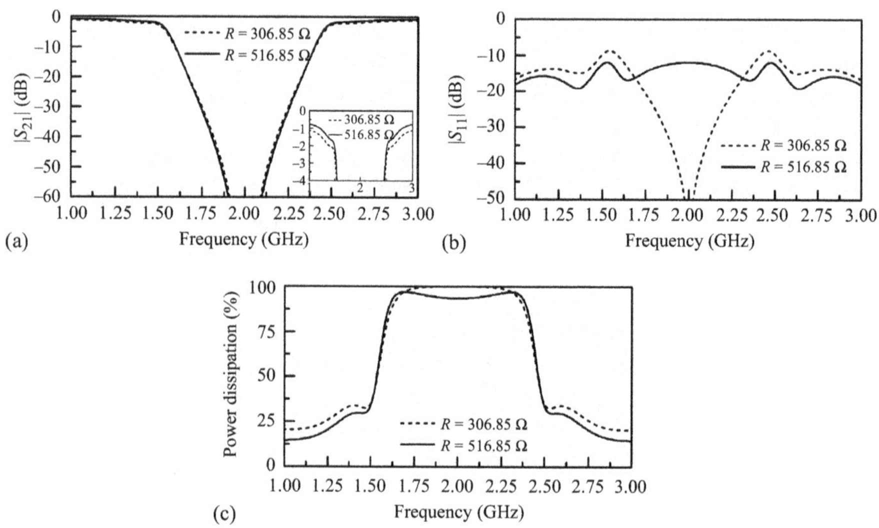

Figure 7.31 Effect of R on the bandstop performance of the wideband ABSF design example ($f_0 = 2$ GHz, FBW = 0.5 and n = 4) for R = 306.85 and 516.85 Ω. Circuit simulated (a) $|S_{21}|$ and (b) $|S_{11}|$ frequency responses. (c) Power dissipated in the ABSF with the input at Port 1

coefficients at the two stopband edges decrease as R increases. Especially, when R increases to 516.85 Ω, the maximal $|S_{11}|$ reduces to a level that is about equal to the reflection coefficient at f_0, and a nearly equal-ripple $|S_{11}|$ response within the stopband is obtained. In addition, the $|S_{11}|$ at all frequencies are now below −11.87 dB. On the other hand, the $|S_{21}|$ response does not change much with the varying resistance. This can be examined in detail by referring to the comparison between the two extreme cases with $R = 306.85$ and 516.85 Ω in Figure 7.31. By increasing R from 306.85 to 516.85 Ω, the maximal $|S_{11}|$ is reduced from −8.62 to

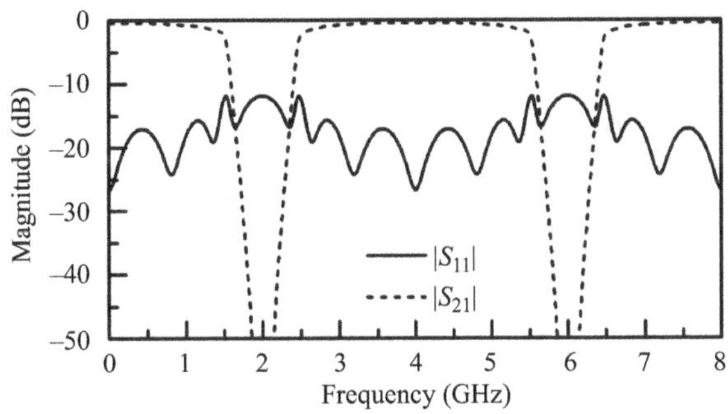

Figure 7.32 Wideband circuit simulated frequency response of the wideband ABSF design example ($f_0 = 2$ GHz, FBW = 0.5 and n = 4) with R = 516.85 Ω

−11.87 dB as expected, while the difference in $|S_{21}|$ between the two cases is negligible as shown in Figure 7.31(a). According to Figure 7.31(c), the power dissipation at f_0 reduced from 100 to 93.5% after R increases to 516.85 Ω. Anyway, the bandwidth for larger than 93.5% power dissipation is from 1.626 to 2.374 GHz (or 37.4% in terms of fractional bandwidth) while the reflected power is less than 6.5% in the same frequency range, which is considered as acceptable for most applications.

In addition to an absorptive stopband, a good passband performance is also obtained for this wideband ABSF after R increases to 516.85 Ω. This is because the value of resistive loading R is now closer to that of the original reflective BSF, i.e., $R = \infty$. According to the wideband frequency response in Figure 7.32, it exhibits a passband with better than 1 dB insertion loss from dc to 1.2 GHz and from 2.8 to 5.2 GHz. The corresponding passband input/output return losses are all better than 15.87 dB, while the input return loss $|S_{11}|$ is better than 11.87 dB at all frequencies. Repeated absorptive stopbands at odd multiples of f_0 are observed due to the periodic property of the original optimal microwave BSF. Therefore, increasing the resistance of R until an equal-ripple $|S_{11}|$ around f_0 is reached is considered as the final step of the design procedure of the proposed wideband ABSF. Still, R is a free variable the designer can use to tradeoff between the return loss at f_0 and the minimum in-band return loss depending on the application requirement.

Although the proposed design method starts from the condition of Z_{in} in Figure 7.27(a) equals infinity at the stopband center frequency only, the resulted ABSF can handle a 30-dB stopband bandwidth from about 12 to 42% [18] according to the following analysis. Figure 7.33 compares the circuit simulated frequency responses of several wideband ABSF designs with four stubs (i.e., $n = 4$) and different FBWs ranging from 0.3 to 0.9. Here, FBW refers to the ripple bandwidth of the optimal microwave BSF designed with a Chebyshev response and a ripple constant $\varepsilon = 0.1005$. The circuit parameters are obtained using (7.16) and the design tables in [20], and the results are listed in Table 7.4. Notably, the resistance of R for all cases have been increased from the initial value given by (7.15) to the

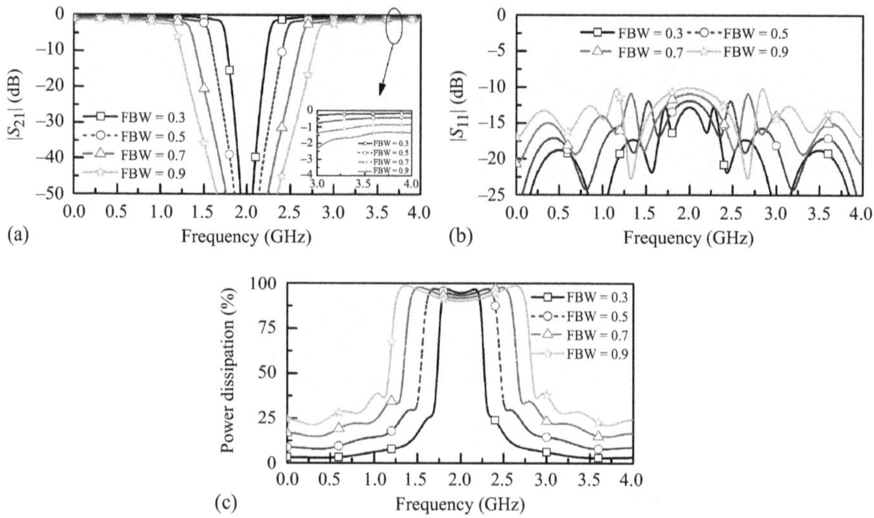

Figure 7.33 Circuit simulated (a) $|S_{21}|$ and (b) $|S_{11}|$ frequency responses of the wideband ABSF with $n = 4$ and $f_0 = 2$ GHz but different FBWs; and (c) power dissipated in the ABSF with the input at Port 1

Table 7.4 Circuit parameters of the wideband ABSFs with different FBWs in Figure 7.33

FBW	R (Ω)	$Z_1 = Z_4$ (Ω)	$Z_2 = Z_3$ (Ω)	$Z_{1,2} = Z_{3,4}$ (Ω)	$Z_{2,3}$ (Ω)
0.3	1,499	216.74	123.78	53.54	54.74
0.5	306.85	123.86	69.33	60.84	63.21
0.7	246.10	82.83	45.54	70.32	74.28
0.9	62.94	34.09	80.61	86.28	151.15

value in Table 7.4 so as to achieve an equal-ripple $|S_{11}|$ response around f_0 according to the analysis in Figure 7.30.

All four designs in Figure 7.33 demonstrate the desired reflectionless bandstop responses, and their performances are summarized in Table 7.5. According to Figure 7.33(a) and 7.33(c), the bandwidths of 30-dB stopband rejection for the four designs with FBW = 0.3, 0.5, 0.7, and 0.9 are equal to 13.6%, 25.8%, 41.2%, and 56.2%, respectively, while the bandwidths for larger than 90% power dissipation are equal to 22.1%, 39.8%, 60.8%, and 80.6%, respectively. However, it can be observed from Figure 7.33(b) that the maximal input reflection coefficient gradually increases as FBW increases. Specifically, the design with FBW = 0.9 may be considered as the upper limit of realizable FBW because its maximal $|S_{11}|$ already reaches −10.1 dB. In addition, the minimum passband insertion loss at dc and $2f_0$ also increases with FBW according to Table 7.5. This also limits the maximal realizable FBW of proposed wideband ABSF.

Table 7.5 Performance summary of the wideband ABSFs with different FBWs in Figure 7.33

FBW	30-dB Stopband bandwidth	Bandwidth for >90% power dissipation	Max. input reflection coeff.	Min. passband insertion loss
0.3	13.6%	22.1%	−12.78 dB	0.14 dB
0.5	25.8%	39.8%	−11.87 dB	0.41 dB
0.7	41.2%	60.8%	−10.92 dB	0.84 dB
0.9	56.2%	80.6%	−10.1 dB	1.33 dB

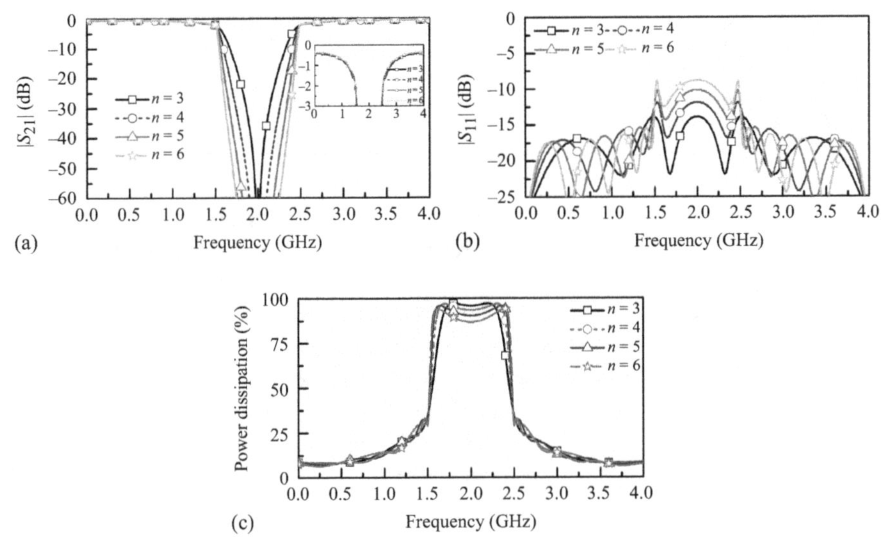

Figure 7.34 Circuit simulated (a) $|S_{21}|$ and (b) $|S_{11}|$ frequency responses of the wideband ABSFs with $f_0 = 2$ GHz, FBW = 0.5 but different number of stubs and (c) power dissipated in the ABSF with the input at Port 1

On the other hand, the problem with smaller FBWs is that the required line impedance becomes abnormally high such that it may not be realizable in practice, which is a common issue for stub BSF structures. For the case with FBW = 0.3 in Table 7.4, the maximal line impedance is equal to 216.74 Ω and it is not realizable using planar transmission line structures. Therefore, a modified design approach will be proposed in the next subsection to deal with this issue so as to extend the lower bound of realizable FBW for the proposed wideband ABSF.

Figure 7.34 compares the circuit simulated results of several wideband ABSFs with the same FBW of 0.5 but different numbers of shunt stubs (i.e., n), which means different filter orders. The circuit parameters are listed in Table 7.6, and the resistance R for each case has been increased again to achieve an equal-ripple input reflection coefficient around f_0. According to Figure 7.34, the

Table 7.6 Circuit parameters for the wideband ABSF designs with different number of stubs in Figure 7.34

n	3	4	5	6
R (Ω)	528.78	516.87	549.74	601.40
Z_1 (Ω)	132.43	123.86	120.36	118.61
Z_2 (Ω)	78.99	69.33	66.40	65.12
Z_3 (Ω)	132.43	69.33	62.24	60.13
Z_4 (Ω)	–	123.86	66.40	60.13
Z_5 (Ω)	–	–	120.36	65.12
Z_6 (Ω)	–	–	–	118.61
$Z_{1,2}$ (Ω)	59.55	60.84	61.41	61.70
$Z_{2,3}$ (Ω)	59.55	63.21	64.42	64.94
$Z_{3,4}$ (Ω)	–	60.84	64.42	65.54
$Z_{4,5}$ (Ω)	–	–	61.41	64.94
$Z_{5,6}$ (Ω)	–	–	–	61.70

frequency selectivity increases with the number of stubs n as expected. However, it is found that the maximal $|S_{11}|$ also increases with n. For $n = 6$, the maximal $|S_{11}|$ already exceeds -10 dB, which may not be considered as usable in practice. For $n = 5$, it is equal to -10.2 dB and this barely passes the common requirement on input return loss. Although it seems that the frequency selectivity of proposed wideband ABSF is therefore limited due to the increase of maximal $|S_{11}|$ with n, a modified design will be given in the next subsection to improve the input reflection coefficient for larger n such that a highly selective wideband ABSF with proper input impedance matching can still be achieved.

7.4.2 Design modifications

The wideband ABSF in Figure 7.27(a) is based on the conventional optimal microwave BSF that employs the stub BSF structure. Therefore, it is more suitable for wideband designs because the line impedance becomes unreasonably high if the required stopband bandwidth is very narrow, which can be told from Table 7.4. On the other hand, due to the lowpass to bandstop transformation used in microwave filter synthesis, the frequency selectivity decreases as the stopband bandwidth increase for the same number of stubs, i.e., the same circuit size. Therefore, in case that the required stopband bandwidth is not large, a moderate bandwidth ABSF is preferred due to the better frequency selectivity. A modified design approach to reduce the required line impedance is developed here so as to ease the realization of proposed ABSF with moderate bandwidth.

Size reduction of transmission line through the use of capacitive or inductive loadings has been widely adopted for the development of compact microwave passive circuits. As pointed out in [25], when a line is inductively loaded, not only the line length but also the characteristic impedance of the line must be decreased in order to maintain the total capacitance and inductance of the original line. This

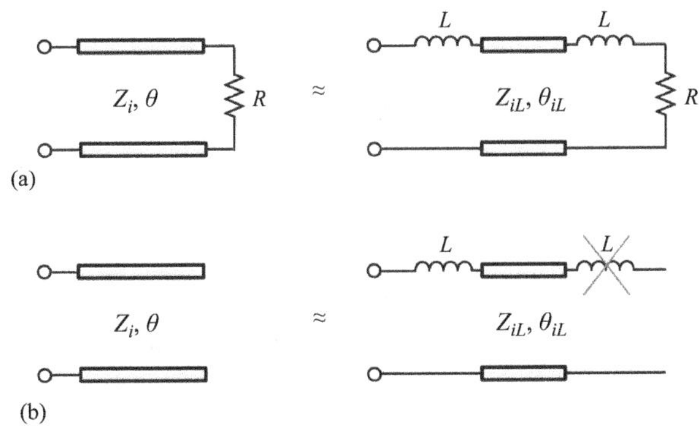

Figure 7.35 Design modifications of the (a) resistor loaded stub and (b) open stub in the wideband ABSF in case that the line impedance Z_i is higher than the maximal realizable impedance

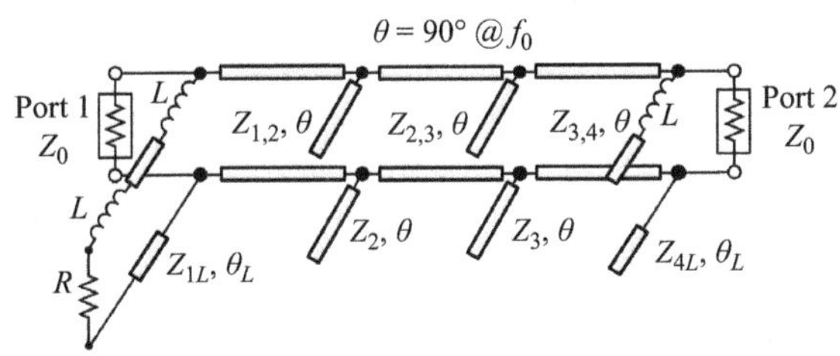

Figure 7.36 Modified design-I of the wideband ABSF of moderate bandwidth for n = 4

property can be used to solve the issue of unreasonably high line impedance for the proposed ABSF with moderate bandwidth. Specifically, the stubs in the proposed ABSF with a line impedance Z_i that is too large can be modified to the inductor-loaded ones shown in Figure 7.35. The required inductance L, line impedance Z_{iL}, and line length θ_{iL} can be determined according to [25] as:

$$\begin{aligned} Z_{iL} &= Z_i \sin\theta_{iL} \\ L &= \frac{Z_i}{\omega_0}\cos\theta_{iL} \end{aligned} \tag{7.17}$$

According to (7.17), the line impedance Z_i can be reduced by a factor of $\sin\theta_{iL}$ with the inductor loadings. However, it should be noted that if θ_{iL} is chosen too small, the bandwidths of the two equivalences in Figure 7.35 will also reduce. Therefore, it is suggested to set Z_{iL} close to the maximal realizable line impedance of the given fabrication process, and calculate the required θ_{iL} and L accordingly.

As a demonstration, shown in Figure 7.36 is the modified design-I of proposed wideband ABSF with four stubs (i.e., $n = 4$) based on the equivalences illustrated in Figure 7.35. According to the circuit parameters given in Table 7.4, when

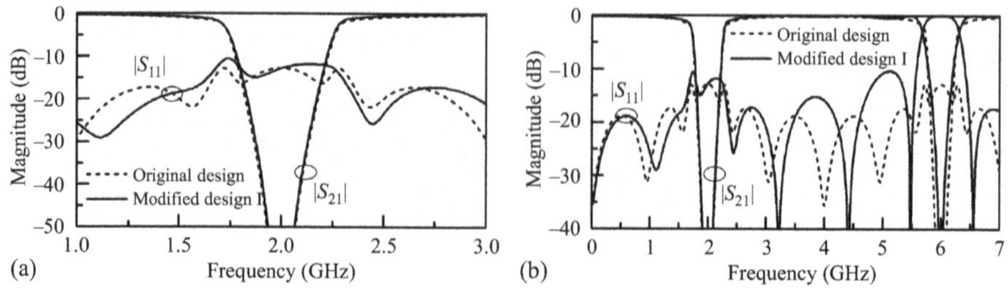

Figure 7.37 Circuit simulated results for the modified design-I of the wideband ABSF in Figure 7.36 and the original design with $f_0 = 2$ GHz, FBW = 0.3, and n = 4: (a) narrowband and (b) wideband S-parameter frequency responses when the input is at Port 1

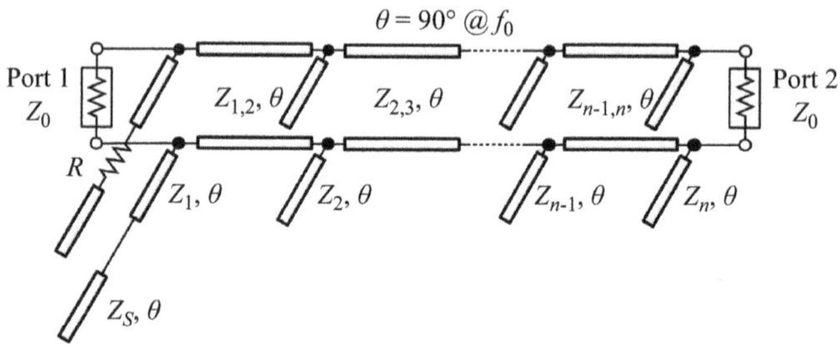

Figure 7.38 Modified design-II of the wideband ABSF for improving the input return loss at Port 1 as well as the passband insertion loss

FBW = 0.3 the first and last shunt stubs require a high line impedance of 216.74 Ω. Therefore, these two stubs are replaced with the inductively loaded ones in Figure 7.35 while all other parts of the filter structure remain unchanged. By setting $Z_{iL} = 139.3$ Ω ($i = 1$ or 4), the required L and θ_{iL} can be obtained using (7.17) as 13.2 nH and 60° at $f_0 = 2$ GHz, respectively. Shown in Figure 7.37 are the circuit simulated S-parameters of this modified wideband ABSF design-I, and those of the original ABSF design based on Figure 7.27(a) are also included for comparison. The two simulated $|S_{21}|$ responses are almost identical from dc to 5 GHz or $2.5f_0$, which implies that the bandwidths of the two equivalences in Figure 7.35 are sufficient for this ABSF design.

Although the simulated $|S_{11}|$ of the modified design-I deviates from that of the original ABSF design for $f \neq f_0$, good input impedance matching is still achieved and the simulated $|S_{11}|$ of the modified design-I is better than −10.5 dB from dc to 5.05 GHz. The problem of high line impedance is thus solved, and this extends the lower limit of realizable bandwidth for the proposed wideband ABSF.

As pointed out in the discussions of Figure 7.33 and 7.34, the maximal input reflection coefficient of the wideband ABSF in Figure 7.27(a) increases with the FBW as well as the number of stubs n. This can be improved by the modified design-II shown in Figure 7.38, in which the short end of the resistor R is replaced by a $\lambda/4$ open stub. According to Figure 7.29, the input impedance Z_{inR} looking into the open end of the first open stub is purely resistive at f_0 only. For frequencies below and above f_0, it becomes complex impedances with inductive and capacitive reactances, respectively. This is why a single resistor R in the original design cannot achieve wideband impedance matching. By changing R into the combination of R and a $\lambda/4$ open stub as shown in Figure 7.38, the bandwidth of impedance matching can be improved because the $\lambda/4$ open stub is capacitive for $0 < f < f_0$ and inductive for $f_0 < f < 2f_0$, while it is equal to an ideal short at f_0. Therefore, by properly setting the line impedance Z_S of the $\lambda/4$ open stub, it is possible to achieve conjugate matching at f_0 as well as two other frequencies below and above f_0. In this way, the maximal input reflection coefficient within the stopband can be reduced.

The performance of the modified design-II in Figure 7.38 is examined by a wideband ABSF design with five stubs (i.e., $n = 5$), and the specifications are $f_0 =$ 2 GHz, $Z_0 = 50\ \Omega$, and FBW $= 0.5$. The circuit parameters are those listed in Table 7.6 for $n = 5$, except that the resistance R is calculated using (7.15) as 289.73 Ω so as to achieve perfect input impedance matching at f_0. Shown in Figure 7.39 are the circuit simulated S-parameters of this ABSF under different line impedances Z_S of the $\lambda/4$ open stub ranging from 200 to 400 Ω. The results for the original wideband ABSF design in which the resistor R is terminated to ground instead is also included for comparison. It is found that the input reflection coefficient $|S_{11}|$ remains equal to 0 at f_0 for all four cases in Figure 7.39 as expected, while the maximal $|S_{11}|$ occurs at the two stopband edges improves slightly for the modified design-II. In addition, the maximal $|S_{11}|$ reduces as Z_S increases. However, even for the case with $Z_S = 400\ \Omega$, the maximal $|S_{11}|$ still exceeds -10 dB. Again, to further improve the overall input impedance matching, the resistor R can be increased until a nearly equal-ripple $|S_{11}|$ within the stopband is reached. Shown in Figure 7.40 are the circuit simulated results of the three modified designs with Z_S equals to 200, 300, and 400 Ω as well as that of the original design, in which the

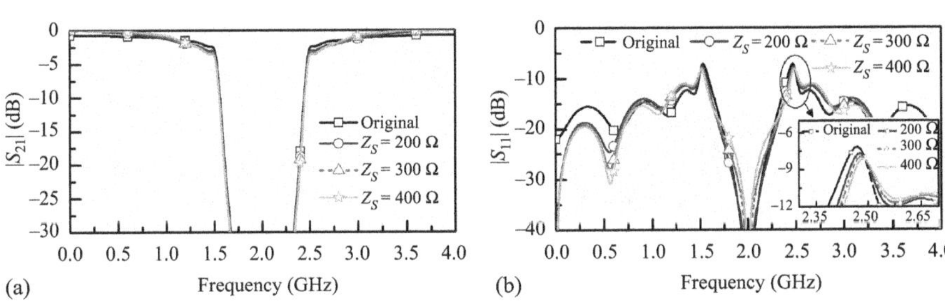

Figure 7.39 Circuit simulated S-parameters of the modified design-II of the wideband ABSF in Figure 7.38 with $f_0 = 2$ GHz, FBW $= 0.5$ and $n = 5$ under different line impedance Z_S of the $\lambda/4$ open stub. The circuit simulated results of the original wideband ABSF design is also included for comparison. (a) $|S_{21}|$. (b) $|S_{11}|$

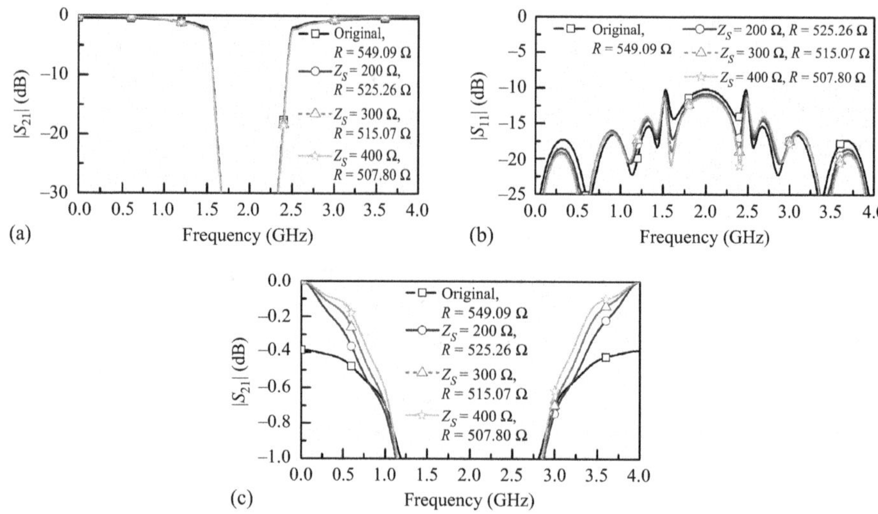

Figure 7.40 Circuit simulated results for the modified design-II of the ABSF in Figure 7.38 with $f_0 = 2$ GHz, FBW = 0.5, and $n = 5$ under different line impedances Z_S of the $\lambda/4$ open stub. The results for the original wideband ABSF design are also included for comparison. The resistance R for each case is increased to the value shown in the figure legend. (a) $|S_{21}|$; (b) $|S_{11}|$; and (c) close-up view of the $|S_{21}|$ responses in the passband

resistances R for them are increased to the value shown in the figure legend. The maximal $|S_{11}|$ of the original wideband ABSF design is now improved to -10.19 dB, while those of the modified designs with a $\lambda/4$ open stub are -10.77, -11.05, and -11.26 dB for Z_S equals to 200, 300, and 400 Ω, respectively, which are all better than that of the original wideband ABSF design.

Although the improvement in $|S_{11}|$ is not large, there are two additional advantages of the modified design-II with a $\lambda/4$ open stub shown in Figure 7.38. First, according to Figure 7.40(c) the passband insertion loss becomes substantially smaller than the original wideband ABSF design. This is because the resistor R becomes floating at dc and $2f_0$ such that the insertion loss becomes zero at these two frequencies. Second, no via is required for the modified design-II with a $\lambda/4$ open stub, which greatly simplifies the circuit fabrication process.

Even though the modified design-II with a $\lambda/4$ open stub can achieve better insertion loss and return loss, a high Z_S larger than 200 Ω is required for a notable improvement in $|S_{11}|$. Such unreasonably high line impedances may not be achievable in practice, so a further modification shown in Figure 7.41 is proposed, in which the equivalent circuit in Figure 7.35(b) is employed. Specifically, the $\lambda/4$ open stub is now replaced by the combination of a series inductor L and an open stub of line impedance Z_{SL} and electrical length θ_{SL} at f_0 ($\theta_{SL} < 90°$). According to the previous discussions, with the introduction of L, one can achieve a higher

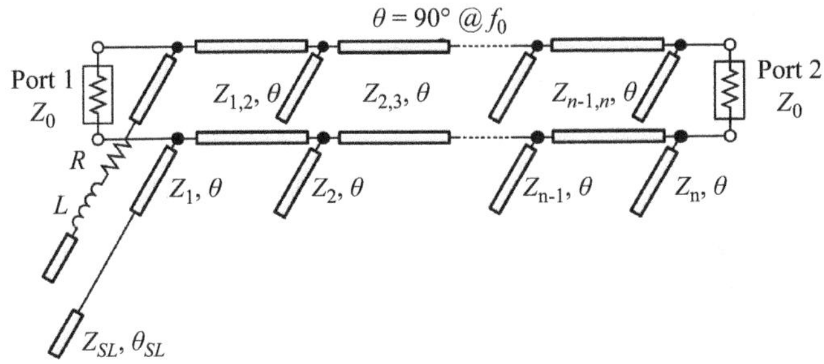

Figure 7.41 Modified design-III of the wideband ABSF for improving the input return loss at Port 1 as well as the passband insertion loss

equivalent line impedance Z_S using a lower line impedance Z_{SL}. Specifically, L, Z_{SL}, and θ_{SL} are related to the equivalent line impedance Z_S of the $\lambda/4$ open stub by (7.18)

$$\begin{aligned} Z_{SL} &= Z_S \sin\theta_{SL} \\ L &= \frac{Z_S}{\omega_0}\cos\theta_{SL} \end{aligned} \tag{7.18}$$

In order to achieve the largest possible Z_S, Z_{SL} can be set as the highest realizable line impedance for a given fabrication process, while L is selected as the largest inductor whose self-resonance frequency is no less than the upper limit of the required passband. The corresponding Z_S and θ_{SL} can then be obtained by

$$\begin{aligned} \theta_{SL} &= \tan^{-1}\left(\frac{Z_{SL}}{\omega_0 L}\right) \\ Z_S &= Z_{SL}\csc\theta_{SL} \end{aligned} \tag{7.19}$$

For the fabrication process considered in this work, the maximal realizable line impedance is $Z_{SL} = 147.9\ \Omega$ and the largest chip inductor with a self-resonance frequency higher than 5 GHz (or $2.5f_0$) is $L = 15$ nH. The corresponding θ_{SL} is 38.1° at $f_0 = 2$ GHz and the equivalent Z_S is 239.6 Ω according to (7.19). Figure 7.42 compares the circuit simulated frequency responses of proposed modified design-II and -III with $n = 5$, $f_0 = 2$ GHz, FBW = 0.5, and $Z_0 = 50\ \Omega$. The modified design-II uses a $\lambda/4$ open stub of line impedance $Z_S = 239.6\ \Omega$ while the modified design-III employs a 15-nH ideal inductor in series with an open stub with $Z_{SL} = 147.9\ \Omega$ and $\theta_{SL} = 38.1°$ at f_0. The resistor R for both designs are 520.89 Ω so as to achieve a nearly equal-ripple in-band $|S_{11}|$ response as shown in Figure 7.42(a), and the maximal $|S_{11}|$ is −10.9 dB. The frequency responses of these two modified designs are almost identical up to 5 GHz, and this validates the effectiveness of the modified design-III in Figure 7.41. The modified design-III can thus be used to improve the $|S_{11}|$ and $|S_{21}|$ response of the wideband ABSF such that the upper limits of realizable FBW and filter order can be further extended.

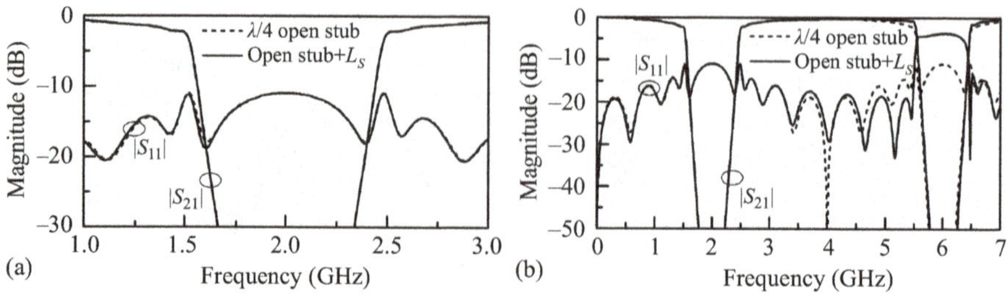

Figure 7.42 Circuit-simulated frequency responses of the modified design-II and -III with $n = 5$, $f_0 = 2$ GHz, $FBW = 0.5$, and $Z_0 = 50\ \Omega$: (a) narrowband and (b) wideband $|S_{11}|$ and $|S_{21}|$ frequency responses

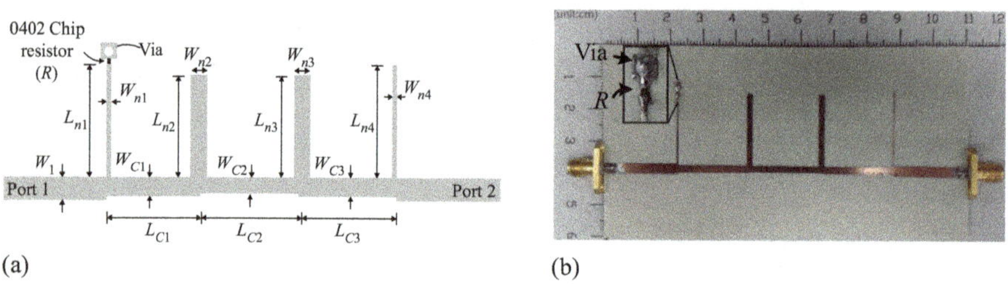

Figure 7.43 (a) Layout and (b) photograph of the microstrip wideband ABSF design example in Figure 7.31 with $f_0 = 2$ GHz, $FBW = 0.5$, and $n = 4$. (Geometrical parameters: $W_1 = 3.3$, $W_{C1} = 2.4$, $W_{C2} = 2.2$, $W_{C3} = 2.4$, $W_{n1} = 0.4$, $W_{n2} = 1.8$, $W_{n3} = 1.8$, $W_{n4} = 0.4$, $L_{C1} = 21.8$, $L_{C2} = 21.8$, $L_{C3} = 21.8$, $L_{n1} = 21$, $L_{n2} = 22.3$, $L_{n3} = 22.3$, and $L_{n4} = 23.3$. Units in mm. 0402 chip component value: $R = 510\ \Omega$.) Reprinted with permission from Reference [18]; © 2017 IEEE

7.4.3 Filter implementation

According to the design method outlined in the previous subsections, a wideband ABSF can be easily designed for a given specification. Three design examples will be implemented in microstrip form and the measured performance will be presented.

First, the wideband ABSF with $f_0 = 2$ GHz, FBW = 0.5, $n = 4$, and $Z_0 = 50\ \Omega$ in Figure 7.31 is implemented in microstrip using a RO4350B substrate ($\varepsilon_r = 3.48$, $\tan\delta = 0.004$, and thickness $h = 1.524$ mm). It is designed based on an optimal microwave BSF with a Chebyshev response and a ripple constant $\varepsilon = 0.1005$. The required Z_i, $Z_{i,j}$, and R have been given in the second row of Table 7.4. By using a computer-aided design tool, e.g. ADS LineCalc, the corresponding line width and line length of each microstrip line section can be easily obtained and they are given in the filter layout in Figure 7.43(a). The resistor R is implemented with a 510-Ω 0402 chip resistor because the required $R = 516.85\ \Omega$ is not available in practice. Shown in Figure 7.43(b) is the photograph of the fabricated filter, and the measured

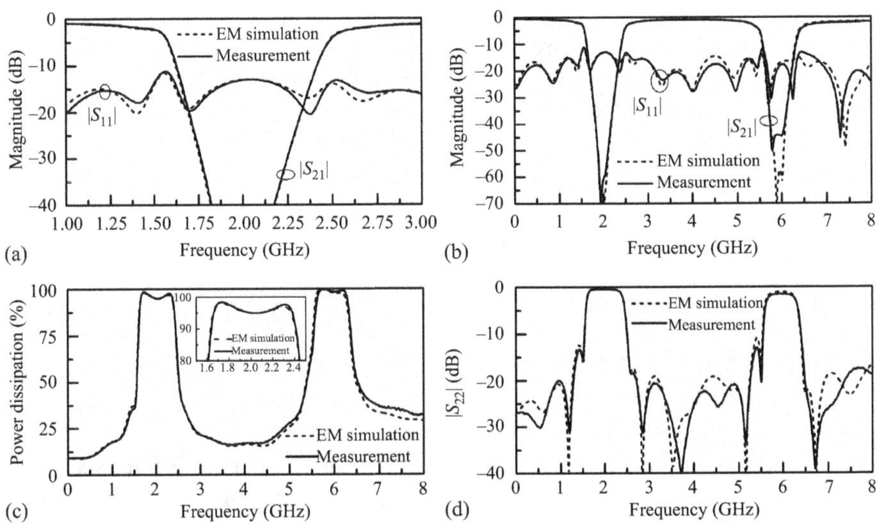

Figure 7.44 Measured and EM simulated results of the microstrip wideband ABSF design example in Figure 7.31 with $f_0 = 2$ GHz, FBW = 0.5, and n = 4: (a) narrowband and (b) wideband S-parameter frequency responses; (c) power dissipation frequency responses when the input is at Port 1; and (d) $|S_{22}|$ frequency response

and EM simulated frequency responses are shown in Figure 7.44. The EM simulation is done by the full-wave simulator Ansys HFSS. Good agreement between measured and EM simulated results is obtained. The measured $|S_{21}|$ exhibits a bandstop frequency response centered at 2 GHz as expected, and the measured stopband rejection is 60.7 dB at 2 GHz. The measured 30-dB rejection bandwidth is from 1.77 to 2.24 GHz or 23.5% in terms of fractional bandwidth. Good stopband selectivity is achieved and the shape factor, i.e., $RBW_{3\text{-dB}}/RBW_{30\text{-dB}}$ is 2.1. Notably, wideband impedance matching is obtained at Port 1 and the measured $|S_{11}|$ is better than −11.2 dB from dc to the highest measurement frequency of 8 GHz. This is because the resistor R can also absorb the input power at the harmonic resonances. In addition, the measured $|S_{11}|$ within the 30-dB rejection bandwidth is better than −13 dB. Shown in Figure 7.44(c) is the power dissipation of this wideband ABSF when the input is at Port 1. The measured power dissipation is better than 95% from 1.66 to 2.38 GHz, and the corresponding fractional bandwidth is 35.6%.

In addition to the wide stopband and high level of stopband power dissipation, good passband performance is also observed. The measured passband insertion loss is within 1 dB from dc to 1.1 GHz and from 3.2 to 4.59 GHz, while the measured passband return loss is better than 16.9 dB in the same frequency range. On the other hand, the measured passband insertion loss is within 1.5 dB from dc to 1.32 GHz and from 2.82 to 5.05 GHz, while the measured passband return loss is greater than 15.3 dB in the same frequency range. As for the output reflection coefficient, the input power at Port 2 is mostly reflected within the stopband as shown in Figure 7.44(d) because the resistor R is only introduced to the first stub near Port 1.

Next, the modified design-I of proposed ABSF with moderate bandwidth in Figure 7.37 is implemented. The filter specifications are: $f_0 = 2$ GHz, FBW = 0.3, $n = 4$, and $Z_0 = 50\ \Omega$. Again, the filter is designed based on the optimal microwave BSF with a Chebyshev response and a ripple constant $\varepsilon = 0.1005$, but the first and last open stubs are replaced by the inductively loaded ones as shown in Figure 7.45. By setting $Z_{iL} = 139.3\ \Omega$, all the other circuit parameters can be calculated accordingly as given in Section 7.4.2. The ABSF is implemented using the same RO4350B substrate and the corresponding microstrip filter layout is shown in Figure 7.45(a). The Murata LQW15AN series 0402 wirewound chip inductor is used to implement the inductive loadings because a high self-resonance-frequency is required to extend the upper passband. The values of lumped chip inductors are properly selected to take the parasitic effects into account, and they are given in the caption of Figure 7.45. In addition, a 0402 resistor of 1.5 kΩ is used for R because the designed value of 1,499 Ω is not available.

Shown in Figure 7.46 are the measured and EM simulated frequency responses. The measured $|S_{21}|$ exhibits a bandstop frequency response with an absorptive stopband, and the measured rejection is 72.77 dB at $f = 2$ GHz. The measured 30-dB rejection bandwidth is from 1.89 to 2.13 GHz or 11.9% in terms of fractional bandwidth. Good frequency selectivity is achieved and the roll-off rate in the transition band from 10 to 30 dB rejection is better than 2 dB per 10 MHz. The measured $|S_{11}|$ is better than −12.9 dB from dc to 5.56 GHz. On the other hand,

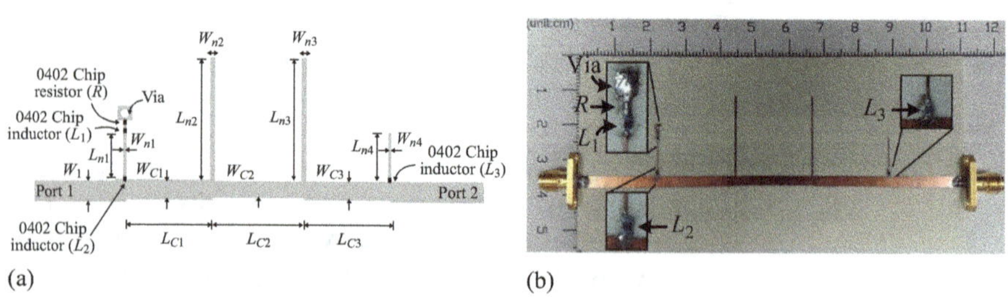

Figure 7.45 (a) Layout and (b) photograph of the modified design-I of the microstrip ABSF design example in Figure 7.37 with $f_0 = 2$ GHz, FBW = 0.3, and $n = 4$. (Geometrical parameters: $W_1 = 3.3$, $W_{C1} = 3$, $W_{C2} = 2.8$, $W_{C3} = 3$, $W_{n1} = 0.26$, $W_{n2} = 0.4$, $W_{n3} = 0.4$, $W_{n4} = 0.26$, $L_{C1} = 21.9$, $L_{C2} = 21.9$, $L_{C3} = 21.9$, $L_{n1} = 8.4$, $L_{n2} = 23.1$, $L_{n3} = 23.1$, and $L_{n4} = 9.4$. Units in mm. 0402 chip component values: $R = 1.5$ kΩ, $L_1 = 8.7$ nH, $L_2 = 8.7$ nH, and $L_3 = 12$ nH.) Reprinted with permission from Reference [18]; © 2017 IEEE

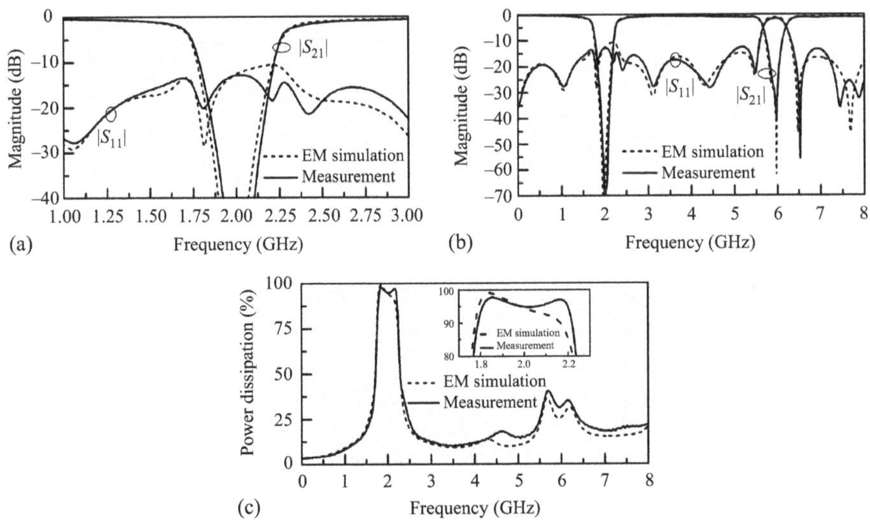

Figure 7.46 Measured and EM simulated results of the modified design-I of the wideband microstrip ABSF in Figure 7.37 with $f_0 = 2$ GHz, FBW = 0.3 and n = 4: (a) narrowband and (b) wideband S-parameter frequency responses; and (c) power dissipation frequency responses when the input is at Port 1

the measured passband insertion loss is within 1 dB from dc to 1.48 GHz and from 2.61 to 5.16 GHz. Shown in Figure 7.46(c) is the power dissipation of this ABSF with the input at Port 1. The measured power dissipation is better than 90% from 1.79 to 2.22 GHz, which corresponds to a fractional bandwidth of 21.4%.

The frequency selectivity of the wideband ABSF can be improved by using more stubs at the expense of a larger circuit area and a higher passband insertion loss. Although the maximal reflection coefficient increases as n increases, the modified design-III in Figure 7.41 can help reduce $|S_{11}|$ and improve $|S_{21}|$. The design example with $f_0 = 2$ GHz, FBW = 0.5 and $n = 5$ in Figure 7.42 is also implemented in microstrip form using the same substrate. The circuit layout is shown in Figure 7.47(a) with the circuit photograph in Figure 7.47(b). The Murata LQW15AN series 0402 wirewound chip inductor of 15 nH is used to realize L, and a 0402 resistor of 500 Ω is chosen for the resistor R. The $\lambda/4$ microstrip open stub behind the resistor R is bent to save some circuit area.

According to the measured and EM simulated results in Figure 7.48, the measured $|S_{21}|$ features a rejection of 65.7 dB at $f_0 = 2$ GHz. The measured 30-dB rejection bandwidth is from 1.68 to 2.32 GHz or 32.4% in terms of fractional bandwidth. The shape factor ($RBW_{3\text{-dB}}/RBW_{30\text{-dB}}$) is 1.7, which is obviously smaller than that of the design example in Figure 7.44 with the same FBW but fewer number of stubs. The roll-off rate in the transition band from 10 to 30 dB rejection is better than 1.67 dB per 10 MHz. An absorptive stopband is again achieved and the

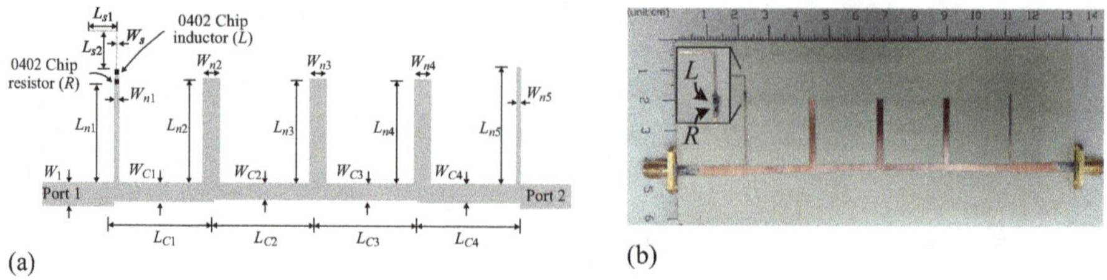

Figure 7.47 *(a) Layout and (b) photograph of the modified design-III of the microstrip wideband ABSF design example in Figure 7.42 with $f_0 = 2$ GHz, FBW = 0.5, and n = 5. (Geometrical parameters: $W_1 = 3.3$, $W_{C1} = 2.3$, $W_{C2} = 2.1$, $W_{C3} = 2.1$, $W_{C4} = 2.3$, $W_{n1} = 0.45$, $W_{n2} = 2$, $W_{n3} = 2.2$, $W_{n4} = 2$, $W_{n5} = 0.45$, $W_s = 0.2$, $L_{C1} = 21.6$, $L_{C2} = 22.8$, $L_{C3} = 22.8$, $L_{C4} = 21.6$, $L_{n1} = 21.5$, $L_{n2} = 22.3$, $L_{n3} = 22.2$, $L_{n4} = 22.3$, $L_{n5} = 23.4$, $L_{S1} = 4.4$, and $L_{S2} = 5$. Units in mm. 0402 chip component values: $R = 500\ \Omega$ and $L = 15$ nH.) Reprinted with permission from Reference [18]; © 2017 IEEE*

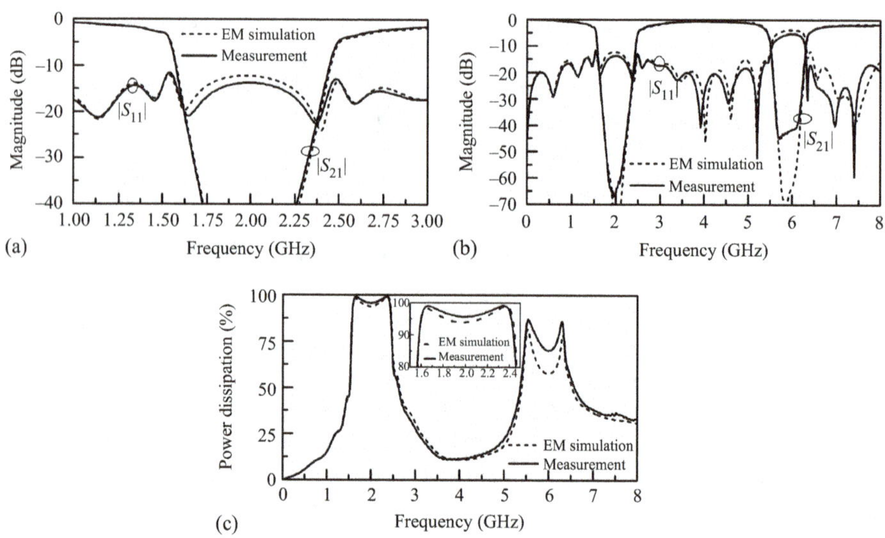

Figure 7.48 *Measured and EM simulated results of the modified design-III of the microstrip wideband ABSF in Figure 7.42 with $f_0 = 2$ GHz, FBW = 0.5, and n = 5: (a) narrowband and (b) wideband S-parameter frequency responses; and (c) power dissipation frequency responses when the input is at Port 1*

measured $|S_{11}|$ is better than −11.5 dB from dc to 5.52 GHz. The measured power dissipation is better than 90% from 1.59 to 2.43 GHz according to Figure 7.48(c), and corresponding fractional bandwidth is 41.8%. As for the passband, the measured insertion loss is within 1 dB from dc to 1.09 GHz and from 3.25 to 4.89 GHz.

7.5 Summary

In this chapter, a simple and very effective way to realize narrowband and wideband ABSFs is presented. Analytical design equations are established to allow the direct synthesis of ABSFs according to the desired specifications. Good stopband rejection, low input reflection coefficient, and high level of input power dissipation within the stopband are achieved as demonstrated by the design examples. In addition, wide passband with low passband insertion loss and good impedance matching is also obtained. Both the narrowband and wideband ABSFs can be easily extended to higher-order ones for better selectivity. They can be used in applications where the absorption of unwanted signal/interference is required.

References

[1] D. R. Jachowski, "Passive enhancement of resonator Q in microwave notch filters," in *IEEE MTT-S Int. Microw. Symp. Dig.*, 2004, pp. 1315–1318.
[2] D. R. Jachowski, "Compact frequency-agile, absorptive bandstop filters," in *IEEE MTT-S Int. Microw. Symp. Dig.*, June 2005, pp. 513–516.
[3] H. W. Bode, "Wave Filter," U.S. Patent 2,035,258, March 24, 1936.
[4] H. W. Bode, "Wave Filter," U.S. Patent 2,002,216, May 21, 1935.
[5] A. C. Guyette, I. C. Hunter, R. D. Pollard, and D. R. Jachowski, "Perfectly-matched bandstop filters using lossy resonators," in *IEEE MTT-S Int. Microw. Symp. Dig.*, June 2005, pp. 517–520.
[6] M. A. Morgan and T. A. Boyd, "Theoretical and experimental study of a new class of reflectionless filter," *IEEE Trans. Microw. Theory Techn.*, vol. 59, no. 5, pp. 1214–1221, 2011.
[7] M. A. Morgan and T. A. Boyd, "Reflectionless filter structures," *IEEE Trans. Microw. Theory Techn.*, vol. 63, no. 4, pp. 1263–1271, 2015.
[8] J. Lee, T. C. Lee, and W. J. Chappell, "Lumped-element realization of absorptive bandstop filter with anomalously high spectral isolation," *IEEE Trans. Microw. Theory Techn.*, vol. 60, no. 8, pp. 2424–2430, 2012.
[9] T. Snow, J. Lee, and W. J. Chappell, "Tunable high quality-factor absorptive bandstop filter design," in *IEEE MTT-S Int. Microw. Symp. Dig.*, 2012.
[10] B. Kim, J. Lee, J. Lee, B. Jung, and W. J. Chappell, "RF CMOS integrated on-chip tunable absorptive bandstop filter using Q-tunable resonators," *IEEE Trans. Electron Device.*, vol. 60, no. 5, pp. 1730–1737, 2013.
[11] T.-C. Lee, J. Lee, E. J. Naglich, and D. Peroulis, "Octave tunable lumped-element notch filter with resonator-Q-independent zero reflection coefficient," in *IEEE MTT-S Int. Microw. Symp. Digest*, June 2014.
[12] T.-H. Lee, B. Kim, K. Lee, W. J. Chappell, and J. Lee, "Frequency-tunable low-Q lumped-element resonator bandstop filter with high attenuation," *IEEE Trans. Microw. Theory Techn.*, vol. 64, no. 11, pp. 3549–3556, 2016.
[13] J.-C. S. Chieh and J. Rowland, "A fully tunable C-band reflectionless bandstop filter using L-resonators," in *Proc. 46th Eur. Microw. Conf.*, October 2016.

[14] Y. Morimoto, T. Yuasa, T. Owada, and M. Miyazaki, "Multi-harmonic absorption filter using quasi-multilayered striplines," in *IEEE MTT-S Int. Microw. Symp. Dig.*, June 2014.

[15] Y. Morimoto, T. Yuasa, T. Owada, *et al.*, "A multi-harmonic absorption circuit using quasi-multilayered striplines for RF power amplifiers," *IEEE Trans. Microw. Theory Techn.*, vol. 65, no. 1, pp. 109–118, 2017.

[16] J.-Y. Shao and Y.-S. Lin, "Millimeter-wave bandstop filter with absorptive stopband," in *IEEE MTT-S Int. Microw. Symp. Dig.*, June 2014.

[17] J.-Y. Shan and Y.-S. Lin, "Narrowband coupled-line bandstop filter with absorptive stopband," *IEEE Trans. Microwave Theory Tech.*, vol. 63, no. 10, pp. 3469–3478, 2015.

[18] S.-H. Chien and Y.-S. Lin, "Novel wideband absorptive bandstop filters with good selectivity," *IEEE Access*, vol. 5, pp. 18847–18862, 2017. DOI: 10.1109/ACCESS.2017.2752903.

[19] G. L. Matthaei, L. Young, and E. M. T. Jones, *Microwave Filters, Impedance-Matching Networks, and Coupling Structures*. Norwood, MA: Artech House, 1980.

[20] J.-S. Hong, Microstrip Filters for RF/Microwave Applications. 2nd. ed., New Jersey: Wiley, 2011.

[21] G. I. Zysman and A. K. Johnson, "Coupled transmission line networks in an inhomogeneous dielectric medium," *IEEE Trans. Microw. Theory Techn.*, vol. MTT–17, no. 10, pp. 753–759, 1969.

[22] M. C. Horton and R. J. Menzel, "General theory and design of optimum quarter wave TEM filters," *IEEE Trans. Microw. Theory Techn.*, vol. MTT–13, no. 3, pp. 316–327, 1965.

[23] M. C. Horton and R. J. Menzel, "The effectiveness of component elements in commensurate line length filters," *IEEE Trans. Microw. Theory Techn.*, vol. MTT-16, no. 8, pp. 555–557, 1968.

[24] O. P. Gupta and R. J. Menzel, "Design tables for a class of optimum microwave bandstop filters," *IEEE Trans. Microw. Theory Techn.*, vol. MTT-18, no. 7, pp. 402–404, 1970.

[25] K. Hettak, G. A. Morin, and M. G. Stubbs, "The integration of thin-film technologies for reduced-size MMICs," *IEEE Trans. Microw. Theory Techn.*, vol. 53, no. 1, pp. 283–291, 2005.

Chapter 8

Acoustic-wave-lumped-element-resonator-based bandpass filters

Dimitra Psychogiou[1] and Roberto Gómez-García[2]

8.1 Introduction

Emerging wireless communication systems, such as 5G, are increasingly calling for new developments in the physical layer. These include the realization of highly versatile RF transceivers with the ability to support multiple standards, frequency bands, and waveforms as well as the development of reconfigurable RF hardware with small *size, weight, and power* (SWaP). *Bandpass filters* (BPFs), and especially the ones that are located in the pre-select stage of the RF front-end chain, are among the most critical elements of these systems due to their ability to select the desired band of interest while at the same time suppressing interference and noise.

Among all the different filtering technologies that have been reported in the open technical literature, BPFs using *acoustic-wave* (AW) resonators, such as those based in *surface-acoustic-wave* (SAW) and *bulk-acoustic-wave* (BAW) phenomena, are the most commonly employed ones in wireless communication transceivers [1,2]. This is attributed to their unique RF performance advantages including: (i) *quality factors* (Qs) in the order of 1,000—e.g., ~10,000–15,000 for SAW resonators for frequencies between 100 MHz and 1 GHz—, (ii) small form factors—1–4 mm^3 per resonator—, and (iii) compatibility with low-cost manufacturing [3,4]. Alternative AW-resonator-based BPF architectures have been reported to date. These include ladder-, lattice-, and self-cascaded configurations [1] which exhibit excellent RF performance characteristics in terms of *insertion loss* (IL), selectivity, and size. Nevertheless, their performance is limited by: (i) narrow *fractional bandwidth* (FBW)—restricted by the *electromechanical coupling coefficient* (k_t^2) of their structural material and is typically between 0.4–0.8k_t^2 [3]—, (ii) the presence of spurious modes, (iii) the need for AW resonators with dissimilar resonant frequency and k_t^2 which makes them sensitive to manufacturing and k_t^2 variations, (iv) large in-band group-delay (τ_g) variation—±59% for commercially available SAW filters—, and (v) static response due to the lack of tunable AW resonators [4].

[1]Department of Electrical, Computer, and Energy Engineering, University of Colorado, USA
[2]Signal Theory and Communications Department, University of Alcala, Spain

Taking into consideration the aforementioned RF performance limitations, this chapter reports on a new type of planar and high-Q BPFs in which AW resonators are effectively combined with *electromagnetic* (EM) elements [4–14] in mixed-technology implementations. The merits of the proposed hybridized high-Q/low-Q AW-resonator based BPF concept include the realization of: (i) FBWs that do not depend on the k_t^2 of the AW resonators and can be even made wider (e.g., ~$5k_t^2$ in [4]), (ii) passbands whose group delay (τ_g) can be designed to be flat [11–14], (iii) transfer functions with multiple levels of continuous analog RF tuning [8,9], and (iv) RF filtering architectures in which identical AW resonators can be employed. Although the design of AW filters is traditionally performed by means of computationally expensive finite-element full-wave simulations, this chapter focuses on a new design approach using *coupling-matrix*-(CM)-based synthesis.

The content of this chapter builds upon the work published in [4–14] and is summarized as follows. Section 8.2 focuses on the CM modeling of two-terminal one-port-type AW resonators and introduces a simple CM-based model for the representation of low- and high-frequency spurious resonances. In Section 8.3, the CM modeling of the most fundamental element of the proposed high-Q/low-Q resonator technology, the so called *acoustic-wave-lumped-element resonator* (AWLR), is presented. Afterwards, Section 8.4 discusses an RF design methodology for quasi-elliptic-type BPFs with high out-of-band isolation levels and FBW $\gg k_t^2$. BPFs with continuous analog RF tuning and transfer functions that can be designed to exhibit either one or multiple passbands—i.e., multi-band behavior—are discussed in Section 8.5. Lastly, Section 8.6 presents an RF design methodology for the realization of flat-in-band-τ_g AWLR-based BPFs and RF diplexers.

8.2 Acoustic-wave resonators

8.2.1 Coupling-matrix-based modeling

8.2.1.1 Coupling matrix representation of the Butterworth–Van Dyke model

Two-terminal one-port-type AW resonators exhibit two high-Q closely spaced resonances (>10,000 for SAW resonators), namely the series resonance f_s and the anti-resonance f_a, and are typically modeled by the *Butterworth–Van Dyke* (BVD) circuit model—*motional resistance* R_M, *motional capacitance* C_M, *motional inductance* L_M, and *parallel capacitance* C_P. The BVD schematic along with an example frequency response of a commercially available SAW resonator is depicted in Figure 8.1(a). For a given AW resonator geometry, f_s and f_a can be specified from the geometrical and material properties of the AW resonator [3] or they can be calculated from the BVD circuit parameters using (8.1).

$$f_S = (2\pi\sqrt{L_M C_M})^{-1}, \qquad f_a = (2\pi\sqrt{L_M C_M C_P/(C_M + C_P)})^{-1} \tag{8.1}$$

Taking into consideration that the frequency response of the AW resonator comprises one series and one parallel-type resonance, a *coupling-matrix diagram*

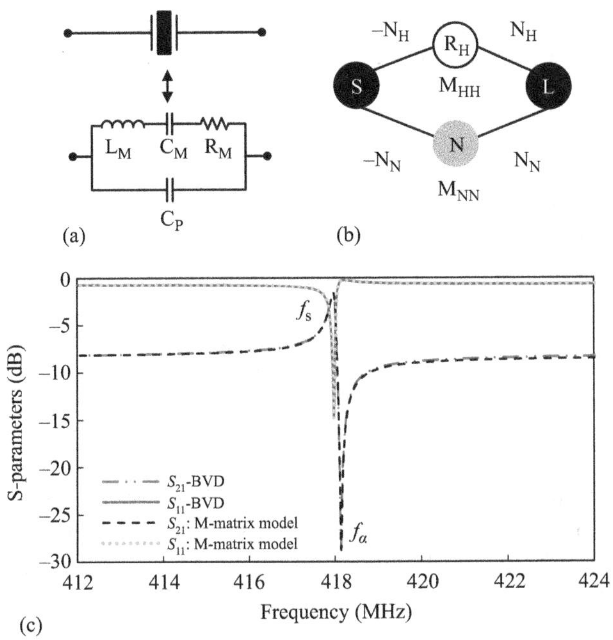

Figure 8.1 *AW resonator. (a) Butterworth–Van Dyke (BVD) model. (b) CMD with $N_H = N_N = 0.707$, $M_{HH} = -0.22i$ (i.e., $Q_{RH} = 13{,}000$), and $M_{NN} = -2.42$. (c) Synthesized S-parameters using the BVD model in (a) for a two-terminal one-port-type AW resonator with $L_M = 119.7\,\mu H$, $C_M = 1.21\,fF$, $R_M = 24\,\Omega$, $C_P = 1.59\,pF$, $k_t^2 = 1 - (f_s/f_a)^2 = 0.077\%$, and the CMD in (b). Reprinted with permission from Reference [5]; © 2015 IEEE*

(CMD) that comprises one resonant node R_H and one non-resonant node N can be used for its representation as shown in Figure 8.1(b) [5]. The self-coupling coefficients M_{HH} and M_{NN} can be calculated using (8.2) whereas N_N and N_H ($N_N = 0.707$ and $N_H = 0.707$ for the example in Figure 8.1) are obtained by fitting the CMD response to the one obtained by RF-measurements or EM-simulations assuming an FBW equal to FBW_{AW}. Note that FBW_{AW} and Q_{AW} in (8.2) respectively refer to the FBW and Q of the series resonant branch of the AW resonator that can be determined using (8.3), in which Z_0 denotes the reference impedance of the two-port network.

$$M_{HH} = -i/(FBW_{AW}Q_{AW}), \quad M_{NN} = 2(f_S - f_a)/(FBW_{AW}f_s) \tag{8.2}$$

$$\begin{aligned} FBW_{AW} &= \frac{\sqrt{Z_0{}^2 + A + 2Z_0\sqrt{A}} - \sqrt{Z_0{}^2 + A - 2Z_0\sqrt{A}}}{2\pi L_M f_S} \\ A &= L_M{}^2(2\pi f_S)^2 + Z_0{}^2 \\ Q_{AW} &= 2\pi f_S L_M / R_M \end{aligned} \tag{8.3}$$

Having defined the CMD of the AW resonator it can be readily used in coupling-matrix-based synthesis for the realization of a wide set of RF filtering transfer functions, including bandpass-type, bandstop-type and multi-band ones as it will be shown in the next sections. In addition, it allows the use of AW resonators together with EM-based ones in mixed-technology configurations.

8.2.1.2 Spurious-mode modeling

In a practical realization scenario, AW resonators exhibit low- and high-frequency spurious modes that appear in a close proximity to their fundamental resonances f_s and f_a and may limit the filter performance. In order to include them at the simulation level and study their effect, the multi-mode CMD in Figure 8.2(a) can be employed [5]. As can be seen, the higher-order modes are taken into consideration by adding extra resonant nodes—S_1–S_K, where K is the number of the existing spurious modes—in-parallel to the primary resonant node. In this case, the self-coupling coefficients M_{11}–M_{KK} and the inter-node coupling coefficients N_1–N_K can be calculated through (8.4), in which FBW_X, Q_X, and f_X are, respectively, the FBW, Q, and resonant frequency of each spurious resonance that can be derived from the RF-measured or EM-simulated response. A comparison between the CM-synthesized and the RF-measured response of a commercially available SAW resonator is depicted in Figure 8.2(b) and verifies the applicability of the multi-mode CMD for spurious-mode representation.

$$
\begin{aligned}
M_{XX}|_{X=1,\ldots K} &= -i/(Q_X FBW_{AW}) + 2(f_S - f_X)/(FBW_{AW} f_s) \\
N_X|_{X=1,\ldots K} &= N_H (FBW_X / FBW_{AW})^{0.5}
\end{aligned}
\tag{8.4}
$$

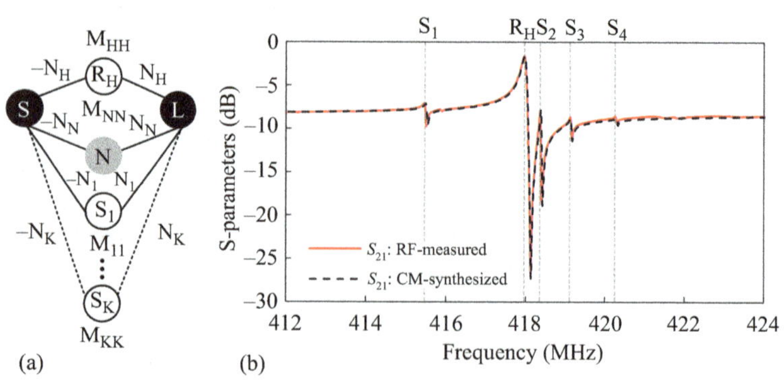

Figure 8.2 CMD of a multi-mode AW resonator. The parallel-cascaded resonant nodes represent the nodes associated to the fundamental series resonant frequency (R_H) and the spurious resonant frequencies (S_K), respectively. (b) Measured response of a commercially available SAW resonator (ASR418S2) and synthesized response for $N_H = N_N = 0.707$, $N_1 = 0.155$, $N_2 = 0.23$, $N_3 = 0.135$, $N_4 = 0.07$, $M_{NN} = -2.42$, $M_{HH} = -0.22i$ (i.e., $Q_{RH} = 13{,}000$), $M_{11} = 37.5 - 0.33i$, $M_{22} = -6.2 - 0.22i$, $M_{33} = -17.8 - 0.3i$, and $M_{44} = -35.5 - 0.23i$. Reprinted with permission from Reference [5]; © 2015 IEEE

8.3 Acoustic-wave-lumped-element resonators

A component of fundamental importance for all filter architectures that will be discussed in this chapter is the *acoustic-wave-lumped-element resonator* (AWLR). It is based on a hybrid implementation scheme in which one high-Q AW resonator is connected in parallel with a low-Q *lumped-element* (LE) resonator and results in a single high-Q—as high as the Q of the AW resonator—resonance in-between two symmetrically or asymmetrically allocated *transmission zeros* (TZs). The operational principles and RF-performance advantages of the AWLR can be better understood by looking into its CMD for both the series (Figure 8.3) and the parallel (Figure 8.4) configurations as described in the next two sections.

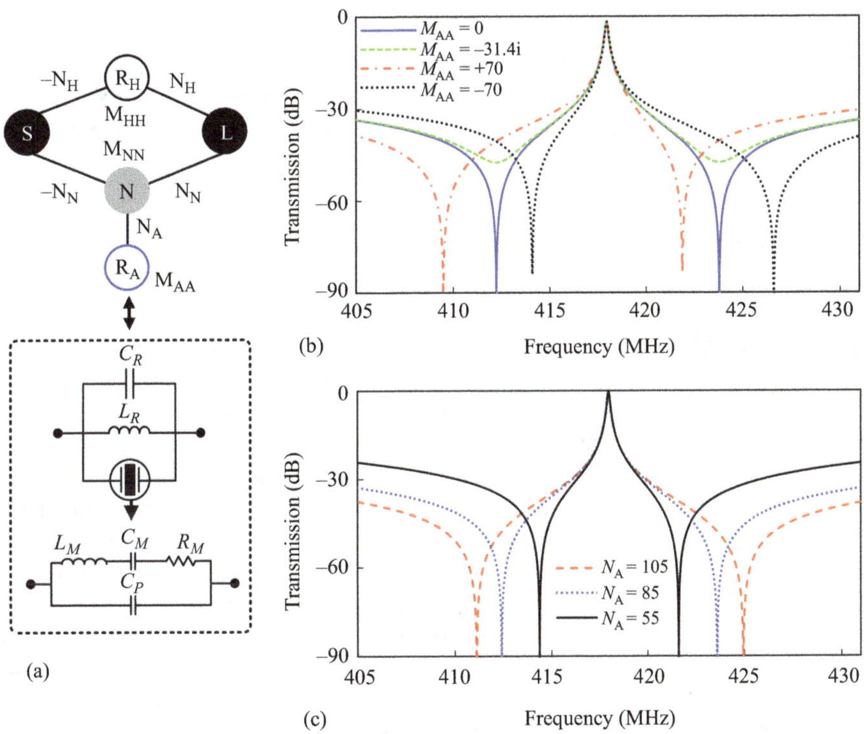

Figure 8.3 *(a) CMD and RF circuit equivalent of the AWLR. (b) Examples of synthesized power transmission responses for various levels of M_{AA}: 70, -70, and $-31.4i$ ($Q_{RA} = 100$). The rest of the coefficients are: $N_H = N_N = 0.707$, $N_A = 87$, $M_{HH} = -0.22i$ ($Q_{RH} = 13{,}000$), and $M_{NN} = 0$. (c) Examples of synthesized power transmission responses for various levels of N_A: 55, 85, and 105. The rest of the coefficients are: $N_H = N_N = 0.707$, $N_A = 87$, $M_{HH} = -0.22i$ ($Q_{RH} = 13{,}000$), $M_{NN} = 0$, and $M_{AA} = 0$. Reprinted with permission from References [5,6]; © 2015 IEEE*

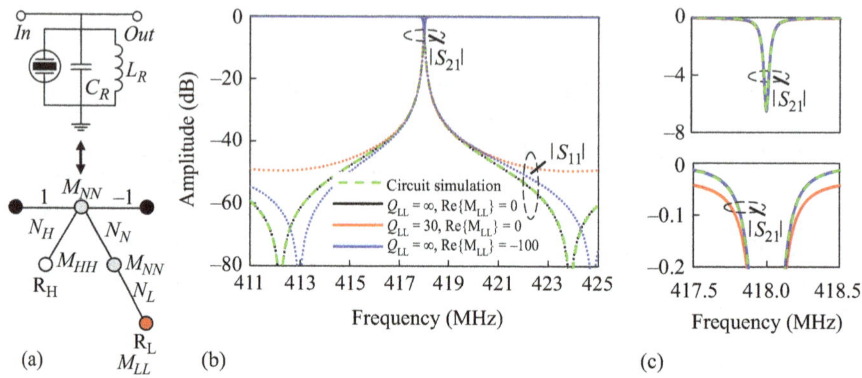

Figure 8.4 *(a) Circuit schematic and CMD. (b) Examples of synthesized power transmission and reflection responses for various Qs (Q_{LL}) and resonant frequencies Re{M_{LL}} of R_L. The rest of the coupling coefficients are: $N_H = 1.41$, $N_N = 1$, $N_L = 250$, $M_{HH} = -0.84i$—i.e., $Q_{HH} = 13{,}000$—, and $M_{NN} = 0$. Reprinted with permission from Reference [7]; © 2016 IEEE*

8.3.1 Series configuration

The CMD and the corresponding circuit schematic equivalent of the AWLR when connected in series to the input/output of a two-port network is shown in Figure 8.3(a) and stems from the CMD in Figure 8.1(b) [5,6]. It can be shown that by adding a resonant node R_A below the non-resonating node N of the CMD of the AW resonator, f_s is effectively decoupled from f_a as shown in the example responses in Figure 8.3(b) and (c). In this manner, a high-Q passband that is located in-between two TZs with FBW that can be calculated through (8.3) can be created. In a practical filter design scenario, the AWLR can be implemented by adding a LE resonator (L_R and C_R) in parallel to the parasitic capacitance C_P of the AW resonators, Figure 8.3(a). Considering that for a given AW resonator topology the coupling coefficients M_{NN}, M_{HH}, N_N, and N_H are fixed, the location of the out-of-band TZs at $\Omega_{Z1,2}$ can be only adjusted by modifying M_{AA} and N_A as described in (8.5) that respectively correspond to altering the resonant frequency of the parallel resonator (through the variation of L_R, C_R, and C_P) or its relevant L_R, C_R, and C_P values—for the same resonant frequency as the high-Q resonant node—in relation to the ones of the series resonant branch L_M, C_M.

$$\Omega_{Z1,2} = \left(-M_{AA}N_N \pm \sqrt{M_{AA}^2\, N_N^2 + 4\, N_H^2\, N_A^2}\right)/2\, N_N \tag{8.5}$$

A set of power transmission response examples are plotted in Figure 8.3(b) and (c), respectively, to illustrate the aforementioned TZ controllability. The dependence of the power transmission response on the finite Q of R_A (Q_{RA}) is also shown at the same figure. It can be observed that the isolation levels at the location of the TZs only depends on Q_{RA}, which is practically defined by the Q of L_R and C_R since the

C_P element of the AW resonator is considered lossless. Noteworthy is that even for a Q_{RA} of the order of 100, the isolation levels at the location of the TZs are over 45 dB. In addition, it should be noted that the passband insertion loss (IL) remains unchanged. This demonstrates the suitability of the AWLR module for the realization of high-Q BPFs.

8.3.2 Parallel configuration

AWLRs can be also realized by connecting one AW resonator and one LE resonator in parallel to the input/output of a two-port network as shown in Figure 8.4. In an analogous manner to the one discussed in the previous section, its CMD can be represented by one high-Q resonator R_{H}, one low-Q resonator R_{L}, and two non-resonating nodes as shown in Figure 8.4(a) [7]. In particular, $\mathrm{R_H}$ is the high-Q series resonance of the AW resonator and R_{L} is the low-Q resonance from C_P, C_R, and L_R. The coupling coefficients of the AWLR CMD can be extracted by fitting the synthesized CMD response to the one obtained by circuit/EM simulations or RF measurements and they are listed in the caption of Figure 8.4 for the example case of a commercially available SAW resonator from Abracon Corporation and a LE resonator with $C_R = 0$. As expected, the in-parallel connected AWLR creates one TZ and two *reflection zeros* (RZs) whose spectral location can be adjusted by modifying the resonant frequency of R_{L} (i.e., $\mathrm{Re}\{M_{LL}\}$). As can be also seen in the same figure, the finite Q of $\mathrm{R_L}$ (Q_{LL}) does not affect the depth (i.e., the effective Q of R_{H} remains almost unchanged) of the TZ. It only results in slightly reduced-magnitude RZs and an insignificant amount of passband IL (0.04 dB) even for Q_{LL}s as low as 30. A comparison between the CM-synthesized and the EM-simulated response of an AWLR that comprises a commercially available (SAW) resonator ($C_M = 1.211$ fF, $L_M = 119.67$ µH, $R_M = 22.2$ Ω, and $C_P = 1.59$ pF) is illustrated in Figure 8.4(b) successfully validating the CMD model for in-parallel connected AWLRs.

8.4 Quasi-elliptic-type bandpass filters using a hybrid combination of high-Q and low-Q resonator modules

8.4.1 Hybrid high-Q/low-Q resonator modules

8.4.1.1 RF designs with enhanced fractional bandwidth

The simplest form of a BPF using AWLRs can be designed by connecting one AWLR in-between the input/output ports of a two-port network as shown in Figure 8.3 [6]. It results in a low-order bandpass-type transfer function shaped by one pole and two TZs—i.e., one-pole/two-TZ BPF. However, its passband exhibits a FBW equal to $\mathrm{FBW_{AW}}$ which is fairly narrow and smaller than the k_t^2 of the AW resonator—$0.4k_t^2$ in this case. In order to increase the FBW of the AWLR module, additional impedance inverters with a normalized coupling coefficient S_r need to be added at the input/output ports as shown in Figure 8.5(a). In this case, the FBW of the passband can be arbitrarily designed to a value FBW_{NEW} by selecting S_r through (8.6). In order to demonstrate the operational principles of

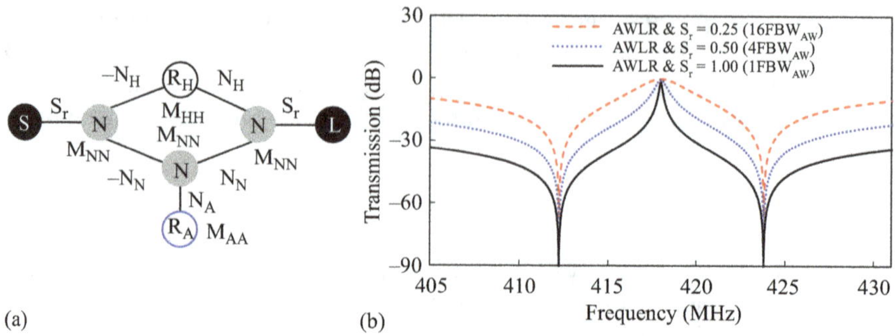

Figure 8.5 (a) CMD of the enhanced-FBW one-pole/two-TZ AWLR-based BPF and (b) power transmission response for various levels of S_r and $N_H = N_N = 0.707$, $N_A = 87$, $M_{HH} = -0.22i$ (i.e., $Q_{RH} = 13{,}000$) and $M_{NN} = M_{AA} = 0$. Reprinted with permission from Reference [5]; © 2015 IEEE

this configuration, Figure 8.5(b) illustrates the synthesized response of the CMD in Figure 8.5(a) for various sets of S_r. As can be seen, the FBW of the obtained BPF response can be arbitrarily increased to theoretically any value and does not depend on the k_t^2 of the AW resonator as opposed to conventional ladder- or lattice-type AW resonator-based BPFs that exhibit FBWs in the range 0.4–0.8k_t^2—e.g., the ones in [1–3].

$$S_r = 1/\sqrt{r}, \quad r = FBW_{NEW}/FBW_{AW} \tag{8.6}$$

8.4.1.2 RF designs with enhanced FBW and out-of-band isolation

In addition to controlling the FBW of a BPF, it is highly desirable to be able to tailor its out-of-band isolation. This is facilitated by cascading two low-Q resonant nodes R_L in-series with the AWLR as shown in the CMD in Figure 8.6. Note that in this configuration both nodes need to resonate at the center frequency of the passband (f_s) while featuring different inductance (L_{RL}) and capacitance (C_{RL}) values than those of the remaining CMD nodes. In this manner, the inter-node coupling coefficients that are adjacent to R_L (i.e., N_N, N_H, and S_r) are scaled by a factor x that represents the inductance scaling among the resonators as well as the nodes transformation in relevance to the CMD in Figure 8.5(a). For a given inductance L_{RL}, the scaling factor x can be calculated using (8.7). A comparison between the synthesized response of the hybrid high-Q/low-Q AWLR-based CMD and that of Figure 8.5 is illustrated in Figure 8.6 for various levels of x. As can be seen, a decrease of x produces an increase of the out-of-band rejection. The finite Q effect of the resonant nodes R_L (with Q equal to Q_{RL}) is also shown in this figure through a comparison between the synthesized transfer function of the CMD in Figure 8.6 for Q_{RL} infinite and Q_{RL} of 100. It can be observed that it comes at the expense of increased IL. However, the corresponding effective quality factor (Q_{eff})

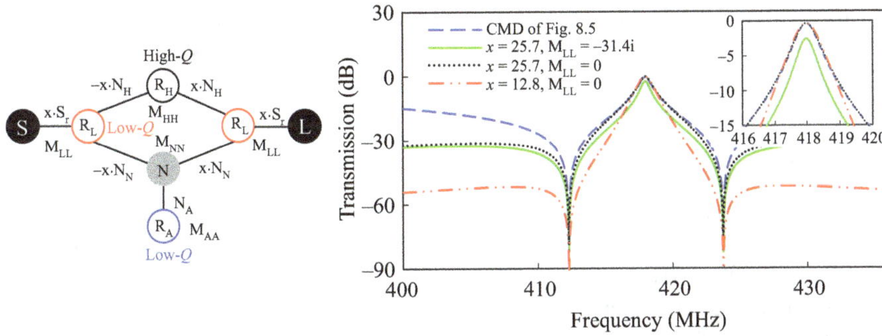

Figure 8.6 CMD and synthesized power transmission response of the high-Q/low-Q AWLR BPF with enhanced FBW and improved isolation for various levels of x (x = 12.8 and 25.7) and $M_{LL} = 0$ and $-31.4i$ (i.e., Q_{RL} equal to infinite and 100). The values of the rest of the coupling coefficients are: $M_{AA} = 0$, $S_r = 0.42$, $N_A = 87$, and $M_{NN} = 0$. Node R_A is considered lossless. Reprinted with permission from References [5,6]; © 2015 IEEE

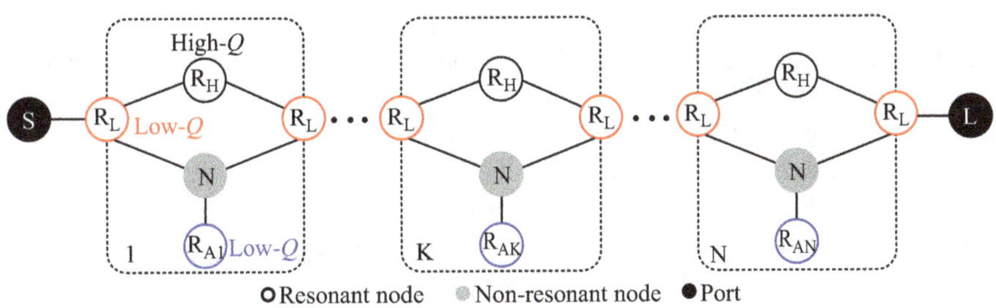

Figure 8.7 Generalized CMD of an Nth-order quasi-elliptic BPF shaped by N high-Q/low-Q resonator modules that exhibit N-poles and 2N TZs. Reprinted with permission from Reference [5]; © 2015 IEEE

of the resulting passband is of the order of 2,300—despite the use of low-Q resonant nodes (R_L) with Q_{RL} around 100.

$$x = \sqrt{\frac{L_{RL} 2\pi f_S}{FBW_{AW} Z_o}} \tag{8.7}$$

8.4.2 RF design of high-order transfer functions

8.4.2.1 Generalized coupling-matrix diagram

Taking into consideration that each high-Q/low-Q resonator module contributes to one pole and two TZs, high-order quasi-elliptic bandpass-type transfer functions shaped by N poles and $2N$ TZs can be designed by cascading in-series N high-Q/low-Q resonator modules through impedance inverters as shown in Figure 8.7.

The illustrated CMD facilitates the synthesis of BPF transfer functions with FBW $\gg k_t^2$ whose TZs can be located around the passband in a symmetric or in an asymmetric fashion. This is done by controlling the coupling coefficients N_A and M_{AA} of each AWLR module as described in (8.5).

8.4.2.2 Second-order design example

This section discusses the synthesis of a second-order transfer function that is shaped by two poles and four symmetrically allocated TZs. For this design, the CMD and its equivalent coupling-matrix representation in Figure 8.8(a) are used and its functionality is summarized as follows. The coupling sections with normalized coefficients R_R and S_r tailor the BPF FBW and power matching levels, respectively. Furthermore, the BPF isolation is controlled by x as in the first-order transfer function in Figure 8.6. An example of a synthesized second-order transfer function that realizes a 3-dB absolute bandwidth of 0.8 MHz (lossless case) and features symmetrically located TZs is depicted in Figure 8.8(b) for various levels of Q_{RL}. Note that, in this implementation, lossless resonant nodes R_A are considered. As it can be seen, the cascade of two AWLR modules with improved isolation gives rise to spurious passbands that are symmetrically located around the BPF's center frequency. However, their spurious transmission can be significantly mitigated by introducing a finite Q in the resonant nodes R_L, which is typically the case in a practical BPF development in which LE resonators are considered. It is apparent that there is a trade-off between passband IL and the magnitude of the spurious modes as observed in Figure 8.8(b). However, in most realistic implementations, the use of resonant nodes R_L with $Q_{RL} < 200$ offers a relatively good compromise in terms of the above parameters. Such an example is shown in Figure 8.8(b), in which a second-order BPF response with Q_{eff} of the order of 1,200 and minimum isolation of 55 dB is obtained by using R_L nodes with Q_{RL} of only 75, making the proposed BPF synthesis approach suitable for LE development.

8.4.2.3 Planar integration scheme and experimental validation

In order to demonstrate the practical realization aspects of the high-Q/low-Q AWLR-based BPF design concept, the RF design of a second-order BPF is discussed in this section. The filter was designed on a Rogers RO 4003C dielectric substrate with the following characteristics: dielectric permittivity $\varepsilon_r = 3.55$, dielectric thickness $H = 1.524$ mm, dielectric loss tangent $\tan(\delta_D) = 0.0027$, and 35-μm-thick Cu-cladding. In particular, a BPF design for a center frequency of 418 MHz, 50-Ω reference impedance, and a 3-dB absolute bandwidth of 0.5 MHz was considered. Commercially available two-terminal one-port-type SAW resonators (ASR418S21) from Abracon Corp. were employed as R_H resonant nodes. Furthermore, for the realization of the low-Q resonant nodes R_A and R_L and their adjacent impedance inverters (coupling coefficients S_r and R_R), LE components from Johanson Tech. and Coilcraft Inc. were utilized as shown in the circuit equivalent in Figure 8.9. It should be noted that the illustrated circuit schematic is a one-to-one representation of the CMD in Figure 8.8. Its RF design is summarized as follows. First, we define the CMD and construct its coupling matrix for the desired FBW and out-of-band isolation levels. Afterwards, having obtained its

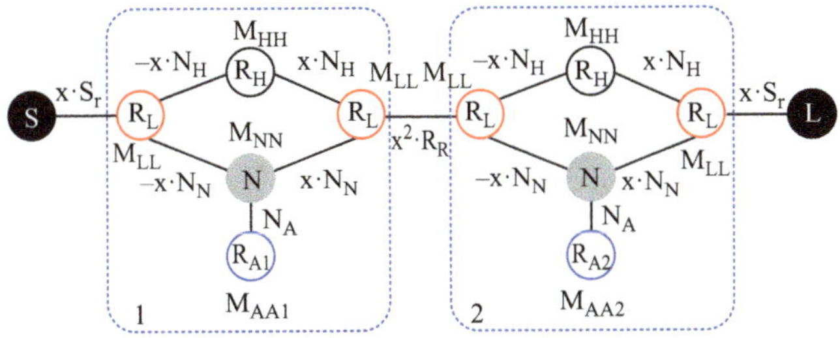

$$
M=\begin{pmatrix}
0 & S_r & 0 & 0 & 0 & 0 & 0 & 0 & 0 & 0 & 0 & 0\\
S_r & M_{LL} & -xN_H & -xN_N & 0 & 0 & 0 & 0 & 0 & 0 & 0 & 0\\
0 & -xN_H & M_{HH} & 0 & 0 & xN_H & 0 & 0 & 0 & 0 & 0 & 0\\
0 & -xN_N & 0 & M_{NN} & N_A & xN_N & 0 & 0 & 0 & 0 & 0 & 0\\
0 & 0 & 0 & N_A & M_{AA1} & 0 & 0 & 0 & 0 & 0 & 0 & 0\\
0 & 0 & xN_H & xN_N & 0 & M_{LL} & x^2R_R & 0 & 0 & 0 & 0 & 0\\
0 & 0 & 0 & 0 & 0 & x^2R_R & M_{LL} & -xN_H & -xN_N & 0 & 0 & 0\\
0 & 0 & 0 & 0 & 0 & 0 & -xN_H & M_{HH} & 0 & 0 & xN_H & 0\\
0 & 0 & 0 & 0 & 0 & 0 & -xN_N & 0 & M_{NN} & N_A & xN_N & 0\\
0 & 0 & 0 & 0 & 0 & 0 & 0 & 0 & N_A & M_{AA2} & 0 & 0\\
0 & 0 & 0 & 0 & 0 & 0 & 0 & xN_H & xN_N & 0 & M_{LL} & S_r\\
0 & 0 & 0 & 0 & 0 & 0 & 0 & 0 & 0 & 0 & S_r & 0
\end{pmatrix}
$$

(a)

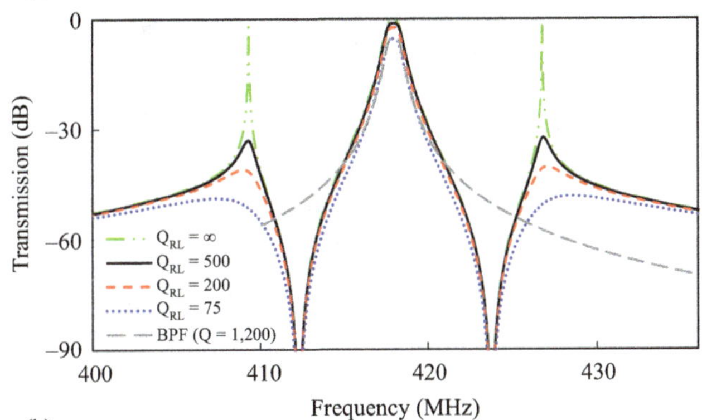

(b)

Figure 8.8 *(a) Second-order high-Q/low-Q BPF with improved isolation. (a) CMD and CM. (b) Frequency response of a second-order transfer function that stems from the CMD in (a) with 3-dB absolute bandwidth of 0.8 MHz and alternative levels of the Q of the resonator* R_L *(lossless* R_A *is considered in these examples) using the following coupling coefficients:* $x=37$, $N_H=N_N=0.707$, $N_A=87$, $M_{AA1}=M_{AA2}=0$, $M_{HH}=0.22i$ *(i.e.,* $Q_{RH}=13{,}000$*),* $S_r=0.132$*, and* $R_R=0.1$*. The frequency response of a second-order BPF with similar bandwidth and Q of 1,200 is also shown. Reprinted with permission from Reference [5]; © 2015 IEEE*

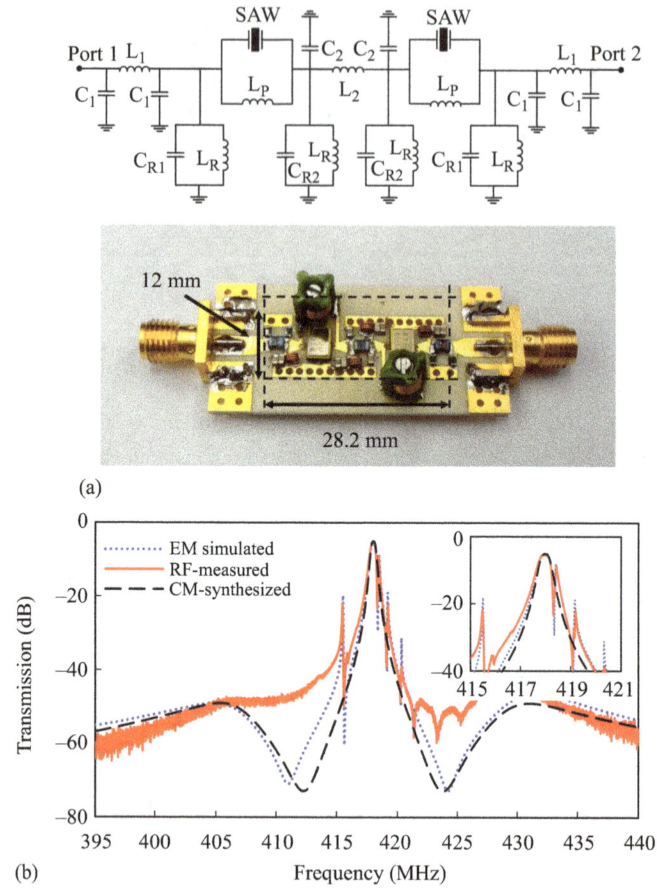

Figure 8.9 (a) Second-order BPF using the CMD in Figure 8.8 and (b) prototype. The utilized LE components are as follows: C_1 = Johanson 251R14S2R0BV4T, L_1 = Coilcraft 0805HQ-39N, C_2 = Johanson 251R14S1R2BV4T, L_2 = Coilcraft 0805CS-6N9, C_{R1} = Johanson 251R14S0R3BV4T//251R14S180JV4T, L_R = Coilcraft 0806SQ-6N9, C_{R2} = Johanson 251R14S0R5BV4T//251R14S160GV4T, L_P = Coilcraft 165-05-A06, and SAW = Abracon ASR418S2. (b) RF-measured, EM-simulated (multi-mode resonator model), and CM-synthesized power transmission responses. Reprinted with permission from Reference [5]; © 2015 IEEE

inter-node coupling coefficients, the coupled-resonator approach in [15] is employed in order to specify the realistic impedance-inverter values (90°-length-at-f_s transmission lines) which are in turn implemented by their first-order π–type LE equivalent circuit as shown in Figure 8.9. Lastly, the final values of the LEs that need to be utilized in the actual BPF prototype are obtained through full-wave post-layout simulation analysis of the realistic filter geometry (landing pads for the LE components and RF excitation interface) using the EM solver of the Advanced Design Systems (ADS) software from Keysight Technologies. An example filter prototype along with its RF-measured, EM-simulated and CM-synthesized responses in which the multi-mode AW resonator module in Figure 8.2 was used are shown in Figure 8.9. The filter RF performance characteristics are summarized as follows: 3-dB absolute bandwidth equal to 0.52 MHz, minimum in-band IL = 5.4 dB (SMA connector loss and RF excitation interface are included), $Q_{\mathrm{eff}} = 1{,}900$, FBW = $0.12\% \approx 1.6kt^2 \approx 3.5FBW_{\mathrm{AW}}$, out-of-band isolation levels higher than 46 dB, and in-band minimum return loss (RL) = 33 dB. They confirm the applicability of the high-Q/low-Q BPF concept for the realization of low-IL quasi-elliptic-type transfer functions with FBWs that are wider than the k_t^2 of the AW resonator.

8.5 Tunable bandpass filters using acoustic-wave resonators

The BPF concept introduced in Section 8.4 demonstrated the potential to realize quasi-elliptic-type AW-resonator-based BPFs with low in-band insertion loss and wideband out-of-band suppression bandwidth by cascading in series high-Q AWLRs and low-Q LEs. Despite facilitating the realization of transfer functions with Q_{eff} in the order of a thousand, the overall Q_{eff} of the hybrid high-Q/low-Q resonator module is 5–10 times smaller than the Q of its constituent AW resonator. In this section, we discuss an alternative design concept for AWLR-based BPFs that overcomes this limitation and enables the synthesis of transfer functions with Q_{eff} that are comparable to the one of the AW resonator but at the expense of narrowed out-of-band isolation bandwidth. In addition, as an important advantage to be highlighted, this topology can be made tunable in terms of various RF performance metrics. These include passband bandwidth, out-of-band isolation bandwidth, and selectivity. Lastly, this filter synthesis technique can be readily extended to the design of multi-band bandpass-type transfer functions that can be reconfigured at will.

8.5.1 Single-band AWLR-based bandpass filters

8.5.1.1 Generalized coupling matrix diagram

The generalized CMD of the tunable AWLR-based BPF along with an example power transmission response are depicted in Figure 8.10. It is based on N high-Q AWLR-based modules that are cascaded in parallel to an all-pass network—constructed by $N\!-\!1$ impedance inverters (coupling coefficients: $K_{1,2}$-$K_{N-1,N}$)—through N impedance inverters (coupling coefficient: N_E) and results in a quasi-elliptic-type transfer function that comprises N poles (ω_{S1}-ω_{SN}) and $2N$ TZs ($\omega_{N:Z1}$-$\omega_{N:Z2}$) [8,9]. As it will be further discussed in this section, this BPF allows the reconfiguration of various transfer-function parameters including bandwidth, out-of-band isolation levels, selectivity, and intrinsically selectable all-pass state. These filter characteristics can be respectively controlled by modifying the coupling coefficients N_E and the location of the AWLR TZs (self-coupling coefficient M_{LL} of the resonant node $\mathrm{R_L}$) through the integration of variable capacitors into the

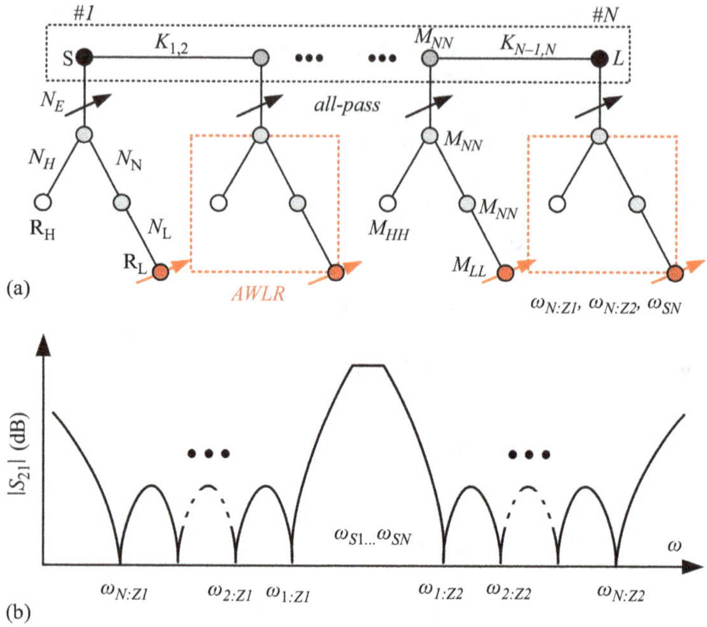

Figure 8.10 Tunable AWLR-based BPF concept. (a) CMD (black circles: source/load; gray circles: non-resonating nodes; white circles: high-Q resonating nodes; red circles: low-Q resonating nodes; and solid lines: inter-node couplings). (c) Illustration of its corresponding transfer function that comprises N poles (ω_{S1} to ω_{SN}) and 2N TZs ($\omega_{1:Z1}$ to $\omega_{N:Z1}$ at the left side of the passband and $\omega_{1:Z2}$ to $\omega_{N:Z2}$ at the right side of the passband). Reprinted with permission from Reference [8]; © 2016 IEEE

corresponding filter parts as it will be discussed in the practical realization example in Section 8.5.1.3.

8.5.1.2 Three-pole/six-transmission zero bandpass filter design

In order to illustrate the operating principles of the tunable AWLR-based BPF, an example three-pole/six-TZ (i.e., $N = 3$) filter architecture is considered. In this example AWLRs with $f_S = 418$ MHz are used as high-Q resonators. Furthermore, the coupling coefficients N_E and $K_{1,2} = K_{2,3}$ are implemented with 90°-long ideal transmission lines at 418 MHz with characteristic impedances Z_E and $Z_1 = Z_2$, respectively. Based on this circuit, various responses are plotted in Figure 8.11. In particular, Figure 8.11(b) and (c) shows that:

(i) The passband bandwidth can be altered by only changing the value of Z_E—i.e., N_E in the CMD of Figure 8.10(a)—so that wider bandwidths are obtained for higher Z_E. In addition, they can be designed to be wider than the k_t^2 of the AW resonators as opposed to conventional ladder or lattice AW resonator-based BPF designs whose bandwidths are limited to 0.4–0.8k_t^2 [1–3].

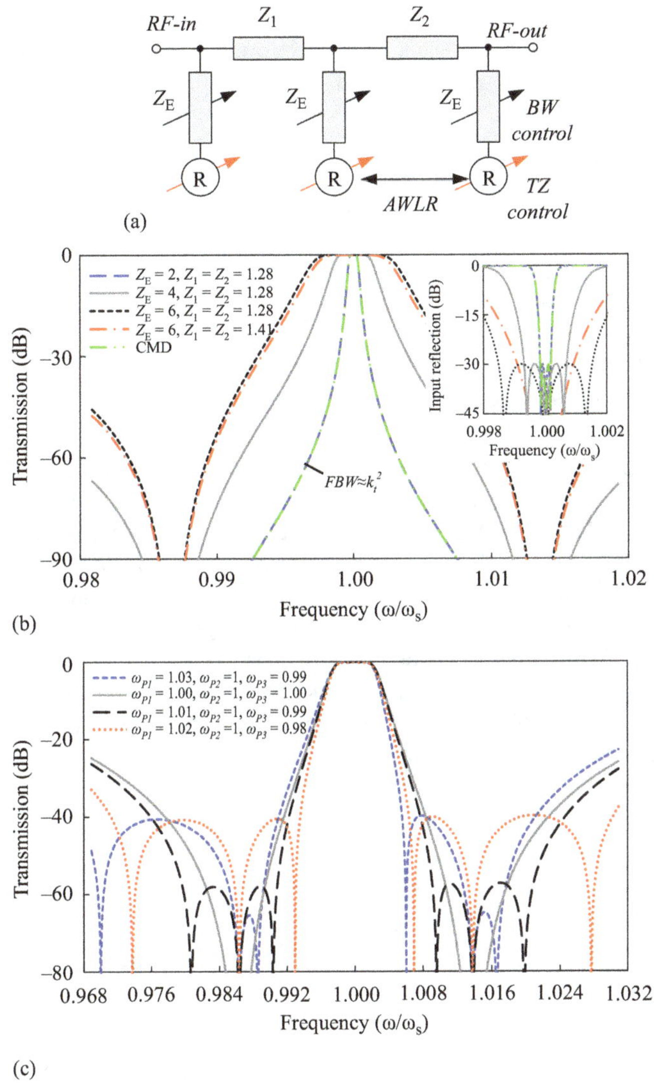

Figure 8.11 Synthesized power transmission responses of a three-pole/six-TZ AWLR BPF ($N = 3$) for: (a) various levels of Z_E and $Z_1 = Z_2$ (normalized to the filter impedance level) and $\omega_S = \omega_{P1} = \omega_{P2} = \omega_{P3}$. The inset shows the input-power-reflection details around the passband. A comparison between a response obtained through linear-circuit simulations and CM synthesis using the CMD in Figure 1(b) for $M_{NN} = 0$, $M_{HH} = 0$, $M_{LL} = 0$, $N_H = 1.4142$, $N_N = 1$, $N_L = 237.5$, $K_{1,2} = K_{2,3} = 0.78$, and $N_E = 0.5$ is also shown; (b) different values of ω_P of its constituent AWLRs (ω_{P1}, ω_{P2}, and ω_{P3}), $Z_E = 6$, and $Z_{1,2} = Z_{2,3} = 1.41$. Reprinted with permission from Reference [8]; © 2016. IEEE

(ii) The input power-reflection characteristics can be adjusted by altering the parameter values of the all-pass network (i.e., $K_{1,2}$ and $K_{2,3}$ in the CMD) that enables both maximally flat (i.e., normalized impedance values $Z_1 = Z_2 = 1.41$) and equi-ripple-type (i.e., normalized impedance values $Z_1 = Z_2 = 1.28$) passbands to be created.

(iii) The out-of-band isolation characteristics can be designed by setting the TZs of each AWLR (six TZs in this case) in a symmetrical or in an asymmetrical fashion (i.e., select ω_{P1}, ω_{P2}, and ω_{P3} or $\mathrm{Re}\{M_{\mathrm{LL}_1}\}$, $\mathrm{Re}\{M_{\mathrm{LL}_2}\}$, and $\mathrm{Re}\{M_{\mathrm{LL}_3}\}$ in the CMD of Figure 8.10(a)) which in turn results in symmetrical or asymmetrical isolation profiles.

8.5.1.3 RF tuning using variable reactive elements

In a practical realization scenario, transfer-function tunability can be realized by incorporating variable reactive elements in the BPF volume. In particular, the out-of-band isolation characteristics are reconfigured by incorporating tunable capacitors into the C_R elements of each AWLR which accordingly change the resonant frequency of each R_{L}. In order to achieve bandwidth reconfiguration, the impedance inverters Z_E are replaced by their first-order π-type (90°-electrically-long transmission-line segment at the filter center frequency) high-pass circuit equivalent (C_E and L_E, inset of Figure 8.12) and tuned through the integration of a variable capacitor into C_E. It should be noticed that, when replacing the ideal inverter Z_E (red trace) by its π-type circuit equivalent (gray trace), the locations of the out-of-band TZs slightly shift towards the center frequency of the passband. This is attributed to the frequency-dependent characteristics of the π-type circuit

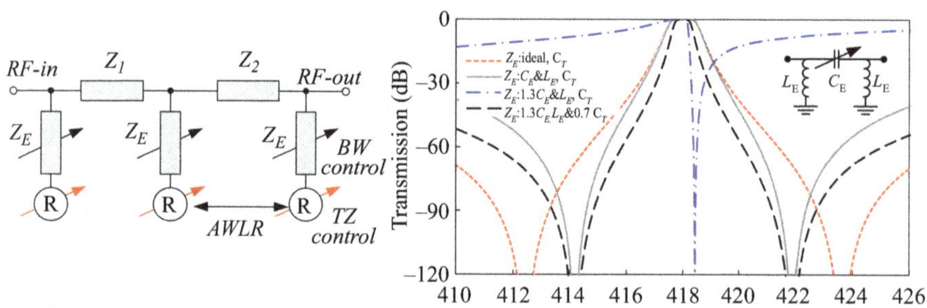

Figure 8.12 Power transmission response of a three-pole/six-TZ AWLR BPF (N = 3) in which Z_E is realized as an ideal 100-Ω impedance inverter (red trace) and as a 90°-electrically long-at-418-MHz transmission-line segment with 100-Ω impedance (gray trace). This transmission-line-based impedance inverter is represented by its high-pass π-type circuit equivalent (L_E, C_E) in order to functionalize the broad bandwidth state and then tuned to a lower bandwidth state by increasing C_E and decreasing C_T. Reprinted with permission from Reference [8]; © 2016 IEEE

equivalent and need to be taken into consideration in the actual filter design. As it can be seen in Figure 8.12, narrower bandwidths can be obtained by increasing C_E. However, the resulting transfer function becomes asymmetrical (blue trace) due to the non-ideal spectral behavior of the reconfigured impedance inverter (electrical length different to 90° at 418 MHz). Considering the dependence of the TZs on the resonant frequency of R_L, this effect can be counteracted by further tuning C_R as shown in the dashed black trace (i.e., decrease its value in terms of the total parallel capacitance $C_T = C_P + C_R$ of the broad bandwidth state, red and gray trace). Consequently, for the realization of symmetrical transfer functions with continuously variable bandwidth both the C_E and C_T—through C_R—elements of each resonant branch need to be tuned.

8.5.1.4 Planar integration scheme and experimental validation

In order to validate the tunable AWLR-based BPF concept, a single-band three-pole/six-TZ BPF was designed and manufactured on a Rogers 4003C dielectric substrate considering a 50-Ω reference impedance, center frequency of 418 MHz, bandwidth of 0.37 MHz, and a maximally flat in-band response. Commercially available two-terminal/one-port-type SAW resonators from Abracon Corporation (ASR418S2 part with series resonant frequency of 418 MHz) were employed as high-Q resonant nodes R_H. The rest of the filter components, including all the impedance inverters and the low-Q resonant nodes R_L, were materialized through LEs from Coilcraft Inc. (inductors) and Johanson Tech. (capacitors). The all-pass-type inverters Z_1 and Z_2 in Figure 8.12 were realized by their second-order low-pass-type circuit equivalent whereas the Z_E ones by their first-order high-pass one in order to allow for bandwidth tuning through the incorporation of a variable trimmer capacitor (1–5 pF from Johanson Tech.) into each branch. An example filter assembly is shown in Figure 8.13. It exhibits a measured bandwidth of 0.37 MHz (FBW around $1.2k_t^2$), minimum in-band IL of 1.8 dB, and RL of 30 dB that correspond to a $Q_{\text{eff}} > 10{,}000$. A comparison between the RF-measured and

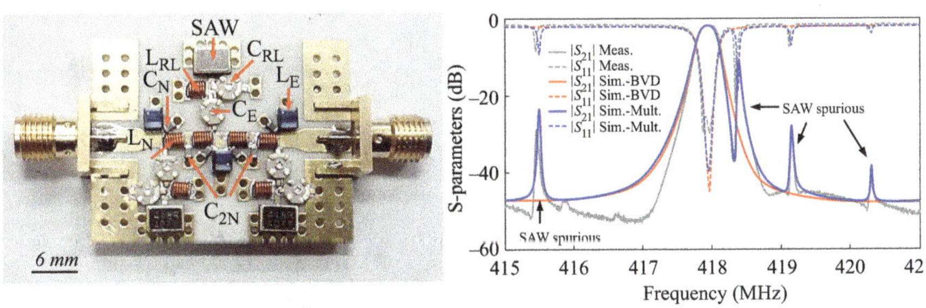

Figure 8.13 Manufactured prototype of a three-pole/six-TZ AWLR-based BPF and comparison between its RF measured and EM simulated response using the single-mode BVD and the multi-mode AW resonator model. Reprinted with permission from Reference [8,9]; © 2016 IEEE

EM-simulated response—single-mode representation of the SAW resonator using the CMD in Section 8.2.1.1 and multi-mode representation of the SAW resonator using the CMD in Section 8.2.1.2—is also shown in Figure 8.13 which, as it can be seen, are in good agreement.

The reconfiguration capabilities of the manufactured prototype are demonstrated in Figure 8.14 [8]. In particular Figure 8.14(a) shows ***analog bandwidth tuning***—obtained by simultaneously varying C_R and C_E in the filter prototype in Figure 8.13, in which larger bandwidths are obtained by decreasing C_E and increasing C_R— in the range 0.16–0.49 MHz (i.e., 3:1 tuning ratio). This corresponds to an FBW tuning in the interval 0.5–1.5k_t^2. For the aforementioned tuning range, the minimum IL was measured in the range 3.3–1.2 dB and the RL below 30 dB, which correspond to $Q_{\mathrm{eff}} > 10{,}000$. Furthermore, Figure 8.14(b) depicts ***selectivity/order tuning*** by altering the number of the active resonant branches within the filter volume. It is implemented by intrinsically switching-off the high-Q resonant branches by decreasing the capacitance C_E of the corresponding impedance inverter Z_E to its minimum value. Note that for all tunable-order states the minimum IL was measured in the interval 1.2–2.7 dB. Furthermore, it can be observed that for a particular bandwidth state the in-band IL remains unchanged when tuning the filter order. This is attributed to the IL of the filter being dominated by the all-pass-type network (shaped by finite Q LEs), which corresponds to a $Q_{\mathrm{eff}} > 3{,}500$, $>8{,}500$, and $>11{,}000$ for the first-, second-, and third-order transfer

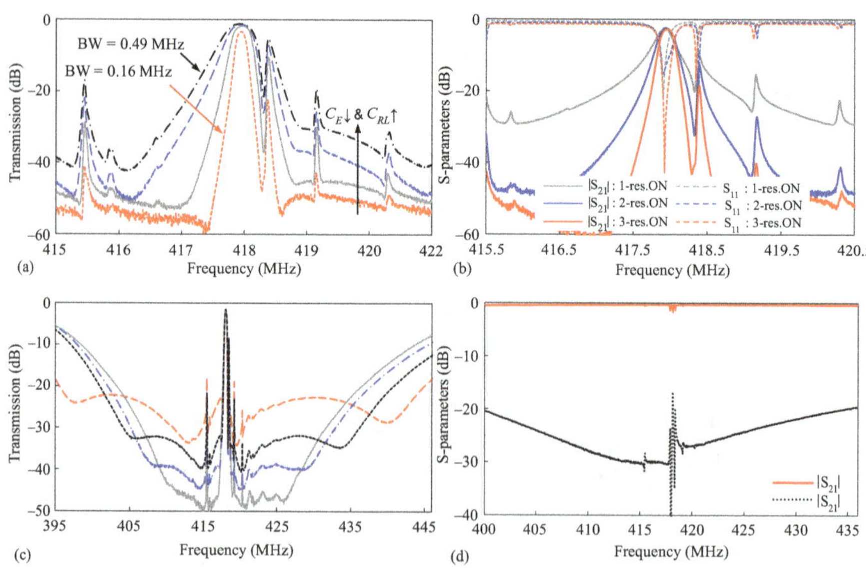

Figure 8.14 RF-measured response of the three-pole/six-TZ BPF prototype: (a) bandwidth tuning in the range 0.16–0.49 MHz; (b) selectivity tuning; (c) out-of-band isolation tuning; and (d) reconfigured all-pass state. Reprinted with permission from Reference [8,9]; © 2016 IEEE

functions, respectively. As also discussed in the previous section, ***out-of-band isolation tuning*** can also be realized by adjusting the location of the out-of-band TZs through the variation of C_R in each AWLR. This feature is shown in Figure 8.14(c) for a bandwidth state of 0.35 MHz. As can be seen, the manufactured BPF allows for tunable isolation between 23 dB and 46 dB at 426 MHz and a variable bandwidth (isolation <20 dB) in the range 18–26.5 MHz (at the upper stopband). Furthermore, an ***all-pass operational mode*** can be obtained by intrinsically switching off—through C_E—all the active resonant branches of the circuit as shown in Figure 8.14(d).

8.5.2 Multi-band AWLR-based bandpass filters

8.5.2.1 Generalized coupling matrix diagram

The generalized CMD and transmission response of the multi-band AWLR-based BPF are illustrated in Figure 8.15. It is based on N high-Q multiresonant AWLR branches that are shaped by a parallel combination of K AW resonators and one LE resonator. Note that each AWLR resonator exhibits K distinct resonances which in turn give rise to K transmission poles ω_{S1} to ω_{SK} that are allocated at the center frequency of each of the K bands. Furthermore, due to the presence of N branches, each passband is shaped by N poles and the overall filter transfer function exhibits $N(K+1)$ TZs. As an advantage to be highlighted when compared to conventional multi-channel/band parallelized-type RF filter architectures, it requires fewer inverters when a large number of passbands needs to be synthesized. This results in

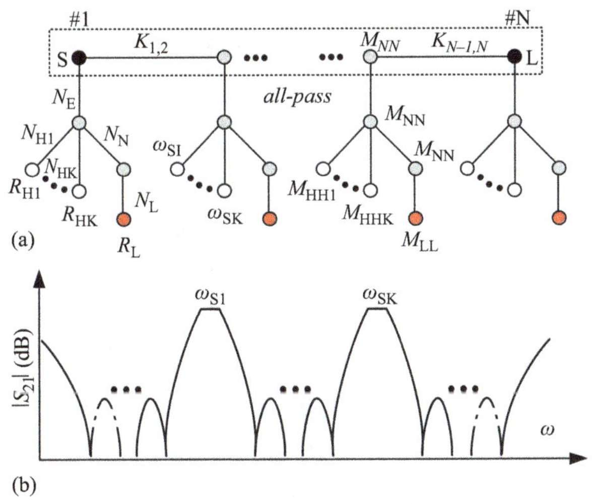

Figure 8.15 Multi-band AWLR-based BPF that exhibits K passbands (ω_{S1} to ω_{SK}), each one shaped by N poles (equal to the number of resonant branches) and N(K + 1) TZs in its out-of-band response. Reprinted with permission from Reference [8]; © 2016 IEEE

lower IL and significantly smaller size. Note that for the synthesis of a K-band transfer function with N poles at each band, $N(K + 1) - 1$ inverters are necessary in the devised filter approach—due the series-resonant nature of each AW resonator, the N_{HK} inverters are not taken into account as they are not physically present in the realistic filter prototype—whereas $K(N + 1)$ inverters are needed in the classic channelized multi-band schemes. As such, the engineered multi-band AWLR filter concept requires less inverters when $K > N - 1$.

8.5.2.2 Dual-band four-pole/six-transmission zero design example

To demonstrate the RF design principles of the multi-band AWLR-based BPF concept, various transfer-function examples of a four-pole/six-TZ dual-band BPF architecture were synthesized and are shown in Figure 8.16. As it can be seen, two transmission bands are created at ω_{S1} and ω_{S2}. Their locations are controlled by the natural resonant frequencies of the high-Q resonant nodes R_{H1} (nodes 3 and 10 with self-coupling coefficients M_{HH_1} in the CMD of Figure 8.16) and R_{H2} (nodes 4 and 9 with self-coupling coefficients M_{HH_2} in the CMD of Figure 8.16). Furthermore, symmetrical and asymmetrical transfer-function profiles can be created by controlling the location of the six TZs in the out-of-band response. This is done by adjusting the natural resonant frequencies of the low-Q resonant nodes R_{L1} (node 6 with self-coupling coefficient M_{LL_1} in the CMD of Figure 8.16) and R_{L2} (node 7 with self-coupling coefficient M_{LL_1} in the CMD of Figure 8.16). It can be observed that symmetrical profiles are obtained for $M_{LL_1} = -M_{LL_2}$ and asymmetrical ones for $M_{LL_1} \neq -M_{LL_2}$, which demonstrates the ability to produce multi-band filtering responses with symmetrical/asymmetrical passband and stopband regions. Although not shown here, the passband bandwidth can be effectively controlled by modifying the inter-node coupling coefficient N_E in a similar manner to the one described for its single-band counterpart (i.e., $K = 1$). Note that in a realistic filter

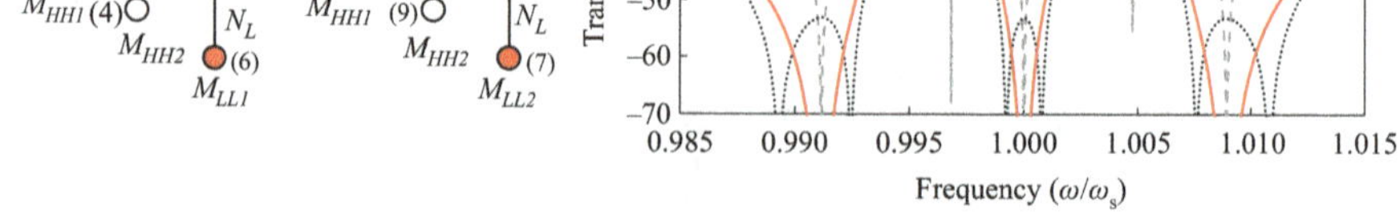

Figure 8.16 CMD and synthesized response of a dual-band BPF that exhibits two transmission bands each one shaped by two poles. The coupling coefficients of the CMD are: $M_{NN} = 0$, $M_{HH1} = -100$, $M_{HH2} = 100$, $N_H = 1.4142$, $N_E = 0.55$, $N_N = 1$, $N_L = 100$ and $K_{12} = 1$. Reprinted with permission from Reference [8]; © 2016 IEEE

development the aforementioned transfer-function metrics can be tuned by incorporating variable-reactance elements into R_L (TZ and bandwidth tuning) and N_E (bandwidth tuning) as it will be experimentally verified in the next section.

8.5.2.3 Experimental validation

For experimental validation purposes, a tunable-bandwidth dual-band six-pole/nine-TZ BPF prototype with two passbands located at 418 and at 433.9 MHz was designed, built, and measured following a similar integration and design process as the one discussed in Section 8.5.1.4. Bandwidth tunability is enabled by inserting variable trimmer capacitors (1–5 pF) at the C_E and C_{RL} locations. The corresponding manufactured prototype and RF-measured responses are illustrated in Figure 8.17. In particular, Figure 8.17(b) depicts various symmetrically tuned bandwidth states that are obtained by simultaneously tuning C_E and C_{RL} at each resonant branch—larger bandwidths are obtained by decreasing C_E and increasing C_{RL}. Note that, in the examples of Figure 8.17(b), the lower-passband bandwidth is tuned in the range 0.3–0.54 MHz and the upper-passband one in the interval 0.6–1.23 MHz with minimum in-band IL from 0.81 dB to 1.75 dB for all bandwidth states. Furthermore, the dual-band BPF prototype allows for continuously tunable asymmetrical-bandwidth profiles that are obtained by only altering the AWLR's C_{RL}. This is illustrated in Figure 8.17(c) for four different transfer-function examples. Summarizing the aforementioned RF-measured performance metrics, the conceived AWLR-based BPF principle allows for multi-band (two in this case) quasi-elliptic-type BPFs with enhanced FBW and smaller size when compared to conventional channelized architectures. Moreover, it enables simultaneous analog transfer-function adaptivity in terms of symmetrical and asymmetrical bandwidth tuning.

8.6 Constant-in-band-group-delay AWLR-based bandpass filters and RF diplexers

Modern wireless-communication systems are increasingly calling for preselect BPFs with as flat as possible in-band group delay (τ_g) due to its importance in the phase distortion of the received RF signals. However, this requirement cannot be met in the currently employed SAW or BAW ladder—or lattice—RF filtering architectures since their τ_g depends on the k_t^2 of its constituent AW resonators. Such an example is illustrated in Figure 8.18 for a ladder-type AW-resonator-based BPF that comprises one and two sections. As can be seen, the τ_g variation becomes more prominent at the passband edges. Furthermore, it worsens with the increase of the number of sections (filter order) which may severely distort the RF signals that pass through the BPF. In order to address the aforementioned limitation, this section discusses an RF design methodology [12–14] that allows the realization of flat-in-band-τ_g AW-resonator-based BPFs by co-integrating AWLRs within a Gaussian-type impedance-inverter network. In addition, the proposed design concept is extended to the realization of RF diplexers with flat-in-band-τ_g characteristics.

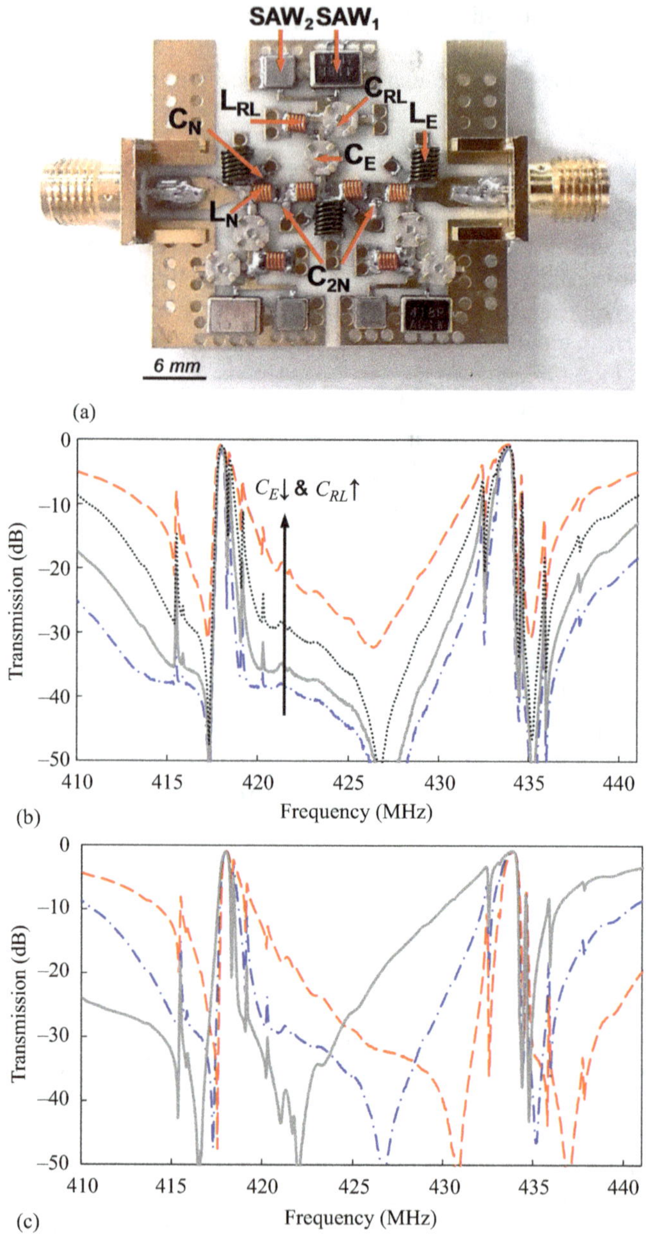

Figure 8.17 Dual-band tunable-bandwidth AWLR-based BPF. (a) Manufactured prototype. (b) RF-measured power transmission response for various symmetrically tunable bandwidth states. (c) RF-measured power transmission response for various asymmetrically tunable bandwidth states. Reprinted with permission from Reference [8]; © 2016 IEEE

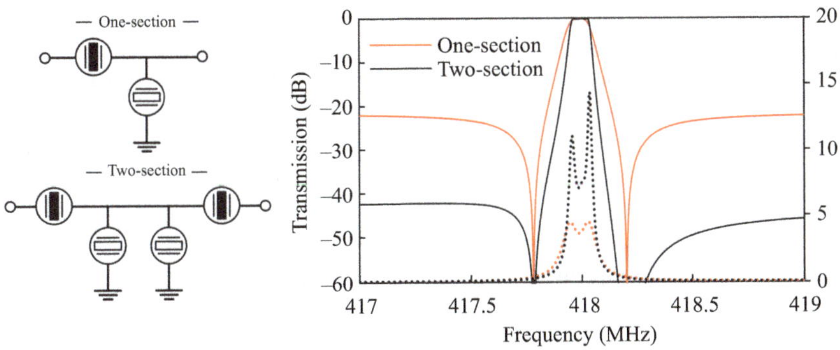

Figure 8.18 Frequency response of conventionally designed ladder-type AW-resonator-based BPFs in terms of power transmission response and τ_g. Lossless one-port-type AW resonators with $k_t^2 = 0.1\%$, series resonant frequency $f_S = 418$ MHz for the series AW resonator, and series resonant frequency $f_S = 417.78$ MHz for the parallel AW resonator are employed. Reprinted with permission from References [13,14]; © 2018 IEEE

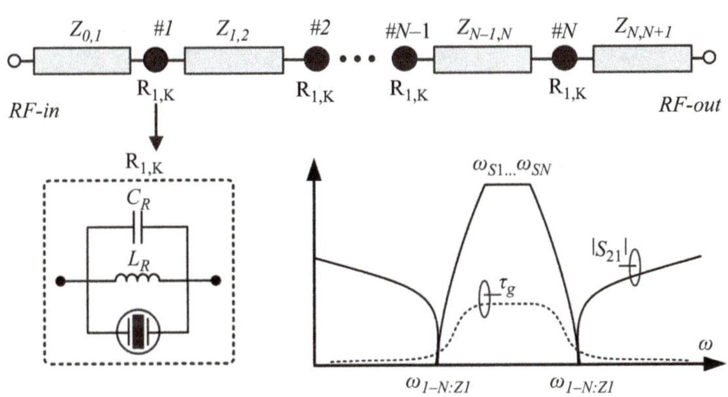

Figure 8.19 Block diagram and conceptual transfer function of the flat in-band τ_g AWLR-based BPF shaped by N + 1 impedance inverters—gray rectangles $Z_{0,1}$ to $Z_{N,N+1}$ and AWLRs—black circles—that contribute to N poles and 2N TZs. Reprinted with permission from References [13,14]; © 2018 IEEE

8.6.1 Bandpass filter with flat in-band group delay

8.6.1.1 Filter concept and design example

The circuit schematic and a conceptual transfer function of the flat-in-band-τ_g AWLR-based BPF is illustrated in Figure 8.19. It is based on N identical AWLRs (Nth-order response) that are in-series cascaded with $N + 1$ impedance inverters and exhibits a quasi-elliptic transfer function that comprises N poles and $2N$ TZs [12–14]. By appropriately selecting the location of the poles through the impedance values of the $N+1$ inverters, $Z_{0,1}$-$Z_{N,N+1}$, a constant in-band τ_g can be obtained in its passband. The location of the TZs is controlled by the resonant frequency of the LE resonators with L_R and C_R elements. As discussed in the previous AWLR-based BPF examples, they can be symmetrically arranged around the passband by assigning the resonant frequency of each LE resonator equal to f_S—series resonant frequency of each AW resonator and center frequency of the BPF passband.

In order to demonstrate the operational principles of the constant-in-band-τ_g filter concept, various examples of conventional Gaussian-type BPFs and the AWLR-BPF are shown in Figure 8.20 for both narrow (Figure 8.20(a)) and broad

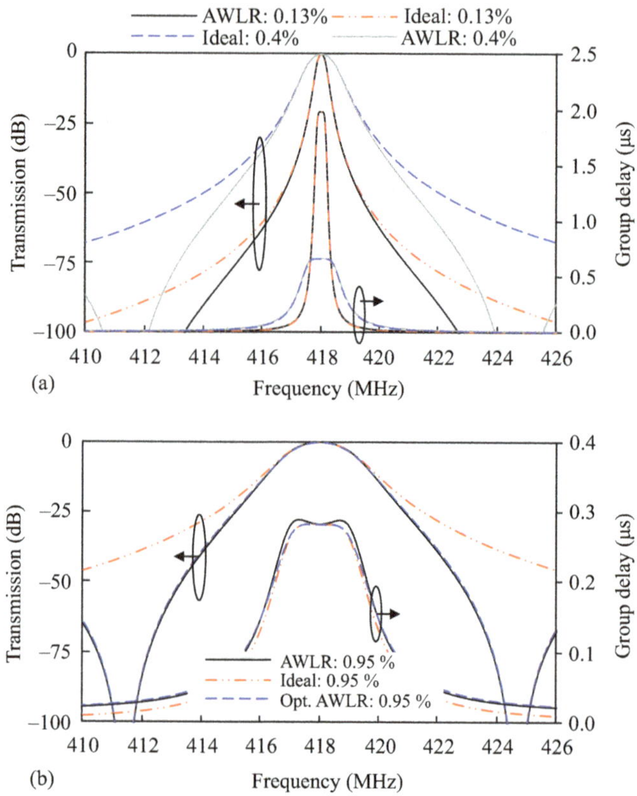

Figure 8.20 Power transmission response and τ_g of an ideal Gaussian-type BPF (three poles) and the proposed AWLR-based BPF (three poles and six TZs) with constant in-band τ_g: (a) narrow-bandwidth states: 0.13% ($1.8k_t^2$)—$Z_{0,1}=1.4$, $Z_{1,2}=2.9$, $Z_{2,3}=7.4$, and $Z_{3,4}=3.5$—and 0.4% ($5.3k_t^2$)—$Z_{0,1}=2.4$, $Z_{1,2}=8.7$, $Z_{2,3}=2.2$, and $Z_{3,4}=6.$; (b) broad-bandwidth states: 0.95% ($12k_t^2$)—Ideal: $Z_{0,1}=3.6$, $Z_{1,2}=20$, $Z_{2,3}=52$, and $Z_{3,4}=9$. Optimized/finely tuned: $Z_{0,1}=3.6$, $Z_{1,2}=20.5$, $Z_{2,3}=56$, and $Z_{3,4}=9.8$. The AWLRs consist of AW resonators with $k_t^2=0.08\%$. The indicated impedance values are normalized to the system reference impedance Z_0. Reprinted with permission from Reference [13]; © 2018 IEEE

(Figure 8.20(b)) bandwidth states and for an AW resonator with $C_M = 1.21$ fF, $L_M = 119.6$ μH, $C_P = 1.2$ pF, and $k_t^2 = 0.08\%$. The illustrated power transmission and τ_g responses have been obtained using circuit-level simulations, in which the component values are specified using (8.8)-(8.10) [15] where g_0–g_N are the normalized element values of the lowpass Gaussian-type prototype and Z_0 is the system reference impedance.

$$Z_{0,1} = \left(\frac{Z_0 FBW\,\omega_S L_M}{g_0 g_1}\right)^{0.5} \tag{8.8}$$

$$Z_{i,i+1} = \frac{FBW\,\omega_S L_M}{(g_i g_{i+1})^{0.5}} \tag{8.9}$$

$$Z_{N,N+1} = \left(\frac{Z_0 FBW\,\omega_S L_M}{g_N g_{N+1}}\right)^{0.5} \tag{8.10}$$

Figure 8.20(a) depicts a comparison between the frequency responses of conventional Gaussian-type BPFs and the AWLR BPF for narrow FBW states—obtained using (8.8)–(8.10). As can be seen, their passband characteristics are in close agreement with each other. On the contrary, when the passband bandwidth becomes wider and its edges get closer to the TZs, small differences in terms of in-band τ_g can be observed. This is illustrated in Figure 8.20(b). However, they can be readily counteracted by finely adjusting the values—2–5% from the ones calculated with (8.8)–(8.10)—of the impedance inverters. As an important advantage to be highlighted in relation to state-of-the-art AW resonators-based BPFs [1–3], this architecture facilitates the realization of FBWs that are not limited by k_t^2 and can be even made wider as shown in the examples in Figure 8.20. Due to the high inductance value of the AW resonators (~μH for SAW resonators) for the realization of wide FBW states, high impedance—proportional to L_M—inverters could be required, which may result in unrealizable values. However, this design aspect can be readily counteracted by lowering the effective series inductance of the AWLR through the parallel cascade of multiple AW resonators [11].

8.6.1.2 Experimental validation

To validate the described flat-in-band-τ_g design principles, a three-pole/six-TZ AWLR-based BPF was designed, manufactured, and measured and is shown in Figure 8.21(a). It was built on a Rogers 4003C substrate using commercially available SAW resonators from Abracon (ASR418S2 with $C_M = 1.211$ fF, $L_M =$ 119.67 μH, $R_M = 24.2$ Ω, and $C_P = 1.59$ pF). Its RF-measured response is shown in Figure 8.21(b) and (c) in terms of *S*-parameters and τ_g. Based on these responses, the measured 3-dB bandwidth, minimum in-band IL, RL, Q_{eff}, and in-band τ_g can be calculated and are respectively equal to 0.3 MHz, 2.1 dB, 32 dB, 9,000, and 1.78 ± 0.02 μs. Note that the measured passband exhibits a 2.4-times larger FBW ($0.95k_t^2$) than conventional ladder-type AW-resonator-based BPFs, for example, the ones in [1–3]. In addition, it exhibits a nearly constant τ_g over the entire passband, which is

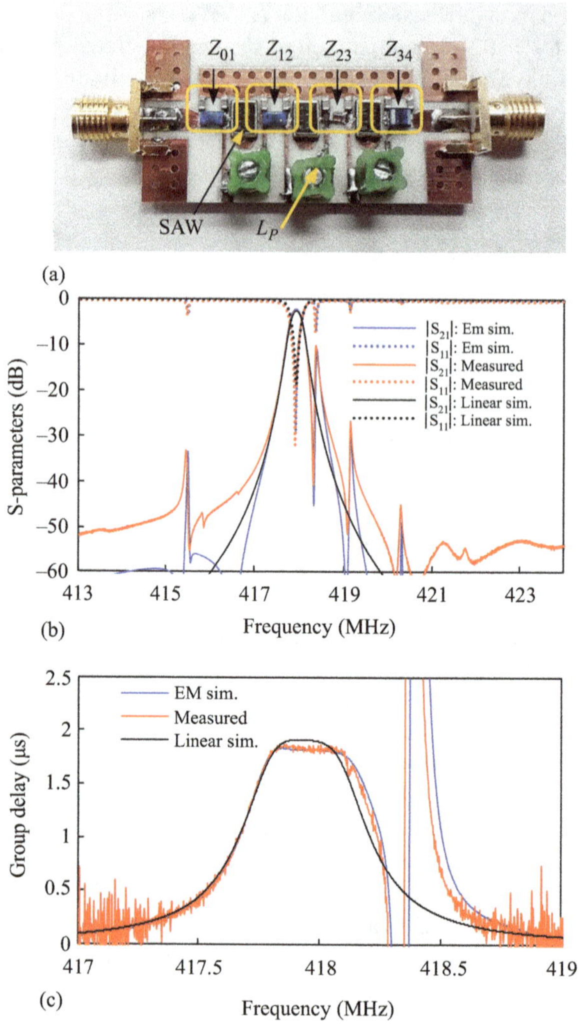

Figure 8.21 Three-pole/six-TZ AWLR BPF with constant in-band τ_g; (a) manufactured prototype; (b) simulated (linear and EM using the multimode AWLR model) and RF-measured power transmission and reflection responses; and (c) simulated (linear and EM using the multimode AWLR model) and RF-measured τ_g. Reprinted with permission from References [13,14]; © 2018 IEEE

much flatter than in AW-resonator-based BPFs for much lower minimum IL levels successfully demonstrating the merits of the proposed AWLR-based BPF with constant in-band τ_g. The spurious resonances that are observed around the passband are due to the multi-resonant nature of the SAW element, as typically observed in all AW-resonator-based BPFs. A comparison between the RF-measured,

EM-simulated—AW resonator modeled by its multi-mode circuit equivalent using the design methodology in Section 8.2.1.2—, and ideally synthesized (linear-circuit simulations) *S*-parameters and τ_g are also shown at the same figure. As can be seen, they are in a fairly close agreement, successfully validating the practical viability of the engineered flat-in-band-τ_g BPF concept.

8.6.2 RF diplexers with flat in-band group delay

8.6.2.1 Filter concept and design example

This section discusses the RF design of AWLR-based diplexers with flat-in-band-τ_g characteristics. As it can be seen in Figure 8.22, the circuit is shaped by two *N*th-order AWLR-based BPFs that are designed using the RF design methodology of Section 8.6.1. After designing each individual BPF, they are connected at the common input port (node 1) through transmission-line segments that are appropriately designed in order to preserve the passband characteristics of each band in the presence of the other. Figure 8.23 depicts ideally synthesized responses through linear circuit simulations of an RF diplexer that comprises two bands that are either widely—(example 1) or closely spaced. In both examples, the connecting transmission lines to the input node are designed as follows. Their characteristic impedances are set equal to the system reference impedance Z_0 and are 90°-long at the center frequency of each band—i.e., f_2 for Z_{inv_1} and f_1 for Z_{inv_2}. As it can be seen throughout these examples, constant in-band τ_g can be obtained for both closely spaced and widely spaced bands. Furthermore, depending on the realized design case, the illustrated responses are obtained by first selecting the impedance values using (8.8)–(8.10) and, if necessary, finely tune them as also discussed in Section 8.6.1.

8.6.2.2 Experimental validation

The manufactured prototype of example AWLR-based RF diplexer geometry is illustrated in Figure 8.24. It consists of two channels, one located at 418 MHz and the other one at 433.9 MHz. Each channel is shaped by a two-pole/four-TZ AWLR BPF. For the 418-MHz channel, the Abracon ASR418S (f_S = 418 MHz, C_M = 1.21 fF,

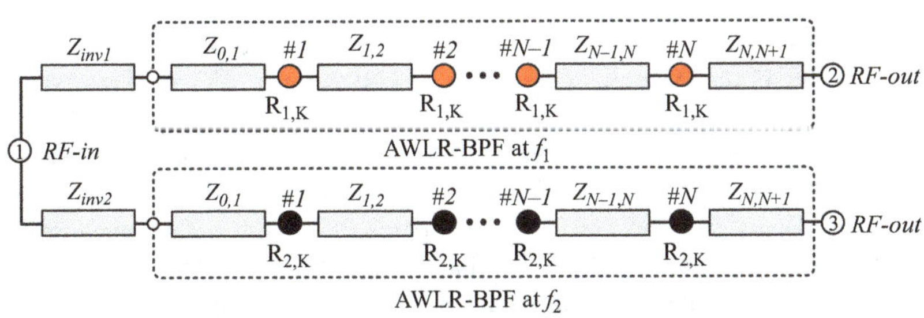

Figure 8.22 Block diagram of an AWLR-based diplexer. Reprinted with permission from Reference [13]; © 2018 IEEE

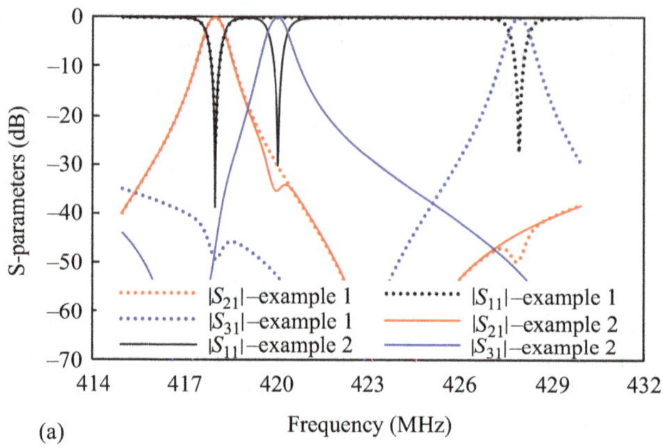

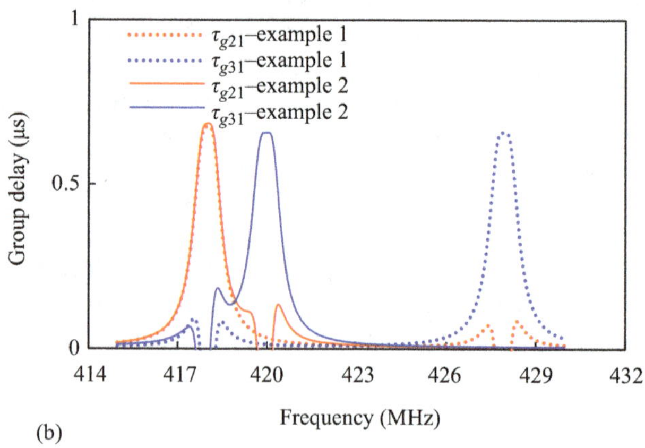

Figure 8.23 (a) Example S-parameter response for two channels each of them shaped by a two-pole/four-TZ AWLR BPF. (b) Example τ_g response for two-pole/four-TZ passbands. Example 1: closely spaced bands. Example 2: widely spaced bands. Reprinted with permission from Reference [13]; © 2018 IEEE

$L_M = 119.67$ μH, $R_M = 24.2$ Ω, and $C_P = 1.59$ pF) SAW resonator was employed, whereas for the 433.9-MHz channel the EPCOS R900 ($f_S = 433.9$ MHz, $C_M = 1.62$ fF, $L_M = 82.984$ μH, $R_M = 19$ Ω, and $C_P = 1.97$ pF) was used. Both filters are connected to the input junction through 90°-long transmission lines with characteristic impedance of 50 Ω. The filter components are shown in Figure 8.24. Its measured response was evaluated in terms of *S*-parameters and τ_g that are depicted in Figure 8.25. In particular, the 418-MHz passband exhibits the following characteristics: measured 3-dB absolute bandwidth of 0.27 MHz, minimum in-band IL of 2 dB, RL of 40 dB, $Q_{\mathrm{eff}} > 7{,}000$, and in-band τ_g around 1.43 μs. The performances of the 433.9-MHz transmission band are as follows: measured 3-dB absolute bandwidth

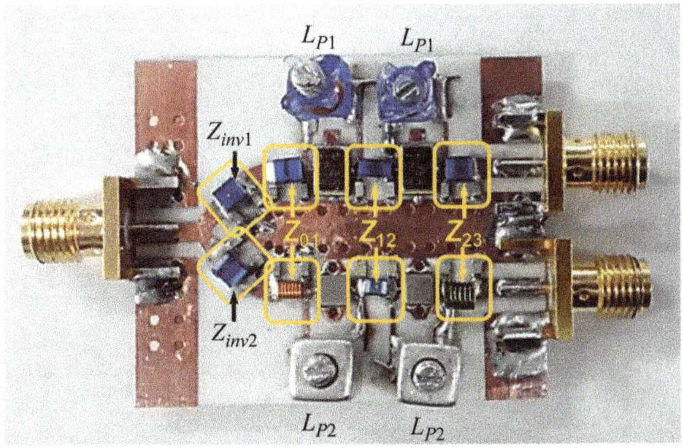

Figure 8.24 Manufactured prototype of a diplexer that comprises channels each of them shaped by two-pole/four-TZ AWLR BPFs built on a Rogers 4003, 1.52 mm-thick substrate. Components for upper channel: Z_{inv_1}*: C = 251R14S6R2CV4T (6.2 pF ceramic capacitor from Johanson Tech. Inc.) and L = 1008CS180 (18 nH inductor from Coilcraft Inc.),* Z_{01}*: C = 251R14S5R6CV4T (5.6 pF ceramic capacitor from Johanson Tech. Inc.) and L = 1008CS270 (27 nH inductor from Coilcraft Inc.),* Z_{12}*: C = 251R14S0R2AV4T (0.2 pF ceramic capacitor from Johanson Tech. Inc.) and L = 1206CS560 (56 nH inductor from Coilcraft Inc.),* Z_{23}*: C = 251R14S1R3BV4T (1.3 pF ceramic capacitor from Johanson Tech. Inc.) and L = 1008CS470 (47 nH inductor from Coilcraft Inc.),* L_{P1} *= 165-06A06L (90 nH trimmer inductor from Coilcraft Inc.), SAW = ASR418SR. Components for lower channel:* Z_{inv_1}*: C = 251R14S2R2CV4T (2.2 pF ceramic capacitor from Johanson Tech. Inc.) and L = 0807SQ10N (10.2 nH inductor from Coilcraft Inc.),* Z_{01}*: C = 251R14S6R8CV4T (6.8 pF ceramic capacitor from Johanson Tech. Inc.) and L = 0806SQ19N (19.4 nH inductor from Coilcraft Inc.),* Z_{12}*: C = 251R14S1R6AV4T (1.6 pF ceramic capacitor from Johanson Tech. Inc.) and L = 0805CS430 (43 nH inductor from Coilcraft Inc.),* Z_{23}*: C = 251R14S1R8BV4T (1.8 pF ceramic capacitor from Johanson Tech. Inc.) and L = 1111SQ39N (39 nH inductor from Coilcraft Inc.),* L_{P2} *= 164-04A06SL (56 nH trimmer inductor from Coilcraft Inc.), SAW: EPCOS R900. Reprinted with permission from Reference [13]; © 2018 IEEE*

of 0.25 MHz, minimum in-band IL of 2.6 dB, RL of 13 dB, $Q_{\mathrm{eff}} > 7{,}000$, and in-band τ_g around 1.65 μs. As can be seen, both bands feature fairly constant in-band τ_g. Note also that while the proposed concept has been tested for second-order channels, it can be readily extended to higher-order realizations or closely spaced channels.

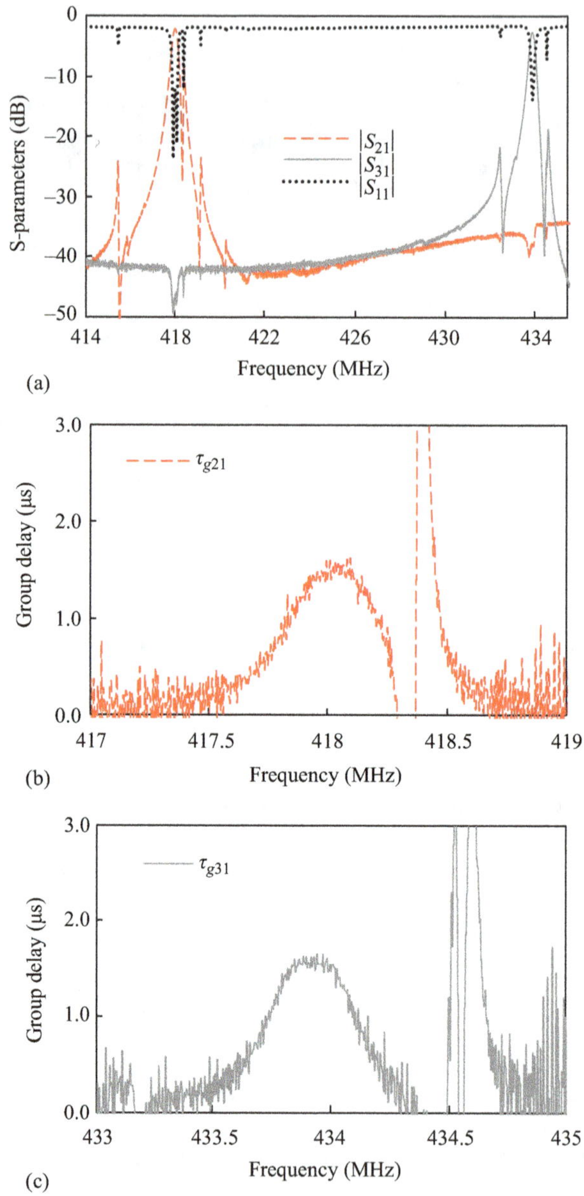

Figure 8.25 RF-measured response of the AWLR-based diplexer with flat in-band τ_g. (a) S-parameters. (b) Group delay of the lower channel (τ_{g21}) (c) Group delay of the upper channel (τ_{g31}). Reprinted with permission from Reference [13]; © 2018 IEEE

8.7 Summary

This chapter has been devoted to the RF design of a new class of planar microwave RF filters in which AW resonators are effectively combined with LEs to create quasi-elliptic bandpass-type transfer functions with the following unique RF performance characteristics: (i) FBWs that do not depend on the k_t^2 of the AW resonators, (ii) high Q_{eff}s of the order of a thousand, (iii) passbands whose in-band group delay (τ_g) can be designed to be flat, and (iv) transfer functions with multiple levels of continuous analog RF tuning. In particular, we focused on a new RF design approach that uses as a basis CM-based synthesis. The extraction of CM-based models for single- and multi-mode AW resonators as well as their equivalents when used in series or parallel configurations have been discussed in detail. Using as a basis these CM AWLR-based models, the synthesis and RF circuit/EM design of high-order BPFs has been presented. In particular, we demonstrated quasi-elliptic transfer functions with maximally flat-, equiripple-, and Gaussian-type in-band response. Furthermore, we additionally addressed the realization of multiband bandpass-type transfer functions with symmetric or asymmetric-passbands. RF tuning methods using variable reactive elements have also been discussed and demonstrated the potential for multiple levels of transfer-function tunability including passband bandwidth, out-of-band isolation, and selectivity. Lastly, this chapter presented practical integration schemes using commercially available SAW resonators and LEs.

References

[1] A. Link and P. Warder, "Golden age for filter design," IEEE Microw. Mag., Vol. 16, Issue 7, pp. 60–72, 2015.

[2] R. Aigner, "SAW and BAW technologies for RF filter applications: A review of the relative strengths and weaknesses," in IEEE Ultrason. Symp., Beijing, China, Nov. 2–5, 2008, pp. 582–589.

[3] S. Gong and G. Piazza, "Design and analysis of Lithium-Niobate-based high electromechanical coupling RF-MEMS resonators for wideband filtering," IEEE Trans. Microw. Theory Techn., Vol. 61, Issue 1, pp. 403–414, 2013.

[4] D. Psychogiou, R. Gómez-García, R. Loeches-Sánchez, and D. Peroulis, "Hybrid acoustic-wave-lumped-element resonators (AWLRs) for high-Q bandpass filters with quasi-elliptic frequency response," IEEE Trans. Microw. Theory Techn. Vol. 63, Issue 07, pp. 2233–2244, 2015.

[5] D. Psychogiou, R. Gómez-García, and D. Peroulis, "Coupling-matrix-based design of high-Q bandpass filters using acoustic-wave lumped-element resonator (AWLR) modules," IEEE Trans. Microw. Theory Techn. Vol. 63, Issue 12, pp. 4319–4328, 2015.

[6] D. Psychogiou, R. Gómez-García, and D. Peroulis, "High-Q bandpass filters using hybrid acoustic-wave-lumped-element resonators (AWLRs) for UHF applications", in IEEE MTT-S International Microwave Symposium Digest (IMS), Phoenix, AZ, USA, May 2015.

[7] D. Psychogiou, R. Gómez-García, and D. Peroulis, "Continuously-tunable-bandwidth acoustic-wave resonator-based bandstop filters and their multi-mode modeling," in Proc. 46th European Microwave Conference (EuMC), 2016, London, UK, Oct. 2016.

[8] D. Psychogiou, R. Gómez-García, and D. Peroulis, "Single- and multi-Band acoustic-wave-lumped-element-resonator (AWLR) bandpass filters with reconfigurable transfer function," IEEE Trans. Microw. Theory Techn., Vol. 64, Issue 12, pp. 4394–4404, 2016.

[9] D. Psychogiou, R. Gómez-García, and D. Peroulis, "Tunable acoustic-wave-lumped-element Resonator (AWLR)-based bandpass filters", in IEEE MTT-S International Microwave Symposium Digest (IMS), San Francisco, CA, USA, May 2016.

[10] D. Psychogiou, R. Gómez-García, and D. Peroulis, "Multi-resonant acoustic-wave-lumped-element resonators (AWLRs) for multi-band band-pass filters with enhanced fractional bandwidth," in 2017 IEEE Int. Conf. on Microw., Commun., Antennas Electron. Syst. (COMCAS), Tel Aviv, Israel, Nov., 2017.

[11] D. Psychogiou, R. Gómez-García, "Switched-bandwidth SAW-based band-pass filters with flat group delay," IET Electron. Lett., Vol. 54, Issue 7, pp. 460–462, May 2018.

[12] D. Psychogiou, R. Gómez-García, and D. Peroulis, "SAW-based bandpass filters with flat in-band group delay and enhanced fractional bandwidth", in IEEE MTT-S International Microwave Workshop Series on Advanced Materials and Processes for RF and THz Applications (IMWS-AMP), September 20–22, Pavia, Italy.

[13] D. Psychogiou, R. Gómez-García, and D. Peroulis, "Constant in-band group-delay acoustic-wave-lumped-element-resonator-based bandpass filters and diplexers," IEEE Trans. Microw. Theory Techn., Vol. 66, Issue 5, pp. 2199–2209, May 2018.

[14] D. Psychogiou, R. Gómez-García, and D. Peroulis, "RF Design of acoustic-wave-lumped-element-resonator-(AWLR)-based bandpass filters with constant in-band group delay", in IEEE MTT-S International Microwave Symposium Digest (IMS), Honolulu, HI, USA, June 4–7 2017.

[15] J.-S. Hong, Microstrip Filters for RF/Microwave Applications, 2nd Ed., Hoboken, NJ, USA: Wiley, 2011.

Chapter 9
Tunable and reconfigurable SIW filters

Juseop Lee[1]

9.1 Introduction

Microwave filters are used in wireless communications systems for selecting specific frequency signals. Designing microwave filters are mainly based on specifications such as the center frequency, the bandwidth, the insertion loss, the selectivity, and so on. In addition to the aforementioned electrical specifications, mechanical ones such as volume also need to be taken into account in the design of microwave filters. Microwave resonators are a critical building block for bandpass and bandstop filters, since the electrical and mechanical features of microwave filters are mainly determined by resonators. One of the most important parameters characterizing microwave resonators is the unloaded quality (Q) factor, and it has a significant impact on electrical performances of filters. For example, a bandpass filter having high-Q resonators has a low insertion loss in the passband and a bandstop filter composed of high-Q resonators has a high attenuation in the stopband. In general, a resonator with a large volume has a higher unloaded Q-factor at microwave frequencies when compared with a smaller resonator. Hence, high-Q resonators such as air-filled metallic waveguide resonators, dielectric resonators, and coaxial cavity resonators are used in designing high-performance static and tunable microwave filters for wireless base stations and satellites [1–5].

In general, planar transmission line resonators such as stripline and microstrip resonators are considered to be small when compared with non-planar resonators such as waveguide and dielectric resonators. In addition, they can be fabricated at relatively low cost. However, the unloaded Q-factor of a planar transmission line resonator is quite low and usually ranges up to a few hundreds [6–8]. On the other hand, the unloaded Q-factor of a waveguide resonator is very high and ranges from a few thousand at least [6–8]. Hence, microwave filters made of waveguide resonators are used in wireless systems in which very low loss is required. However, they are bulky and require sophisticated fabrication processes, which results in high fabrication cost. Since a microwave resonator does not have a high unloaded Q-factor and a small form factor at the same time, both the unloaded Q-factor and

[1]College of Informatics, Korea University, South Korea

size must be taken into account when choosing a resonator type in designing a microwave filter.

A substrate-integrated waveguide (SIW) resonator is a semi-planar version of a waveguide resonator. It is implemented in a substrate and filled completely or in part with a dielectric material from the substrate. As a result, it is typically smaller than a classical waveguide resonator and has a lower unloaded Q-factor. In addition, its unloaded Q-factor and size is typically higher and larger than those of a planar transmission line. Hence, the SIW resonator can be considered to be a compromise between a planar transmission line resonator and a conventional waveguide resonator in terms of the unloaded Q-factor and the size. Since the SIW resonator has a higher Q-factor and a smaller form factor when compared with a planar transmission line resonator and a classical waveguide resonator, respectively, it has found tremendous amount of applications in designing filters operating at microwave frequencies. The static filter designs using the SIW resonators can be found widely in the literature. For example, an SIW bandpass filter having single-mode direct-coupled rectangular resonators is a fundamental structure [9], and advanced bandpass filter structures have also been designed extensively by making use of, for example, and multiple coupling paths and multiple resonant modes [10–16]. In addition, the SIW resonators have been used in designing bandstop filters [17,18].

The main purpose of this chapter is to introduce the recent advances in frequency-tunable and reconfigurable SIW filter designs based on the use of frequency-tunable SIW resonators. Overall, it covers the design of bandpass, bandstop, and multifunction filters. This chapter first deals with a frequency-tunable bandpass filter design in Section 9.2. The basic concepts for a conventional frequency-tunable SIW resonator are briefly reviewed, and a SIW resonator structure suited for the design of filters operating at high frequency bands such as K-band is introduced. A K-band frequency-tunable filter design is also included. Section 9.3 then describes bandstop filter designs using the frequency-tunable SIW resonators. Bandstop filter topologies with and without interresonator couplings are discussed. The bandstop filter topologies containing the interresonator couplings are useful for the design of multifunction filter capable of producing both bandstop and bandpass responses. Section 9.4 introduces the concepts and theories for the design of multifunction filters based on using the bandstop filter topologies containing the interresonator couplings. Section 9.5 provides the summary of this chapter.

9.2 Frequency-tunable bandpass filter

9.2.1 Introduction

In this section, a design method for a frequency-tunable SIW bandpass filter is presented. Since studies on designing frequency-tunable filters using SIW resonators operating at low frequency bands have been extensively carried out and reported [19–22], this section specifically focuses on a design approach for a frequency-tunable SIW filter operating at K-band. In general, the designs of

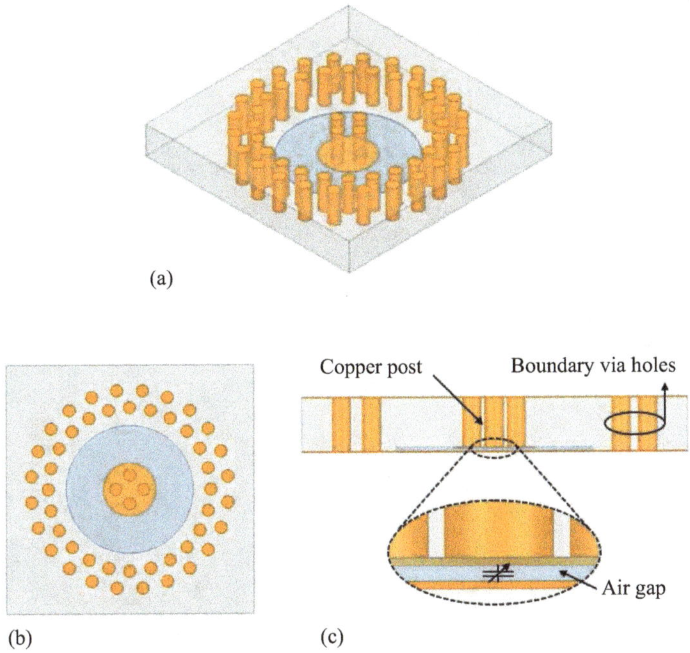

Figure 9.1 General frequency-tunable SIW resonator with a post: (a) three-dimensional view (conductor sheets on the top and bottom sides are not shown for clarity); (b) top view; and (c) cross-sectional side view

frequency-tunable SIW filters operating at low frequency bands are mainly based on using the fundamental mode of a frequency-tunable SIW resonator with a single post at the center. Figure 9.1 shows a simple example of a frequency-tunable resonator integrated in a dielectric substrate. This resonator has conductive via holes that define its side wall. At its center, there exists a single post with conductive via holes and a circular patch. An air gap is formed between the patch of the post and a flexible conductor sheet on the bottom side of the substrate, since an adhesive layer with a circular hole is used for attaching the conductor sheet to the bottom side of the substrate. The resonant frequency of the resonator can be adjusted by changing the thickness of the air gap formed between the post and the conductor sheet called membrane on the bottom side. The detailed operating principles and design methods of the single-post frequency-tunable SIW resonator and its applications to filter designs can be found in [19–22]. In this section, we introduce a recent advance in a *K*-band frequency-tunable SIW resonator and filter technology.

9.2.2 Resonator

The simple model of a frequency-tunable SIW resonator having two posts is shown in Figure 9.2 [23]. The substrate containing the resonator is Rogers TMM3 that has the dielectric constant of 3.27. Each post is made with a conductive via hole and a

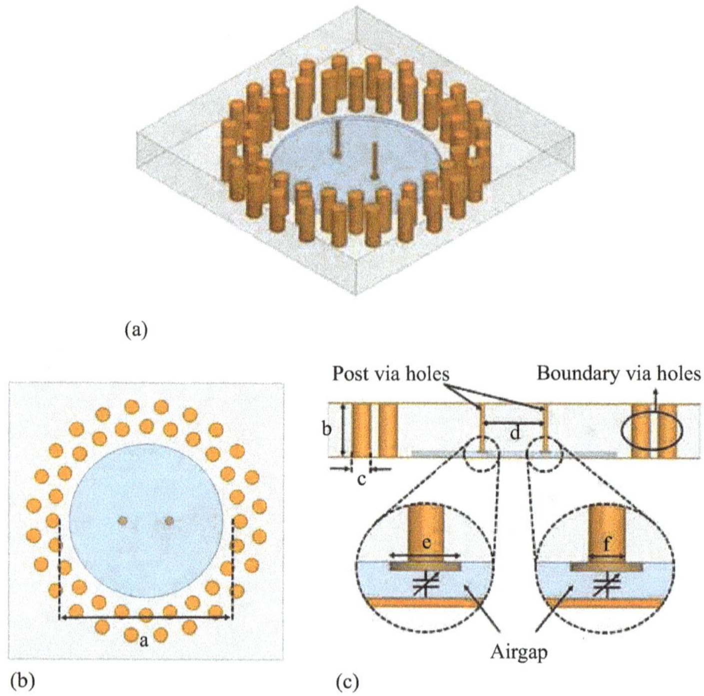

Figure 9.2 Frequency-tunable SIW resonator: (a) three-dimensional view (conductor sheets on the top and bottom sides are not shown for clarity); (b) top view; and (c) cross-sectional side view. Reprinted with permission from [23]; © 2015 IEEE

circular patch. The existence of the two posts results in generating non-fundamental resonant modes that do not exist in a resonator structure with a single post. This type of resonator can be designed such that one of the non-fundamental modes resonates in *K*-band. An air gap exists between the plane containing the patches of the two posts and the conductor plane on the bottom side called membrane. Hence, the air-gap thickness changes when the membrane is adjusted in the vertical direction. This in turn results in the variation of the capacitance between the patches of the posts and the membrane, which leads to the variation of the resonant frequency of each resonant mode. Two-layer via holes are employed in forming the sidewall of the resonator for reducing the radiation loss.

The approximate magnetic field distributions of the first three resonant modes are shown in Figure 9.3. The fundamental mode has the magnetic field circulating around the two posts as if there were a single post. The second resonant mode has the magnetic field circulating around each post and has a symmetric field distribution. On the other hand, the third one has an antisymmetric field distribution.

Figure 9.4 shows the variation of the resonant frequencies of the first three resonant modes of the resonator shown in Figure 9.2. In obtaining the results shown

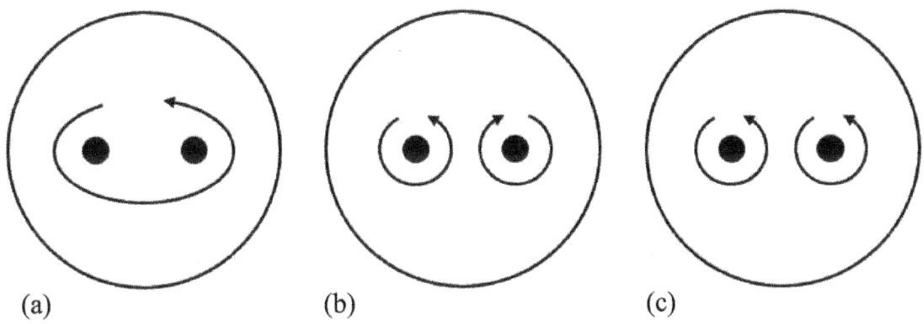

Figure 9.3 Magnetic fields of the first three resonant modes

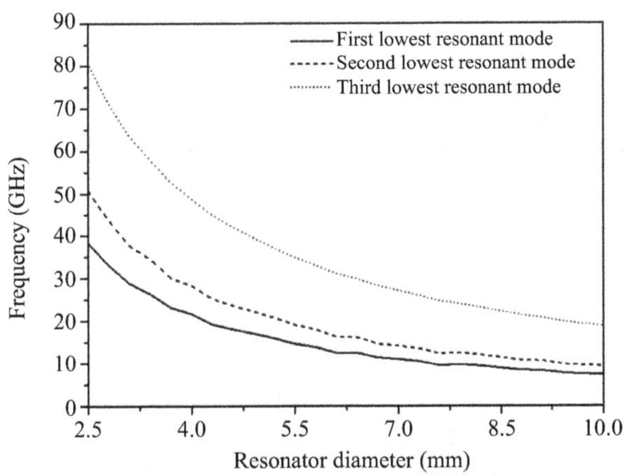

Figure 9.4 Resonant frequencies of the first three resonant modes of the frequency-tunable SIW resonator with two posts. Reprinted with permission from [23]; © 2015 IEEE

in Figure 9.4, the physical dimension ratios given by $b = a/4.462$, $d = a/3.7$, $e = a/21.25$, and $f = a/42.5$ were used. In addition, it was assumed that the air-gap thickness was 25 μm and the diameter of the via holes forming the side wall of the resonator was 0.7 mm. In general, it can be concluded that the resonant frequencies increase when the size of the resonator decreases. In addition, a higher-order resonant mode is required to have larger resonator dimensions in order to resonate at a specific resonant frequency in comparison with a lower-order mode. For example, if the fundamental mode is used in the design of a filter operating at 22.2 GHz, Figure 9.4 indicates that the diameter of the resonator is required to be around 3.9 mm. In addition, it must be 4.8 mm and 8.5 mm when the second and the third modes are in use, respectively. This implies that using the first resonant mode requires accurate fabrication processes with high cost for implementing a physically small resonator in comparison with the second and third resonant modes. According to the full-wave simulations, the unloaded Q-factors of the first three resonant modes of the resonator shown in Figure 9.2 are 573, 449, and 850. Since

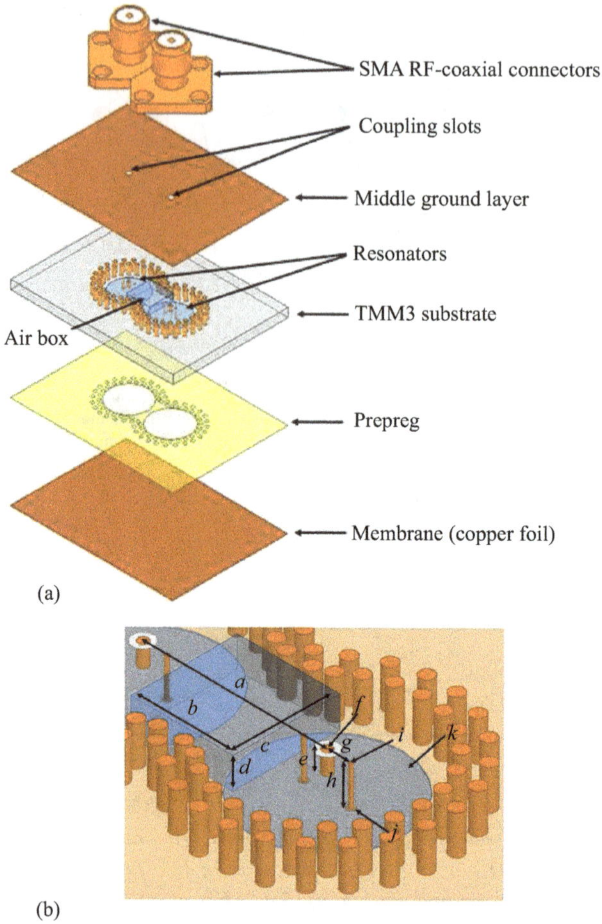

Figure 9.5 (a) Layer-by-layer view of the designed second-order filter and (b) physical dimensions (a = 9 mm, b = 5 mm, c = 5.2 mm, d = 1.5 mm, e = 1 mm, f = 0.25 mm (radius), g = 1.15 mm, h = 1.845 mm, i = 0.1 mm (radius), j = 0.2 mm (radius), k = 3.6 mm (radius)). Reprinted with permission from [23]; © 2015 IEEE

the third resonant mode has a higher *Q*-factor in comparison with the first two modes, we will demonstrate a design technique for *K*-band filter designs using the third resonant mode.

9.2.3 K-band bandpass filter

A second-order frequency-tunable SIW resonator filter operating in *K* band is presented in Figure 9.5 [23]. The resonators with two posts are embedded in a 1.905-mm-thick Rogers TMM3 substrate. The radius of the via holes forming the sidewalls of the resonators is 0.35 mm. The slots are formed at the center of each resonator so that the center conductor of each coaxial connector can be inserted into the resonator without being short to the ground. The two resonators are coupled to each other by virtue of the inductive iris formed by removing a number of via holes between the two resonators. A part of dielectric between the two resonators is removed for improving the upper stopband, and the details will be demonstrated. The conductor plane on the bottom side is called membrane and is made with a flexible copper foil. It can deflect in the vertical direction, and this makes it possible to adjust the resonant frequency of the resonators.

The frequency-tunable filter was designed to have a second-order Butterworth response with 400-MHz bandwidth at 22.2 GHz. The coupling matrix can be obtained by using filter synthesis methods [6,24], and the one for the second-order Butterworth response is given by

$$\mathbf{M} = \begin{bmatrix} 0 & M_{s,1} & 0 & 0 \\ M_{s,1} & 0 & M_{1,2} & 0 \\ 0 & M_{1,2} & 0 & M_{2,l} \\ 0 & 0 & M_{2,l} & 0 \end{bmatrix} \tag{9.1}$$

where the normalized coupling values are $M_{s,1} = M_{2,l} = 0.841$ and $M_{1,2} = 0.707$. In general, the denormalized coupling values are used in determining the physical dimensions of the coupling structures, and they are related to the normalized values as

$$k_{\text{ex}} = M_{\text{ex}}\sqrt{\Delta} \tag{9.2}$$

$$k_{\text{in}} = M_{\text{in}}\Delta \tag{9.3}$$

where k_{ex} and M_{ex} are the denormalized and normalized external coupling values, respectively, and k_{in} and M_{in} are the denormalized and normalized interresonator coupling values, respectively. Δ in (9.2) and (9.3) is the fractional bandwidth. In the design of the second-order filter, M_{ex} refers to $M_{s,1}$ and $M_{2,l}$, and M_{in} refers to $M_{1,2}$. Hence, $k_{\text{ex}} = k_{s,1} = k_{2,l}$ and $k_{\text{in}} = k_{1,2}$.

The external coupling structure must be designed in a way to have the coupling value in (9.2), and it is 0.113 in this design. Figure 9.6 shows the side view of the external coupling structure and the resonator used in the filter design. A circular slot is formed at the center of the top conductor plane. The center conductor of the coaxial connector passes through the slot and is inserted in the resonator. The external coupling structures in [25–30] also use a slot, but the port is of microstrip structure. Hence an additional substrate must be used for implementing the microstrip structure. It is evident that attaching the coaxial connector directly to the substrate containing the resonator as shown in Figure 9.6 can avoid using an additional substrate.

The physical external coupling value of the coupling structure shown in Figure 9.6 can be found by full-wave simulation, and it is given by

$$k_{\text{ex}} = \sqrt{\frac{f_2 - f_1}{f_0}} \tag{9.4}$$

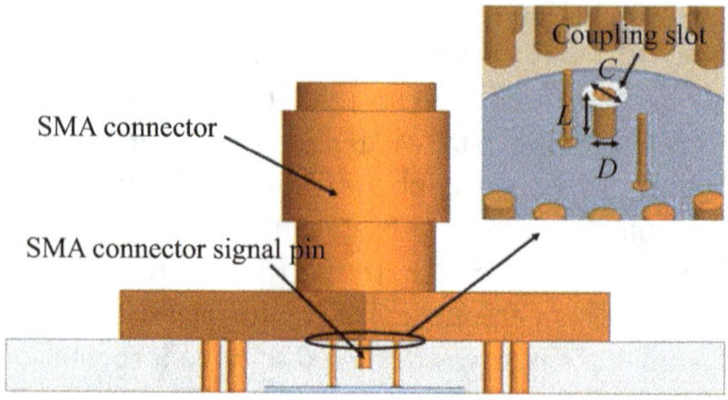

Figure 9.6 Frequency-tunable SIW resonator and external coupling structure. Reprinted with permission from [23]; © 2015 IEEE

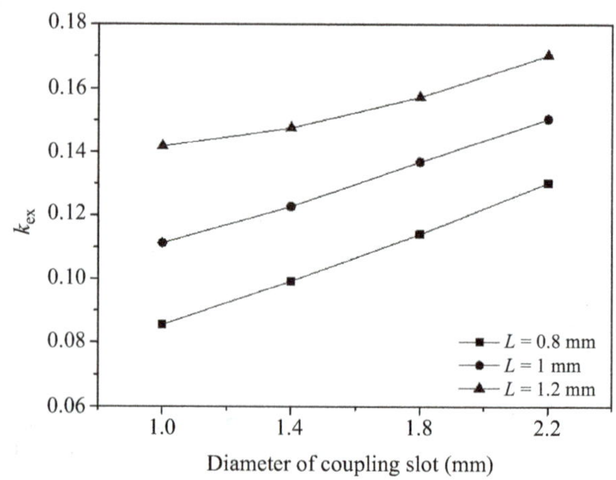

Figure 9.7 External coupling value. Adapted with permission from [23]; © 2015 IEEE

where f_0 is the resonant frequency and f_1 and f_2 are the frequencies where the phase of the reflection coefficient deviates by 90° from the phase at f_0. The detailed procedure for extracting the external coupling value from the reflection coefficient can be found in [31]. The physical coupling value of the external coupling structure shown in Figure 9.6 varies with a number of physical dimensions. For example, the length and the diameter of the center conductor of the coaxial connector inserted in the resonator (L and D in Figure 9.6) determine the physical coupling value. In addition, the external coupling value depends on the diameter of the circular slot (C in Figure 9.6). It has been found that the coupling value increases with the aforementioned three physical parameters. Although the external coupling value are mainly determined by the three physical parameters, it is recommended to examine how the coupling value varies with the length of the center conductor of the coaxial connector and the diameter of the circular slot since the diameter of the center conductor of the coaxial connector is defined by a connector manufacturer. Figure 9.7 shows the external

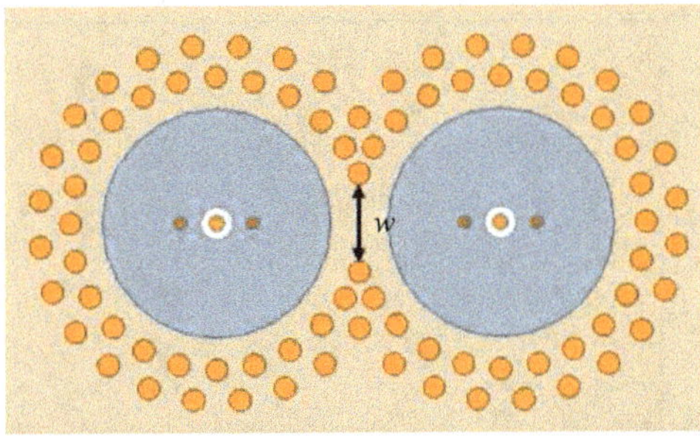

Figure 9.8 Interresonator coupling structure. Adapted with permission from [23]; © 2015 IEEE

coupling value variation with respect to the length of the center conductor of the coaxial connector and the diameter of the circular slot when the diameter of the center conductor is 0.5 mm. Figure 9.7 confirms that the external coupling values increases as the length of the center conductor of the coaxial connector and the diameter of the circular slot increase. By reading Figure 9.7, we can find the length of the center conductor and the diameter of the slot. Among many possible dimensions, 1.0 mm and 1.0 mm can be chosen for the length of the center conductor and the diameter of the slot, respectively, for getting the theoretical external coupling value.

Next, we design the interresonator coupling structure. Figure 9.8 shows the top view of the two coupled SIW resonators. The two resonators are coupled by the inductive iris, and its width and the distance between the two resonators mainly determine the interresonator coupling value. The coupling value of the interresonator coupling structure can be calculated by using the transmission response of the weakly excited coupled resonators. The interresonator coupling value is given by

$$k_{\text{in}} = \frac{f_2^2 - f_1^2}{f_2^2 + f_1^2} \tag{9.5}$$

where f_1 and f_2 are the first and second resonant frequencies of the weakly excited coupled resonator structure, respectively [24]. In general, the coupling value increases with the iris width, and the calculated value using (9.5) is shown in Figure 9.9 when the distance between the two resonators is 9 mm. By reading Figure 9.9, one can find that the iris width giving the interresonator coupling value needed in this filter design is 2.5 mm.

Using the physical dimensions of the external and interresonator coupling structures discussed previously, we can design a K-band second-order bandpass filter and its structure is shown in Figure 9.10. Figure 9.11 shows the simulated transmission and reflection responses of the filter shown in Figure 9.10. The second-order Butterworth response appears at the predefined center frequency at 22.2 GHz. However, the performance of the upper stopband is poor due to the existence of the undesired resonances. It has been found that these resonances are newly generated in and around the iris when the two resonators are coupled. Hence,

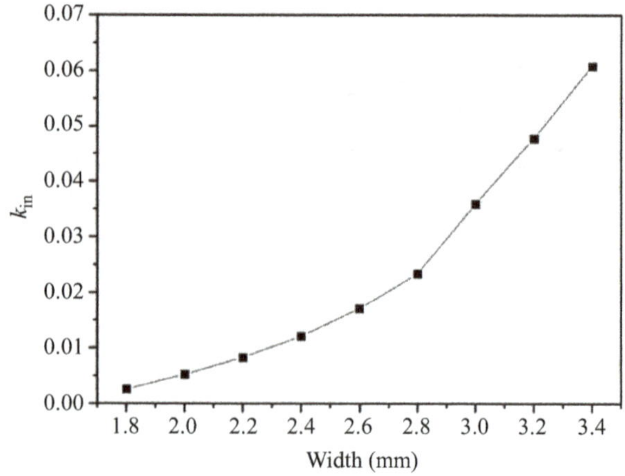

Figure 9.9 Interresonator coupling value. Adapted with permission from [23]; © 2015 IEEE

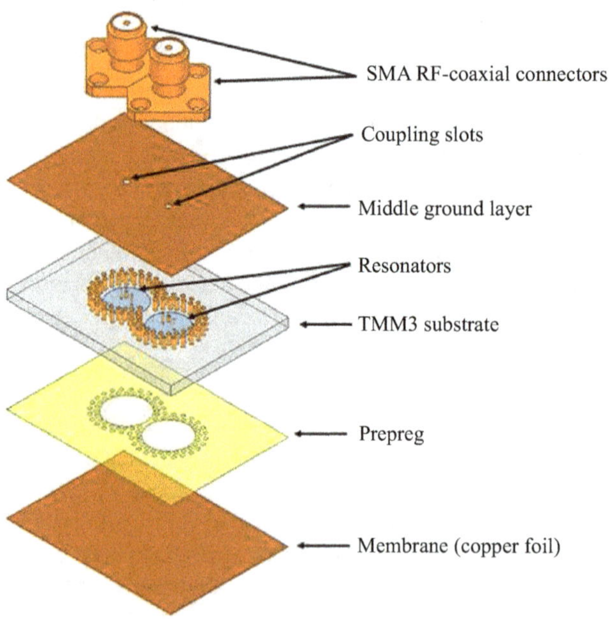

Figure 9.10 Layer-by-layer view of the second-order filter employing the coupling structures shown in Figures 9.6 and 9.8. Adapted with permission from [23]; © 2015 IEEE

one can predict that these undesired resonances can be moved to either higher or low frequencies by altering the inductive iris structure. Various methods for improving the stopband performance for SIW filters have been presented [16,32,33]. However, they cannot be used in this filter design since they are

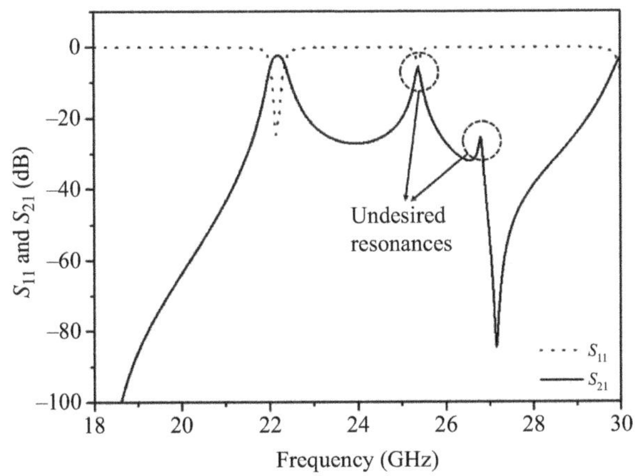

Figure 9.11 Simulated frequency response of the filter shown in Figure 9.10. Adapted with permission from [23]; © 2015 IEEE

intended to suppress the undesired resonances generated in the resonator structures, not in the interresonator coupling structures. Hence, it is preferred to modify the interresonator structure for the purpose of enhancing the upper stopband.

Since the undesired resonances in the upper stopband are generated in the coupling structure, we can increase the frequencies of them by simply removing a part of the dielectric material in the iris structure. This is because the resonant frequency of a resonant structure is in general inversely proportional to the permittivity of the material filling the resonant structure. Since removing a part of the dielectric material is equivalent to replacing it with an air box, the permittivity effectively decreases. This approach was employed in the filter design, and Figure 9.12 shows the top view of the modified interresonator coupling structure and its coupling value variation with respect to the iris width. In this design, the width of the air box is set to be 0.4 mm smaller than the iris width and the length of the air box is 5 mm. One can find that the iris width of 5.6 mm corresponds to the required coupling value.

Using the design results for the external and interresonator coupling structures, we can design the entire filter structure and the physical dimensions are summarized in Figure 9.5. Figure 9.13 shows the comparison between the simulated responses of the two filters shown in Figure 9.5 and Figure 9.10. One filter contains the modified interresonator coupling structure shown in Figure 9.12 and the other has the original coupling structure shown in Figure 9.8. Figure 9.13 confirms that the filter with the modified iris structure has an improved upper stopband performance.

The filter shown in Figure 9.5 was fabricated. The via holes for the posts and the side walls of the resonators were fabricated by creating holes by drilling followed by copper plating. The patches of the copper posts were fabricated by a milling process. The copper foil was attached to the substrate by using a prepreg in between. Since the prepreg has two cylindrical holes such that it does not cover the

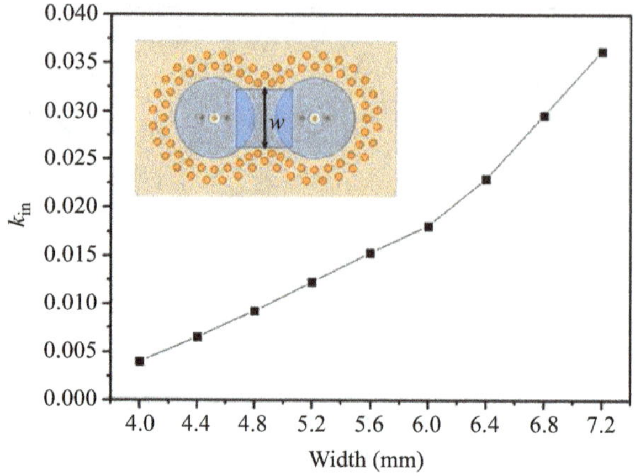

Figure 9.12 Modified interresonator coupling structure and its coupling value. Adapted with permission from [23]; © 2015 IEEE

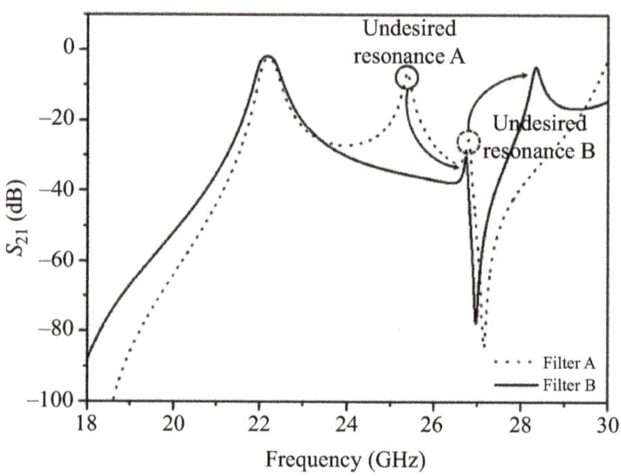

Figure 9.13 Simulated frequency responses of the two filters shown in Figures 9.5 and 9.10. Adapted with permission from [23]; © 2015 IEEE

resonator area, the air gap exists in each resonator. Since the prepreg is not conductive, the copper foil and the conductor of the substrate were not originally connected electrically. For electrical connection between the copper foil and the conductor of the substrate, we created holes in the copper foil and the prepreg, filled the holes with solder paste followed by thawing the paste. Finally, the coaxial connectors were inserted at the center of each resonator. Figure 9.14(a) shows the top and bottom views of the fabricated filter before mounting the tuning devices. Figure 9.14(b) shows the fabricated filter after mounting the tuning devices for adjusting the air gaps for tuning the resonant frequencies of the resonators. In this

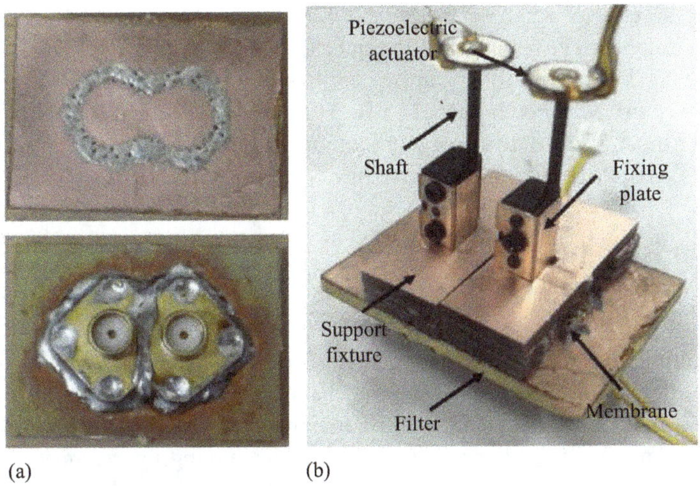

Figure 9.14 *(a) Top and bottom views of the fabricated filter before installing the tuning devices and (b) three-dimensional view of the fabricated filter after installing the tuning devices. Adapted with permission from [23]; © 2015 IEEE*

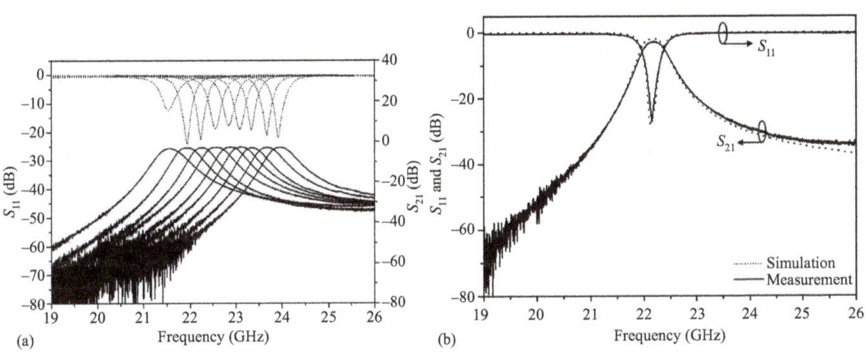

Figure 9.15 *(a) Comparison between simulation and measurement of the filter centered at 22.2 GHz and (b) measured frequency tuning responses of the filter. Adapted with permission from [23]; © 2015 IEEE*

work, two piezoelectric actuators from Piezoelectric Technology were used for controlling the resonators. The fixing plates are attached to the support fixtures, and the shafts can move in the vertical direction by virtue of the actuators. Since the shafts are attached to the membrane, the membrane can be adjusted leading to the variation of the resonant frequency of each resonator.

The comparison between the simulated and measured results of the filter centered at 22.2 GHz is shown in Figure 9.15(a). It shows an excellent agreement between the simulation and the measurement of the filter. In addition, both return losses are larger than 25 dB at the center frequency. Since the filter contains the frequency-tunable SIW resonators, its center frequency can be adjusted. It is shown in Figure 9.15(b) that the filter can be continuously tuned from 21.6 to 24.0 GHz having the return loss better than 12 dB. The measured return loss is better than 17 dB over most of the entire frequency tuning range. The measured insertion loss is from 2.7 to 3.4 dB over the tuning range.

9.3 Bandstop filter

9.3.1 Introduction

In this section, we provide the fundamental design theories for SIW resonator bandstop filters. We discuss conventional bandstop filter topologies that have no interresonator couplings, and they are compared with the ones containing interresonator couplings. The bandstop filter topologies having interresonator coupling is important in the sense that it allows us to design reconfigurable filters capable of producing bandstop and bandpass responses. The discussion employs the coupling matrix theories and it is shown that the bandstop filter topologies containing interresonator couplings can generate the responses of the classical bandstop filter topologies with no interresonator couplings.

9.3.2 Second-order bandstop filter

Second-order bandstop filter topologies are shown in Figure 9.16. The numbered filled circles represent the resonators. The classical topology has no interresonator couplings whereas the other topology has an interresonator coupling between resonators 1 and 2 [25]. The coupling matrix for the classical bandstop filter topology is given by

$$\mathbf{M} = \begin{bmatrix} 0 & M_{s,1} & 0 & M_{s,l} \\ M_{s,1} & 0 & 0 & 0 \\ 0 & 0 & 0 & M_{2,l} \\ M_{s,l} & 0 & M_{2,l} & 0 \end{bmatrix} \tag{9.6}$$

where s and l represent the source and load, respectively, and $M_{s,1} = M_{2,l}$ when the external couplings for resonators 1 and 2 are the same. The coupling matrix can be obtained by means of using well-known filter synthesis processes [6,24]. For example, the coupling matrix values for the Butterworth bandstop responses in the normalized frequency domain shown in Figure 9.17 can be found and they are $M_{s,1} = M_{2,l} = 1.189$ and $M_{s,l} = \pm 1.000$. The concept of the sign of the coupling values is relative [24,34]. In other words, if a coupling is positive, then a negative coupling means that its phase response is opposite to that of the positive coupling. For example, if we assign the positive sign to a coupling with $+90°$ phase shift, then a coupling with $-90°$ phase shift has the negative sign. In this discussion, we assign the negative sign to a coupling with $-90°$ phase shift, and it is assumed that

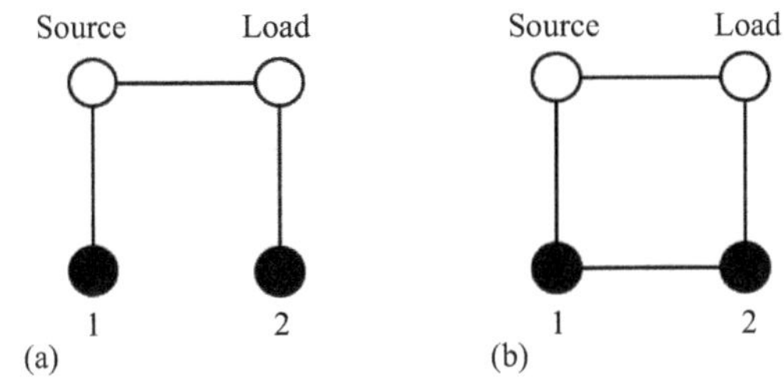

Figure 9.16 *(a) Second-order bandstop filter topology without interresonator couplings and (b) second-order bandstop filter topology with an interresonator coupling. Adapted with permission from [25]; © 2010 IEEE*

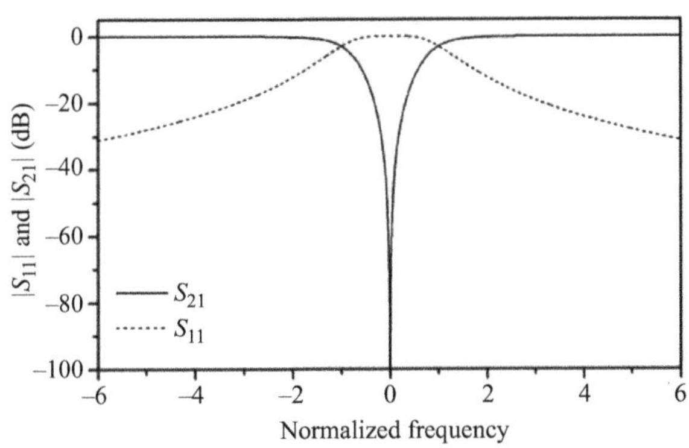

Figure 9.17 *Butterworth bandstop responses in the normalized frequency domain*

$M_{s,l}$ has $-90°$ phase shift leading to $M_{s,l} = -1.000$ for demonstrating the relationship between the two bandstop filter topologies.

The coupling matrix for the bandstop filter topology with an interresonator coupling is given by

$$\mathbf{M}' = \begin{bmatrix} 0 & M'_{s,1} & 0 & M'_{s,l} \\ M'_{s,1} & M'_{1,1} & M'_{1,2} & 0 \\ 0 & M'_{1,2} & M'_{2,2} & M'_{2,l} \\ M'_{s,l} & 0 & M'_{2,l} & 0 \end{bmatrix} \tag{9.7}$$

where $M'_{s,1} = M'_{2,l}$ assuming that the topology is symmetric with respect to the center of it. It is worth noting that the coupling matrix has a nonzero coupling value $M'_{1,2}$ and the self-coupling values, $M'_{1,1}$ and $M'_{2,2}$. We can find the coupling values in (9.7) corresponding to a prescribed filter response using filter synthesis techniques. Alternatively, they can be found by using the relationship between (9.6) and (9.7) for having the same response. It has been found that the bandstop filter

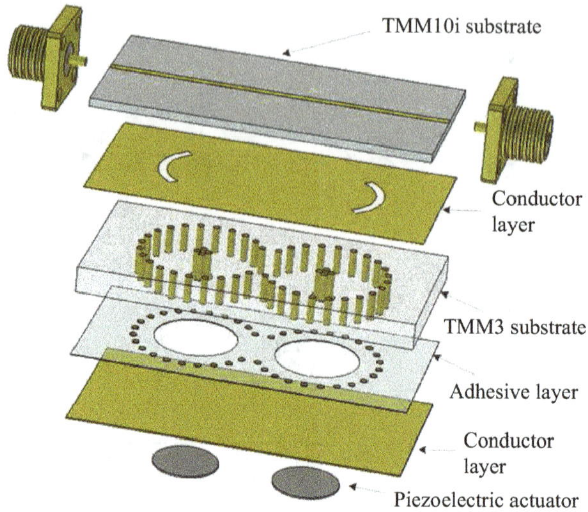

Figure 9.18 Layer-by-layer view of the second-order frequency-tunable SIW resonator bandstop filter. Adapted with permission from [25]; © 2010 IEEE

topology with an interresonator coupling (Figure 9.16(b)) can reproduce the responses of the conventional topology when [25]

$$\begin{aligned} M'_{s,1} &= M_{s,1} \\ M'_{s,l} &= -M_{s,l} = 1.000 \\ M'_{1,1} &= -M'_{2,2} = \pm\sqrt{M'^2_{s,1} M'_{1,2} - M'^2_{1,2}} \end{aligned} \tag{9.8}$$

where

$$M'^2_{s,1} > M'_{1,2}. \tag{9.9}$$

$M'_{1,2}$ is allowed to have an arbitrary value under the condition that (9.9) is satisfied. Using (9.8) and $M'_{1,2}$ satisfying (9.9), we can find the coupling values of (9.7). For example, a Butterworth bandstop filter with $M'_{1,2} = 0.700$ is required to have $M'_{1,1} = -M'_{2,2} = 0.707$. According to (9.9), the maximum $M'_{1,2}$ we can choose is 1.414 for the Butterworth response shown in Figure 9.17. Equations (9.8) and (9.9) indicate that a bandstop filter can have an interresonator coupling of an arbitrary value if the resonators can be tuned independently such that they resonate at the resonant frequency corresponding to $M'_{1,1}$ and $M'_{2,2}$.

A frequency-tunable SIW resonator bandstop filter structure designed using the topology containing the interresonator coupling (Figure 9.16(b)) is shown in Figure 9.18. The filter mainly consists of two substrates. Two frequency-tunable cylindrical SIW resonators are implemented in a 3.175-mm-thick Rogers TMM3 substrate ($\varepsilon_r = 3.27$). Each resonator has a post at its center made with 4 via holes and a circular patch. The side walls of the resonators are formed by conducting via holes of which diameter is 0.8 mm. The radius of the cavity is defined by the distance between the center of the post and the center of the one of via holes forming the side walls, and it is 7.5 mm in this design. There exists an air gap between the plane containing the circular patch of the post of each resonator and the conductor layer on the bottom side of the filter. The resonant frequency of each resonator can be controlled by adjusting the thickness of the air gap. Two piezoelectric actuators from Piezo Systems are attached to the bottom conductor layer for the purpose of adjusting the air-gap thickness by deforming the conductor layer. The iris is made by removing a number of the via holes between the two resonators, and it forms the interresonator coupling. The width of the iris can be chosen arbitrarily since $M'_{1,2}$ can be arbitrary as long as (9.9) is satisfied. In this design, the iris width is 5.5 mm.

The microstrip line is fabricated in a 1.27-mm-thick Rogers TMM10i substrate ($\varepsilon_r = 9.8$). The two coupling slots in the conductor layer between the two substrates are for implementing the external couplings for the two resonators. The size of the coupling slots was determined by the normalized coupling values $(M'_{s,1}, M'_{2,l})$ the bandwidth, and the center frequency. The predefined response used in this filter design is the Butterworth response with the bandwidth of 30 MHz when centered at 3.25 GHz. The normalized external coupling values, $M'_{s,1}$ and $M'_{2,l}$, can be denormalized into

$$k'_{s,1} = (k'_{2,l}) = M'_{s,1}\sqrt{\frac{\Delta}{2}}\left(= M'_{2,l}\sqrt{\frac{\Delta}{2}}\right). \tag{9.10}$$

Hence, the two coupling slots were designed such that they have the denormalized coupling value given in (9.10). The coupling value of an external coupling structure in a bandstop filter configuration can be obtained by means of simulating a single resonator coupled to a transmission line [35]. The length of the microstip between the two coupling slots is three-quarter wavelength for implementing $M'_{s,l} = 1$.

Figure 9.19 shows the measured frequency response of the bandstop filter. Since the filter employs the frequency-tunable SIW resonators, its center frequency can be tuned. In addition, the filter is capable of having a Butterworth response while being centered at different frequencies since the iris is allowed to have an arbitrary coupling value. It is shown that the measured responses agree well with the theoretical responses.

9.3.3 Third-order bandstop filter

Third-order bandstop filter topologies are shown in Figure 9.20 [36]. The numbers represent the resonators and NRN stands for nonresonating nodes. The concept of NRN is introduced and detailed in [37]. The conventional bandstop filter topology (Figure 9.20(a)) has no interresonator couplings. A bandstop filter may have interresonator couplings and a topology having interresonator couplings between

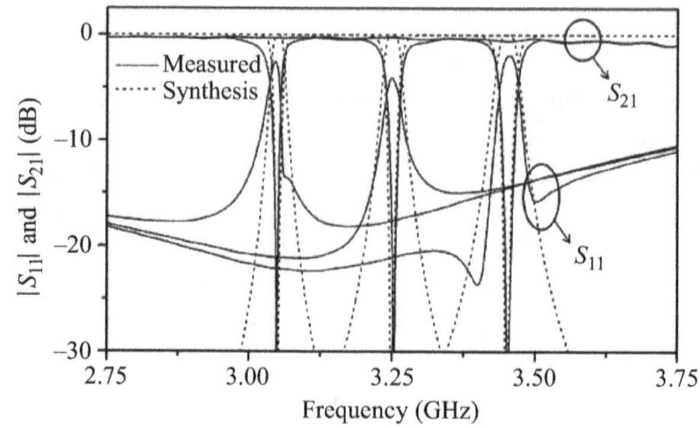

Figure 9.19 Comparison between the synthesized and measured responses of the frequency-tunable bandstop filter shown in Figure 9.18. Reprinted with permission from [25]; © 2010 IEEE

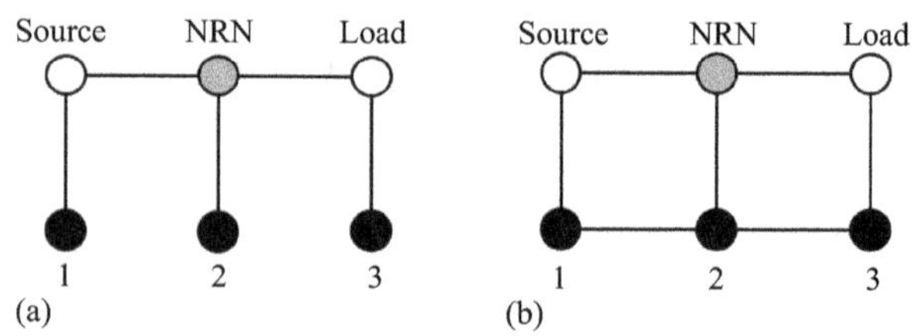

Figure 9.20 (a) Third-order bandstop filter topology with no interresonator couplings and (b) third-order bandstop filter topology with two interresonator couplings. Adapted with permission from [36]; © 2016 IEEE

the pairs of consecutively-numbered resonators is shown in Figure 9.20(b). Assuming that the external couplings for resonators 1 and 3 have the same value and self coupling values for resonators 1 and 3 have same magnitude but opposite sign, one can express the coupling matrix for the conventional topology as

$$\mathbf{M} = \begin{bmatrix} 0 & M_{s,1} & 1 & 0 & 0 & 0 \\ M_{s,1} & M_{1,1} & 0 & 0 & 0 & 0 \\ 1 & 0 & 0 & M_{n,2} & 0 & 1 \\ 0 & 0 & M_{n,2} & M_{2,2} & 0 & 0 \\ 0 & 0 & 0 & 0 & -M_{1,1} & M_{s,1} \\ 0 & 0 & 1 & 0 & M_{s,1} & 0 \end{bmatrix} \tag{9.11}$$

when the coupling values between the source and NRN and between NRN and the load are 1, which is usually the case. The coupling matrix for producing a pre-defined filtering response can be obtained by filter synthesis methods [6,24]. For example, the coupling values in (9.11) for the third-order Butterworth bandstop

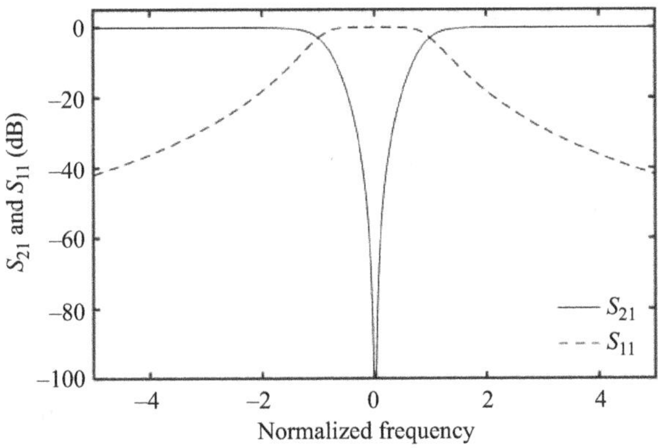

Figure 9.21 Third-order Butterworth response. Adapted with permission from [36]; © 2016 IEEE

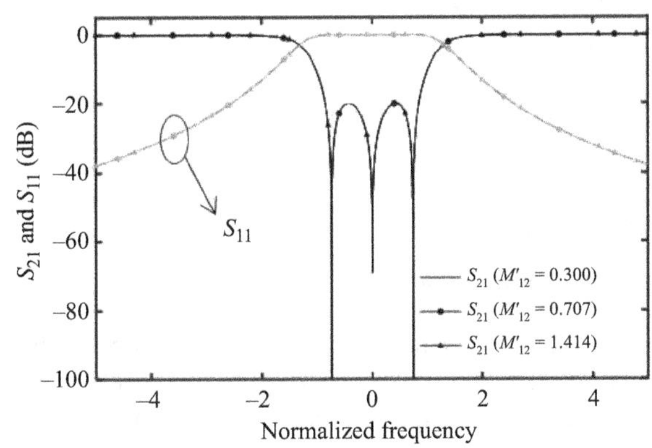

Figure 9.22 20-dB equiripple response. Adapted with permission from [36]; © 2016 IEEE

filter response in the normalized frequency domain (Figure 9.21) are $M_{s,1} = 1.000$, $M_{n,2} = 1.414$, $M_{1,1} = 0.000$, and $M_{2,2} = 0.000$. In addition, the conventional topology is capable of producing equiripple responses. An equiripple response can be obtained by adjusting resonators 1 and 3. For example, by changing $M_{1,1} = 0.000$ to $M_{1,1} = 0.739$, we can have the 20-dB equiripple response shown in Figure 9.22. It is worth noting that the 3-dB and equirripple stopbands range from −1.313 to 1.313 and from −0.853 to 0.853, respectively, in the normalized frequency domain. It indicates that we can increase the 3-dB bandwidth by changing the response from the Butterworth response to an equiripple response. Since both Butterworth and equiripple responses can be obtained with $M_{2,2} = 0$, we let the self coupling value for resonator 2 be 0 in this section.

With the assumption used in (9.11) and the one that the interresonator couplings between resonators 1 and 2 and between resonators 2 and 3 are the same, the coupling matrix for the topology with the interresonator couplings can be written as

$$\mathbf{M}' = \begin{bmatrix} 0 & M'_{s,1} & 1 & 0 & 0 & 0 \\ M'_{s,1} & M'_{1,1} & 0 & M'_{1,2} & 0 & 0 \\ 1 & 0 & 0 & M'_{n,2} & 0 & 1 \\ 0 & M'_{1,2} & M'_{n,2} & M'_{2,2} & M'_{1,2} & 0 \\ 0 & 0 & 0 & M'_{1,2} & -M'_{1,1} & M'_{s,1} \\ 0 & 0 & 1 & 0 & M'_{s,1} & 0 \end{bmatrix}. \tag{9.12}$$

The transmission coefficients (S_{21}) of the two bandstop filter topologies in Figure 9.20 can be obtained from the coupling matrices and expressed in terms of the coupling values. By comparing the two transmission coefficients, the coupling values of the filter topology containing the interresonator couplings for duplicating the frequency responses of the conventional one can be obtained. The coupling value relationships are given by

$$\begin{aligned} M'_{s,1} &= M_{s,1} \\ M'_{n,2} &= M_{n,2} \\ M'_{2,2} &= M_{2,2} = 0 \\ M'_{1,1} &= \sqrt{M_{1,1}^2 + 2M'_{1,2}(M_{s,1}M_{n,2} - M'_{1,2})} \\ M'_{1,2} &= \text{arbitrary value} < M_{s,1}M_{n,2}. \end{aligned} \tag{9.13}$$

It is worth noting that all coupling values for the bandstop filter topology with interresonator couplings can be derived from the coupling values for the conventional bandstop filter topology. In addition, the normalized coupling value $M'_{1,2}$ can be chosen arbitrarily but must not be larger than $M_{s,1}M_{n,2}$. It can be concluded from (9.13) that the filter topology containing the interresonator couplings can reproduce the frequency responses of the conventional topology by having different self-coupling values for resonators 1 and 3. For example, the topology with the inter-resonator couplings can produce the Butterworth response shown in Figure 9.21 when $M'_{1,1} = 0.818$ and $M'_{1,2} = 0.300$. The 20-dB equiripple response shown in Figure 9.22 can also be produced by having $M'_{1,1} = 1.102$ with $M'_{1,2} = 0.300$. The same response can be obtained with different values of $M'_{1,2}$ as shown in Figure 9.22.

It has been found that the topology shown in Figure 9.20(b) is also capable of having the all-pass response provided that

$$\begin{aligned} M'_{s,1} &= \text{arbitrary value} \\ M'_{n,2} &= \sqrt{2}M'_{s,1} \\ M'_{1,2} &= \frac{M'_{s,1}M'_{n,2}}{2} \\ M'_{1,1} &= M'_{2,2} = 0. \end{aligned} \tag{9.14}$$

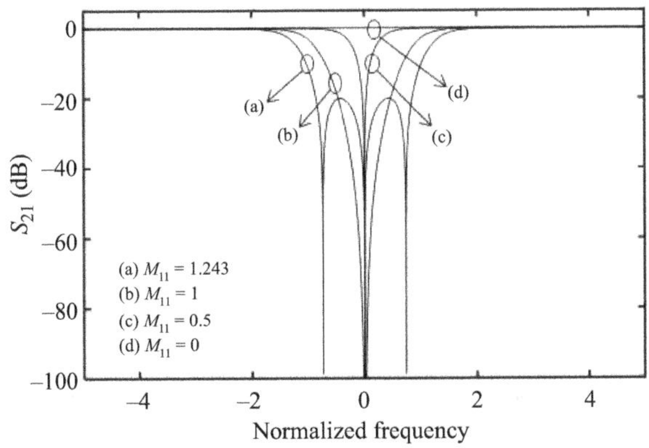

Figure 9.23 Frequency responses of the third-order bandstop filter topology with interresonator couplings

Hence, by considering the condition for producing the conventional bandstop responses (Butterworth and equiripple responses) given in (9.13) and the one for the all-pass response given in (9.14), we can design a bandstop filter using the topology shown in Figure 9.20(b) capable of having both the conventional bandstop responses (Butterworth and equiripple responses) and the all-pass response. For example, we can find all coupling values except the self-coupling values satisfying (9.13) and (9.14) simultaneously, and they are

$$\begin{aligned} M'_{s,1} &= 1.000 \\ M'_{n,2} &= 1.414 \\ M'_{1,2} &= 0.707. \end{aligned} \tag{9.15}$$

According to (9.13), the filter having the coupling values in (9.15) along with $M'_{1,1} = 1.000$ and $M'_{2,2} = 0.000$ has the Butterworth response shown in Figure 9.21. The filter response changes to the equiripple response shown in Figure 9.22 when $M'_{1,1}$ is adjusted to 1.243. In addition, according to (9.14), the filter with (9.15) has the all-pass response when its resonators are tuned to have $M'_{1,1} = M'_{2,2} = 0.000$. In short, the filter having the normalized coupling values in (9.15) is capable of having Butterworth, equiripple, and all-pass responses, and the filter response is determined by the self-coupling values for resonators 1 and 3. Figure 9.23 shows four examples of the frequency responses that can be produced by the filter topology shown in Figure 9.20(b) having the coupling values given in (9.15). Plots (a) and (b) in Figure 9.23 can be obtained from the filter topologies shown in Figures 9.20(a) and 9.20(b) while plots (c) and (d) can be produced only by the filter topology containing the interresonator couplings. It is worth noting that the filter can achieve a variable bandwidth by simply adjusting the resonant frequency of resonators 1 and 3. For example, the 3-dB stopband is from -0.298 to 0.298 when $M'_{1,1} = 0.500$ while it is from -1.000 to 1.000 when $M'_{1,1} = 1.000$. In summary, the bandstop filter topology containing the interresonator couplings can exhibit various frequency responses by

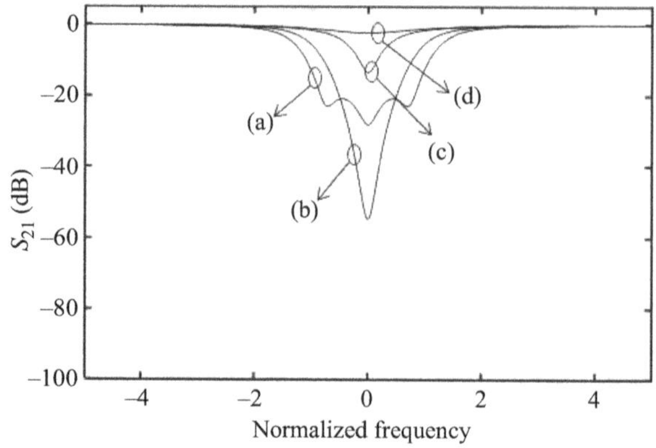

Figure 9.24 Effect of resonator loss on frequency responses of the third-order bandstop filter topology with interresonator couplings. Plots (a), (b), (c), and (d) in this figure correspond to those in Figure 9.23. Adapted with permission from [36]; © 2016 IEEE

tuning the self-coupling coefficients without the changes of external and interresonator couplings when (9.15) is satisfied.

We have discussed a number of important features of the bandstop filter topology with the interresonator couplings based on the assumption that there is no loss in a filter structure. The loss of the resonators can be taken into account in predicting the response of the filter topology. It can be executed by having an negative imaginary part in the self-coupling value for each resonator. For example, Figure 9.24 shows the frequency responses of the filter topology having the coupling values in (9.15) when the imaginary part of the self-coupling values of the resonators is $-j0.133$. It can be observed that the attenuation of the Butterworth-like response decreases at the center frequency as the bandwidth becomes smaller.

Two bandstop filters have been designed based on the two aforementioned filter topologies, and their structures are shown in Figure 9.25. Both filters are designed using frequency-tunable SIW resonators. Both filters consist of two substrates, a 0.508-mm-thick Rogers RO4350B and a 3.175-mm-thick Rogers TMM3, for the microstrip line and frequency-tunable SIW resonators, respectively. The frequency-agile feature is achieved by virtue of the air gap formed in each SIW resonator. The copper sheet shown in Figure 9.25 is laminated to the TMM3 substrate using a prepreg (not shown in Figure 9.25) for forming the air gap between the conductive post and the copper sheet. The coupling between the microstrip line and each resonator is achieved via the coupling slot etched on the ground plane, and the microstrip line includes the patches placed above the coupling slots for improving the passband matching performance [38].

The coupling matrix in (9.11) with $M_{s,1} = 1.000$ and $M_{2,n} = 1.414$ is used for designing a bandstop filter with no interresonator couplings, and the one in (9.12) with the coupling values in (9.15) is used in the design of a bandstop filter with the interresonator couplings. Both filters are designed to have 1.5% bandwidth at

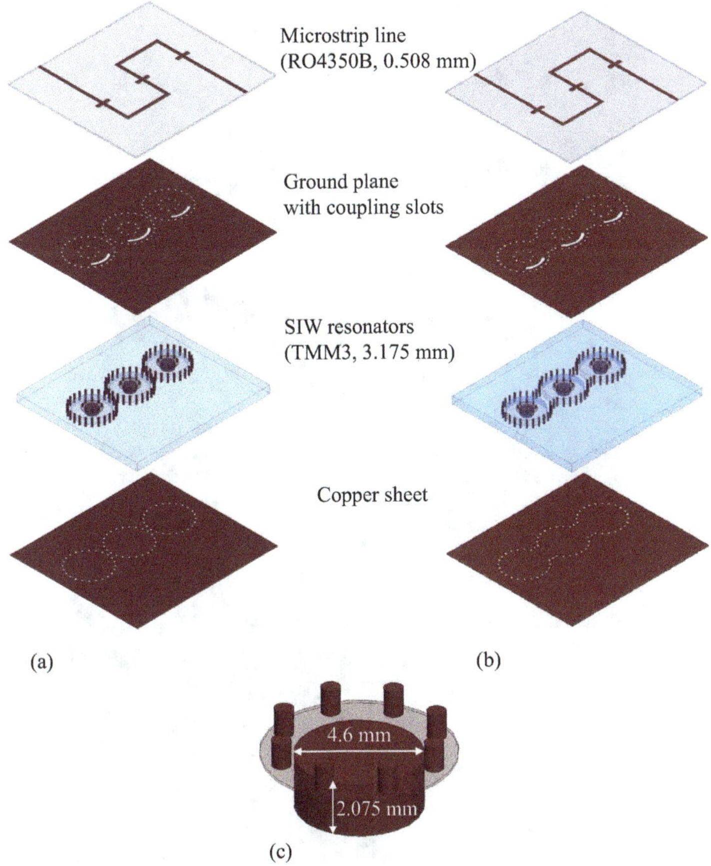

Figure 9.25 Third-order frequency-tunable SIW resonator bandstop filters: (a) filter with no interresonator couplings; (b) filter with two interresonator couplings; and (c) detailed view of the conductive post in each resonator. Reprinted with permission from [36]; © 2016 IEEE

3.3 GHz. In designing the physical coupling structures, the denormalized coupling values were used. More specifically, the coupling slots were designed such that they have the coupling values given by

$$k = M\sqrt{\frac{\Delta}{2}} \tag{9.16}$$

where M refers to M_{s1} and M_{n2} for the filter without interresonator couplings and it is M'_{s1} and M'_{n2} for the filter with the two interresonator couplings. The irises of the filter with the two interresonator couplings were designed in a way to have the coupling values given by

$$k = M\Delta \tag{9.17}$$

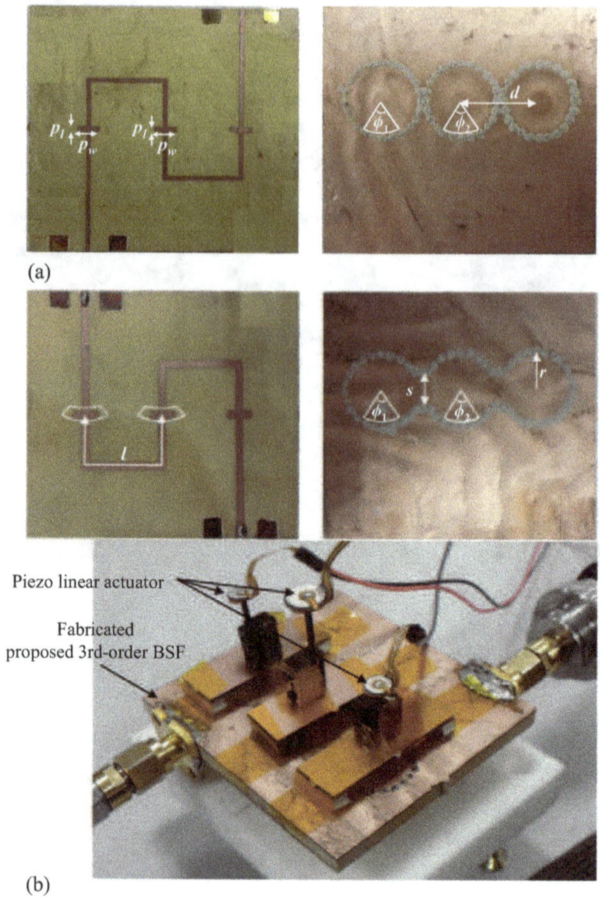

Figure 9.26 (a) Top and bottom views of the filter with no interresonator couplings before mounting the tuning devices and (b) top and bottom views of the filter with the two interresonator couplings before mounting the tuning devices and its three-dimensional view after mounting the tuning devices. Reprinted with permission from [36]; © 2016 IEEE

where M is M'_{12}. Figure 9.26 shows the fabricated filters along with the detailed physical dimensions. The coupling slot dimensions of one filter are identical to those for the other filter. The electrical length between two coupling slots is set to be 270 degrees at 3.3 GHz, and it includes the electrical length generated by the coupling slots. Both filters have frequency-agile features by virtue of the air gap formed in each resonator, and its resonant frequency can be tuned by means of using an actuator. In this work, a linear actuator from Piezoelectric Technology Company is used for each resonator. Detailed operating principle can be found in [23]. In short, one side of a shaft is attached to the copper sheet and the other side to

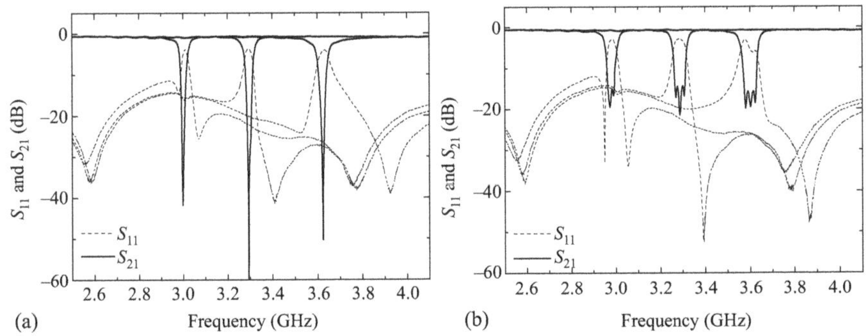

Figure 9.27 Measured frequency responses of the filter with no interresonator couplings: (a) Butterworth and (b) equiripple. Reprinted with permission from [36]; © 2016 IEEE

the actuator, and it moves in the vertical direction by means of the voltage pulses applied to the actuator. The air-gap thickness can be controlled by the amplitude and the number of the pulses of the voltage. Hence, the resonant frequency of each resonator can be adjusted independently.

Figure 9.27 shows the measured results of the fabricated bandstop filter with no interresonator couplings. The filter can be tuned from 3.0 to 3.6 GHz having a Butterworth response with the attenuation level larger than 40 dB at the center frequency. This filter can also have equiripple stopband responses by adjusting the resonant frequency the resonators. As an example, Figure 9.27(b) shows the measurements with 15-dB equiripple responses in the frequency tuning range.

The measured results of the bandstop filter designed using the topology with the interresonator couplings are shown in Figure 9.28. Since the filter was designed using the coupling matrix in (9.12) with the coupling values in (9.15), it is capable of reproducing the conventional responses (Butterworth and equiripple responses) and varying the attenuation level in the stopband. One of the advantages of using the filter topology shown in Figure 9.20(b) in a bandstop filter design is that it is capable of controlling the attenuation level by means of adjusting the resonant frequencies of resonators 1 and 3. Having various responses can find its application to controlling or equalizing the signal amplitude from system design perspective.

9.3.4 Fourth-order bandstop filter

In addition to the second- and third-order bandstop filter topologies we discuss fourth-order bandstop filter topologies in this section. Figure 9.29 shows two fourth-order bandstop filter topologies [29,39]. The classical bandstop filter topology (Figure 9.29(a)) has no interresonator couplings whereas the other one has the interresonator couplings between resonators 1 and 2 and between resonators 3 and 4. Assuming that the coupling values between the source/load and NRN and the

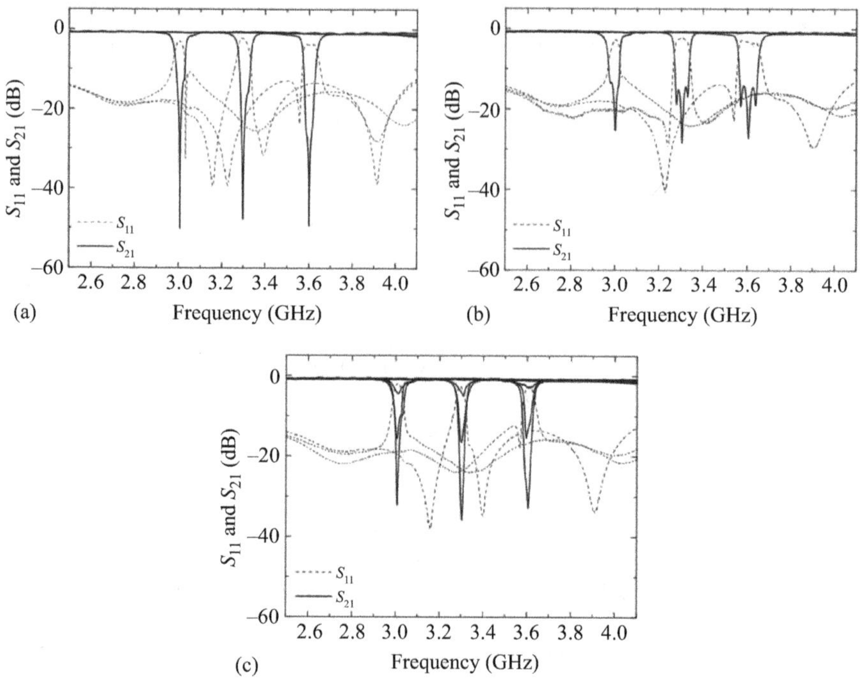

Figure 9.28 Measured frequency responses of the filter with two interresonator couplings: (a) Butterworth; (b) equiripple; and (c) variation of attenuation. Adapted with permission from [36]; © 2016 IEEE

one between the NRNs are 1, we can write the coupling matrix corresponding to the topology with no interresonator couplings as

$$\mathbf{M} = \begin{bmatrix} 0 & M_{s,1} & 0 & 1 & 0 & 0 & 0 & 0 \\ M_{s,1} & M_{1,1} & 0 & 0 & 0 & 0 & 0 & 0 \\ 0 & 0 & M_{2,2} & M_{2,n1} & 0 & 0 & 0 & 0 \\ 1 & 0 & M_{2,n1} & 0 & 1 & 0 & 0 & 0 \\ 0 & 0 & 0 & 1 & 0 & M_{2,n1} & 0 & 1 \\ 0 & 0 & 0 & 0 & M_{2,n1} & -M_{2,2} & 0 & 0 \\ 0 & 0 & 0 & 0 & 0 & 0 & -M_{1,1} & M_{s,1} \\ 0 & 0 & 0 & 0 & 1 & 0 & M_{s,1} & 0 \end{bmatrix} \quad (9.18)$$

In (9.18), the fourth and fifth rows and columns correspond to the first and second NRNs, respectively. In addition, the first, second, sixth, and seventh rows and columns are for resonators 1, 2, 3, and 4, respectively. In addition, it is assumed that $M_{n2,3} = M_{2,n1}$, $M_{4,l} = M_{s,1}$, $M_{3,3} = -M_{2,2}$, and $M_{4,4} = -M_{1,1}$ in (9.18) for simplicity. The topology shown in Figure 9.29(a) can produce the Butterworth response (Figure 9.30) when the coupling values are $M_{s,1} = 0.875$, $M_{2,n1} = 1.359$,

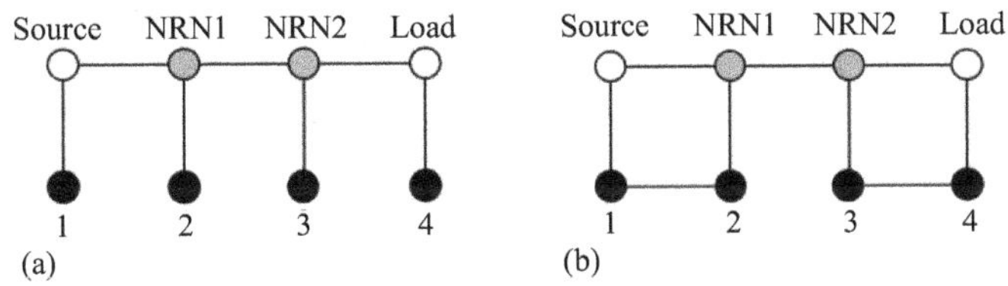

Figure 9.29 (a) Fourth-order bandstop filter topology with no interresonator couplings; (b) fourth-order bandstop filter topology with two interresonator couplings. Adapted with permission from [29]; © 2013 IEEE

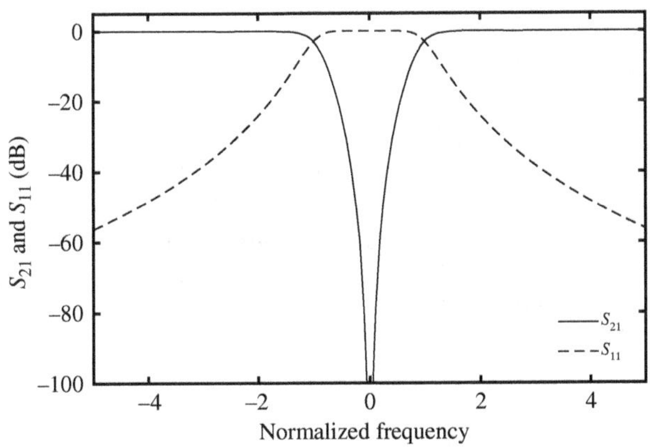

Figure 9.30 Fourth-order Butterworth response. Adapted with permission from [29]; © 2013 IEEE

and $M_{1,1} = M_{2,2} = M_{3,3} = M_{4,4} = 0.000$. By adjusting the self-coupling values, one can obtain various frequency responses. For example, $M_{1,1} = -M_{4,4} = 0.529$ and $M_{2,2} = -M_{3,3} = 0.219$ give the 40 dB equiripple response shown in Figure 9.31.

The coupling matrix for the topology with the interresonator couplings can also be formulated, and it is given by

$$\mathbf{M}' = \begin{bmatrix} 0 & M'_{s,1} & 0 & 1 & 0 & 0 & 0 & 0 \\ M'_{s,1} & M'_{1,1} & M'_{1,2} & 0 & 0 & 0 & 0 & 0 \\ 0 & M'_{1,2} & M'_{2,2} & M'_{2,n1} & 0 & 0 & 0 & 0 \\ 1 & 0 & M'_{2,n1} & 0 & 1 & 0 & 0 & 0 \\ 0 & 0 & 0 & 1 & 0 & M'_{2,n1} & 0 & 1 \\ 0 & 0 & 0 & 0 & M'_{2,n1} & -M'_{2,2} & M'_{1,2} & 0 \\ 0 & 0 & 0 & 0 & 0 & M'_{1,2} & -M'_{1,1} & M'_{s,1} \\ 0 & 0 & 0 & 0 & 1 & 0 & M'_{s,1} & 0 \end{bmatrix} \tag{9.19}$$

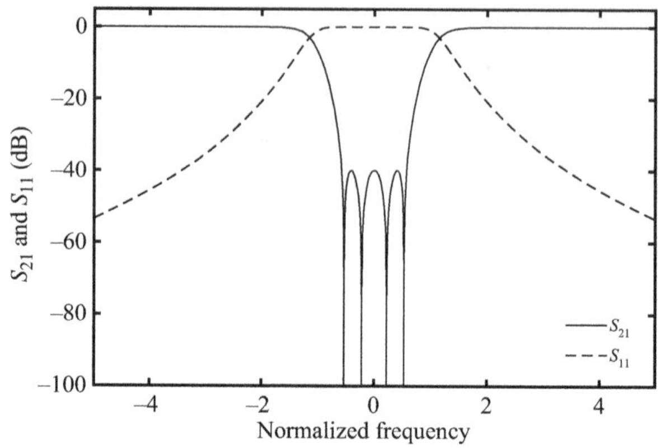

Figure 9.31 Fourth-order 40-dB equiripple response

In (9.19), it is assumed that the interresonator coupling between resonators 3 and 4 is the same as the one between resonators 1 and 2. In addition, it is assumed that $M'_{n2,3} = M'_{2,n1}$, $M'_{4,l} = M'_{s,1}$, $M'_{3,3} = -M'_{2,2}$, and $M'_{4,4} = -M'_{1,1}$, as was the case with the topology with no interresonator couplings. The filter topology shown in Figure 9.29(b) can produce the frequency responses of the filter topology shown in Figure 9.29(a) provided that [39]

$$\begin{aligned} M'_{s,1} &= M_{2,n1} \\ M'_{2,n1} &= M_{s,1} \\ M'_{1,1} &= M_{2,2} \\ M'_{2,2} &= M_{1,1} \\ M'_{1,2} &= M_{s,1}M_{2,n1} \end{aligned} \tag{9.20}$$

For example, in the case that $M'_{s,1} = 1.359$, $M'_{2,n1} = 0.875$, $M'_{1,2} = 1.189$ and $M'_{1,1} = M'_{2,2} = M'_{3,3} = M'_{4,4} = 0.000$, the topology shown in Figure 9.29(b) produces the Butterworth response shown in Figure 9.30. In addition, the equiripple response shown in Figure 9.31 can be obtained by making small adjustment to the self-couplings such that $M'_{1,1} = -M'_{4,4} = 0.219$ and $M'_{2,2} = -M'_{3,3} = 0.529$.

Figure 9.32 shows the two frequency-tunable bandstop filter structures designed by using the two topologies shown in Figure 9.29. Each filter is comprised of two substrates, a Rogers RO4350B substrate with the thickness of 0.508 mm for implementing the microstrip structure and a Rogers TMM3 substrate with the thickness of 3.175 mm for accommodating the SIW resonators. For comparison, the physical dimensions of both filters were determined such that they have the Butterworth response with 1.5% fractional bandwidth when centered at 3.3 GHz. The detailed processes for finding the dimensions are the same as those for the second-order and third-order bandstop filter designs discussed previously. It is interesting to note that the slot size for resonators 1 and 4 of the filter with interresonator couplings shown in Figure 9.32(b) is identical to that for resonators 2 and 3 of the filter with no interresonator couplings shown in Figure 9.32(a). In addition,

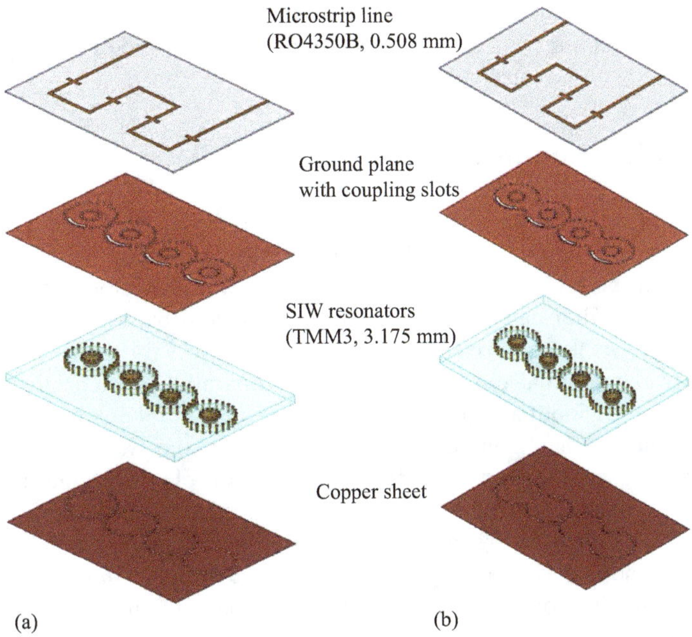

Figure 9.32 Fourth-order frequency-tunable SIW resonator bandstop filters: (a) filter with no interresonator couplings and (b) filter with two interresonator couplings

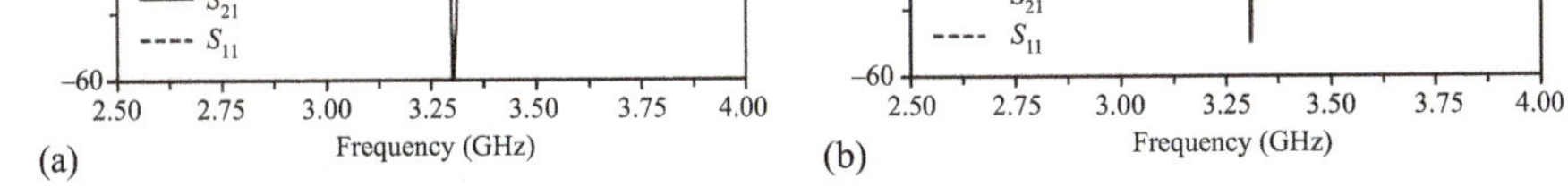

Figure 9.33 Simulated responses of fourth-order frequency-tunable SIW resonator bandstop filters: (a) filter with no interresonator couplings and (b) filter with two interresonator couplings

the slot size for resonators 2 and 3 of the filter shown in Figure 9.32(b) is the same as that for resonators 1 and 4 of the filter shown in Figure 9.32(a). This is because the filter shown in Figure 9.32(b) is required to satisfy $M'_{s,1} = M_{2,n1}$ and $M'_{2,n1} = M_{s,1}$ in (9.20) for producing the same Butterworth response.

Figure 9.33 shows the simulated frequency responses of the two filters shown in Figure 9.32 when they are centered at 3.3 GHz and tuned in a way to have the prescribed Butterworth response. It shows that the bandstop filter containing interresonator couplings can reproduce the frequency response of the conventional bandstop filter with no interresonator couplings.

9.4 Multifunction filter

In this section, we describe a design method for reconfigurable filters capable of having multiple functionalities. More specifically, we focus on implementing filters that can be reconfigured to have bandpass and bandstop modes at least. A dual-function filter capable of having bandpass and bandsotp responses can replace a filter bank comprised of two filters (Figure 9.34). In the filter bank structure, two individual circuits are positioned parallel to each other and two switches are at the input and output ports. Hence, the filter bank has a number of downsides since it has a large volume taken up by the two circuits and two switches add loss. Hence, a multifunction filter can play an important role in reducing the volume and the loss of a system in which a filter bank is in use. However, designing such reconfigurable filter is challenging mainly because microwave bandpass filters are generally required to have interresonator couplings whereas traditional bandstop filters do not have them. Due to this fundamental difference between the bandpass filters and the traditional bandstop filters in terms of the existence of the interresonator couplings, designing a bandpass-to-bandstop reconfigurable filter requires using tunable interresonator couplings whose values can be tuned down to zero.

A different way of designing reconfigurable filters is to use the bandstop filter topologies described in the previous section. For example, a second-order reconfigurable filter can be designed by using the bandstop filter topology shown in Figure 9.16(b). By comparing the bandstop filter topology shown in Figure 9.16(b) with a second-order direct-coupled bandpass filter topology shown in Figure 9.35, one can conclude that a bandstop filter designed using the filter topology containing the interresonator coupling can be switched to a bandpass filter by removing the

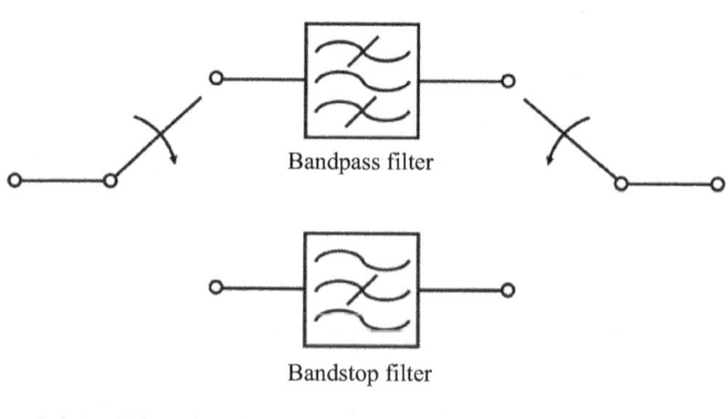

Figure 9.34 Filter bank comprised of bandpass and bandstop filters

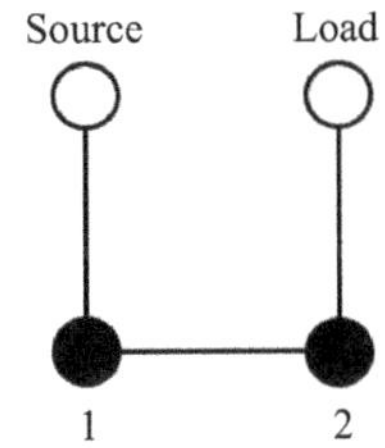

Figure 9.35 Second-order direct-coupled bandpass filter topology

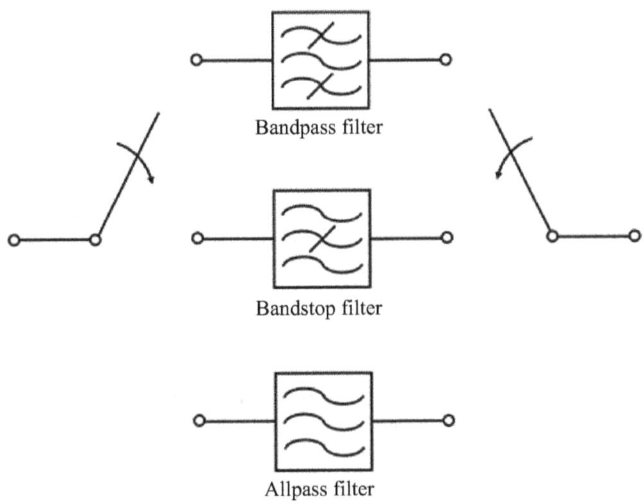

Figure 9.36 Filter bank comprised of bandpass, bandstop, and allpass filters. Adapted with permission from [41]; © 2017 IEEE

source-load coupling. This design approach was applied to second-order bandpass-to-bandstop reconfigurable filters and the details of the designs can be found in [26] and [40].

The aforementioned design approach can also be applied to higher-order reconfigurable filter designs. In this section, let us consider the design of a third-order reconfigurable filter. More specifically, we discuss a filter capable of having three modes: bandpass, bandstop, and allpass [41]. In short, the tri-function filter is equivalent to the filter bank shown in Figure 9.36 in terms of the functionalities. Since it can replace the filter bank, it is capable of reducing the size of a system in which the filter bank is used.

In the previous section, the third-order bandstop filter topology containing the interresonator couplings shown in Figure 9.20(b) was discussed. Using the bandstop filter topology with the interresonator couplings, we can design a filter capable of having both bandstop and all-pass responses. In view of the classical third-order direct-coupled bandpass filter topology shown in Figure 9.37, it can be concluded

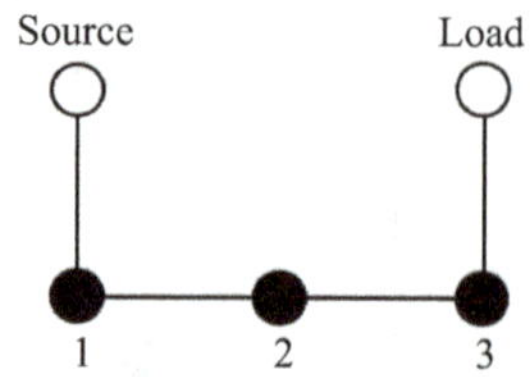

Figure 9.37 Third-order direct-coupled bandpass filter topology

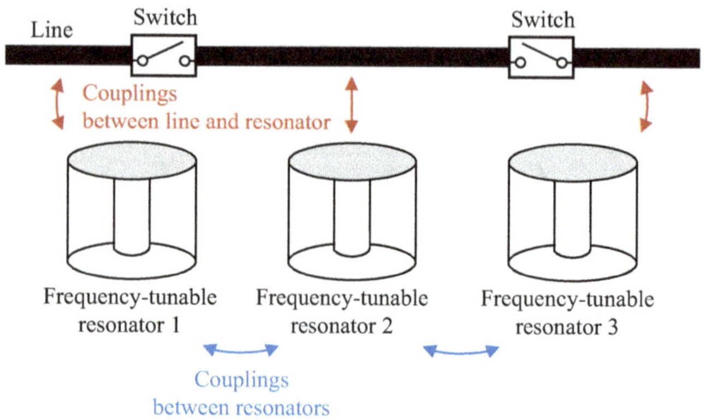

Figure 9.38 Circuit schematic of third-order reconfigurable filter. Adapted with permission from [41]; © 2017 IEEE

that a third-order filter designed using the topology shown in Figure 9.20(b) can be reconfigured in a way to operate as a bandpass filter by removing the couplings between the source and the load.

Figure 9.38 shows a simplified circuit schematic of a third-order reconfigurable filter. The transmission line connects the input port and the output port, and it is coupled to three frequency-agile resonators. There are two switches embedded in the line. For the purpose of designing a filter capable of implementing the topologies shown in Figure 9.20(b) and Figure 9.37, the filter schematic has the inter-resonator couplings between resonators 1 and 2 and between resonators 2 and 3.

When the two switches are closed, the topology of the filter comes to be the one shown in Figure 9.20(b) and the corresponding coupling matrix is given in (9.12). In this case, as discussed in the previous section, the filter can produce bandstop and all-pass responses. For example, the filter produces the all-pass response shown in Figure 9.39 when the normalized coupling values are

$$\begin{aligned} M'_{s,1} &= 1.000 \\ M'_{n,2} &= 1.414 \\ M'_{1,2} &= 0.707 \\ M'_{1,1} &= M'_{2,2} = 0.000. \end{aligned} \tag{9.21}$$

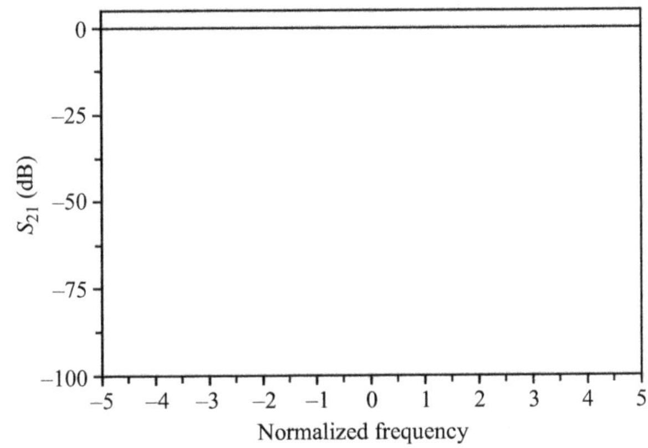

Figure 9.39 Allpass response. S_{11} (dB) is not shown since it is $-\infty$. Adapted with permission from [41]; © 2017 IEEE

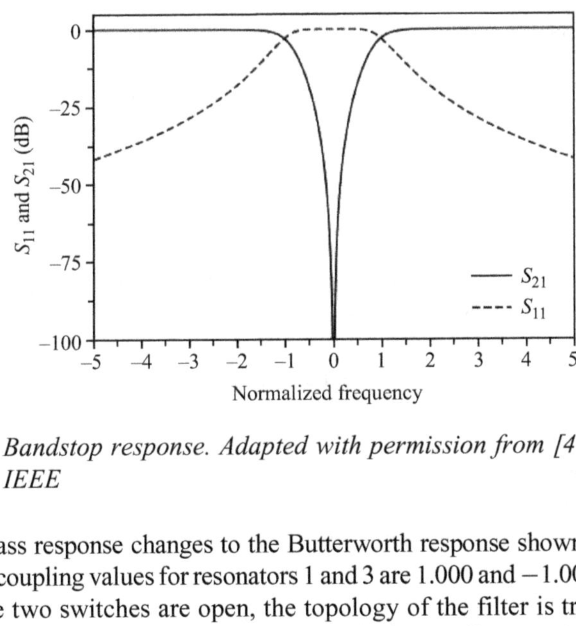

Figure 9.40 Bandstop response. Adapted with permission from [41]; © 2017 IEEE

The all-pass response changes to the Butterworth response shown in Figure 9.40 when the self-coupling values for resonators 1 and 3 are 1.000 and −1.000, respectively.

When the two switches are open, the topology of the filter is transformed into the bandpass topology shown in Figure 9.37 whose coupling matrix is given by

$$\mathbf{M} = \begin{bmatrix} 0 & M^p_{s,1} & 0 & 0 & 0 \\ M^p_{s,1} & 0 & M^p_{1,2} & 0 & 0 \\ 0 & M^p_{1,2} & 0 & M^p_{1,2} & 0 \\ 0 & 0 & M^p_{1,2} & 0 & M^p_{s,1} \\ 0 & 0 & 0 & M^p_{s,1} & 0 \end{bmatrix} \tag{9.22}$$

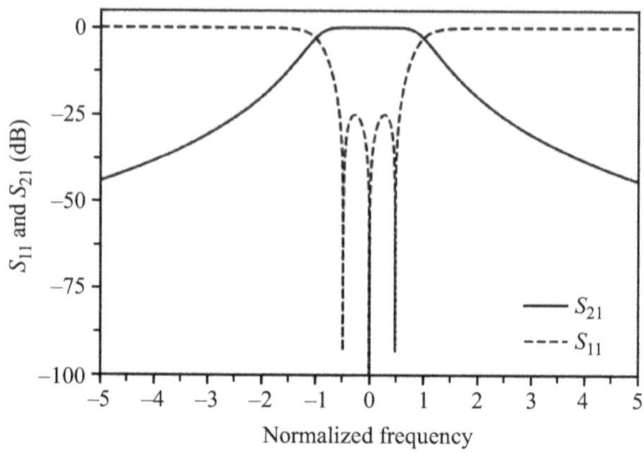

Figure 9.41 Bandpass response. Adapted with permission from [41]; © 2017 IEEE

In (9.22), it is assumed that the topology is symmetric with respect to resonator 2 and all the self coupling values are 0. The coupling values in (9.22) can be found for a predefined third-order bandpass response. For example, the coupling values for the 25-dB equiripple bandpass response shown in Figure 9.41 are

$$\begin{aligned} M^p_{s,1} &= 0.910 \\ M^p_{1,2} &= 0.679 \end{aligned} \tag{9.23}$$

It is worth noting that the coupling values in (9.23) are for the 25-dB equiripple bandpass response whose 3-dB passband is from -1 to 1 in the normalized frequency domain.

In this section, we demonstrate the design example of a tri-function filter capable of producing three different responses shown in Figures 9.39, 9.40, and 9.41. The filter must be designed such that its coupling structures have the denormalized coupling values required by the predefined frequency responses. When switches are closed for implementing the all-pass and bandstop responses, the physical coupling values are related to the normalized values by

$$\begin{aligned} k'_{\mathrm{ex}} &= M'_{\mathrm{ex}}\sqrt{\frac{\Delta_s}{2}} \\ k'_{\mathrm{in}} &= M'_{\mathrm{in}}\Delta_s \end{aligned} \tag{9.24}$$

where k'_{ex} and M'_{ex} represent the denormalized and normalized external coupling values, respectively, and k'_{in} and M'_{in} represent the denormalized and normalized interresonator coupling values, respectively. Δ_s denotes the fractional bandwidth of the bandstop response. The normalized external coupling values are associated with one resonator and they are represented by M'_{ex} which includes $M'_{s,1}$ and $M'_{n,2}$. The normalized interresonator coupling, M'_{in}, is $M'_{1,2}$ in this filter design.

When the switches are open for producing a bandpass response, the denormalized coupling values are related to the normalized ones by

$$\begin{aligned} k_{\text{ex}}^{p} &= M_{\text{ex}}^{p}\sqrt{\Delta_p} \\ k_{\text{in}}^{p} &= M_{\text{in}}^{p}\Delta_p \end{aligned} \tag{9.25}$$

where k_{ex}^{p} and M_{ex}^{p} represent the denormalized and normalized external coupling values, respectively, for a bandpass response, and k_{in}^{p} and M_{in}^{p} represent the denormalized and normalized interresonator coupling values, respectively. Δ_p denotes the fractional bandwidth of the bandpass response.

For simple demonstration, static coupling structures were adopted in the filter design and they were designed such that they have the desired denormalized coupling values given in (9.24) and (9.25) at the design frequency of 3.3 GHz. Since the interresonator coupling values do not change when turning on and off the switches, the filter must be designed such that k_{in}' equals to k_{in}^{p}. Hence, in this design example, the design fractional bandwidths, Δ_s and Δ_p were set to 0.020 and 0.021 for demonstration, respectively, so that

$$k_{1,2}' = k_{1,2}^{p} = 0.014 \tag{9.26}$$

when the filter is centered at 3.3 GHz. Using the predefined design fractional bandwidths, we can find the other coupling values using (9.24) and (9.25). When the switches are closed, they are

$$\begin{aligned} k_{s,1}' &= 0.100 \\ k_{n,2}' &= 0.141. \end{aligned} \tag{9.27}$$

When the switches are open, the denormalized external coupling values are

$$k_{s,1}^{p} = 0.131. \tag{9.28}$$

Now, we describe physical design of the frequency-agile tri-function filter having the tuning range from 3.0 to 3.6 GHz by means of using the physical coupling values in (9.26)–(9.28) at the design frequency of 3.3 GHz. Figure 9.42 shows the frequency-tunable SIW resonator structure used in the filter design. The application of this type of resonator structure to frequency-tunable filter designs are discussed extensively in the literature [23,27,30,38,42]. Hence, this section gives a brief description on it. The resonator in this filter design is implemented in a 3.175-mm-thick Rogers TMM3 substrate. The resonant frequency of the resonator is mainly determined by the resonator radius, the post radius, and the thickness of the air gap formed between the post and the flexible conducting layer called membrane on one side of the substrate. Hence, tuning the resonant frequency can be carried out by changing the air-gap thickness by bending the membrane. Considering the frequency tuning range, the resonator radius was chosen to be 7.7 mm, and Figure 9.43 shows the resonant frequency variation of the resonator shown in Figure 9.42 with respect to the air-gap thickness.

Figure 9.44 shows the top view and cross-sectional side view of the frequency-tunable tri-function filter designed to have the physical coupling values given in

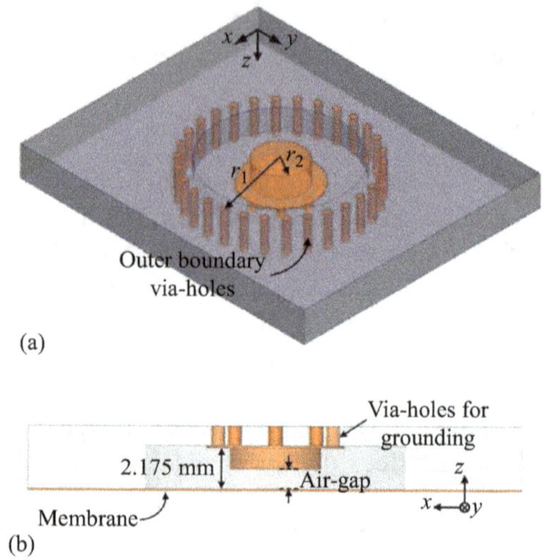

Figure 9.42 Resonator structure: (a) three-dimensional view (two conductor planes on top and bottom sides of the substrate are not shown for clarity) and (b) cross-sectional side view. Reprinted with permission from [41]; © 2017 IEEE

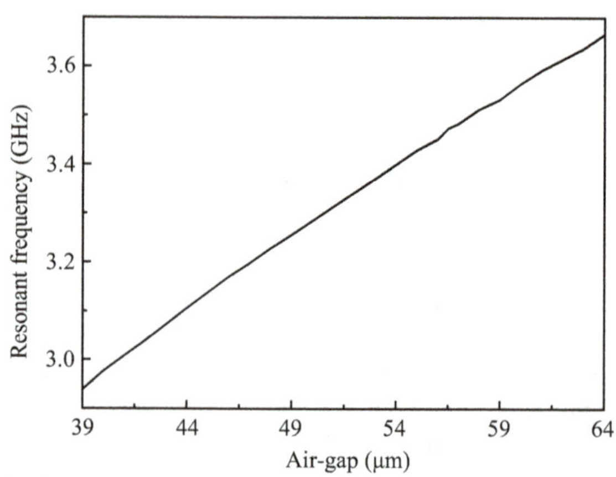

Figure 9.43 Simulated resonant frequency of the resonator. Reprinted with permission from [41]; © 2017 IEEE

(9.26)–(9.28). In this design, the filter has two dielectric substrates, and they are a 0.508-mm-thick Rogers RO4350B substrate and a 3.175-mm-thick Rogers TMM3. A 50-Ω microstrip line is used for connecting the input port and the output port. There are two pads for mounting the two switches.

The physical design of the filter starts with designing the coupling slots. The microstrip line is coupled to the frequency-tunable SIW resonators through the coupled slots on the ground plane between the two dielectric substrates. When the

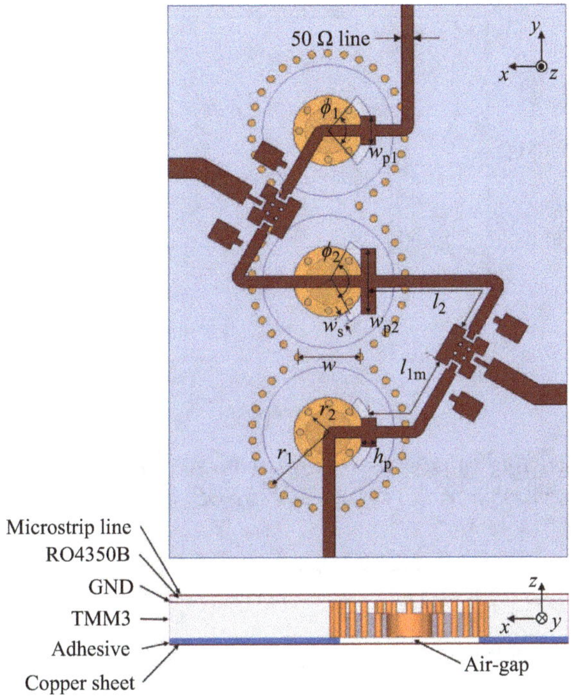

Figure 9.44 Top and side view of the proposed reconfigurable filter. ($\phi_1 = 100°$, $\phi_2 = 124°$, $w = 6.2$ mm, $l_{1m} = 11.1$ mm, $l_2 = 18.9$ mm, $w_{p1} = 2.8$ mm, $w_{p2} = 6.5$ mm, $w_s = 1$ mm, $h_p = 1.5$ mm, $r_1 = 7.7$ mm, and $r_2 = 2.3$ mm). Reprinted with permission from [41]; © 2017 IEEE

switches are closed, the external coupling values are mainly determined by the size and location of each slot. In this design, the coupling value of the coupling slot between the microstrip line and the resonator was computed by full-wave simulation for various values of the slot angle when the other dimensions are the ones given in Figure 9.44. Figure 9.45 shows the variation of the coupling value with respect to the angle of the slot. In view of (9.27) and Figure 9.45, we can conclude that the slot angle for resonators 1 and 3 must be 100° and the one for resonator 2 is 124°.

When the switches are open, the external coupling value of the slot for implementing the predefined bandpass response depends on the distance between the slot for the first (or third) resonator and the open end of the line as well as its size and location. Since the size and location of each slot was already determined, we need to find the line length from the slot and its open end giving the coupling value in (9.28). Figure 9.46 shows the variation of the coupling value versus the line length from the slot to its open end. According to Figure 9.46, the line length of 14 mm gives the desired coupling value.

Next, the interresonator coupling value is mainly determined by the width of the iris between a pair of two coupled resonators. Full-wave simulation was carried

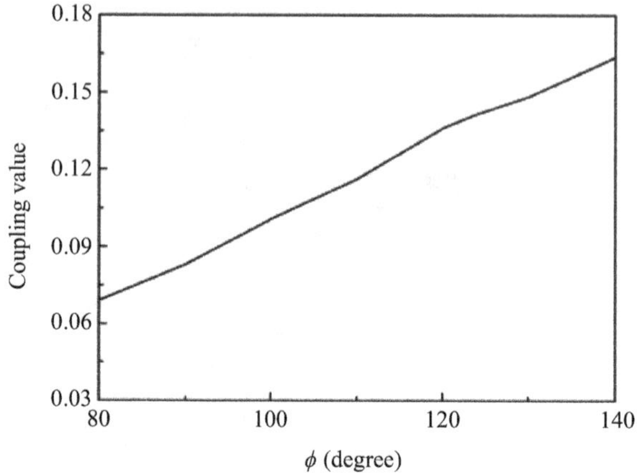

Figure 9.45 Coupling value of the slot when the switches are closed. Adapted with permission from [41]; © 2017 IEEE

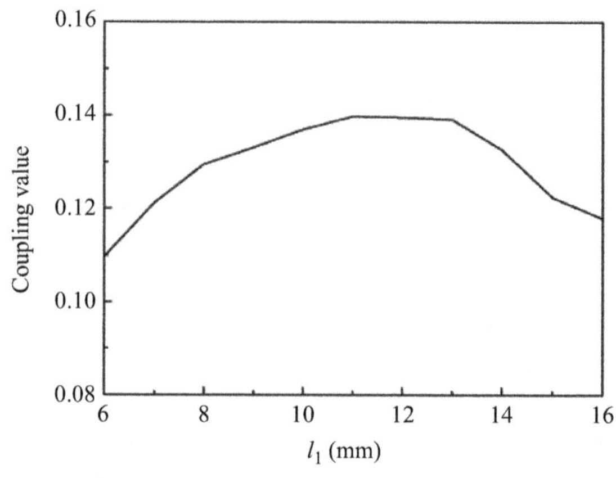

Figure 9.46 Coupling value of the slot when the switches are open. Adapted with permission from [41]; © 2017 IEEE

out for calculating the interresoantor coupling value between the coupled resonators used in the filter structure, and the coupling value is shown in Figure 9.47. Figure 9.47 and (9.26) mandate that the iris width is 6.2 mm.

Previously, the length of the line from the slot to its open end was found for realizing the desired external coupling value for the bandpass response. In this case, the open end is implemented by the off-state switch, but the off-state switch does not provide the ideal open circuit in practice. Hence, the actual switch parameters need to be incorporated in finding the line length. The switches from Radant Company, RMSW-200hp, were used in this filter design, and one of them was measured to finalize the design of the microstrip line. The measurement of the switch showed that the phase of the reflection coefficient was −38° when it was off

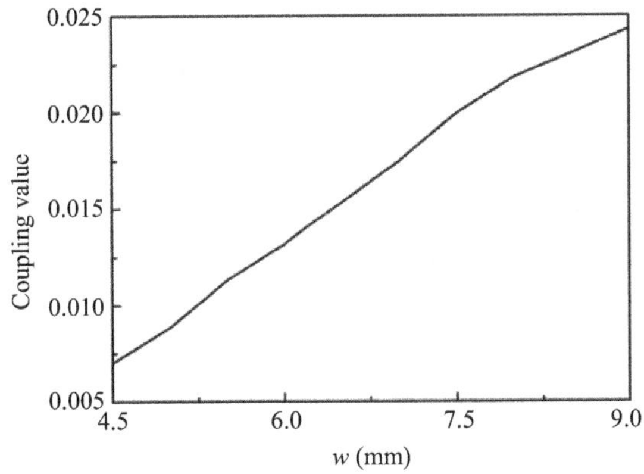

Figure 9.47 Coupling value of the iris between two resonators. Adapted with permission from [41]; © 2017 IEEE

and the phase of the transmission coefficient was −49° when it was on. Since the reflection coefficient of −38° is equivalent to the line length of 2.9 mm, the line length from the coupling slot for resonator 1 (or resonator 3) to the open end was adjusted from 14 mm to 11.1 mm.

Designing the lines between neighboring coupling slots must be carried out in a way to satisfy the desired coupling values. According to (9.12), the normalized coupling value between the source (or load) and the NRN is 1. In addition, the normalized coupling value between resonators 1 and 2 (or resonators 2 and 3) is a positive one. Since they have the same sign, they are required to have the same phase shift. Each inter-resonator coupling structure is implemented by an inductive iris used for realizing an inverter having +90° phase shift. Hence, the electrical length of the line between the two neighboring coupling slots must be +270° including the phase shifts of the switches and the two coupling slots. Hence, l_2 in Figure 9.44 was determined to be 18.9 mm. The rectangular patches placed above the coupling slots are intended to compensate the impedance mismatch caused by the coupling slots in the ground plane [38].

The filter shown in Figure 9.44 was fabricated and the fabricated filter is shown in Figure 9.48. The microstrip line structure contains the dc-bias lines and the paths to the ground for the two switches. In this work, the fabrication technique described in [43] was adopted. Piezoelectric actuators from PiezoElectric Technology (TULA series) are employed in the filter structure for controlling the resonant frequency of each resonator.

Figure 9.49 shows the measured filter responses when the two switches are closed. In this case, the filter can produce the all-pass and bandstop responses. When all resonators are tuned to resonate at the same frequency, the filter produces the all-pass response shown in Figure 9.49(a). The measured insertion loss over the frequency range of interest (3.0 to 3.6 GHz) is about 1.0 dB. The loss can be attributed to the two switches, the connectors, and the microstrip line. A small notch is observed at 3.67 GHz, and it is due to the finite Q-factor of the SIW

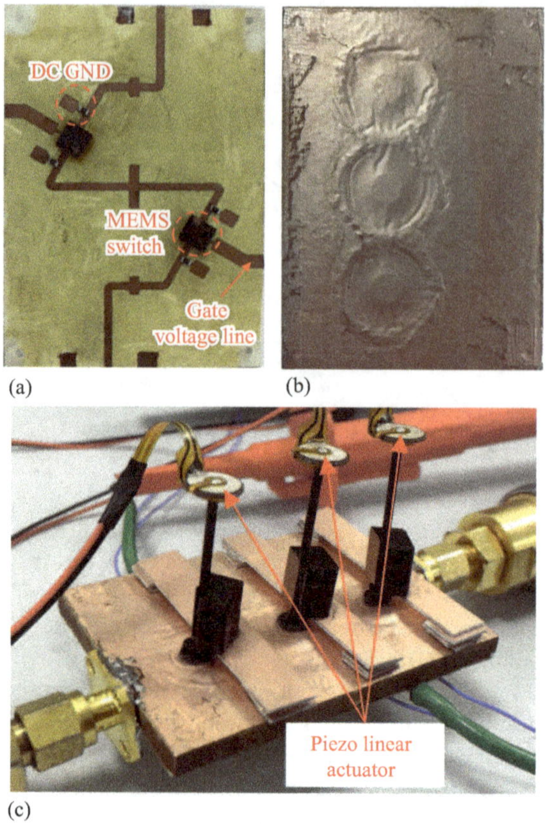

Figure 9.48 Fabricated filter: (a) top view; (b) bottom view before mounting tuning devices; and (c) three-dimensional view after mounting tuning devices. Reprinted with permission from [41]; © 2017 IEEE

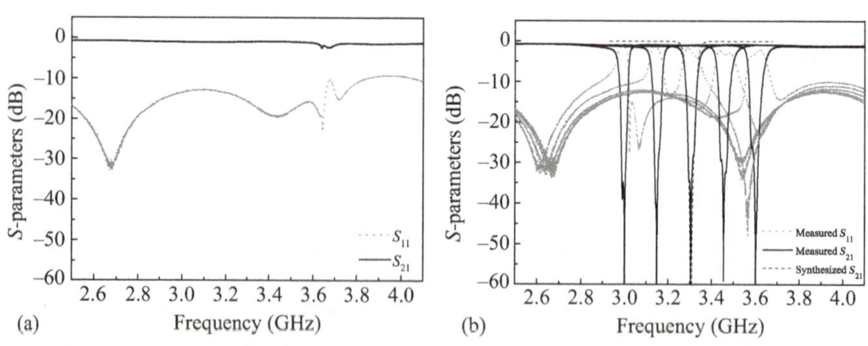

Figure 9.49 Measured frequency responses of the fabricated filter: (a) allpass mode; and (b) bandstop mode. Reprinted with permission from [41]; © 2017 IEEE

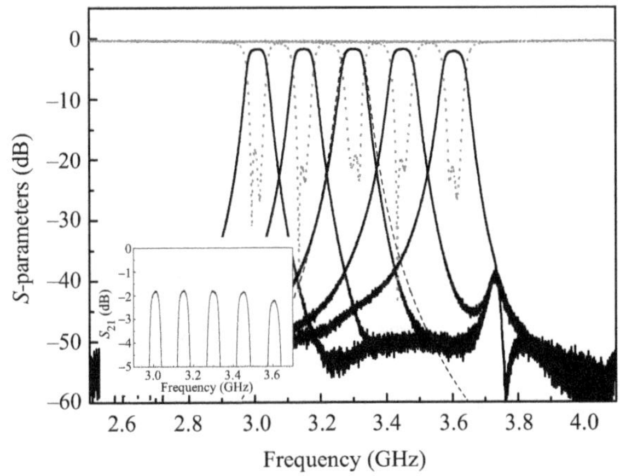

Figure 9.50 Measured bandpass responses of the fabricated filter. Reprinted with permission from [41]; © 2017 IEEE

resonators. Since, this small notch that appears in the all-pass response is inevitable, it is placed outside of the frequency range of interest by virtue of the frequency-tunable feature of each resonator.

One can obtain a Butterworth bandstop response at the resonant frequency of resonator 2 by tuning resonators 1 and 3 such that one resonates at a frequency lower than the resonant frequency of resonator 2 and the other does at a frequency higher than the resonant frequency of resonator 2. The resonant frequencies of resonators 1 and 3 are determined by the self-coupling values in the coupling matrix, the center frequency, and the design bandwidth. The measured responses provided in Figure 9.49(b) shows that the stopband can be tuned from 3.0 to 3.6 GHz and the attenuation level can reach at least 50 dB at the center frequency of the stopband.

Figure 9.50 shows the measured bandpass responses when the two switches are open. The measured insertion loss of the filter is smaller than 2.4 dB when the filter is tuned from 3.0 to 3.6 GHz. For comparison, a theoretical response centered at 3.3 GHz is included in Figure 9.50. The theoretical response is based on the assumption that the Q-factor of each resonator is 450. It can be concluded that there is a good agreement between the measured and prescribed bandpass responses.

In conclusion, the fabricated filter can have three different responses over the 1.2:1 frequency range from 3.0 to 3.6 GHz, and it indicates that the filter can replace a filter bank comprised of three building blocks namely a bandstop filter, a bandpass filter, and an all-pass circuit.

9.5 Summary

This chapter is devoted to provide an overview of the techniques for designing microwave filters using frequency-tunable SIW resonators. First, the basic frequency-tunable SIW resonator structure having a single post is reviewed, and a modified resonator structure containing two posts is described. The discussion on the characteristics of the first three resonant modes of the modified resonator is provided, and a design approach for a *K*-band frequency-tunable SIW resonator filter using the third resonant mode of the resonators is described. In addition, this chapter includes a methodology to improve the upper stopband of the *K*-band frequency-tunable filter.

Frequency-tunable SIW resonators can also be employed in designing frequency-tunable bandstop filters. For understanding filter designs, bandstop filter topologies have been explored. Second-order, third-order, and fourth-order conventional bandstop filter topologies with no interresonator couplings are described. Since it is preferable to have interresonator couplings in designing multifunction filters capable of having bandstop and bandpass responses, bandstop filter topologies containing interresonator couplings are also discussed. The bandstop filter topologies containing interresonator couplings can reproduce the frequency responses of the topologies with no interresonator couplings, and the relationships between them are summarized in this chapter. They allow us to find the coupling values of the bandstop filter topologies containing the interresonator couplings directly from the ones of the conventional topologies. This chapter illustrates second-order, third-order, and fourth-order frequency-tunable bandstop filter design examples using the conventional topologies and the topologies containing the interresonator couplings.

This chapter also includes a design example of a third-order frequency-tunable multifunction filter capable of having three different responses: bandpass, bandstop, and all-pass responses. The filter structure has two switches and they determine the operation mode of the filter. When the switches are open, it is in bandpass mode. On the other hand, the filter can produce bandstop and all-pass responses when switches are closed. Since frequency-tunable multifunction filters can replace filter banks comprised of multiple filters, it is believed that they are useful in reducing the size of multifunction wireless systems.

References

[1] Mansour RR. Filter technologies for wireless base stations. IEEE Microwave Magazine. 2004;5(1):68–74.

[2] Yu M, Yassini B, Keats B, *et al.* The sound the air makes: High-performance tunable filters based on air-cavity resonators. IEEE Microwave Magazine. 2014;15(5):83–93.

[3] Yassini B, Yu M, Keats B. A Ka-band fully tunable cavity filter. IEEE Trans Microw Theory Techn. 2012;60(12):4002–4012.

[4] Arnold C, Parlebas J, Zwick T. Reconfigurable waveguide filter with variable bandwidth and center frequency. IEEE Trans Microw Theory Techn. 2014;62(8):1663–1670.

[5] Nam S, Lee B, Koh B, *et al.* K-band fully reconfigurable pseudo-elliptic waveguide resonator filter with tunable positive and negative couplings. IEICE Trans Commun. 2016;99(10):2136–2145.

[6] Cameron RJ, Kudsia CM, Mansour RR. Microwave filters for communication system. New York: Wiley; 2007.

[7] Mansour RR. High-Q tunable dielectric resonator filters. IEEE Microwave Magazine. 2009;10(6):84–98.

[8] Chen XP, Wu K. Substrate integrated waveguide filter: Basic design rules and fundamental structure features. IEEE Microwave Magazine. 2014;15(5): 108–116.

[9] Deslandes D, Wu K. Single-substrate integration technique of planar circuits and waveguide filters. IEEE Trans Microw Theory Techn. 2003;51(2):593–596.

[10] Chen X, Hong W, Cui T, *et al.* Substrate integrated waveguide elliptic filter with transmission line inserted inverter. Electronics Letters. 2005;41(15):851–852.

[11] Chen X, Hong W, Cui T, *et al.* Substrate integrated waveguide (SIW) linear phase filter. IEEE Microw Wireless Compon Lett. 2005;15(11):787–789.

[12] Potelon B, Favennec JF, Quendo C, *et al.* Design of a substrate integrated waveguide (SIW) filter using a novel topology of coupling. IEEE Microw Wireless Compon Lett. 2008;18(9):596–598.

[13] Shen W, Wu LS, Sun XW, *et al.* Novel substrate integrated waveguide filters with mixed cross coupling (MCC). IEEE Microw Wireless Compon Lett. 2009;19(11):701–703.

[14] Shen W, Sun XW, Yin WY, *et al.* A novel single-cavity dual mode substrate integrated waveguide filter with non-resonating node. IEEE Microw Wireless Compon Lett. 2009;19(6):368–370.

[15] Deslandes D, Wu K. Substrate integrated waveguide dual-mode filters for broadband wireless systems. In: Radio and Wireless Conf. Boston, MA; Aug. 2003. p. 385–388.

[16] Chen XP, Wu K, Drolet D. Substrate integrated waveguide filter with improved stopband performance for satellite ground terminal. IEEE Trans Microw Theory Techn. 2009;57(3):674–683.

[17] Han S, Wang XL, Fan Y. Analysis and design of multiple-band bandstop filters. Progr Electromagn Res. 2007;70:297–306.

[18] Ahmad BH, Hunter IC. Design and fabrication of a substrate integrated waveguide bandstop filter. In: European Microwave Conference. Amsterdam, Netherlands; Oct. 2008. p. 40–42.

[19] Joshi H, Sigmarsson HH, Moon S, *et al.* High Q narrow-band tunable filters with controllable bandwidth. In: IEEE MTT-S Int. Microwave Symp. Dig. Boston, MA; Jun. 2009. p. 629–632.

[20] Moon S, Sigmarsson HH, Joshi H, *et al.* Substrate integrated evanescent-mode cavity filter with a 3.5 to 1 tuning ratio. IEEE Microw Wireless Compon Lett. 2010;20(8):450–452.

[21] Sigmarsson HH, Lee J, Peroulis D, *et al.* Reconfigurable-order bandpass filter for frequency agile systems. In: IEEE MTT-S Int. Microwave Symp. Dig. Anaheim, CA; May. 2010. p. 1756–1759.

[22] Koh B, Lee B, Nam S, *et al.* Integration of interresonator coupling structures with applications to filter systems with signal route selectivity. IEEE Trans Microw Theory Techn. 2016;64(9):2790–2803.

[23] Lee B, Nam S, Koh B, *et al.* K-band frequency tunable substrate-integrated-waveguide resonator filter with enhanced stopband attenuation. IEEE Trans Microw Theory Techn. 2015;63(11):3632–3640.

[24] Hong JS, Lancaster ML. Microstrip Filters for RF/Microwave Applications. New York: Wiley; 2001.

[25] Lee J, Naglich EJ, Chappell WJ. Frequency response control in frequency-tunable bandstop filters. IEEE Microw Wireless Compon Lett. 2010;20(12): 669–671.

[26] Naglich EJ, Lee J, Peroulis D, *et al.* A tunable bandpass-to-bandstop reconfigurable filter with independent bandwidths and tunable response shape. IEEE Trans Microw Theory Techn. 2010;58(12):3770–3779.

[27] Lee J, Naglich EJ, Sigmarsson HH, *et al.* Tunable inter-resonator coupling structure with positive and negative values and its application to the field-programmable filter array (FPFA). IEEE Trans Microw Theory Techn. 2011;59(12):3389–3400.

[28] Naglich EJ, Lee J, Peroulis D, *et al.* Switchless tunable bandstop-to-all-pass reconfigurable filter. IEEE Trans Microw Theory Techn. 2012;60(5): 1258–1265.

[29] Lee J, Naglich EJ, Sigmarsson HH, *et al.* New bandstop filter circuit topology and its application to design of a bandstop-to-bandpass switchable filter. IEEE Trans Microw Theory Techn. 2013;61(3):1114–1123.

[30] Lee K, Lee TH, Ahn CS, *et al.* Reconfigurable dual-stopband filters with reduced number of couplings between a transmission line and resonators. IEEE Microw Wireless Compon Lett. 2015;25(2):106–108.

[31] Kwok RS, Liang JF. Characterization of high-Q resonators for microwave filter applications. IEEE Trans Microw Theory Techn. 1999;47(1):111–114.

[32] Lee B, Lee TH, Lee K, *et al.* K-band substrate-integrated waveguide resonator filter with suppressed higher-order mode. IEEE Microw Wireless Compon Lett. 2015;25(6):367–369.

[33] Mehrshahi E, Salehi M, Rezaiesarlak R. Substrate integrated waveguide filters with stopband performance improvement. In: in Int. Microw. Millim. Wave Technol. Conf. Chengdu, China; May. 2010. p. 2018–2020.

[34] Jarry P, Beneat J. Design and realizations of miniaturized fractal microwave and RF filters. New Jersey: Wiley; 209.

[35] Kajfez D, Guillon P. Dielectric resonators. Norwood, MA: Artech House; 1986.

[36] Lee TH, Lee K, Park GC, *et al.* Bandstop filter (BSF) topology with variable attenuation. IEEE Trans Microw Theory Techn. 2016;64(2):467–474.

[37] Amari S, Rosenberg U. New building blocks for modular design of elliptic and self-equalized filters. IEEE Trans Microw Theory Techn. 2004;52(2): 721–736.

[38] Naglich EJ, Lee J, Peroulis D, *et al.* Extended passband bandstop filter cascade with continuous 0.85–6.6-GHz coverage. IEEE Trans Microw Theory Techn. 2012;60(1):21–30.

[39] Lee TH, Ahn CS, Kim YS, *et al.* Extension of bandstop filter topology with inter-resonator coupling structures to higher-order filters. IEEE Microw Wireless Compon Lett. 2013;23(8):403–405.

[40] Lee K, Lee TH, Park GC, *et al.* Frequency-tunable bandstop-bandpass dual-function microwave filter. IEICE Electronics Express. 2015;12(11): 20150313.

[41] Lee TH, Lee B, *et al.* Frequency-tunable tri-function filter. IEEE Trans Microw Theory Techn. 2017;65(11):4584–4592.

[42] Joshi H, Sigmarsson HH, Chappell WJ. Analytical modeling of highly loaded evanescent-mode cavity resonators for widely tunable high-Q filter applications. In: in Proc. Union Radio Sci. Int. (URSI). Chicago, IL, USA; Aug. 2008. p. 1–4.

[43] Nam S, Koh B, Lee B, *et al.* Low-cost high-performance frequency-tunable substrate-integrated waveguide filter structure and fabrication method. IEICE Electronics Express. 2016;13(9):20160211.

Chapter 10
Superconducting dual-band filters

Haiwen Liu[1,2] *and Baoping Ren*[1]

10.1 Introduction

In 1986, high-temperature superconductors (HTS) were discovered [1]. Compared with low-temperature superconductors, the cooling environment for HTS operating is easier to achieve, which leads to a lower cost for industries applications. So, there has been an enormous effort worldwide to develop and characterize HTS materials. As well-known, HTS thin films deposited on substrates have the advantage of low surface resistant and making them attractive in designing microwave filters with ultralow loss and excellent out-of-band performance.

With the continuous appearance of various wireless equipments having multi-functional services, compact and high-performance HTS dual-band bandpass filters (BPFs) are highly desired. Owing to the requirement of more design freedoms for controlling two passbands simultaneously, the design of HTS dual-band BPFs is more difficult than single-band BPFs. Although some attempts have been made on dual-band HTS BPFs [2–5], very few effective methods for dual-band HTS filters with design flexibility of controlling the center frequency and bandwidths have been presented. In Reference [6] and [7], the design of fourth-order and eighth-order HTS dual-band BPFs with dual-mode stub-loaded resonators and dual-feeding structure technique are proposed for adjusting both the center frequencies and bandwidths of each passband. However, the four-pole filter has only two reflection zeros in each of the passband and occupies a relative large circuit size, which means that the proposed structure and design are not optimal [6]. In Reference [7], the proposed eighth-order dual-band HTS BPF has no transmission zeros in the finite frequency scope and its coupling scheme should be improved for high-selectivity.

Herein, this chapter presents several outcomes of an extensive research on the HTS dual-band filters based on multimode resonators with compact size and high performance [8,9,12–14]. In Section 10.2, a series of compact two-pole HTS dual-band BPFs based on various multimode resonators are presented with the operating

[1]School of Information Engineering, East China Jiaotong University, China
[2]School of Electronics and Information Engineering, Xi'an Jiaotong University, China

mechanism. To achieve steep sideband and flat passband, a novel dual-mode resonator is proposed and utilized to design high-order HTS dual-band BPF with controllable bandwidths and multi-transmission zeros in Section 10.3. A general dual-path coupling scheme is introduced for high-order coupling realization. The designed dual-band filters are fabricated on MgO substrate with YBCO thin films and measured at 77 K. Finally a summary is given in Section 10.4.

10.2 Compact two-pole dual-band HTS filters

10.2.1 Miniaturization design using split ring resonator

A type of HTS filter employing octagonal half-wavelength split-ring resonators (SRRs) is designed to generate two passbands with compact size [8]. The schematic view of this type of HTS BPF is depicted in Figure 10.1. This BPF includes two sets of embedded octagonal SRRs, two side coupled stubs, and an interdigital coupling structure between the two inner resonators. The outer resonators of the SRR structure, resonator 1 and resonator 4 with long perimeters, are used to generate the lower passband at f_1. Whereas the inner resonators of embedded resonator 2 and resonator 3 with short perimeters, are utilized for producing the upper passband at f_2. Resonator 1 and resonator 4 are connected with two 50 Ω input/output (I/O) lines, respectively. The interdigital structure between resonator 2 and resonator 3 is used to control the coupling of two inner resonators, and thus control the bandwidth of the lower passband. A pair of side-coupled stubs is used to adjust the coupling between the two outer resonators, and thus adjust the bandwidth of the upper passband.

Straight-line resonators and hairpin resonators are two common structures used in superconducting BPFs. To accurately determine the unwounded length and the sizes of these resonators, electromagnetic simulations are carried out and the simulated results are presented in Table 10.1. All resonators are designed at the same frequency with the same line widths. Obviously, it can be observed from Table 10.1 that the octagonal resonator has the most compact size among the three resonator structures.

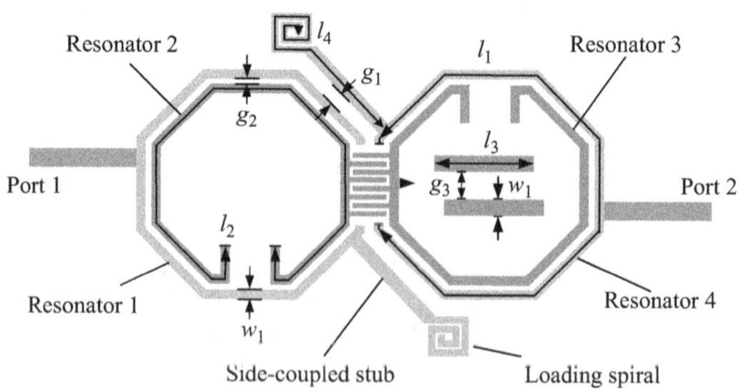

Figure 10.1 Schematic view of the proposed HTS filter. Reprinted with permission from Reference [8]; © 2013 IEEE

Table 10.1 Comparisons between different resonator structures

Resonator structure	Unwounded length (mm)	Circuit size (mm × mm)
Straight line	15.6	15.6 × 0.2
Hairpin	16.1	5.2 × 5.7
Octagonal	15.7	5.2 × 5.2

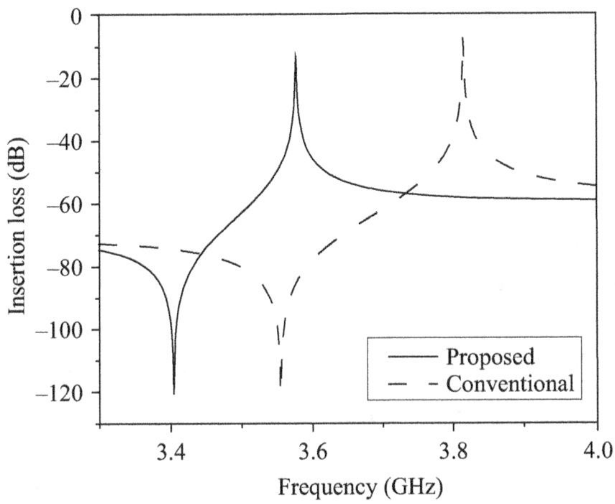

Figure 10.2 Comparison of the simulated insertion losses between the proposed and the conventional octagonal filters. Reprinted with permission from Reference [8]; © 2013 IEEE

For further miniaturization design, side coupled stubs and loading spiral are adopted. As seen from Figure 10.1, side-coupled stub and spiral are loaded to each outer resonator, which results in the corresponding increment of self-inductance and self-capacitance. Figure 10.2 shows a comparison of the simulated insertion losses between the proposed and the conventional octagonal resonators. It can be found that the resonant frequency of the octagonal resonator using the loaded elements can be reduced.

Therefore, in the proposed filter, the center frequency of each passband and the coupling between each resonator can be controlled independently. Small size and high performance filter for dual-band application can be realized. Following the discussion above, a dual-band HTS filter with central frequencies of 2.45 GHz for WLAN and 3.5 GHz for WiMAX is designed, simulated and measured. The filter was fabricated on a 2-in-diameter 0.5-mm-thick MgO wafer with double-sided YBCO films, and measured at the temperature of 77 K. Figure 10.3 shows the frequency responses of the experimental data. Its maximum insertion losses, including SMA connector loss, are measured to be 0.37 and 0.35 dB, respectively. The return losses within the passbands are greater than 12 and 13 dB, respectively.

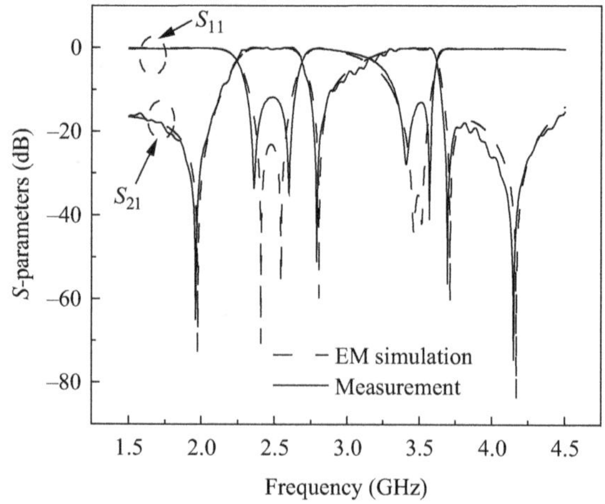

Figure 10.3 Simulated and measured results of the dual-band HTS filter. Reprinted with permission from Reference [8]; © 2013 IEEE

Four transmission zeros located at 1.98, 2.81, 3.71, and 4.17 GHz are realized, which greatly improve the skirt selectivity.

10.2.2 Compact stepped-impedance resonators filter with multiple transmission zeros

In this work, a compact dual-band high-temperature superconducting (HTS) BPF constructed by spirally asymmetric stepped-impedance resonators (SIRs) and a pair of parallel-coupled feed lines is presented in Figure 10.4(a) [9]. The resonant characteristics of the asymmetric SIR can be easily controlled and the first two resonant modes are utilized to realize a dual-band performance at 2.4/5.8 GHz for WLAN applications. In order to achieve the compact size and high selectivity, some improvements are conducted on the asymmetric SIR with spiral high-impedance transmission-line. Besides, owing to its intrinsic characteristics and the multipath propagation effect, multiple transmission zeros (TZs) are created to improve the passband selectivity and the rejection level of stopband. The coupling scheme of the designed dual-band filter can be expressed as Figure 10.4(b). The proposed dual-band HTS filter is designed with the target specifications of the central frequencies of $f_{01} = 2.4$ GHz and $f_{02} = 5.8$ GHz, the corresponding fractional bandwidths of $\mathrm{FBW}^{\mathrm{I}} = 4.1\%$ and $\mathrm{FBW}^{\mathrm{II}} = 1.6\%$, respectively. The circuit is fabricated on a 2-in-diameter 0.5-mm-thick MgO wafer with double-sided $YBa_2Cu_3O_y$ (YBCO) films, which has a relative dielectric constant of 9.78. In addition, five transmission zeros can be observed both in simulation and measurement. Figure 10.5 shows the frequency responses of the fabricated HTS filter. Five TZs located at 2.06, 2.81, 5.33, 5.88 GHz, and 6.55 GHz can greatly improve the passband selectivity and the out-of-band rejection skirts, as observed from Figure 10.5.

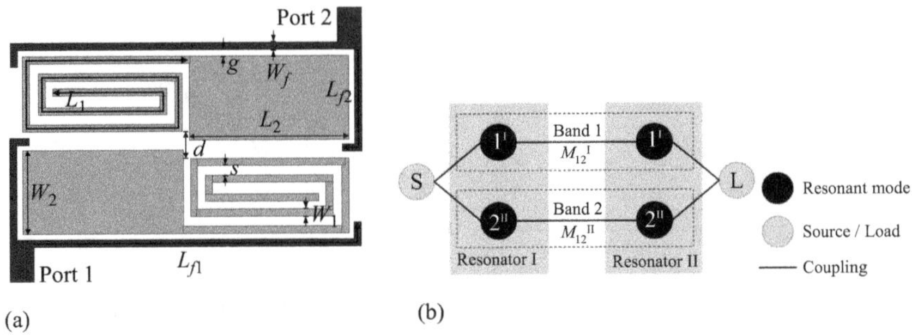

Figure 10.4 (a) Layout of the proposed dual-band HTS filter using the spiral asymmetric SIRs. (b) Its coupling scheme. Reprinted with permission from Reference [9]; © 2016 IEEE

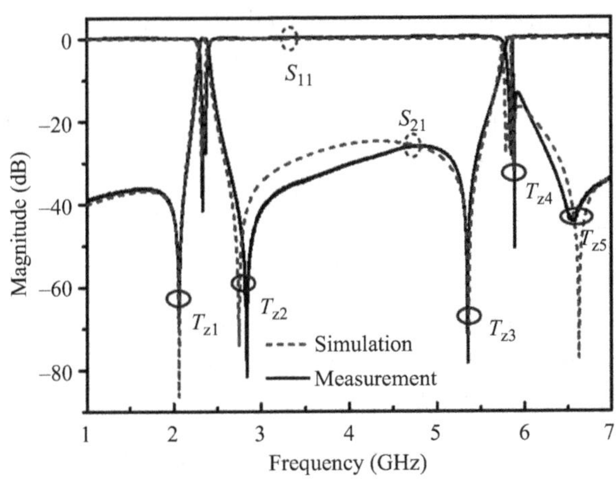

Figure 10.5 Simulated and measured (at 77 K) results of the fabricated HTS filter. Reprinted with permission from Reference [9]; © 2016 IEEE

Moreover, measured temperature dependence and input-power dependence of microwave nonlinear behaviors for the dual-band HTS filter are also presented and discussed. To further clarify microwave properties of HTS filter, it is essential to understand the equivalent circuits as displayed in Figure 10.6, where R_s', L', R_{sk}, and ΔL_k (T) stand for the conventional surface resistance inductance, surface dynamic resistance and the kinetic inductance, respectively. It can be expressed as:

$$\Delta R_{sk} = \frac{\mu_0^2 \omega^2 \sigma_n \left[\lambda_p(T)\right]^3}{2} \tag{10.1}$$

$$\Delta L_k(T) \approx \mu_0 \frac{l}{\omega} \lambda_p(T) \coth \frac{t}{\lambda_p(T)} \tag{10.2}$$

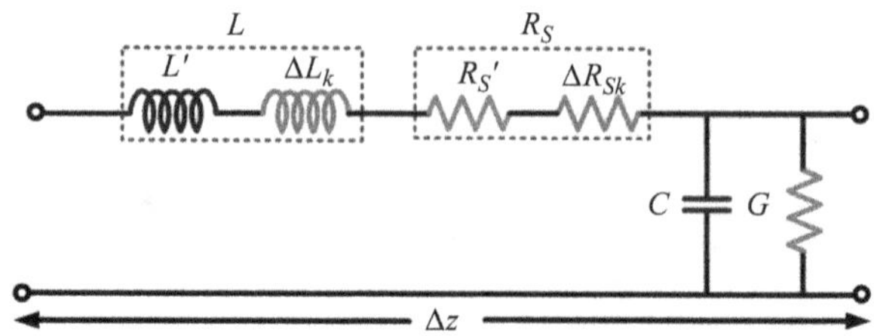

Figure 10.6 Equivalent circuit for an elemental unit of a superconducting transmission line, where $\Delta L_k(T)$ and ΔR_{sk} are nonlinear elements. Reprinted with permission from Reference [9]; © 2016 IEEE

where μ_0 is the permeability of vacuum, σ_n is the normal-state conductivity of the superconductor, t is the thickness of the superconducting film, and l is the lengths of the strip. $\lambda_p(T)$ is magnetic penetration depth [10,11].

From the wideband view in Figure 10.7(a), the frequency shift of the HTS filter is dominated by the temperature. The filter has both excellent dual-band bandpass performance (temperature < 90 K) and good bandstop performance (temperature > 100 K). Because the property of superconducting will disappear when the working temperature is large than the critical temperature (90 K). Besides, when temperature is changed from 65 to 100 K, the insert loss of S_{21} continues to increase, while the passband shift to a lower frequency both of the two bands from the enlarged view of two inset in Figure 10.7(a), which is well described by (10.1) and (10.2). It is evident that the higher temperature lead to a bigger magnetic penetration depth $\lambda_p(T)$, which directly enhanced the dynamic resistance and the kinetic inductance.

Additionally, the transmission spectrum at different input power levels when the operating temperature is fixed at 77 K for the fabricated HTS filter has also been tested. As shown in Figure 10.7(b), it can be observed that as the input power is increased from −6 dBm to 9 dBm, the transmission performance of the HTS filter remains nearly unchanged. That is to say, the proposed dual-band HTS filter has a relatively high power handling capacity. From the enlarged scale of insertion loss results of the two passbands in embedding graph of Figure 10.7(b), the slightly increased insertion loss can be observed with the increasing of input power. This can be explained by the power (or current) dependence of the resistive parts of the surface impedance. At a large input power level, the microwave surface impedance will be increased, resulting in the increase of insertion loss.

10.2.3 Multimode-resonator dual-band HTS filters

To meet the miniaturized and high-performance demand of microwave dual-band BPFs, there has been a growing interest in dual-mode and multimode resonators because of their advantages such as small size, low mass, and low loss. Triple- and quadruple-mode stub-loaded resonators have been successfully proven in the design of multimode filters. Resonant modes of the resonators can be divided into

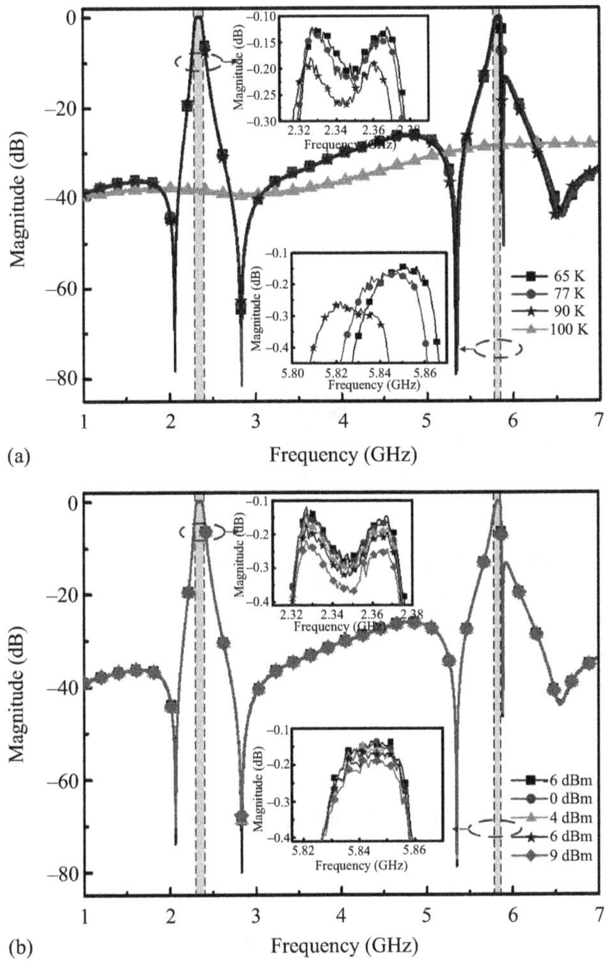

Figure 10.7 Experimental transmission spectrum of the HTS filter: (a) at various operating temperatures and (b) at 77 K with different input powers. Reprinted with permission from Reference [9]; © 2016 IEEE

two pairs, and each pair can form a passband. Therefore, a BPF with a dual-passband response can be realized.

Herein, a dual-band HTS BPF using a quadruple-mode resonator is introduced, as shown in Figure 10.8(a) [12]. This quadruple-mode resonator consists of a half-wavelength uniform impedance resonator (UIR) loaded by one open stub in a symmetrical plane, and two spiral open stubs at the two sides of the symmetrical plane. The quadruple-mode resonator excites four modes to produce dual-band characteristics. Even/odd mode method can be applied to explain its mode-splitting characteristics for its symmetry.

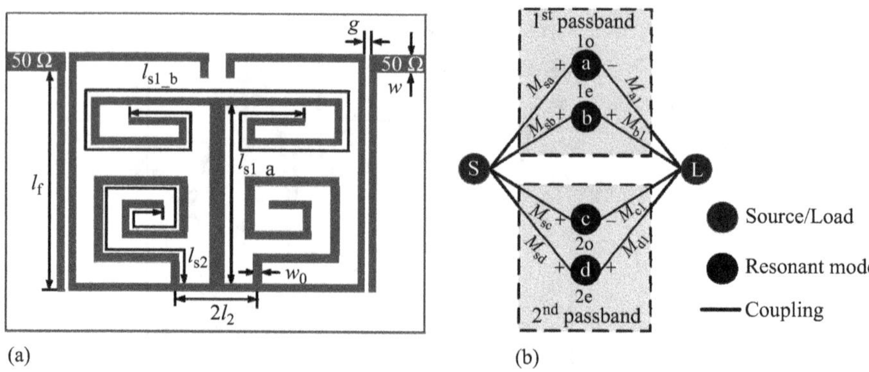

Figure 10.8 (a) Layout of the proposed dual-band HTS filter and (b) coupling scheme of the dual-band filter. Reprinted with permission from Reference [12]; © 2014 IEEE

For this structure, the lower passband frequency f_L is mainly determined by the half-wavelength UIR and the open stub in the symmetrical plane, whereas the higher passband frequency f_H is controlled by both the half-wavelength UIR and the two spiral open stubs at the two side of the symmetrical plane. Moreover, the coupling scheme of the dual-band filter is given in Figure 10.8(b). The gray round dots represent source and load, respectively. The black round dots represent four resonant modes. The source (input) is coupled to the four resonant modes by the admittance inverter, represented by the solid lines, and so is the load (output). The corresponding coupling matrix [M] for the coupling scheme in Figure 10.8(b) is represented as follows:

$$[M] = \begin{bmatrix} 0 & M_{sa} & M_{sb} & M_{sc} & M_{sd} & 0 \\ M_{as} & M_{aa} & 0 & 0 & 0 & M_{al} \\ M_{bs} & 0 & M_{bb} & 0 & 0 & M_{bl} \\ M_{cs} & 0 & 0 & M_{cc} & 0 & M_{cl} \\ M_{ds} & 0 & 0 & 0 & M_{dd} & M_{dl} \\ 0 & M_{la} & M_{lb} & M_{lc} & M_{ld} & 0 \end{bmatrix}. \tag{10.3}$$

Note that a symmetrical coupling structure indicates the equality of the coupling magnitude between source (or load) and each resonant mode. That is to say, the elements $M_{sa} = M_{as} = M_{la} = M_{al}$, $M_{sb} = M_{bs} = M_{lb} = M_{bl}$, $M_{sc} = M_{cs} = M_{lc} = M_{cl}$ and $M_{sd} = M_{ds} = M_{ld} = M_{dl}$. The obtained coupling matrix is shown as (10.4). Afterwards, measured and simulated frequency responses are illustrated in Figure 10.9. Measured results show that the two passbands of the filter are centered at 1.8 (GSM band) and 2.4 GHz (WLAN band) with the fractional bandwidth of 8% and 4.3%, respectively. The first TZ is attributed to the loaded T-shaped open stub, the second TZ is caused by virtual ground due to two spiral

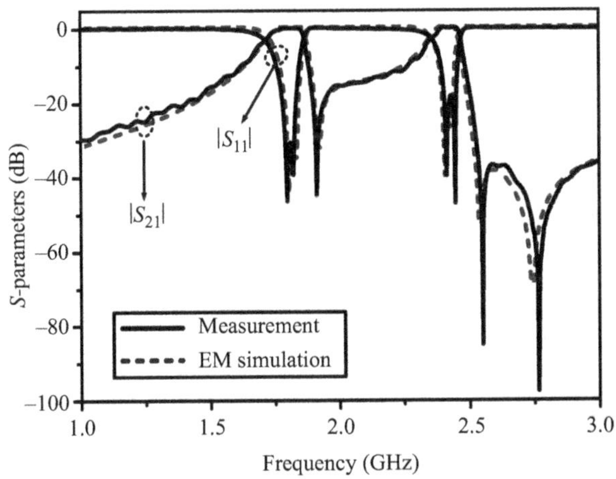

Figure 10.9 EM simulated and measured results of the designed filter. Reprinted with permission from Reference [12]; © 2014 IEEE

loading stubs of l_{s2}, and the third TZ stems from out-of-phase cancellation of two dissimilar signal paths.

$$[M] = \begin{bmatrix} 0 & -0.0968 & -0.1848 & -0.5679 & 0.2493 & 0 \\ -0.0968 & -0.9980 & 0 & 0 & 0 & 0.0968 \\ -0.1848 & 0 & 0.2297 & 0 & 0 & -0.1848 \\ -0.5679 & 0 & 0 & 0.7165 & 0 & 0.5679 \\ 0.2493 & 0 & 0 & 0 & -0.9374 & 0.2493 \\ 0 & 0.0968 & -0.1848 & 0.5679 & 0.2493 & 0 \end{bmatrix}. \tag{10.4}$$

Hairpin resonator and stub-loaded resonator are also commonly used to construct compact and simple multiband BPFs. Figure 10.10(a) shows a dual-band HTS microstrip filter using nondegenerate-mode hairpin resonator [13]. This multimode resonator consists of two hairpin resonators loaded with a common T-shaped stub. The coupling scheme of the dual-band filter is shown in Figure 10.10(b), where the gray nodes represent source/load ports. The black dots, a and b, represent nondegenerate modes of the outer resonator (which form the lower passband I), and the black dots, c and d, represent nondegenerate modes of the inner resonator (which form the higher passband II). Two pairs of nondegenerate modes are explained by even–odd mode method. The T-shaped stub tapped from inside excites one pair of nondegenerate modes for the inner hairpin resonator. Similarly, the inner hairpin resonator loaded with the T-shaped stub produces another pair of nondegenerate modes for the outer hairpin resonator. Moreover, the former generates the higher passband, and the latter produces the lower passband, which can be controlled independently by the corresponding

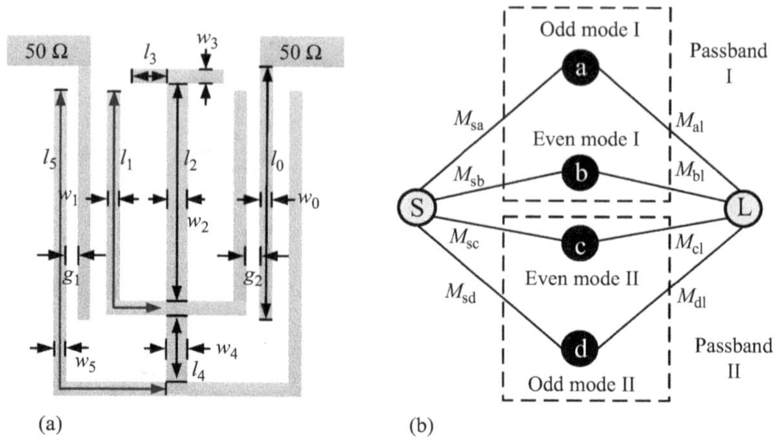

Figure 10.10 (a) Schematic view of the proposed dual-dual HTS filter based on two pairs of nondegenerate mode and (b) coupling schemes of the proposed dual-band filter. Reprinted with permission from Reference [13]; © 2015 IEEE

hairpin resonator. Simulated and measured results of the proposed dual-band filter are compared in Figure 10.11. Results show that the designed dual-band filter is centered at 2.4 GHz (WiMAX-band) and 3.5 GHz (WLAN-band) with the 3-dB fractional bandwidth of 6.88 and 8.57%, respectively. Also, several transmission zeros are realized for high selectivity and stopband suppression. Transmission zeros T_{Z1} (@2.35 GHz) and T_{Z3} (@3.75 GHz) are due to self-coupling. Moreover, transmission zero T_{Z2} (@3.15 GHz) generated by the transversal interference between the two signal paths from one port to the other port.

10.3 High-order HTS dual-band BPFs on dual-mode HRR

The above described two-pole HTS dual-band BPFs have the advantage of compact size and ultra-low insertion loss, but the selectivity of the passband and attenuation in stopband are undesirable and should be improved. One of effective method is the implementation of high-order filtering response. In this section, an eighth-order HTS dual-band filter will be presented to achieve high selectivity and deep attenuation for rejecting interference based on a novel dual-mode hairpin-ring resonator (HRR).

10.3.1 Properties of dual-mode HRR

A compact dual-mode resonator, namely HRR, with two controllable resonant modes is developed [14]. Its configuration is shown in Figure 10.12(a). L_1 to L_4 and w denote the physical lengths and width of the corresponding microstrip line segments, respectively. Compared with the conventional hairpin resonator firstly reported in [15], an additional horizontal line section (in green color) is added to excite dual-mode characteristics.

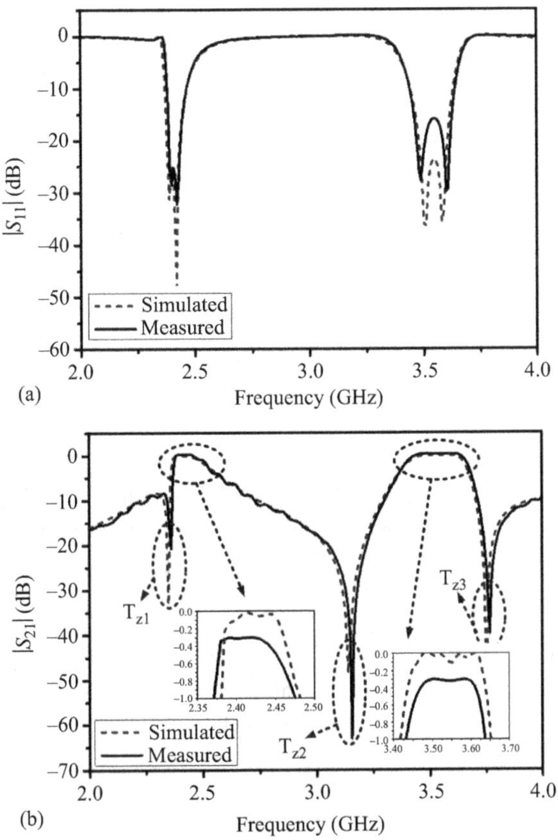

Figure 10.11 *(Dashed line) Simulated and (solid line) measured (a) S_{11} and (b) S_{21}. Reprinted with permission from Reference [13]; © 2015 IEEE*

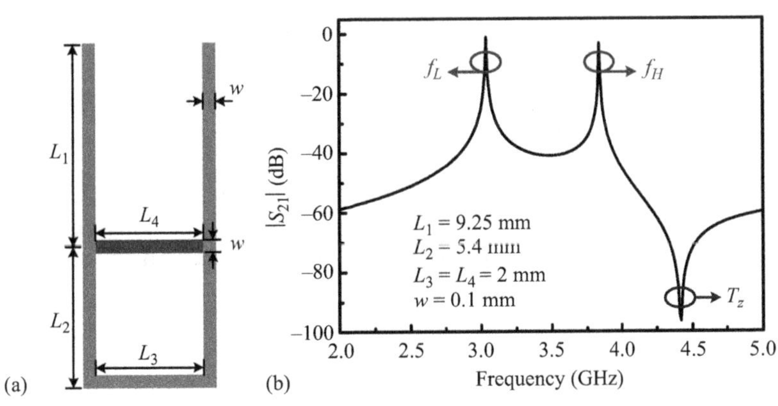

Figure 10.12 *(a) Layout of dual-mode HRR unit and (b) its frequency response of S_{21} magnitude under weak coupling. Reprinted with permission from Reference [14]; © 2017 IEEE*

Owing to the inserted horizontal line, one more signal transmission path will be provided and such that an extra resonant mode could be produced. That is to say, the symmetry of the conventional hairpin structure could be perturbed by the inserted horizontal line which connects two arms of the hairpin resonator as a perturbation and makes it possible for excitation of two orthogonal modes for dual-mode operation. So, this is a new method to construct the dual-resonance characteristics within a hairpin resonator or open-loop resonator without extra circuit size occupation [16]. It is well suited to high-order microwave circuit design and applications. On the other hand, the inserted horizontal line can provide a new signal interference way between the upper and lower horizontal lines so that the transmission zero may be produced and applied for high selectivity performance design.

To explore the resonant characteristics of the proposed dual-mode HRR, a full-wave electromagnetic (EM) simulation is conducted using *Feko Suite* 7.0. The substrate MgO with a relative dielectric constant of 9.78 and a thickness of 0.5 mm for YBCO film is used in the design. Figure 10.12(b) depicts the simulated frequency response of the dual-mode HRR under weak coupling. It is shown that two resonant modes, defined as f_L and f_H, are excited in a specified frequency scope. It is also found interestingly that there is a finite-frequency transmission zero (TZ) when the two modes split.

To clarify the resonance mechanism, the current density distributions of the proposed dual-mode HRR is checked. Figure 10.13(a) and (b) depicts the current distributions at two resonant modes, f_L and f_H, respectively. It is shown that the current is mainly concentrated on the line section of L_1 and L_4 at f_L whereas the current distributes on the line section of L_1, L_2, and L_3 at f_H. From Figure 10.13(a), it reveals that the proposed dual-mode HRR at f_L has a similar current distribution as that of the typical half-wavelength ($\lambda/2$) hairpin resonator. Moreover, the lower horizontal line (L_3) work actually as an open terminal for the mode at f_L. At the same time, the proposed dual-mode HRR at fH has a similar current distribution as that of the typical full-wavelength (λ) hairpin resonator. It is noticed that the upper horizontal line (L_4) is a virtual open terminal for the mode at f_H. Thus, the two resonant modes of the proposed dual-mode HRR can be adjusted independently by lower and upper horizontal lines of L_3 and L_4. So, the discrepancy between two modes are tunable and useful for inter-stage coupling of high-order filter design with excellent performances.

Based on the discussion aforementioned and the transmission line theory, the two modes can be considered as the results of two resonance tanks, i.e., one tank with line section L_1 and L_4 and the other tank with line section L_1, L_2, and L_3. Then, two resonant frequencies of the proposed dual-mode HRR could be estimated as:

$$f_L \approx \frac{c}{(4L_1 + 2L_4)\sqrt{\varepsilon_{\text{eff}}}} \tag{10.5}$$

$$f_H \approx \frac{c}{(2L_1 + 2L_2 + L_3)\sqrt{\varepsilon_{\text{eff}}}} \tag{10.6}$$

where c is the speed of light in the free space and ε_{eff} is the effective dielectric constant of substrate.

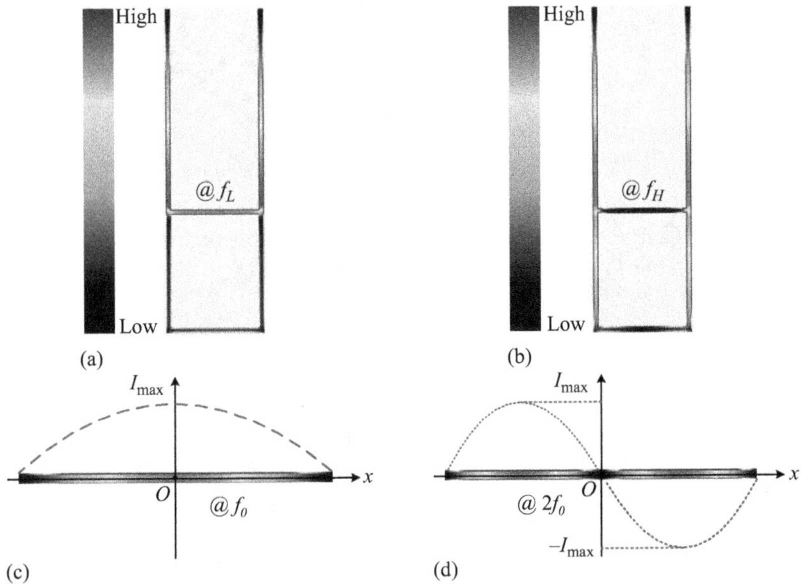

Figure 10.13 *(a) Simulated current density distributions of the HRR at f_L; (b) simulated current density distributions of the HRR at f_H; (c) current distribution along the half-wavelength resonator at f_0; and (d) current distribution along the half-wavelength resonator at $2f_0$. Reprinted with permission from Reference [14]; © 2017 IEEE*

Furthermore, Figure 10.13(c) and (d) depicts the current distributions along the half-wavelength ($\lambda/2$) resonator under the fundamental mode (@f_0) and its second harmonic (@$2f_0$), respectively. It is interestingly found that the current distribution of dual-mode HRR at f_L in Figure 10.13(a) is the similar as that of $\lambda/2$ resonator at the fundamental frequency f_0 in Figure 10.13(c). On the contrary, the current distribution of dual-mode HRR at f_H in Figure 10.13(b) consists with that of $\lambda/2$ open-loop resonator at the second harmonic $2f_0$ in Figure 10.13(d). Therefore, one distinct characteristic of the proposed dual-mode HRR resonator is that the two modes might not be coupled to each other after the modes are split. This non-coupled dual-mode characteristic is entirely different from that of the dual-mode loop resonator with degenerate modes [17].

In addition, no coupling between the split modes of the dual-mode HRR can be explained by the theory of asynchronously tuned coupled resonators. It is confirmed from the obtained equality between the self-resonant frequencies and split-mode frequencies after simulated by placing a magnetic or electric wall along the symmetric axis of the proposed HRR, respectively, as implemented in [18]. Therefore, it can be treated as doublet with a coupling structure shown in Figure 10.14, which can be used as a building block for designing high-order dual-band components such as filters, which are demonstrated in the following part.

For further discussion about the non-coupled mode operation of the proposed dual-mode HRR, the simulated results for different values of L_3 are plotted

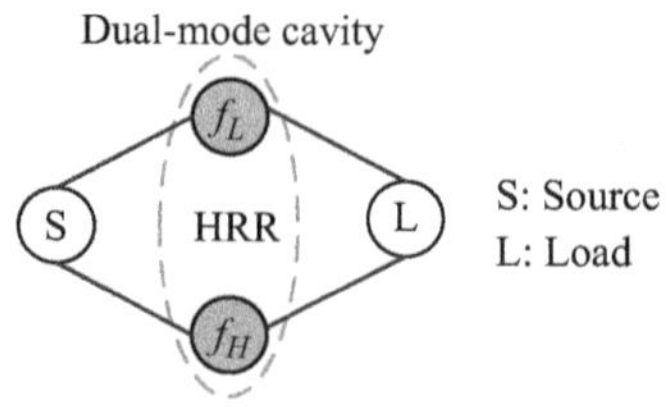

Figure 10.14 Coupling scheme of dual-mode HRR. Reprinted with permission from Reference [14]; © 2017 IEEE

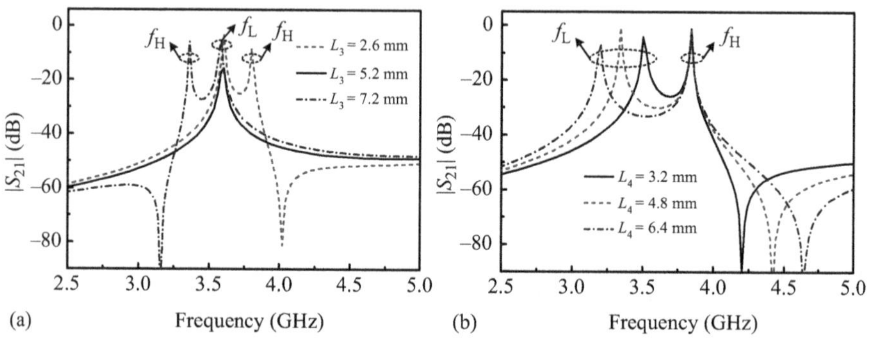

Figure 10.15 (a) L_3-dependence of modal resonant characteristics of the proposed dual-mode HRR for $L_1 = 7.6$ mm, $L_2 = 7$ mm and $L_4 = 2$ mm and (b) L_4-dependence of modal resonant characteristics of the proposed dual-mode HRR for $L_1 = 7.6$ mm, $L_2 = 7$ mm and $L_3 = 2$ mm. Reprinted with permission from Reference [14]; © 2017 IEEE

in Figure 10.15(a). In this simulation, w is set to 0.1 mm. L_1, L_2 and L_4 are kept constant. It is seen from Figure 10.15(a) that, for $L_3 = 5.2$ mm, the proposed HRR exhibit one resonant frequency, i.e., $f_L = f_H$. When L_3 is decreased or increased from this value, two resonant modes split. In this case, f_H is effectively shifted whereas f_L is hardly unchanged. On the contrary, the resonant characteristics of two modes for the varied L_4 are illustrated in Figure 10.15(b). L_1, L_2 and L_3 keep constant. It shows that the change of L_4 has influence on f_L whereas f_H is almost no changed. Hence, f_H and f_L can be independently controlled by L_3 and L_4. Also, it is observed that another distinct characteristic of the dual-mode HRR is that the tunable transmission zero position appears to be closely associated with f_H.

Due to the inserted horizontal line of L_4, there exist two signal paths in HRR structure. As described in [19], the transmission zero will be generated because of the transversal signal interference between two signal paths. The resultant

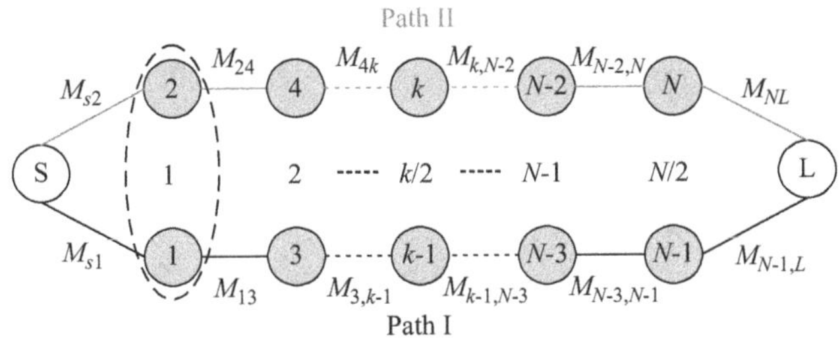

Figure 10.16 Dual-path coupling structure for dual-band transversal filter design. N is the total number of resonators. S and L represents the source and load, respectively. Reprinted with permission from Reference [14]; © 2017 IEEE

transmission zeros can be determined by

$$2\theta_2 + \theta_3 - \theta_4 = (2m + 1) \times 180^\circ, \quad \text{for } m = 0, 1, 2, \ldots \tag{10.7}$$

where θ_2, θ_3, and θ_4, are the electrical lengths with respect to microstrip line section L_2, L_3, and L_4, respectively. TZ indicated in Figure 10.15 is ascertained by (10.7) with $m = 0$. Note that this is one of the advantage that multi-mode resonators always have intrinsic transmission zeros under the condition of no any non-adjacent coupling structures [16,19]. Thus, the band-edge selectivity of microwave components could be enhanced.

10.3.2 Design of third-order HTS dual-band filter

For demonstration, a third-order dual-band BPF is firstly designed by the proposed dual-mode HRR. Then, a general dual-path coupling scheme is introduced and discussed for dual-band implementation. Based on the proposed dual-mode HRR and the coupling diagram, a three-order dual-band HTS filter is designed and simulated.

Figure 10.16 depicts the proposed dual-path coupling structure for high-order dual-band response implementation. It can be obtained from the $N+2$ fully canonical transversal topology network after being simplified with a series of similarity transformations and residue grouping operations [20]. In this dual-path coupling structure, each path (Path I or II) governs the transmission performances of one passband and the solid lines represent the direct coupling routes. Also, each gray node represents a resonator and the total number of resonators for dual-band response implementation is denoted by N.

As shown in Figure 10.16, the total number (N) of resonators can be reduced if each group of two separated resonators, just as the two nodes in elliptic dashed line box, is represented by one dual-mode resonator with two individual modes. In theory, the circuit size can be reduced by 50% at least. Based on the proposed dual-path coupling structure, the dual-mode E-shaped resonator is used to design dual-band filter by the analytic coupling matrix synthesis procedure [21]. But, the

	s	2	4	··	*k*	··	*N*-2	*N*	*L*
s		M_{s2}							
2	M_{2s}	M_{22}	M_{24}						
4		M_{42}	M_{44}	⋱					
⋮			⋱	⋱	⋱				
k				⋱	M_{kk}	⋱			
⋮					⋱	⋱	⋱		
N-2						⋱	$M_{N\text{-}2,N\text{-}2}$	$M_{N\text{-}2,N}$	
N							$M_{N,N\text{-}2}$	M_{NN}	M_{NL}
L								M_{LN}	

Figure 10.17 N/2 + 2 canonical coupling matrix [M] for Path II of the proposed coupling structure in Figure 10.1. The matrix is symmetric about the principal diagonal, i.e., $M_{ij} = M_{ji}$. *Reprinted with permission from Reference [14]; © 2017 IEEE*

synthesis process is complex and the synthesized circuit size is still large for HTS fabrication.

To simplify the design process above and achieve a compact size, a simple design theory and procedure is proposed. That is, two paths are implemented individually with circuit theory of low-pass prototype. Figure 10.17 shows the corresponding coupling matrix of path I in Figure 10.16, where M_{kk} is self-coupling and represents the frequency offset from center frequency.

To realize the desired high-order dual-band performance, the following points should be addressed as:

1. Two non-coupled resonant modes of dual-mode resonator are designed independently and its unit structure is convenient for high-order development.
2. Enough design freedoms for inter-stage coupling of two adjacent resonators is desired for good performance in both in-bands simultaneously.
3. Two individually external quality factors in a wide range for two passbands is required for the arbitrary bandwidths of the dual-band filter.

Due to the size reduction, the dual-mode HRR unit is well reshaped for circuit miniaturization. Herein, two open lines (L_1) of HRR are folded inner to make full use of the space inside and a large patch is inserted to increase the capacitance effect. Also, the spiral open lines can weaken the inter-stage couplings between HRRs, which will further miniaturize the overall filter size. This new and compact

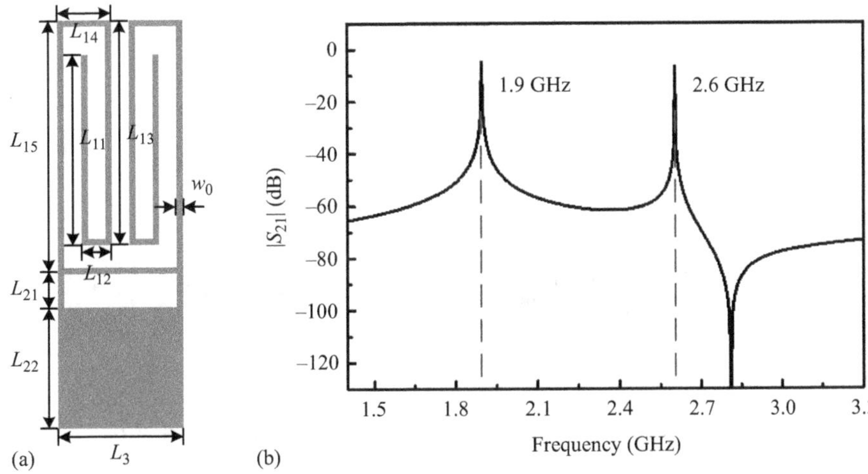

Figure 10.18 (a) Spiral dual-mode HRR and (b) simulated frequency responses with two designed resonant frequencies. Reprinted with permission from Reference [14]; © 2017 IEEE

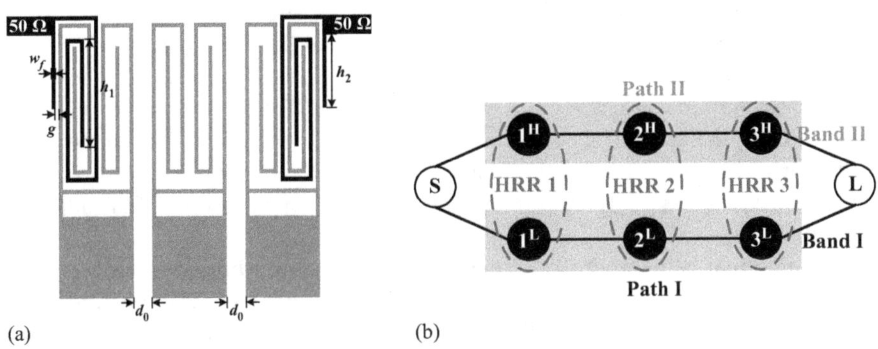

Figure 10.19 (a) Layout of the third-order dual-band HTS filter with spiral dual-mode HRR and (b) its coupling diagram. Reprinted with permission from Reference [14]; © 2017 IEEE

spiral dual-mode HRR is presented in Figure 10.18(a). Herein, two modes of the spiral dual-mode HRR are designed to locate at 1.9 GHz (f_L) and 2.6 GHz (f_H). After optimized by EM simulation software, the values of all variables marked in Figure 10.18(a) are obtained as follows: $L_{11} = 5.05$, $L_{12} = 0.5$, $L_{13} = 5.45$, $L_{14} = 0.9$, $L_{15} = 6$, $L_{21} = 1$, $L_{22} = 2.6$, $L_3 = 2.2$, $w_0 = 0.1$ (unit: mm). The corresponding frequency response is shown in Figure 10.18(b), which meets the design specification.

Figure 10.19(a) presents the geometrical structure of the third-order dual-band HTS filter using the spiral dual-mode HRR, which is designed at 1.9 GHz and

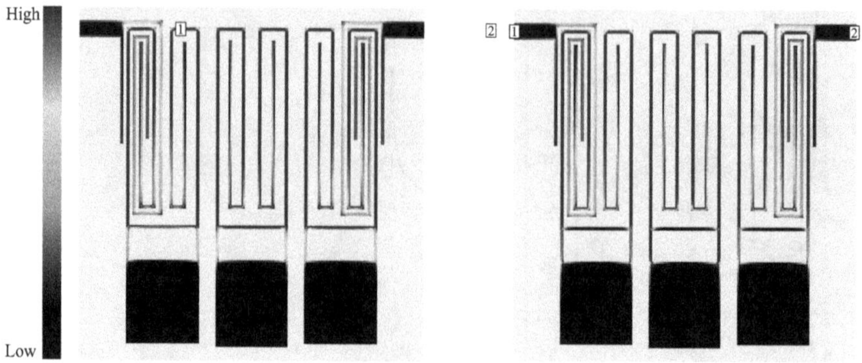

(a) (b)

Figure 10.20 Simulated current distributions at resonant frequencies of (a) 1.9 GHz and (b) 2.6 GHz. Reprinted with permission from Reference [14]; © 2017 IEEE

2.6 GHz, respectively. Based on the coupling scheme discussed in section II, the coupling diagram of the designed third-order dual-band filter using the proposed spiral dual-mode HRR is presented in Figure 10.19(b). It is shown that two coupling paths contribute the formation of two passbands, where each path consists of three resonant nodes provided by three identical dual-mode HRRs. Figure 10.20 shows the current distributions of the third-order dual-band filter at $f_L = 1.9$ GHz and $f_H = 2.6$ GHz. As we can see, the third-order dual-band filter has the same current distribution characteristic as one dual-mode HRR shows in Figure 10.13(a) and (b). Thus, the two non-coupled resonant modes of the spiral dual-mode HRR can be tuned independently so that the desired inter-stage couplings for two passbands can be realized easily by adjusting the coupling gap and offset of the adjacent coupled HRRs.

For the dual-band filter prototype, two passbands are both developed to the Chebyshev frequency response with 0.043-dB ripple level and designed individually. The fractional bandwidth (FBW) of lower and higher passbands is 3.4 and 2.9%, respectively. Based on the general procedure of filter design in [18], the actual coupling coefficients m_{ij} and external quality factors Q_{ex} can be computed. The finally obtained m_{ij} and Q_{ex} are $m_{12} = m_{23} = 0.0353$, $Q_{ex} = 24.9$ for the first passband, and $m_{12} = m_{23} = 0.0298$, $Q_{ex} = 29.5$ for the second passband.

The following is to determine the dimensions of both the feed-line structure and internal coupling region. The first step is to design the external coupling to meet the desired Q_{ex}. To provide flexible choices in designing the bandwidth for two passbands simultaneously, two parallel-coupled microstrip lines are used for the feeding structure, as shown in Figure 10.21(a), which includes one vertical microstrip line indicated as h_2 and the other meander microstrip line indicated as h_1.

Figure 10.21(b) depicts the extracted external quality factor Q_{ex} of two passbands versus coupling length h_1 when h_2 is kept to 0 mm for demonstration. It can be seen that both Q_{ex} of the first and second passbands decrease as h_1 increases.

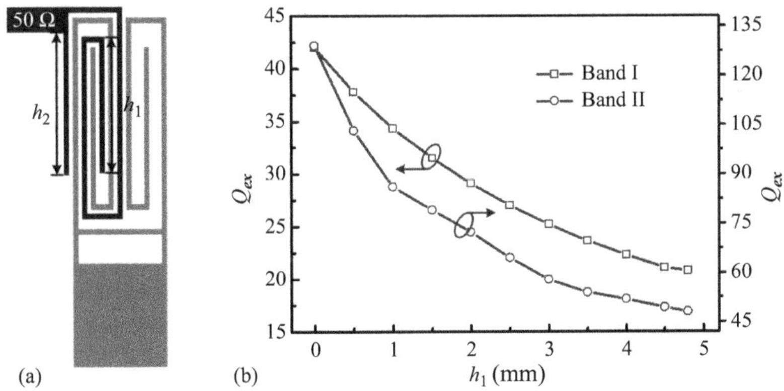

Figure 10.21 *(a) Parallel-coupled feeding structure of the dual-mode HRR and (b) the external quality factors Q_{ex} as a function of coupling length h_1, where $h_2 = 0$ and $w_f = g = 0.1$ mm. Reprinted with permission from Reference [14]; © 2017 IEEE*

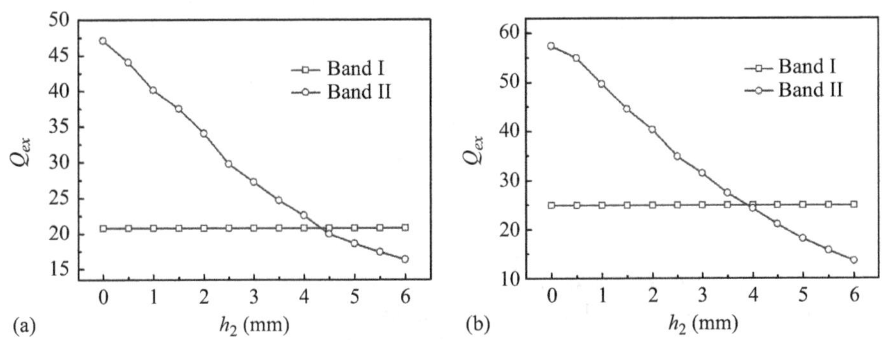

Figure 10.22 *(a) The external quality factors Q_{ex} as a function of coupling length h_2, where $h_1 = 4.8$ and $w_f = g = 0.1$ mm and (b) the external quality factors Q_{ex} as a function of coupling length h_2, where $h_1 = 3.1$ mm. Reprinted with permission from Reference [14]; © 2017 IEEE*

Figure 10.22(a) plots the external quality factor Q_{ex} of two passbands versus coupling length h_2 when h_1 is remained as 4.8 mm. It is shown that Q_{ex} of the second passband decreases with h_2 increasing while Q_{ex} of the first passband is almost unchanged, indicating the independently controllable for Q_{ex} of two passbands. According to the relationship between the coupling lengths (h_1 and h_2) and the external quality factor Q_{ex}, a process to design the parallel-coupled feeding structure is given as follows. Based on the desired Q_{ex} for the first passband, the coupling length h_1 can be firstly determined from Figure 10.22 when $h_2 = 0$. Then, h_2 can be determined by choosing the desired Q_{ex} for the second passband from Figure 10.22(a).

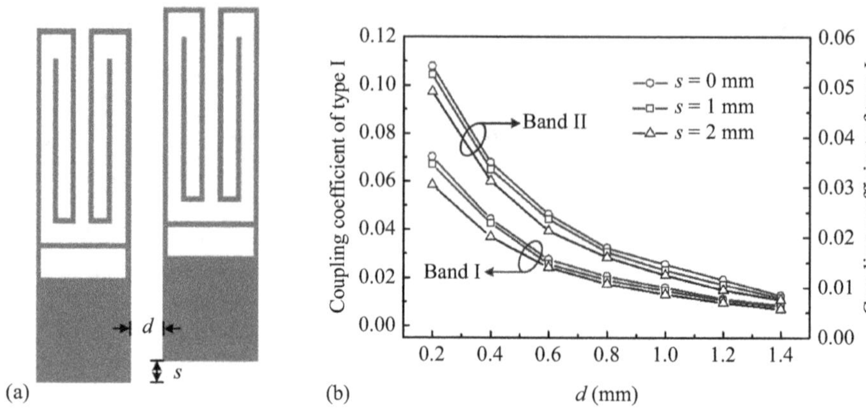

Figure 10.23 (a) The coupling structure between two adjacent HRR with same orientation (Type I) and (b) the corresponding coupling coefficients as a function of coupling space d. Reprinted with permission from Reference [14]; © 2017 IEEE

According to this design process, h_1 of the three-order dual-band filter in Figure 10.19(a) can be firstly ascertained to 3.1 mm from Figure 10.21(b) based on the calculated $Q_{ex} = 24.9$ of the first passband. The following work is to adjust h_2 to satisfy the desired Q_{ex} for the second passband of the designed three-order dual-band filter. Figure 10.22(b) depicts the variation of Q_{ex} versus h_2 when $h_1 = 3.1$ mm. It is shown that Q_{ex} of the first passband is kept unchanged and Q_{ex} of the second passband decreases as h_2 enlarges. Based on the desired $Q_{ex} = 29.5$ for the second passband, h_2 can then be found to be 3.2 mm.

The following step is to adjust the coupling lengths and spaces between the adjacent HRRs to meet the desired coupling coefficients simultaneously for two passbands. Figure 10.23(a) shows the coupling structure of two adjacent HRRs with the same orientation (named as Type I), where d and s are the coupling distance and offset in the vertical direction, respectively. The simulated coupling curves of two passbands are extracted and drawn in Figure 10.23(b). It is observed that the coupling coefficients of two passbands are both decreased monotonously as d and s become enlarged. Based on the desired coupling values, d can be explicitly determined to be 0.5 mm and any two of coupled resonators has no offset in vertical part, i.e., $s = 0$.

After simulated and optimized by *Feko* Suite 7.0, the frequency response of the designed dual-band filter is depicted in Figure 10.24. Three poles can be observed in both two passbands from the figure and the return losses are better than 20 dB. The two passbands are centered at 1.903 and 2.602 GHz and their corresponding 3-dB FBW are 5.78% and 4.46%, respectively, which agree well with the desired specifications. In addition, there are four obvious TZs near the passbands, the first two TZs (1.35 GHz, 1.645 GHz) are in the lower skirt of the first passband and the fourth one (2.85 GHz) is in the upper side of the second passband, which enhance the out of bands rejection of the filter. The third TZ (2.345 GHz) is locate

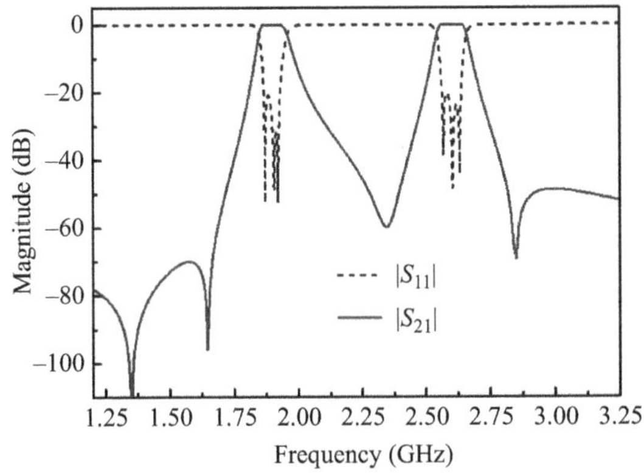

Figure 10.24 Simulated frequency responses of the designed third-order filter. Reprinted with permission from Reference [14]; © 2017 IEEE

in the middle of two passbands, which improve the band-to-band isolation. Among them, the first and second TZs are explained by the mechanism described in Part 10.3.1 and the latter two TZs are provided by the feeding structure. However, only one TZ exists between two passbands and the band-to-band isolation is unsatisfactory. Therefore, the isolation of two bands as well as the steepness of slope of the passbands should be enhanced to meet the potential high-performance requirement of the future wireless system.

10.3.3 Design of eighth-order HTS dual-band filter

In this part, an eighth-order dual-band HTS filter with sharp rate of attenuation near the passband and low insertion loss is designed. To achieve a good isolation of two passbands, a different coupling structure for adjacent coupled resonators is proposed. As shown in Figure 10.25(a), two identical dual-mode HRRs are approximately placed in opposite orientation with a coupling spacing d and offset s to facilitate the coupling. Here, this coupling structure is indicated as Type II. The corresponding frequency responses in a finite frequency scope under weak excitation are shown for comparison in Figure 10.25(b). It is shown that two transmission zeros (TZ_1, TZ_2) between two passbands are produced for Type II, which can improve greatly the band-edge selectivity and band-to-band isolation.

Figure 10.26 presents the layout of eighth-order dual-band HTS filter. As can be seen, any two adjacent resonators are in opposite orientations. $d_1 \sim d_4$ indicate the coupling spaces. The parallel-coupled feeding structure is used as the same way in Figure 10.21(a), which will provide enough design freedoms for two bands. Here, two passbands are developed to the Chebyshev frequency response with 0.01-dB ripple level. The center frequency and ripple FBW of the lower and higher passband is 1.9 GHz, 2.6 GHz and 2.1, 1.9%, respectively. The actual coupling coefficients m_{ij} and external quality factors Q_{ex} can be obtained [18] and listed in Table 10.2.

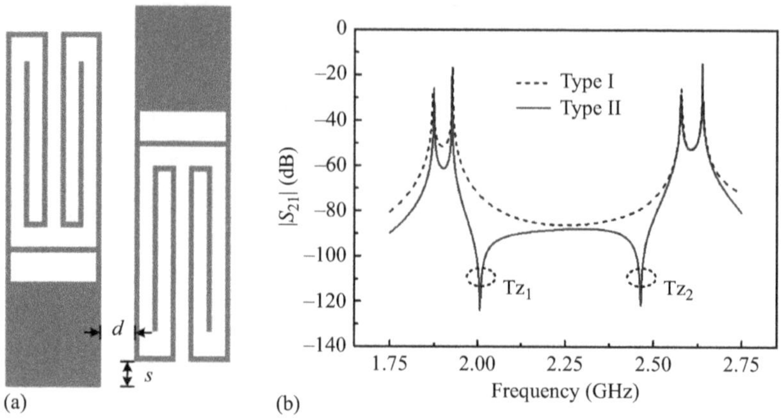

Figure 10.25 (a) The coupling structure between two HRR units with opposite orientation (Type II) and (b) comparison of the frequency responses between two coupling structures under the weak excitation. Reprinted with permission from Reference [14]; © 2017 IEEE

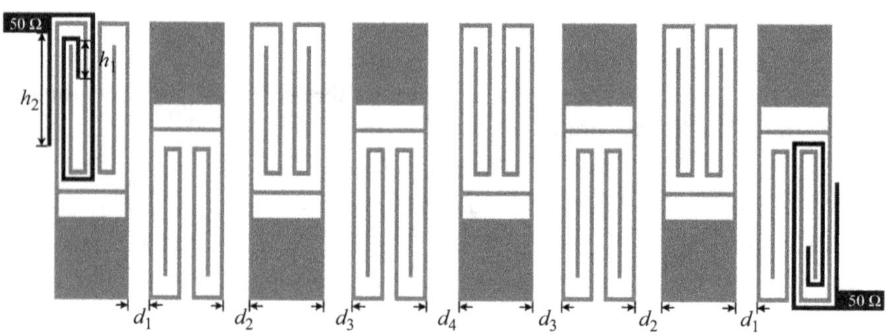

Figure 10.26 Layout of the designed eighth-order dual-band HTS filter. Reprinted with permission from Reference [14]; © 2017 IEEE

Table 10.2 Coupling coefficients of eighth-order dual-band filter

Bands	$M_{12} = M_{78}$	$M_{23} = M_{67}$	$M_{34} = M_{56}$	M_{45}	Q_e
Band I	0.0197	0.0133	0.0122	0.0119	38.4
Band II	0.0180	0.0121	0.0111	0.0109	42.0

Figure 10.27(a) shows the extracted coupling coefficients of Type II. It is seen that the coupling coefficients of two passbands are both decreased monotonously as d enlarges. According to the desired coupling values, the coupling distances in Figure 10.26, $d_1 \sim d_4$, can be explicitly ascertained successively as

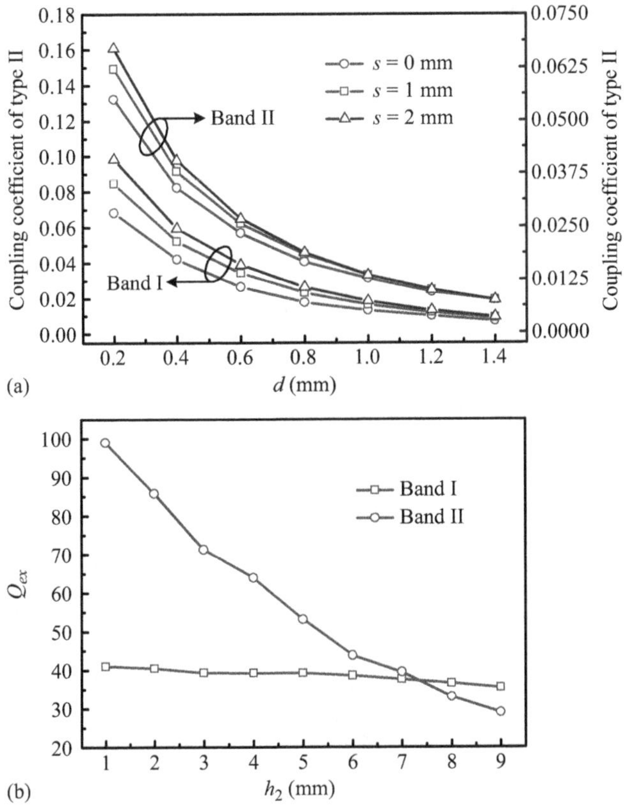

Figure 10.27 *(a) The coupling coefficients of Type II as a function of coupling space d and (b) the external quality factors Q_{ex} as a function of coupling length h_2, where $h_1 = 0.45$ mm. Reprinted with permission from Reference [14]; © 2017 IEEE*

0.65, 0.95, 1.05, and 1.1 mm, respectively. As described in Section 10.2, based on the required Q_{ex} of the first passband, h_1 in Figure 10.26 can be firstly determined to 0.45 mm from Figure 10.21(b). The extracted external quality factors with respect to h_2 is plotted in Figure 10.27(b). From the figure, h_2 can be quickly found to be 7 mm.

After optimized by EM simulation, the simulated frequency response of the designed eighth-order dual-band filter is depicted as black dashed lines in Figure 10.28. The lower passband is centered at 1.9 GHz with a 3-dB FBW of 2.3% and the higher passband is at 2.603 GHz with a 3-dB FBW of 2.2%. Moreover, it is shown that five TZs (1.85, 2.03, 2.43, 2.72, and 3.1 GHz) around two passbands are created for sharp passband slops and high band-to-band isolation.

The simulated $|S_{21}|$ of two designed filters, i.e., third-order one and eighth-order one, are compared and illustrated in Figure 10.29. Several key parameters are

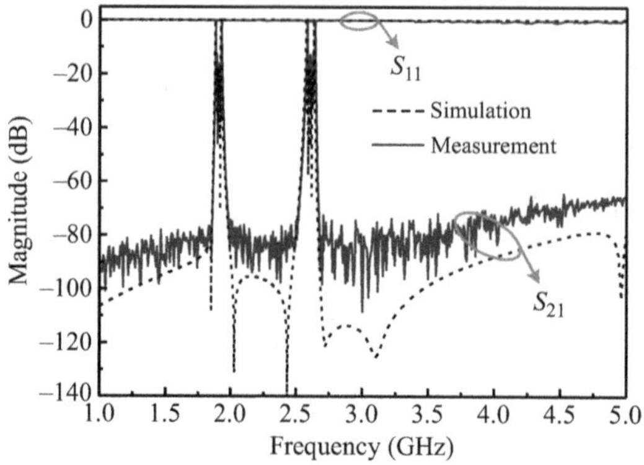

Figure 10.28 Results of the eighth-order dual-band HTS filter. Reprinted with permission from Reference [14]; © 2017 IEEE

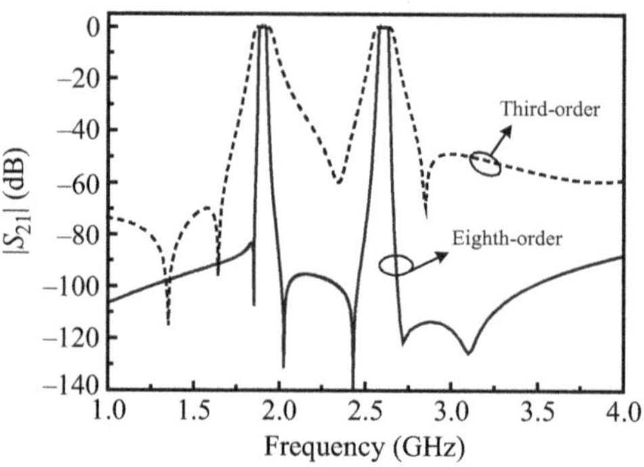

Figure 10.29 Comparison of simulated $|S_{21}|$ between the third-order filter and the eighth-order filter. Reprinted with permission from Reference [14]; © 2017 IEEE

Table 10.3 Performance comparison between the dual-band filters

Filter type	ζ_{ROR} (dB/GHz)	Isolation (dB)	Number of TZs
Third-order	447/221/370/333	<60	4
Eighth-order	4,250/2,833/1,889/3,400	>95	5

listed in detail for performance comparison in Table 10.3, where ζ_{ROR} is the roll-off rate of passband and defined as follows:

$$\zeta_{ROR} = \frac{|\delta_{-20\ dB} - \delta_{-3\ dB}|}{|f_{-20\ dB} - f_{-3\ dB}|} \tag{10.8}$$

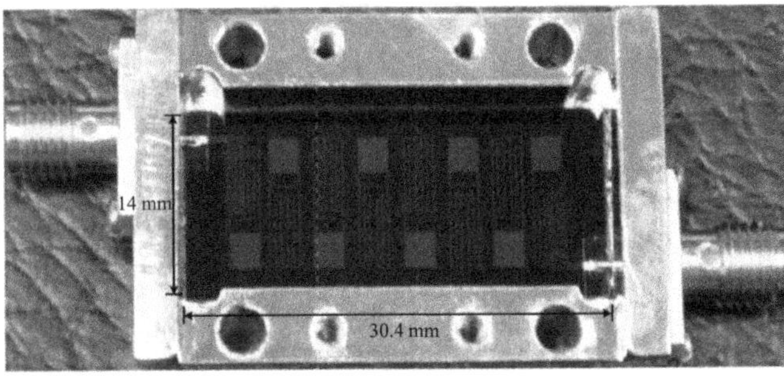

Figure 10.30 Photograph of the fabricated HTS filter with SMA connectors. Reprinted with permission from Reference [14]; © 2017 IEEE

where $\delta_{-20\ \text{dB}}$ and $\delta_{-3\ \text{dB}}$ is the 20-dB and 3-dB attenuation point, respectively. $f_{-20\ \text{dB}}$ and $f_{-3\ \text{dB}}$ is the 20-dB and 3-dB stopband frequency, respectively. Compared with the designed third-order filter, there is a great improvement on the band-edge selectivity and band-to-band isolation of the eighth-order filter, which satisfies the desired performance.

10.3.4 Fabrication and experimental verification

To validate the proposed resonator and design method, the designed eighth-order dual-band filter is fabricated on a 2-in-diameter 0.5-mm-thick MgO wafer with double-sided YBCO films. The film is patterned by the standard photo-lithography. The ion etching technology is used to etch the front-side film to form the circuit structure, and the circuit is mounted on a gold-plated metal carrier and then carefully packaged into a shield box. Figure 10.30 shows the photograph of the fabricated dual-band HTS filter with the cover opened. The overall size of the filter is 24 mm × 9.6 mm (not include feed lines), which amounts to 0.39 $\lambda_g \times 0.156\ \lambda_g$ (λ_g is the guided wavelength of the 50-Ω line in the substrate at the center frequency of the first passband).

The packaged HTS dual-band filter is cooled down to a temperature of 77 K in a cryogenic cooler and measured by HP8753ES network analyzer, after the full two-port calibration for reflection and transmission measurements is performed at ambient temperature. The measured wideband results are depicted as the solid lines in Figure 10.28, which agree well with the simulated frequency responses. Figure 10.31 depicts the enlarged view of the simulated and measured performances in two passbands. The measured low- and high-bandpass center frequencies are 1.91 GHz and 2.615 GHz with 3-dB FBWs of 2.39% and 2.29%, respectively. The maximum values of two in-band insertion loss are approximately 0.29 and 0.45 dB, which shows the superiority of insertion loss by using the HTS film material. Then, the unloaded quality factors of the designed HTS filter in two passbands can be obtained as 7,173 and 4,824. The in-band return loss and out-of-band rejection is better than 12 and 70 dB, respectively. Some slight deviations between the measurements and simulations are mainly due to the potential differences in the dielectric constant and thickness of the substrate.

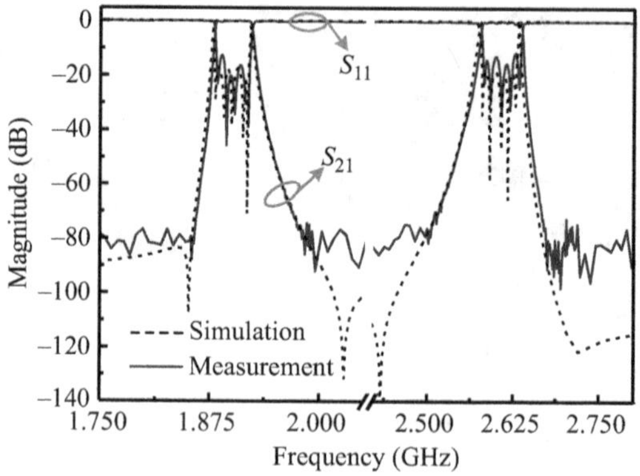

Figure 10.31 Enlarged scale of the simulated and measured results in two passbands. Reprinted with permission from Reference [14]; © 2017 IEEE

10.4 Summary

Several compact and high-performance HTS dual-band BPF based on multi-mode resonators have been reviewed in this chapter. Firstly a series of novel and compact dual-mode or quadruple-mode resonators are presented for constituting the low-order dual-band filters with the electrically small size and ultra-low insertion loss. Then, a novel dual-mode HRR is demonstrated and developed to an eighth-order HTS dual-band filter with sharp passband slope and deep out of band attenuation. In light of the development of more flexible resonant structures and efficient design theory, more compact and high-performance HTS dual-band, even the triple-band or quad-band filter can be designed in the future so as to meet the requirement for the high-quality service of various multi-functional wireless system.

References

[1] Bednorz I. G., Muller K. A., 'Possible high-Tc superconductivity in Be-La-Cu-O system'. *Z. Phys. B.* 1986, vol. 64(2), pp. 189–191.

[2] Wang L. M., Chang M. C., Li S. F., *et al.*, 'Quarter-wavelength stepped-impedance YBCO resonators for miniaturized dual-band high-Tc superconducting filters'. *IEEE Trans. Appl. Supercond.* 2009, vol. 19(3), pp. 895–898.

[3] Yeo K. S., Lancaster M. J., '8 pole high temperature superconductor microstrip dual-band bandpass filter design'. *IEEE MTT-S IMS*, Baltimore, MD, USA, 5–10 June 2011. pp. 1–4.

[4] Seliya N., Sugiyama S., 'Design of miniaturized HTS dual-band bandpass filters using stub-loaded meander line resonators and their applications to

tri-band bandpass filters'. *IEEE Trans. Appl. Supercond.* 2015, vol. 25(3), p. 1500805.

[5] Song F., Wei B., Zhu L., Cao B. S., Lu X. L., 'Dual-band high-temperature superconducting bandpass filter using quint-mode stub-loaded resonators'. *IEEE Trans. Appl. Supercond.* 2015, vol. 25(4), p. 1501410.

[6] Heng Y., Guo X. B., Cao B. S., Wei B., Zhang G. Y., Song X. K., 'Dual-band superconducting bandpass filter using stub-loaded resonators with controllable coupling and feeding structure'. *IEEE Microw. Wireless Compon. Lett.* 2013, vol. 23(8), pp. 400–402.

[7] Lu X. L., Cao B. S., Guo X. B., *et al.*, 'Design of a high-order dual-band superconducting filter with controllable frequencies and bandwidths'. *IEEE Trans. Appl. Supercond.* 2014, vol. 24(2), p. 1500205.

[8] Liu H. W., Fan Y. C., Zhang Z. C., *et al.*, 'Dual-band superconducting bandpass filter using embedded split ring resonator'. *IEEE Trans. Appl. Supercond.* 2013, vol. 23(3), p. 1300304.

[9] Liu H. W., Liu F., Qin F., Ren B. P., Wen P., Guan X. H., 'Compact dual-band HTS bandpass filter using spirally asymmetric stepped-impedance resonators'. *IEEE Trans. Appl. Supercond.* 2016, vol. 26(7), pp. 1501805, 2016.

[10] Mateu J., Collado C., O'Callaghan J. M., 'Modeling superconducting transmission line bends and their impact on nonlinear effects'. *IEEE Trans. Microw. Theory Techn.* 2007, vol. 55(5), pp. 822–828.

[11] Dahm T., Scalapino, D. J., 'Theory of intermodulation in a superconducting microstrip resonator'. *J. Appl. Phys.*, 1997, vol. 81(4), pp. 2002–2009.

[12] Liu H. W., Wen P., Zhao Y. L., Ren B. P., Wang X. M., Guan X. H., 'Dual-band superconducting bandpass filter using quadruple-mode resonator'. *IEEE Trans. Appl. Supercond.* 2014, vol. 24(2), p. 4901204.

[13] Liu H. W., Wen P., Wang X. M., *et al.*, 'Dual-band high-temperature superconducting hairpin-resonator bandpass filter based on two pairs of nondegenerate modes'. *IEEE Trans. Appl. Supercond.* 2015, vol. 25(3), p. 1500104.

[14] Liu H. W., Ren B. P., Hu S. X., Guan X. H., Wen P., Tang J. M., 'High-order dual-band superconducting bandpass filter with controllable bandwidths and multi-transmission zeros'. *IEEE Trans. Microw. Theory Techn.* 2017, vol. 65(10), pp. 3813–3823.

[15] Cristal E. G., Frankel S., 'Hairpin-line and hybrid hairpin-line/half-wave parallel-coupled-line filters'. *IEEE Trans. Microw. Theory Techn.* 1972, vol. 20(11), pp. 719–728.

[16] Hong J. S., Shaman H., Chun Y. H., 'Dual-mode microstrip open-loop resonators and filters'. *IEEE Trans. Microw. Theory Techn.* 2007, vol. 55(8), pp. 1764–1770.

[17] Görur A., 'Description of coupling between degenerate modes of a dual-mode microstrip loop resonator using a novel perturbation arrangement and its dual-mode bandpass filter applications'. *IEEE Trans. Microw. Theory Techn.* 2004, vol. 52(2), pp. 671–677.

[18] Hong J. S., Lancaster M. J., *Microwave Filter for RF/Microwave Application*. New York, NY, USA: Wiley, 2001.

[19] Luo S., Zhu L., Sun S., 'A dual-band ring-resonator bandpass filter based on two pairs of degenerate modes'. *IEEE Trans. Microw. Theory Techn.* 2010, vol. 58(12), pp. 3427–3432.

[20] Cameron R. J., 'Advanced coupling matrix synthesis techniques for microwave filters'. *IEEE Trans. Microw. Theory Techn.* 2003, vol. 51(1), pp. 1–10.

[21] Kuo Y. T., Chang C. Y., 'Analytical design of two-mode dual-band filters using E-shaped resonators'. *IEEE Trans. Microw. Theory Techn.* 2012, vol. 60(2), pp. 250–260.

Chapter 11
Balun bandpass filters

Jianpeng Wang[1], *Feng Huang*[1], *and Xuedao Wang*[1]

11.1 Introduction

In recent years, great advancement of various modern wireless communication systems has put forward higher requirements on radio-frequency (RF) devices with high performance and miniaturized size. To meet these requirements, one effective approach is to design a variety of multifunction embedded components by integrating multiple functions into one component. A typical work to accommodate to this tendency is the balun bandpass filter, which is initially constructed by the proper integration of a balun and a filter. It can not only maintain the function of a balun but also can gain the frequency selectivity of a bandpass filter. Attributing to these specific features, the balun filter can be utilized to effectively simplify the circuit design and reduce the overall size of the RF front-end system as shown in Figure 11.1.

Over the past few years, several approaches have been successfully proposed to deal with various balun bandpass filters. Firstly, the most popular method for the implementation of balun bandpass filter is based on the conventional Marchand-type balun topology by utilizing printed circuit board (PCB) technology [1–3], low-temperature co-fired ceramic technology [4,5]. Secondly, the three-port balun bandpass filters can be straightforwardly realized based on the symmetrical four-port network by keeping one of the four ports to be open or short circuited. The works in [6–9] are typical examples for this method. Thirdly, as another interesting method, alternative balun bandpass filters can be realized by making use of the inherent anti-phase property of the dual-mode ring resonators [10–12] or patch resonators [13,14]. Moreover, some other studies about the realization of balun filters including the multi-coupled lines construction [15], the slotline resonator [16], the different transition structures [17], and the substrate integrated waveguide (SIW) based coupled-resonator model [18] have also been reported.

Recently, the half-wavelength open-ended transmission line has been receiving much attention in generating desired differential signals in the design of balun bandpass filters. In [19,20], balun filters with single-band and dual-band responses

[1]School of Electronic and Optical Engineering, Nanjing University of Science and Technology, China

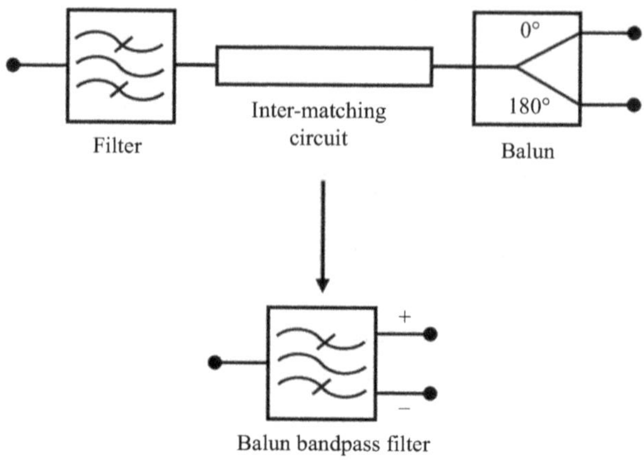

Figure 11.1 Flow diagram for illustrating the transformation from a cascaded filter/line/balun circuit block to an incorporated balun bandpass filter. Reprinted with permission from Reference [20]; © 2016 IEEE

are presented on a planar-type balun topology, whose working principle can be explained by the analysis of standing wave pattern on a half-wavelength open-circuited transmission line. Moreover, the proposed balun topology has an additional advantage for exploration of alternative class of wideband balun bandpass filters. Based on this balun topology, planar wideband balun filters have been successful achieved as discussed in [21] and [22]. In this context, this balun topology exhibits a great flexibility in designing balun bandpass filters.

In this chapter, we will firstly deal with a coupled line based balun topology. Then based on this balun topology, a new method to design a class of planar-type balun bandpass filters will be covered utilizing widely used open-type multi-mode resonators. To extend the application of this simple and effective method, a dual-mode balun diplexer is also introduced. In addition, several planar balun bandpass filters based on different transition structures have been also introduced in this chapter. All designs are verified by simulation and experiment, and they are found in good agreement with each other.

11.2 Balun topology

11.2.1 Basic design concept

First, let us consider that an incident wave of $V_0^+ e^{-j\beta z}$ is generated from a source at $z < 0$ and it is transmitted along a lossless transmission line with the characteristic impedance Z_0 and the terminated load impedance Z_L. We know that a reflected wave must be excited if $Z_L \neq Z_0$. Considering the special case of open circuited termination shown in Figure 11.2(a), where the load impedance $Z_L = \infty$, we can find that the reflection coefficient is $\Gamma = 1$ and the standing wave ratio is infinite for the lossless transmission line. Therefore, a stable standing wave pattern can be

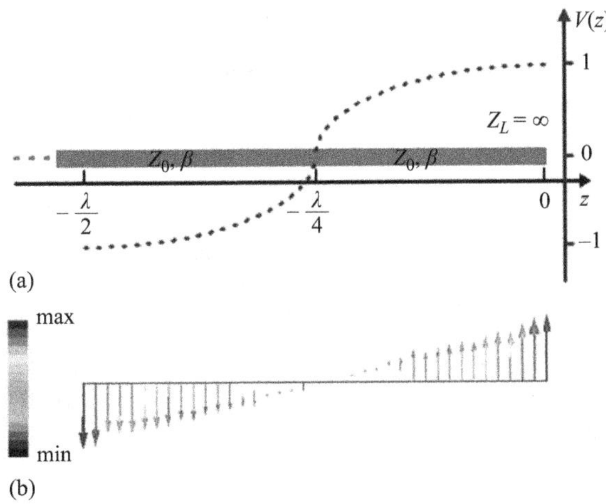

Figure 11.2 Sketch of longitudinal voltage/field distributions along an open-circuited transmission line: (a) normalized voltage wave and (b) electric field intensity. Reprinted with permission from Reference [20]; © 2016 IEEE

formed along the open-circuited transmission line. According to the fundamental transmission line theorem, the voltage distribution along the line consists of a superposition of an incident and reflected wave, such that

$$V(z) = V_0^+(e^{-j\beta z} + e^{j\beta z}) = 2V_0^+\cos(\beta z) \tag{11.1}$$

where β is the propagation constant of the transmission line. According to the formulation (11.1), the normalized fundamental voltage standing wave along the open-circuited line is illustrated by the dash line in Figure 11.2(a). We can see the voltage has a maximum value at the open-end and has a minimum value at a position that is one quarter-wavelength away from the end. Figure 11.2(b) shows the corresponding electric field distribution along this transmission line, which verifies the properties of standing wave distribution. Thus, if the electrical length of the line is half-wavelength, electric field distribution between the input port and open-circuited termination exhibits two distinctive zones, with identical magnitude and out-of-phase characteristics, as highly demanded in the design of a balun circuit.

11.2.2 Theoretical analysis

Now, let us apply the field distribution properties around a half-wavelength open-circuited transmission line for the demonstration and design of a proposed balun [20]. Figure 11.3 depicts the original prototype of this planar-type balun with two quarter-wavelength microstrip coupled line networks, and it can be viewed as a backward wave directional coupler individually, whose layout is shown in Figure 11.4. It should be noted that the two coupler sections are connected to each other through their respective direct ports (2–2′). In addition, one input port (1) is

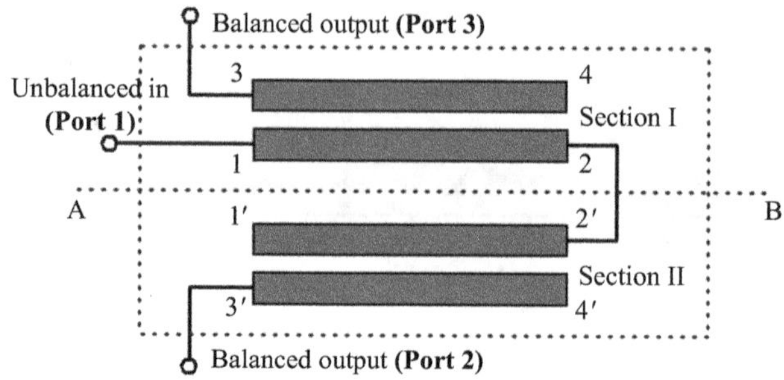

Figure 11.3 Block diagram of the proposed planar-type balun consisting of two identical couplers. Reprinted with permission from Reference [20]; © 2016 IEEE

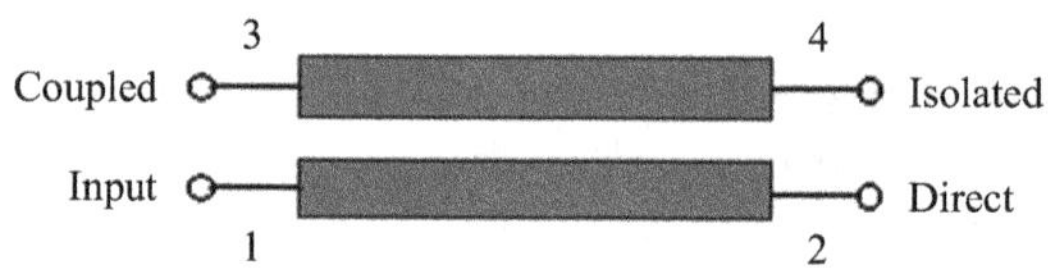

Figure 11.4 Schematic of the backward-wave directional coupler. Reprinted with permission from Reference [20]; © 2016 IEEE

formed as the unbalanced input of the balun and the other input port (1′) is set as an open circuit. Moreover, two coupled ports (3–3′) are considered to be a pair of balanced outputs, while two isolated ports (4–4′) are both open circuited. To facilitate the theoretical analysis of the balun topology, the two following assumptions need to be made:

(a) The characteristic impedance of the coupled line is equal to the identical source and load impedances;
(b) The phase velocities of the even and odd modes are kept identical within the allowed error margin of balun performance.

According to [23], the S-parameters of a four-port coupler are given as:

$$S_{\text{coupler}} = \begin{bmatrix} 0 & x/z & y/z & 0 \\ x/z & 0 & 0 & y/z \\ y/z & 0 & 0 & x/z \\ 0 & y/z & x/z & 0 \end{bmatrix} \tag{11.2}$$

where

$$x = \sqrt{1 - k^2}$$
$$y = jk \sin\theta$$
$$z = \sqrt{1 - k^2}\cos\theta + j\sin\theta$$

where θ and k represent the electrical length and coupling coefficient, respectively. With reference to the port assignment in Figure 11.3, the S-parameters of the couplers in Sections I and II can be expressed by the following two scattering matrices:

$$S_{\mathrm{I}} = \begin{bmatrix} s_{11} & s_{12} & s_{13} \\ s_{21} & s_{22} & s_{23} \\ s_{31} & s_{32} & s_{33} \end{bmatrix} = \begin{bmatrix} 0 & x/z & y/z \\ x/z & y^2/z^2 & xy/z^2 \\ y/z & xy/z^2 & x^2/z^2 \end{bmatrix} \tag{11.3a}$$

$$S_{\mathrm{II}} = \begin{bmatrix} s_{2'2'} & s_{2'3'} \\ s_{3'2'} & s_{3'3'} \end{bmatrix} = \begin{bmatrix} (x^2+y^2)/z^2 & 2xy/z^2 \\ 2xy/z^2 & (x^2+y^2)/z^2 \end{bmatrix} \tag{11.3b}$$

Since the two sections share one common direct port, the whole S-parameters of the proposed balun can be derived by combining (11.3a) and (11.3b), as given in (11.4a). In the design, the electrical length of the coupled line section is equal to $\pi/2$, so that (11.4a) can be simplified to (11.4b). Obviously, we can deduce the relationship of $S_{\text{balun-21}} = -S_{\text{balun-31}}$ from (11.4b), which demonstrates that the proposed balun does provide two perfect differential outputs at ports 2 and 3.

$$S_{\text{balun}} = \begin{bmatrix} \dfrac{x^2(x^2+y^2)}{z^4-y^4-x^2y^2} & \dfrac{x^4y-y^5+yz^4}{z(z^4-y^4-x^2y^2)} & \dfrac{2x^2yz}{z^4-y^4-x^2y^2} \\ \dfrac{x^4y-y^5+yz^4}{z(z^4-y^4-x^2y^2)} & \dfrac{x^2z^2}{z^4-y^4-x^2y^2} & \dfrac{2x^2y^2}{z^4-y^4-x^2y^2} \\ \dfrac{2x^2yz}{z^4-y^4-x^2y^2} & \dfrac{2x^2y^2}{z^4-y^4-x^2y^2} & \dfrac{x^2z^4+y^2z^4-x^2y^2+y^4}{z^2(z^4-y^4-x^2y^2)} \end{bmatrix} \tag{11.4a}$$

$$S_{\text{balun}} = \frac{1}{2k^2+1} \begin{bmatrix} 1-2k^2 & -2k & 2k \\ -2k & -1 & -2k^2 \\ 2k & -2k^2 & -1 \end{bmatrix} \tag{11.4b}$$

11.2.3 Coupled line prototypes

Figure 11.5 shows two types of coupled line prototypes, namely, types A and B, which are applied herein to the implementation of the proposed balun structure. Type A is formed by an anti-parallel coupled line with its coupled port terminated by an open-circuited stub, whereas type B is in a parallel coupled line with a transmission line cascaded in its input port. In what follows, much effort is made to characterize and design these two proposed coupled line prototypes.

For the type A prototype, the input impedance of its open-circuited stub Z_L is derived as:

$$Z_L = -jZ_1 \cot\theta_1 \tag{11.5}$$

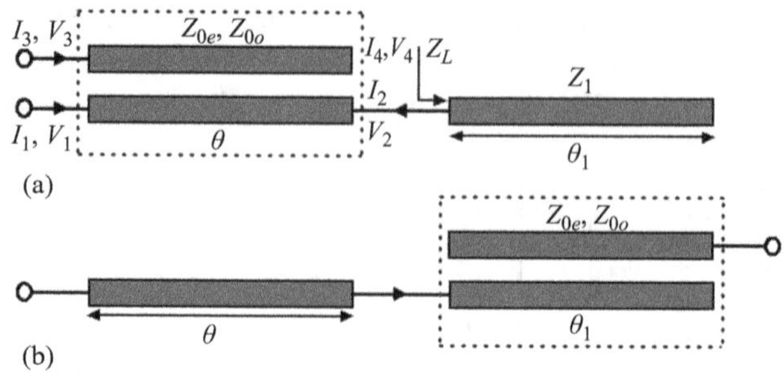

Figure 11.5 Schematics of two coupled line prototypes: (a) Type A and (b) Type B. Reprinted with permission from Reference [20]; © 2016 IEEE

where Z_1 and θ_1 stand for the characteristic impedance and electrical length of the terminated stub, respectively. In order to analytically derive the overall *ABCD* matrix of this two-port circuit, the even- and odd-mode analysis approaches are applied for simplification of our analysis. Based on the previous assumptions and according to the work in [24], two equations can be formulated as follows:

$$\begin{bmatrix} V_1 - V_3 \\ I_1 - I_3 \end{bmatrix} = \begin{bmatrix} \cos\theta & jZ_{0o}\sin\theta \\ jY_{0o}\sin\theta & \cos\theta \end{bmatrix} \begin{bmatrix} V_2 - V_4 \\ -(I_2 - I_4) \end{bmatrix} \tag{11.6a}$$

$$\begin{bmatrix} V_1 + V_3 \\ I_1 + I_3 \end{bmatrix} = \begin{bmatrix} \cos\theta & jZ_{0e}\sin\theta \\ jY_{0e}\sin\theta & \cos\theta \end{bmatrix} \begin{bmatrix} V_2 + V_4 \\ -(I_2 + I_4) \end{bmatrix} \tag{11.6b}$$

$$\begin{aligned} u_0 &= \cos\theta \\ u_1 &= Z_L\cos\theta \\ u_2 &= Z_L Y_{0e} Y_{0o}\sin^2\theta \\ u_3 &= j[(Y_{0e} - Y_{0o})\sin\theta]/2 \\ u_4 &= j[(Y_{0e} + Y_{0o})\sin\theta]/2 \\ u_5 &= j[(Z_{0e} - Z_{0o})\sin\theta]/2 \\ u_6 &= Z_L\cos\theta + j[(Z_{0e} + Z_{0o})\sin\theta]/2 \\ u_7 &= \cos\theta + j[Z_L(Y_{0e} + Y_{0o})\sin\theta]/2 \\ u_8 &= jZ_0\cos\theta \\ u_9 &= j(1/Z_0)\cos\theta \end{aligned} \tag{11.7}$$

For simplification, the port conditions with $I_4 = 0$ and $V_2 = -I_2Z_L$ are taken into account. Besides, with resorting to the relevant parameters denoted in (11.7), the final *ABCD* matrix for type A can be deduced as (11.8a).

$$\begin{bmatrix} A & B \\ C & D \end{bmatrix} = \frac{1}{u_1u_3 - u_4u_5} \begin{bmatrix} -u_4u_6 & -u_0u_6 \\ -u_1u_4 + u_2 & u_3u_5 - u_0u_7 \end{bmatrix} \tag{11.8a}$$

$$\begin{bmatrix} A & B \\ C & D \end{bmatrix} = \begin{bmatrix} u_0 & u_8 \\ u_9 & u_0 \end{bmatrix} \begin{bmatrix} u_3u_8 - \dfrac{u_0^2u_4 + u_4^2u_8}{u_3} & u_0u_5 - \dfrac{u_0^2u_4 + u_0u_4u_8}{u_3} \\ u_0u_3 - \dfrac{u_0^2u_4 + u_0u_4u_9}{u_3} & u_5u_9 - \dfrac{u_0^2u_4 + u_0^2u_9}{u_3} \end{bmatrix} \tag{11.8b}$$

As for the type B prototype, it is formed by cascading a transmission line with a parallel coupled line section. To analyze the type B prototype, the *ABCD* matrix of this two-port circuit can be derived by multiplying the *ABCD* matrices of two different sections, one is a transmission line and the other is a parallel coupled line. With the known parameters described in (11.7), the resultant transmission matrix of this two-port type B prototype can be derived as (11.8b).

Thus, the transmission coefficients of these two coupled line prototypes can be obtained as indicated in (11.9) according to the conversion formula in [24], where Z_0 is the system impedance and its value is set as 50 Ω as usual.

$$S_{\text{type\_A}} = \frac{2Z_0(u_1u_3 - u_4u_5)}{-u_0u_6 + Z_0(u_3u_5 - u_0u_7 - u_4u_6) + Z_0^2(u_2 - u_1u_4)} \tag{11.9a}$$

$$S_{\text{type\_B}} = \frac{2Z_0u_3}{u_0u_3u_5 - u_0^2u_8 - u_0u_4u_8 + Z_0(u_3^2u_8 + u_3u_5u_9 - 2u_0^2u_4 - u_0^2u_9 - u_0^2u_8) + Z_0^2(u_0^2u_3 - u_0^2u_4 - u_0u_4u_9)} \tag{11.9b}$$

Obviously, $S_{\text{type\_A}}$ varies as a function of the electric length θ and its maximum value is reached at $\theta = \pi/2$, and it indicates that type A prototype performs a bandpass behavior near the central frequency of $\theta = \pi/2$. Since the denominator of $S_{\text{type\_A}}$ is finite, the condition for emergence of the transmission zeros in the stopband can be analytically determined by enforcing the numerator of $S_{\text{type\_A}}$ to be equal to zero, which leads to

$$u_1u_3 = u_4u_5 \tag{11.10}$$

After theoretical simplification of (11.10), the following equation is deduced to determine all the transmission zeros:

$$\tan\theta = \frac{2Z_1}{Z_{0e} + Z_{0o}} \cot\theta_1 \tag{11.11}$$

Figure 11.6(a) shows the first two transmission zeros versus the electrical length ratio as $\alpha = \theta_1/\theta$ with $Z_{0e} = 104.7$ Ω, $Z_{0o} = 31.7$ Ω, and $Z_1 = 78.6$ Ω. As we can see, the locations of two transmission zeros change at different values of α. Specifically, there is only one transmission zero which occurs at $\theta = \pi/2$ when $\alpha = 0$.

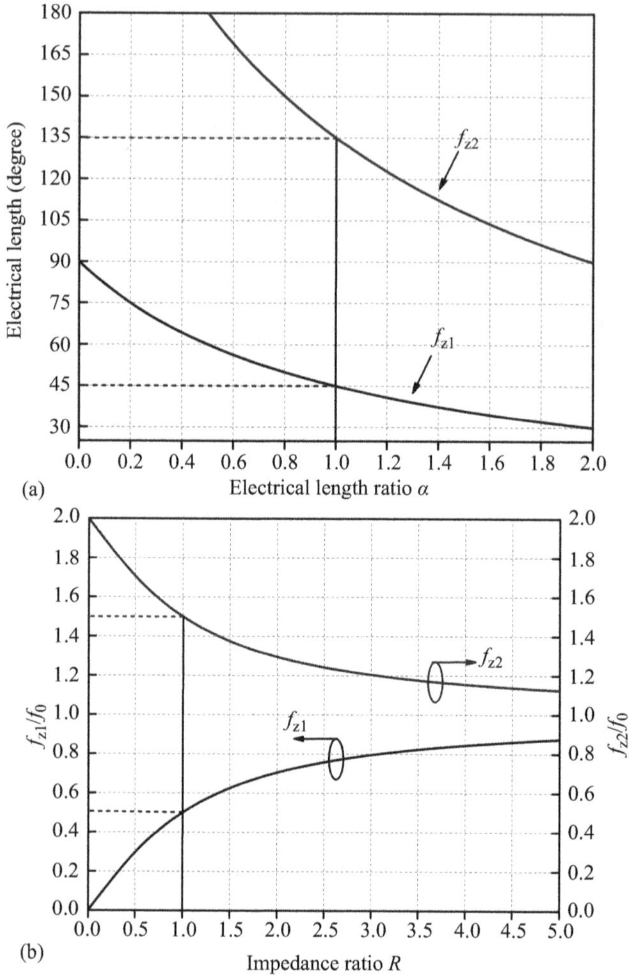

Figure 11.6 (a) Locations of the first two TZs versus electrical length ratio α and (b) locations of the first two TZs versus impedance ratio R. Reprinted with permission from Reference [20]; © 2016 IEEE

Under this circumstance, the type A coupled line prototype is referred to as the anti-parallel coupled-line. What is more, two transmission zeros emerging at the electrical length of $\pi/4$ and $3\pi/4$ could be clearly observed when $\alpha = 1$. Considering $\theta = \theta_1 = \pi/2$ at the specified center frequency f_0 is readily selected, thus the frequencies of the first two transmission zeros can be found in the closed-form manner such that

$$f_{z1} = \frac{2f_0}{\pi}\arctan\sqrt{\frac{2Z_1}{Z_{0e} + Z_{0o}}} \tag{11.12a}$$

$$f_{z2} = \frac{2f_0}{\pi}\left[\pi - \arctan\sqrt{\frac{2Z_1}{Z_{0e} + Z_{0o}}}\right] \tag{11.12b}$$

By defining the impedance ratio of $R = 2Z_1/(Z_{0e} + Z_{0o})$, the relationship between R and the first two transmission zero frequencies is established through (11.12) and graphically expressed in Figure 11.6(b) for direct demonstration. It can be observed from Figure 11.6(b) that these two transmission zeros are symmetrical with respect to the operating center frequency f_0 with a mutual relationship of $0 < f_{z_1} < f_0 < f_{z_2} < 2f_0$. Moreover, these two transmission zeros have the trend to move toward to the central frequency when R gets increased. In other words, these two transmission zeros can be flexibly controlled by varying the characteristic impedance of the terminated open-circuited stub. In our design, R is readily set as 1, thus two transmission zeros would emerge near $0.5f_0$ and $1.5f_0$. Furthermore, these inherent zeros keep emerging periodically in every frequency range $(2nf_0, 2(n+1)f_0)$, where n is an integer.

Similarly, according to the expression of transmission coefficient $S_{\text{type\_B}}$, the type B coupled line prototype discussed herein exhibits a bandpass response as well under the condition where θ is equal to $\pi/2$. By comparatively investigating the type A and B prototypes, no extra one can be found within the frequency range from 0 to $2f_0$ for type B coupled line prototype except for two fixed transmission zeros at 0 and $2f_0$.

11.3 Dual-mode balun bandpass filters

For demonstration, two dual-mode balun bandpass filters are implemented in this section by adopting widely used open-type resonators [19,20]. The practical balun filters are fabricated on a single layer substrate, i.e., Rogers RO4003C with a dielectric constant of 3.55, a loss tangent of 0.0027, and a thickness of 0.508 mm. Simulation is carried out using the commercial full-wave simulator, i.e., ANSOFT HFSS, while the measurement is accomplished using the Agilent N5244A four-port vector network analyzer.

11.3.1 Geometric construction

Based on the balun topology discussed above, a new dual-mode balun bandpass filter is designed and its geometrical layout is displayed in Figure 11.7. In the design, an open-stub-loaded resonator, shown in Figure 11.8, is adopted to achieve the desired dual-mode response. Here, port 1 is the unbalanced input port, while ports 2 and 3 make a differential or balanced output of the proposed balun bandpass filter. As we can observe, two resonators, namely, resonator A and A′ are rotationally symmetrical with respect to the center of the transmission line, and they are quarter- and half-wavelength distances away from the open-circuited termination. In addition, open-stub-loaded resonators are geometrically folded into an E-shaped configuration for size compactness. Using the well-known even- and odd-mode

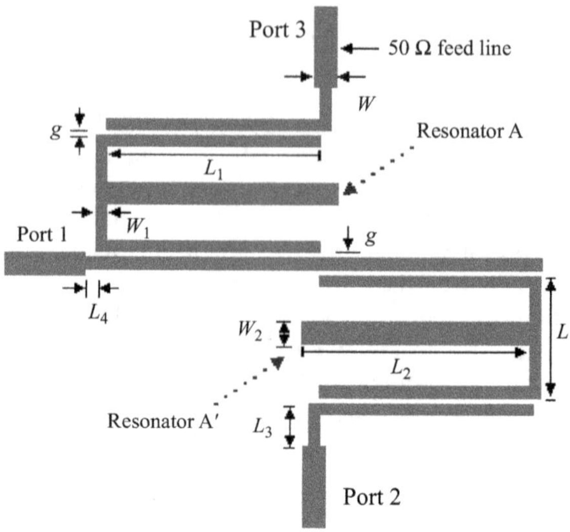

Figure 11.7 Layout of the dual-mode balun bandpass filter with E-shaped resonators

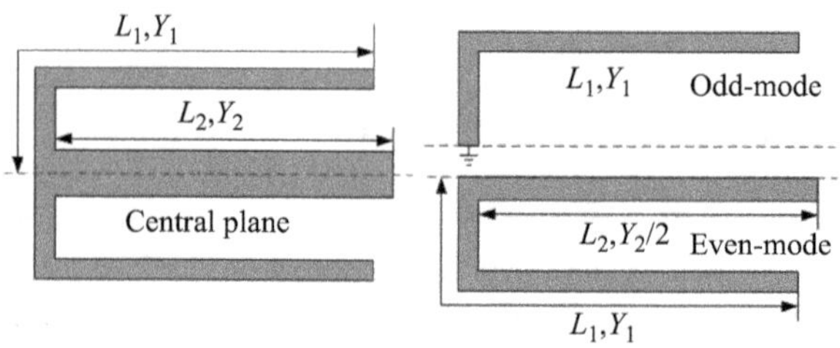

Figure 11.8 E-shaped dual-mode resonator with its half bisections under odd- and even-mode excitations

methods [25], the fundamental resonant frequencies of the dual-mode resonator can be derived as:

$$f_{\text{odd}} = \frac{c}{4L_1\sqrt{\varepsilon_e}} \tag{11.13a}$$

$$f_{\text{even}} = \frac{c}{2(L_1 + L_2)\sqrt{\varepsilon_e}} \tag{11.13b}$$

where c is the speed of light in free space and ε_e is the effective dielectric constant of the substrate. In the design, the stub length L_2 is tuned to properly adjust the even–mode resonant frequency while keeping the odd-mode one fixed. This property can be utilized to achieve a filtering circuit with a certain bandwidth, so as to form up a dual-mode balun bandpass filter as will be discussed herein.

11.3.2 Design

Figure 11.9 shows the coupling scheme of the dual-mode balun bandpass filter, where S, L_+ or L_-, E or E′ and O or O′ stand for source, load, even mode, and odd mode, respectively. As we can see, inside the balun bandpass filter, there exist two main coupling routes sharing a common input feed line, and each route can support two resonant modes of the resonator, i.e., odd and even modes. Based on the previous discussions, two main filtering paths should be achieved with identical magnitude and opposite phase in the desired passband. Thus, the coupling coefficients labeled in Figure 11.9 should satisfy the relationship of $M_{SE} = -M_{SE'}$, $M_{SO} = -M_{SO'}$. In the design, both filtering paths are purposely designed with identical in-band specifications, i.e., a 2 GHz central frequency, a 25 dB in band return losses, and an 11% fractional bandwidth, respectively. In order to achieve the above-specified frequency response, synthesized coupling matrix for each route is given as:

$$M_I = \begin{bmatrix} 0 & 0.9166 & -1.4353 & 0 \\ 0.9166 & -2.763 & 0 & 0.9166 \\ -1.4353 & 0 & 2.6858 & 1.4353 \\ 0 & 0.9166 & 1.4353 & 0 \end{bmatrix} \tag{11.14a}$$

$$M_{II} = \begin{bmatrix} 0 & -0.9166 & 1.4353 & 0 \\ -0.9166 & -2.763 & 0 & -0.9166 \\ -1.4353 & 0 & 2.6858 & -1.4353 \\ 0 & -0.9166 & -1.4353 & 0 \end{bmatrix} \tag{11.14b}$$

The detailed design procedure for the determination of physical dimensions of the balun bandpass filter is summarized as follows.

The first step is to obtain the geometric parameters of the E-shaped resonator. The desired parameters of resonant frequencies f_{odd} and f_{even} can be obtained from the coupling matrix in (11.14), where $f_{odd} = 1.86$ GHz and $f_{even} = 2.16$ GHz. Once these frequencies are confirmed, the relevant physical parameters are initially chosen as $L_1 = 18$ mm, $L = 8$ mm, $L_2 = 17$ mm, and $W_1 = 0.4$ mm. Next step is to

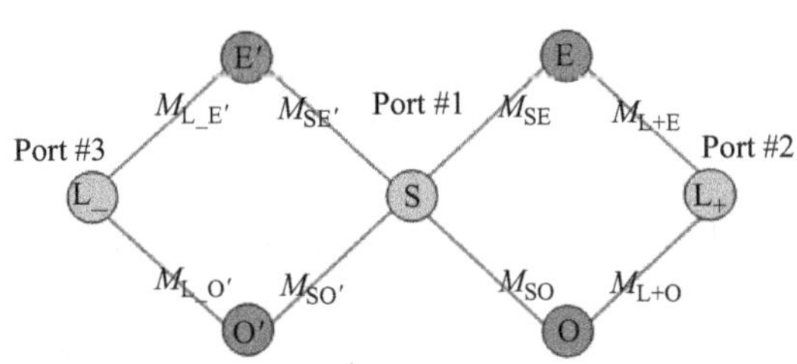

Figure 11.9 Coupling scheme of the proposed dual-mode balun filter

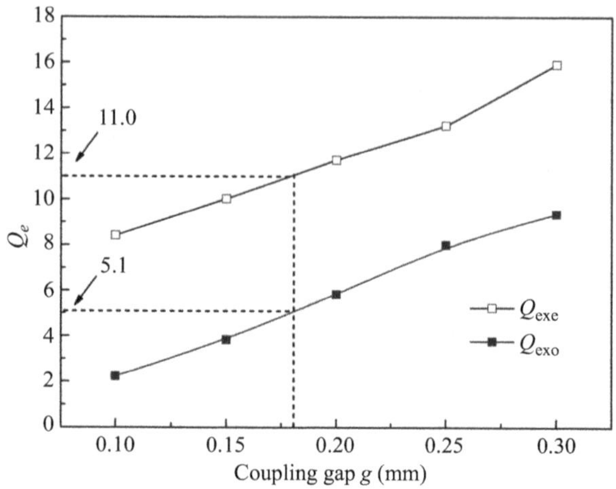

Figure 11.10 External quality factors Q_{exe} and Q_{exo} versus coupling gap spacing g

realize the input/output (I/O) coupling strengths to satisfy the required bandwidth. In general, I/O couplings to even and odd modes can be characterized in terms of external quality factors Q_{exe} and Q_{exo}, which can be extracted from the following [26] through the two group delays, $\tau_{s11}(f_{\text{even}})$ and $\tau_{s11}(f_{\text{odd}})$:

$$Q_{\text{exe}} = \pi f_{\text{even}} \cdot \tau_{s11}(f_{\text{even}}) \tag{11.15a}$$

$$Q_{\text{exo}} = \pi f_{\text{odd}} \cdot \tau_{s11}(f_{\text{odd}}) \tag{11.15b}$$

Figure 11.10 illustrates the extracted Q_e under even- and odd-mode excitations as a function of gap spacing g. These two desired Q-factors can also be derived from the coupling matrix with the values of $Q_{\text{exe}} = 10.8$ and $Q_{\text{exo}} = 4.4$. Thus, the selections of $g = 0.18$ mm can give us these two desired values simultaneously as far as possible.

Finally, the fine tuning process is executed to obtain the definitive dimensions of the overall balun bandpass filter, formed on the microstrip line structure, relying on the full-wave simulation.

11.3.3 Experimental verification

Following the above-described procedure, a dual-mode balun bandpass filter has been designed, fabricated, and measured. Realizable design geometric parameters are determined as: $g = 0.2$ mm, $L_1 = 19$ mm, $W_1 = 0.6$ mm, $L_2 = 20$ mm, $W_2 = 1.2$ mm, $L = 9.2$ mm, $W = 1.18$ mm, $L_3 = 4$ mm, $L_4 = 1$ mm. The overall size of the fabricated circuit shown in Figure 11.11 is $0.6\lambda_g \times 0.46\lambda_g$, where λ_g is the guided wavelength at the central frequency of the balun filter.

The simulated and measured results, i.e., S-parameters, magnitude imbalance, and phase difference, are plotted in Figure 11.12, showing good agreement with each other. The balun filter operates at the central frequency of 2.0 GHz with a

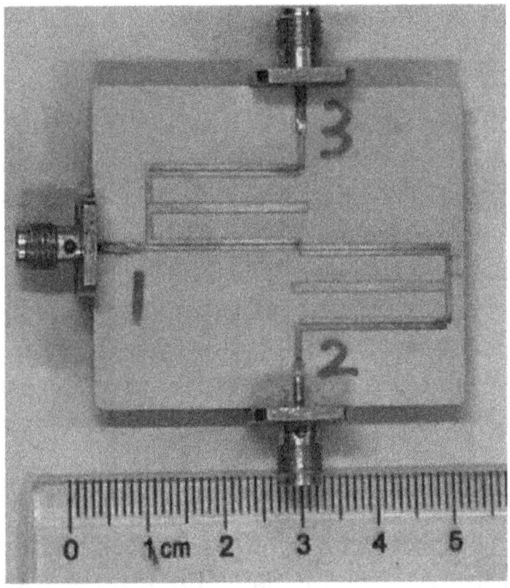

Figure 11.11 Photograph of the fabricated dual-mode balun bandpass filter

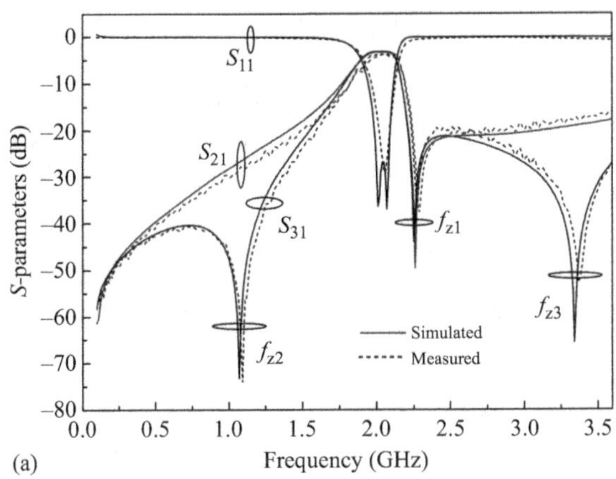

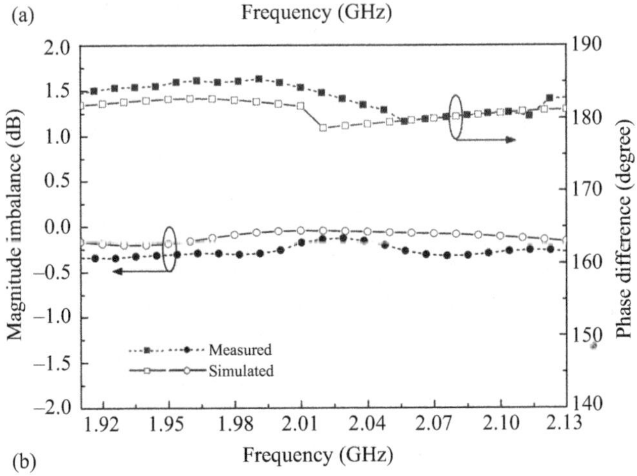

Figure 11.12 Simulated and measured results of the dual-mode balun filter: (a) S-parameters and (b) phase difference and magnitude imbalance

fractional bandwidth of 11%. The measured in-band insertion loss is smaller than 0.8 dB and the return loss is better than 23 dB. In the stopband, one transmission zero caused by the resonance of the central loaded stub in E-shaped resonator emerges at 2.26 GHz for S_{21}. While for S_{31}, besides the transmission zero at 2.26 GHz, there are two extra transmission zeros at 1.07 and 3.38 GHz caused by type B coupled line prototype. These transmission zeros help a lot to enhance the rejection skirt near cutoff frequencies of the desired passband and out-of-band attenuation, so as to improve its frequency selectivity. As illustrated in Figure 11.12(b), the balun bandpass filter exhibits good balanced performance with an amplitude imbalance of less than 0.35 dB and a phase imbalance of better than 5° in the desired passband.

11.3.4 Improvement for miniaturization and stopband rejection

In order to provide a compact configuration, another improved dual-mode balun bandpass filter is proposed [20] and its layout is displayed in Figure 11.13. Similarly, port 1 is the unbalanced input port, and ports 2 and 3 are the balanced outputs. In this design, two identical resonator A and A′ are arranged in the face-to-face format and they are symmetrical with respect to the central plane P-P′. In this context, the overall size of the new balun bandpass filter design can be significantly reduced. What is more, another attractive feature of this balun bandpass filter is embodied in its output feeding line structure. Specifically, as shown in Figure 11.13, a pair of differential signals can be coupled to two balanced ports using the type A coupled line prototype, so as to create extra transmission zeros for both S_{21} and S_{31} responses.

Based on the provided design procedure, an improved dual-mode balun bandpass filter is designed at 2.4 GHz with a 3 dB fractional bandwidth of 5.8% and in-band return loss better than 20 dB. With these requirements, the desired parameters of resonant frequencies f_{odd} and f_{even} then can be obtained from a targeted coupling matrix, where $f_{odd} = 2.33$ GHz and $f_{even} = 2.46$ GHz. Once these frequencies are confirmed, relevant physical parameters are chosen as $L_1 = 17$ mm, $L = 6.8$ mm, and $W_1 = 0.4$ mm. Two desired Q-factors can also be derived from a targeted coupling matrix with the values of $Q_{exe} = 30.77$ and $Q_{exo} = 17.8$. Thus, the

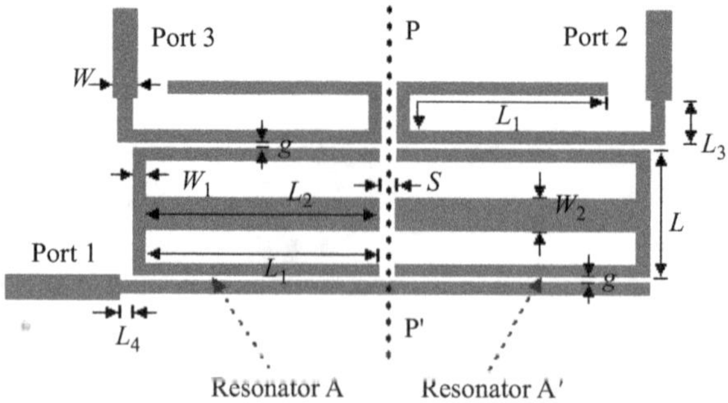

Figure 11.13 Configuration of the improved dual-mode balun bandpass filter. Reprinted with permission from Reference [20]; © 2016 IEEE

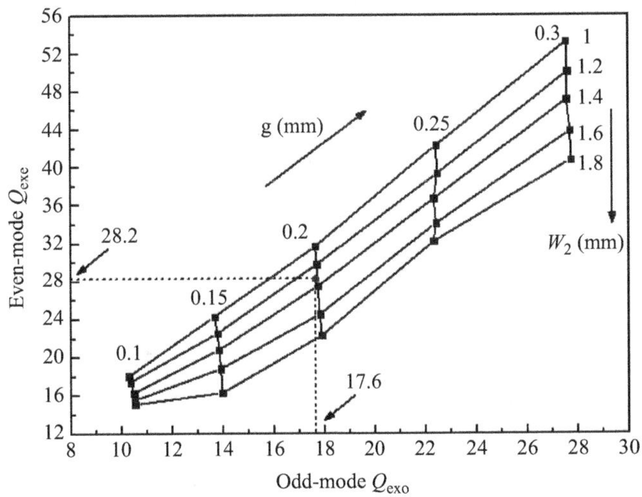

Figure 11.14 External quality factors Q_{exe} and Q_{exo} versus coupling gap spacing g with different stub widths W_2. Reprinted with permission from Reference [20]; © 2016 IEEE

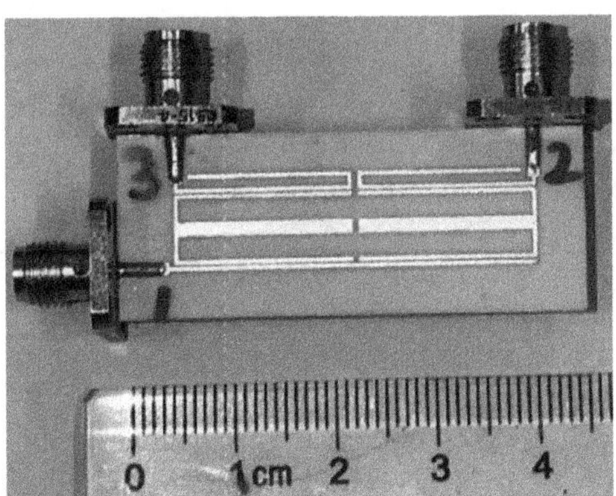

Figure 11.15 Photograph of the fabricated improved dual-mode balun bandpass filter. Reprinted with permission from Reference [20]; © 2016 IEEE

selections of $g = 0.2$ mm and $W_2 = 1.35$ mm are chosen from Figure 11.14. Final geometric parameters are determined as: $g = 0.2$, $W = 1.18$, $W_1 = 0.4$, $W_2 = 1.4$, $L = 6.8$, $L_1 = L_2 = 17$, $L_3 = 0.5$, $L_4 = 0.5$, $S = 0.5$, all in millimeters. Figure 11.15 shows the fabricated improved balun filter. The core occupied area of the improved circuit is about $0.61\lambda_g \times 0.24\lambda_g$, reduced by about 50% when compared with the original dual-mode balun filter as shown in Figure 11.11.

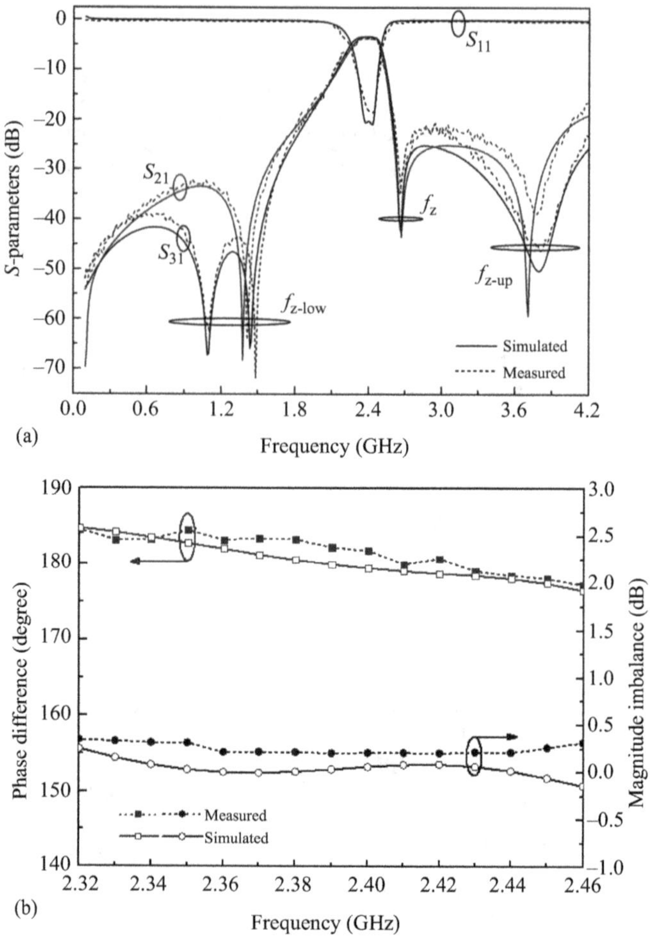

Figure 11.16 Simulated and measured results of the improved balun filter: (a) S-parameters and (b) magnitude and phase differences in the passband. Reprinted with permission from Reference [20]; © 2016 IEEE

Figure 11.16 illustrates the simulated and measured results of the improved dual-mode balun filter. The balun filter operates at 2.4 GHz with a fractional bandwidth of 5.8%. The measured in-band insertion loss is smaller than 0.7 dB and the return loss is better than 18 dB. In the stopband, the transmission zeros, $f_{z\text{-low}}$ (1.1 and 1.45 GHz) and $f_{z\text{-up}}$ (3.8 GHz), are successfully created as expected. Moreover, the common transmission zero at f_z (2.66 GHz) emerging on the higher side of the passband can also be observed clearly. These transmission zeros help a lot to enhance the rejection skirt near cutoff frequencies of the desired passband and out-of-band attenuation, so as to improve its frequency selectivity. In addition, the balun bandpass filter exhibits good balanced performance with an amplitude imbalance of less than 0.3 dB and a phase imbalance of better than 3° in the desired passband.

11.4 Dual-mode dual-band balun bandpass filter

11.4.1 Geometric construction

To further verify the proposed balun topology, its application to the design of dual-mode dual-band balun bandpass filter is presented [20]. The configuration primarily consists of two pairs of dual-mode resonators, namely, resonators A and B, A′ and B′, and three microstrip coupled feed lines, as shown in Figure 11.17. Herein, the dual operating bands are properly achieved with resorting to the coupling between each pair of resonators and the input feed line. Obviously, resonators A and A′ or B and B′ are rotationally symmetrical with regard to the center of the input feed line and have to be operated simultaneously at the same frequency. The space S_1 for separating these two resonators is set to reduce the capacitive coupling between the two adjacent open ends. In addition, S_1 and the offset S_2 are designed to compensate for the phase imbalance caused by the unequal even- and odd-mode phase velocities of the microstrip coupled line. Besides, we can figure out from Figure 11.17 that the center loaded stub is folded into a T-type configuration for flexible tuning of even-mode resonant frequency. Thus, the center frequency and operating bandwidth of each passband can be controlled individually.

11.4.2 Design

The coupling scheme of this dual-mode dual-band balun bandpass filter is depicted in Figure 11.18. Note that nodes 1 and 2 (1′ and 2′) represent the even and odd

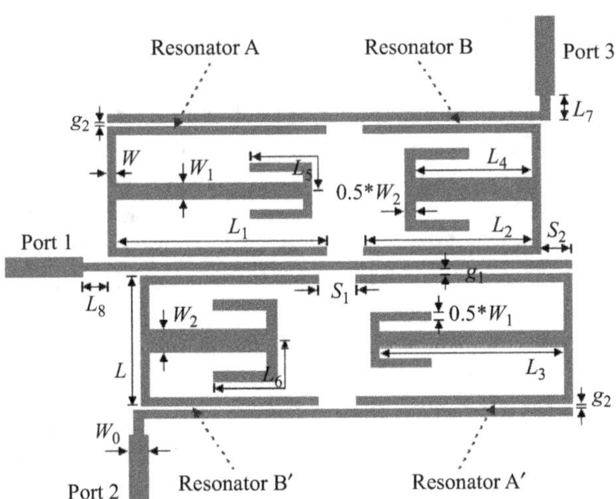

Figure 11.17 Configuration of the dual-mode dual-band balun bandpass filter. Reprinted with permission from Reference [20]; © 2016 IEEE

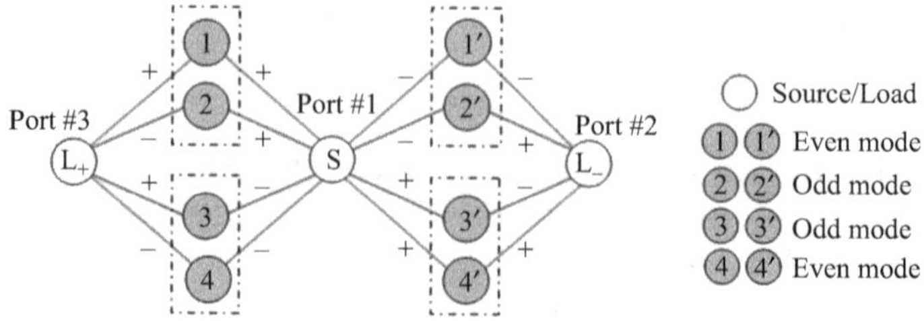

Figure 11.18 Coupling scheme of the dual-mode dual-band balun bandpass filter. Reprinted with permission from Reference [20]; © 2016 IEEE

resonant modes of resonators A and A′, and they are utilized to form up the lower passband. Meanwhile, 3 and 4 (3′ and 4′) denote those resonant modes of the resonators B and B′, and they make up the upper passband.

Following the above design procedure, we first determine the physical dimensions of resonators A and B. Herein, the resonant frequencies with respect to the two passbands are arranged as $f_{\text{even}_1} < f_{\text{odd}_1}$ and $f_{\text{even}_2} > f_{\text{odd}_2}$, and thus two transmission zeros would emerge on the lower side of the first passband and the upper side of the second passband so as to improve its frequency selectivity. For demonstration, a dual-mode dual-band balun bandpass filter operating at 2.28 and 2.72 GHz with the fractional bandwidths of 5.3 and 4.4% is designed. Similarly, its design parameters can be deduced from a targeted coupling matrix according to the specified response, i.e., $f_{\text{even}_1} = 2.23$ GHz, $f_{\text{odd}_1} = 2.32$ GHz, $f_{\text{even}_2} = 2.68$ GHz, and $f_{\text{odd}_2} = 2.76$ GHz. With these design parameters, resonators A and B are initially determined based on (11.13). The strip width on the coupled line section is chosen as 0.5 mm.

Next, the I/O couplings are considered after the determination of the resonators. Similarly, the required external Q-factors from a targeted coupling matrix with respect to the even and odd modes are given as $Q_{\text{exe}_1} = 48.45$, $Q_{\text{exo}_1} = 15.48$, $Q_{\text{exo}_2} = 19.2$, and $Q_{\text{exe}_2} = 55.42$, respectively. Figure 11.19 depicts the extracted Q-factors under varied coupling gap spacing g_2 as well as varied central stub widths W_1 and W_2. Herein, the initial values of g_2, W_1 and W_2 under these desired Q-factors are found to be $g_2 = 0.2$ mm, $W_1 = 0.8$ mm, and $W_2 = 1.2$ mm. After the fine-tuning process is executed, the definitive values of these dimensions are determined as $g_1 = 0.3$ mm, $g_2 = 0.2$ mm, $W_1 = 1$ mm, and $W_2 = 1.4$ mm. In order to achieve the required bandwidth, the two coupling gaps g_1 and g_2 become different due to the three-conductor coupled line sections in the input section.

11.4.3 Experimental verification

Based on the preceding analysis, the above-described dual-mode dual-band balun bandpass filter is designed and fabricated on the Rogers RO4003C substrate ($\varepsilon_r = 3.55$, $\tan\delta = 0.0027$) with a thickness of 0.508 mm. The photograph of the fabricated circuit is depicted in Figure 11.20. The physical dimension is provided as: $g_1 = 0.3$, $g_2 = 0.2$, $W_0 = 1.2$, $W = 0.5$, $W_1 = 1$, $W_2 = 1.4$, $L = 9$, $L_1 = 16$, $L_2 = 13.5$,

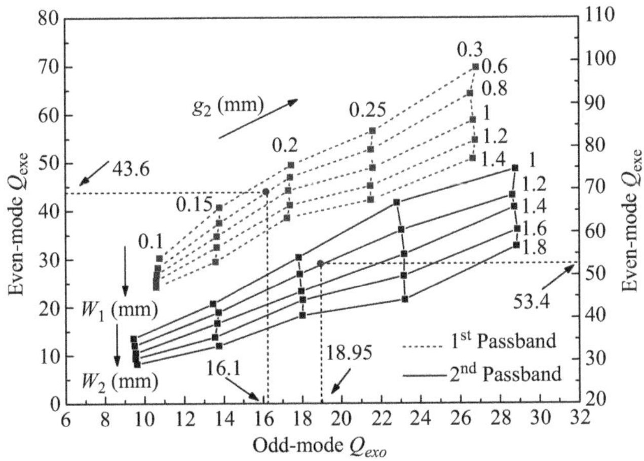

Figure 11.19 External quality factors Q_{exe} and Q_{exo} for dual passbands versus coupling gap spacing g_2 under varied stub widths W_1 and W_2. Reprinted with permission from Reference [20]; © 2016 IEEE

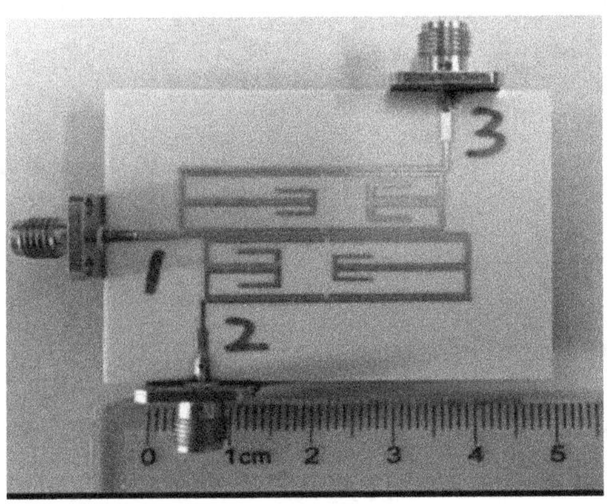

Figure 11.20 Photograph of the fabricated dual-mode dual-band balun bandpass filter. Reprinted with permission from Reference [20]; © 2016 IEEE

$L_3 = 15$, $L_4 = 7.7$, $L_5 = 5.8$, $L_6 = 7.25$, $L_7 = 1$, $L_8 = 0.7$, $S_1 = 2.2$, $S_2 = 2.1$, all in millimeters, and it has the overall area of about $0.63\lambda_g \times 0.46\lambda_g$, where λ_g is the guided wavelength at the central frequency of the lower passband.

The simulated and measured results are plotted in Figure 11.21. Centering at 2.28 and 2.72 GHz, two passbands have the fractional bandwidths of 5.3 and 4.4%, respectively. Inside each passband, two transmission poles can be observed clearly. The insertion losses are smaller than 0.9 and 1.0 dB in the lower and upper passbands, and both return losses are better than 20 dB. Transmission zeros

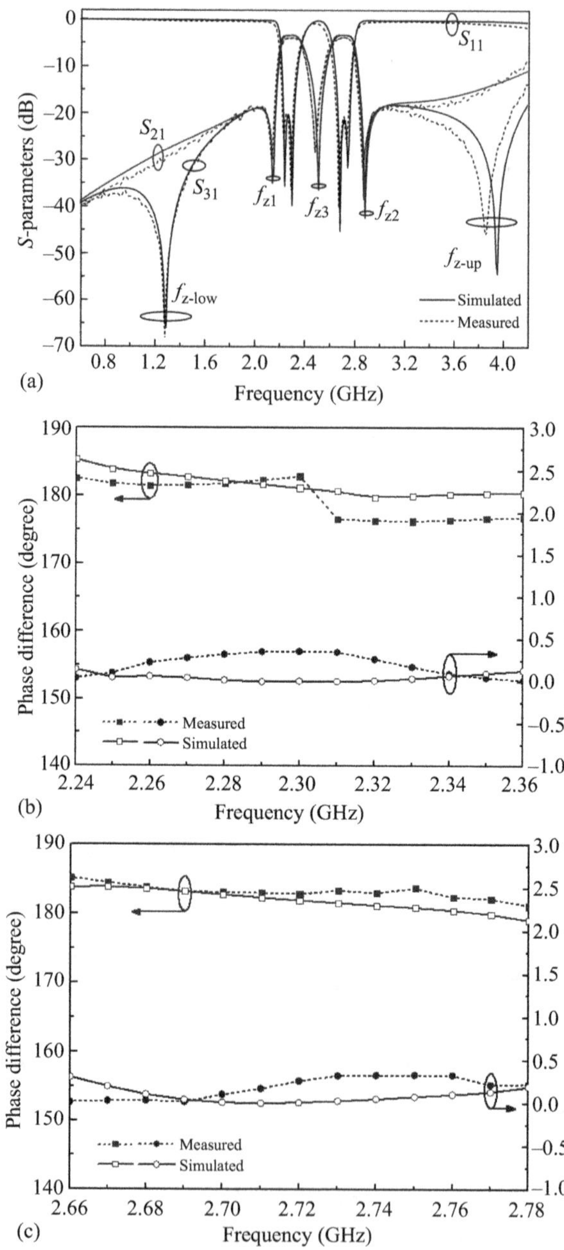

Figure 11.21 Simulated and measured results of the dual-mode dual-band balun filter: (a) S-parameters; (b) magnitude and phase differences in the first passband; and (c) magnitude and phase differences in the second passband. Reprinted with permission from Reference [20]; © 2016 IEEE

$f_{z\text{-low}}$ and $f_{z\text{-up}}$ (1.29 and 3.8 GHz) are as expected resulted from the type A coupled line structure, while f_{z_1} and f_{z_2} (2.15 and 2.89 GHz) are due to the resonance of the centrally loaded stub in the dual-mode resonator. In addition, an extra transmission zero f_{z_3} (2.52 GHz) emerges between two passbands, thus helping us a lot in out-of-band rejection enhancement. Figure 11.21(b) and Figure 11.21(c) depict the phase difference and magnitude imbalance at each operating band. It can be observed that both passbands exhibit good balanced performance with an amplitude imbalance of less than 0.34 dB and a phase imbalance of better than 4.4°.

11.5 Wideband balun bandpass filters

Following the narrow band balun filters with single- /dual-band operation introduced in the previous section, this section will deal with balun bandpass filters for wideband operation. Specifically, two planar wideband balun bandpass filters are introduced, and they are realized by utilizing multi-mode resonators [21,22] based on the above discussed coupled line balun topology. Similarly, the designed circuits are fabricated on Rogers RO4003C substrate with a thickness of 0.508 mm, a relative permittivity of 3.55, and a loss tangent of 0.0027.

11.5.1 Tri-mode balun filter

As shown in Figure 11.22, the tri-mode resonator employed in the design [21] mainly consists of a half-wavelength microstrip transmission line with a stepped-impedance stub loaded on the central plane. Meanwhile, the odd- and even-mode analysis method is utilized to analyze it due to the symmetrical topology of the resonator.

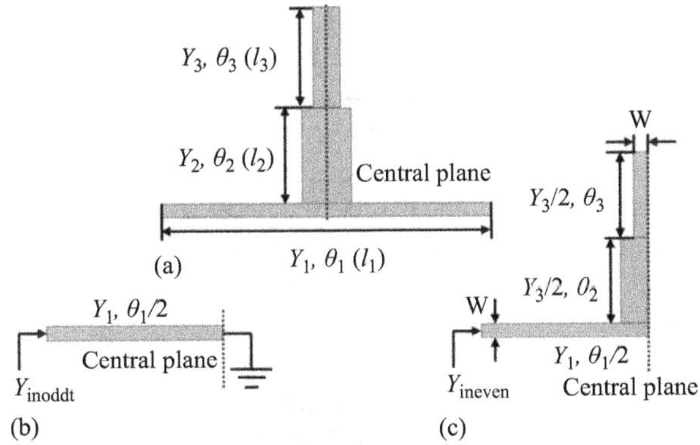

Figure 11.22 (a) Employed tri-mode resonator; (b) odd-mode equivalent circuit and (c) even-mode equivalent circuit. Reprinted with permission from Reference [21]; © 2016 IEEE

The central plane in Figure 11.22(a) is equivalent to an electric and a magnetic wall when the odd- and even-mode are excited, respectively. Therefore, the input admittance of odd- and even-mode equivalent circuit can be derived as:

$$Y_{\text{in}_{\text{odd}}} = \frac{Y_1}{j\tan\left(\frac{\theta_1}{2}\right)} \tag{11.16a}$$

$$Y_{\text{in}_{\text{even}}} = Y_1 \frac{Y_L + jY_1 \tan\left(\frac{\theta_1}{2}\right)}{Y_1 + jY_L \tan\left(\frac{\theta_1}{2}\right)} \tag{11.16b}$$

where

$$Y_L = j\frac{Y_2}{2}\frac{Y_2 \tan\theta_2 + Y_3 \tan\theta_3}{Y_2 - Y_3 \tan\theta_2 \tan\theta_3} \tag{11.16c}$$

For simplification purpose, we set $Y_3 = 2Y_1$ and $\theta = \theta_1/2 = \theta_3$ at the center frequency. By enforcing $Y_{\text{in}_{\text{odd}}} = 0$ and $Y_{\text{in}_{\text{even}}} = 0$, the odd- and even-mode resonant condition can be determined by

$$\frac{1}{j\tan\theta} = 0 \tag{11.17a}$$

$$R^2 \tan\theta_2 + 2R\tan\theta - \tan^2\theta \tan\theta_2 = 0 \tag{11.17b}$$

where $R = Y_2/Y_3$ is the admittance ratio of the stepped impedance stub. Based on equation (11.17a), the odd-mode resonant frequency can be solved as $\theta = \pi/2$. Meanwhile, assuming $K = \theta_2/\theta$, the even-mode resonant condition can be simplified as:

$$R^2 \tan K\theta + 2R\tan\theta - \tan^2\theta \tan K\theta = 0 \tag{11.18}$$

According to equation (11.18), the even-mode resonant frequency can be mathematically achieved. The relationship between the resonant frequency and varied R, K is illustrated in Figure 11.23(a). As we can see, three resonant modes including two even modes locating at both sides of the odd mode can be obtained. And it is clear that the distance of the two even-mode frequencies which mainly determines the bandwidth of the balun bandpass filter can be controlled by R and K. Based on the frequency response of the proposed resonator shown in Figure 11.23(b), two transmission zeros outside of the passband can be achieved, since a transmission zero may occur simultaneously with even mode due to the resonance of the stub loaded at the central plane.

Figure 11.24 shows the configuration of the proposed wideband balun bandpass filter. As we can see, the proposal consists of two identical above discussed tri-mode resonators, a pair of balanced outputs, and an open-circuited $\lambda/2$ transmission line excited by an unbalanced port. Figure 11.25 shows the coupling scheme of the designed wideband balun filter. According to the scheme, three independent coupling paths for the tri-mode resonators can be concluded. To apply our proposed technique, a high selectivity wideband tri-mode balun filter with a center frequency

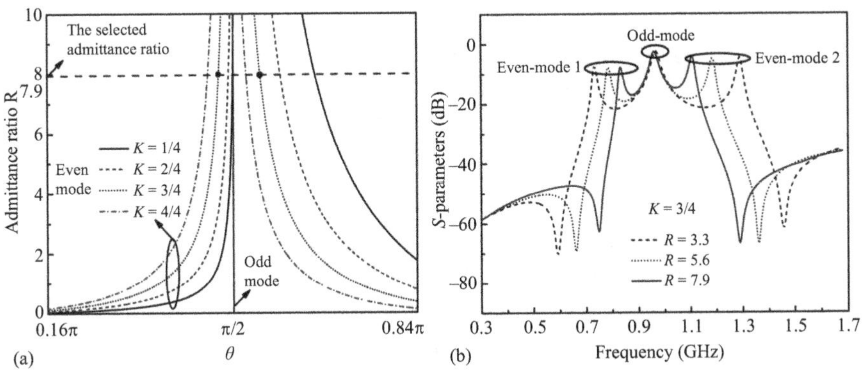

Figure 11.23 Resonance properties of the tri-mode resonator: (a) solved resonant frequency with varied K and R and (b) frequency response of the tri-mode resonator with varied R. Reprinted with permission from Reference [21]; © 2016 IEEE

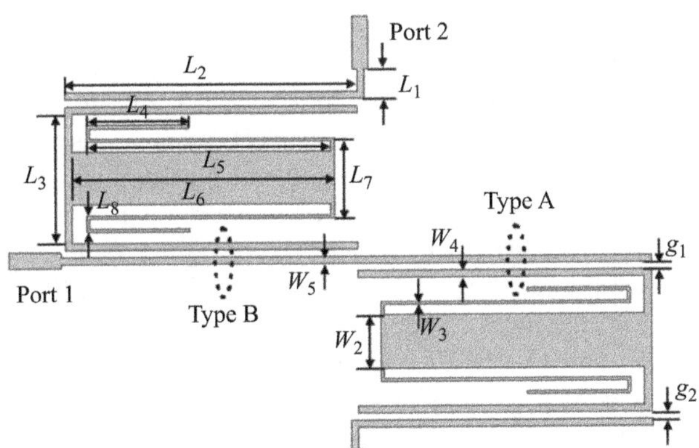

Figure 11.24 Layout of the tri-mode wideband balun filter. Reprinted with permission from Reference [21]; © 2016 IEEE

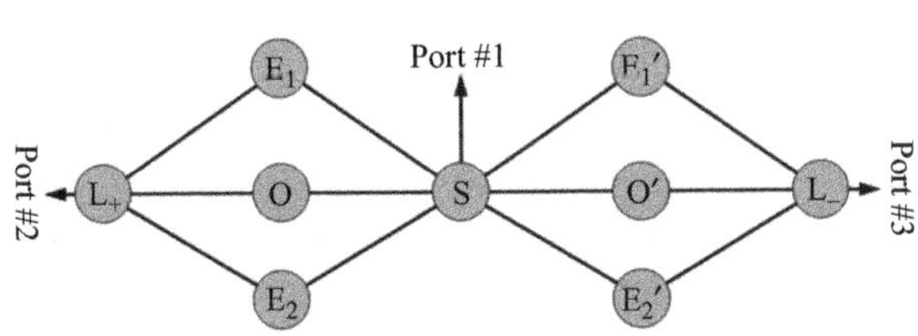

Figure 11.25 Coupling scheme of the proposed wideband balun bandpass filter. Reprinted with permission from Reference [21]; © 2016 IEEE

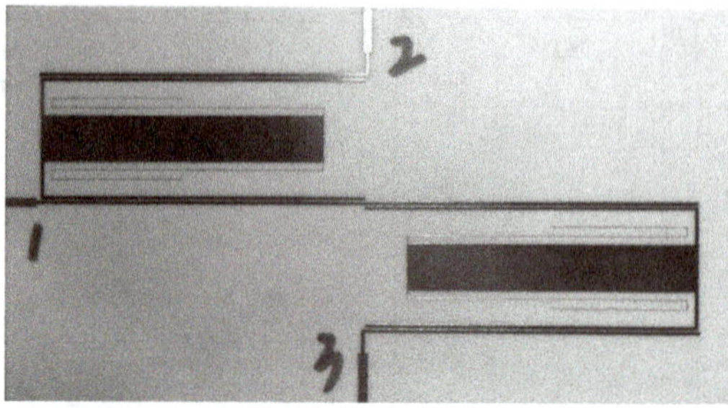

Figure 11.26 Photograph of the fabricated tri-mode wideband balun bandpass filter. Reprinted with permission from Reference [21]; © 2016 IEEE

of 1.0 GHz, a 3 dB fractional bandwidth of 30%, and a good balance performance is designed for show.

The fabricated circuit is shown in Figure 11.26. Here, $K = 3/4$ and $R = 7.9$ are selected to demonstrate the feasibility of the balun filter. The geometric parameters are shown as follows: $L_1 = 3$ mm, $W_1 = 1.2$ mm, $L_2 = 42.1$ mm, $W_2 = 6$ mm, $L_3 = 14.5$ mm, $W_3 = 0.125$ mm, $L_4 = 17$ mm, $W_4 = 0.5$ mm, $L_5 = 35.1$ mm, $W_5 = 0.5$ mm, $L_6 = 36$ mm, $L_7 = 7.8$ mm, $L_8 = 1.3$ mm, $g_1 = 0.1$ mm, $g_2 = 0.1$ mm. Meanwhile, the overall circuit size is $0.586\lambda_g \times 0.317\lambda_g$, where λ_g is the guided wavelength of the center frequency. The simulation and measurement results including the scattering parameters, phase difference, and amplitude imbalance are shown in Figure 11.27. The proposed wideband balun filter operating at the central frequency of 1 GHz with a 3 dB fractional bandwidth of 30% exhibits three transmission poles in the passband. The measured in-band insertion loss is less than 0.8 dB and the return loss is better than 19 dB. Two common transmission zeros located at 0.79 and 1.36 GHz outside the passbands of S_{21} and S_{31} can be clearly observed from the results. Similarly, another two transmission zeros located at 0.5 and 1.61 GHz of S_{21} are the inherent transmission zeros caused by the type A prototype. Besides, the measured amplitude imbalance and phase difference are shown in Figure 11.27(b). It can be concluded that the phase difference between two balanced ports is within $180 \pm 5°$ and the amplitude imbalance is less than 0.35 dB in the passband. These indicate that the proposed wideband balun filter has a good phase, amplitude balance, and a high selectivity filtering performance.

11.5.2 Quad-mode balun filter for dipole antenna application

In order to extend the applications of the proposed balun topology, a new wideband balun bandpass filter is achieved herein by adopting quadruple-mode resonators. In addition, the designed quad-mode balun filter is used to feed a dipole antenna for practical application [22].

The quadruple-mode resonator employed in the design is shown in Figure 11.28(a), and it is symmetrical with respect to its central plane. The even-odd mode method is

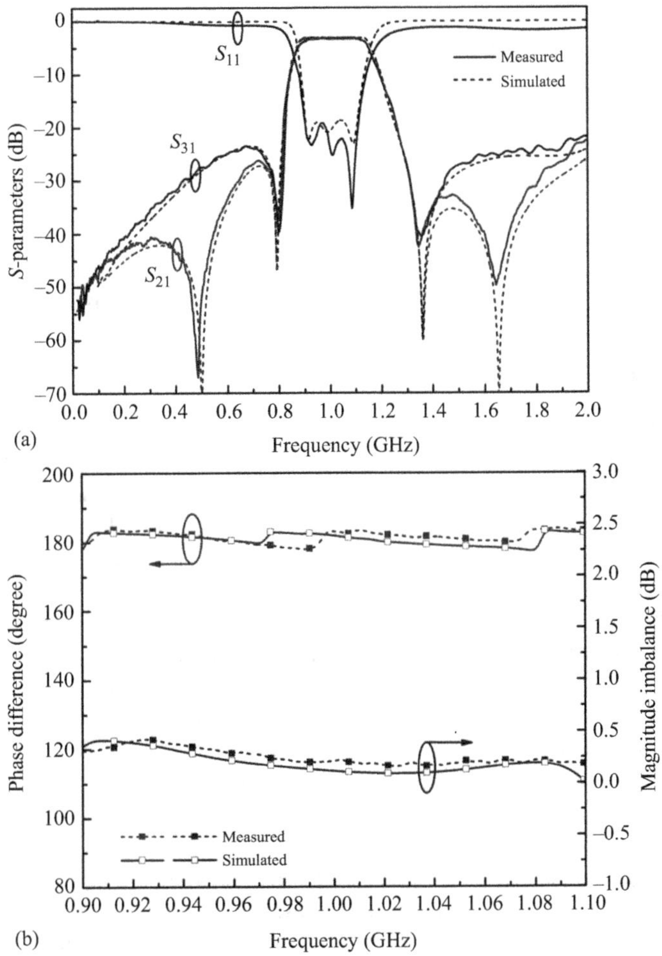

Figure 11.27 *Simulated and measured results of the tri-mode wideband balun filter: (a) S-parameters; and (b) magnitude and phase differences in the passband. Reprinted with permission from Reference [21]; © 2016 IEEE*

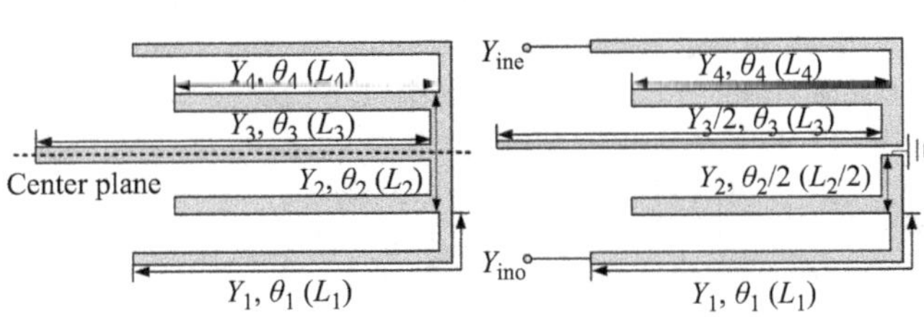

Figure 11.28 *(a) Configuration of the quadruple-mode resonator and (b) even- and odd-mode equivalent circuits. Reprinted with permission from Reference [22]; © 2016 IEEE*

applied to analyze its resonance property. Figure 11.28(b) depicts the corresponding equivalent circuits. Therefore, the input admittance of the even- and odd mode circuits can be calculated as

$$Y_{\text{in}_e} \quad \text{or} \quad Y_{\text{in}_o} = Y_1 \frac{K + jY_1 \tan\theta_1}{Y_1 + jK \tan\theta_1} \tag{11.19a}$$

where

$$K = jY_4 \tan\theta_4 - jY_2 \frac{(Y_3/2)\tan\theta_3 + Y_2 \tan(\theta_2/2)}{Y_2 - (Y_3/2)\tan\theta_3 \tan(\theta_2/2)} \quad \text{(even-mode)} \tag{11.19b}$$

$$K = jY_4 \tan\theta_4 - jY_2 \cot(\theta_2/2) \quad \text{(odd-mode)} \tag{11.19c}$$

In the meantime, two transmission zeros (TZs) can be calculated under the condition of transfer admittance $Y_{12} = 0$, i.e., $Y_{\text{in}_{\text{even}}} = Y_{\text{in}_{\text{odd}}}$. Based on this analysis, a quadruple-mode resonator can be realized and designed. The first four resonant modes are queued as even-, odd-, even-, and odd-mode, and they are expressed as f_{1e}, f_{2o}, f_{3e} and f_{4o}, respectively. For simplification, the assumption $Y_1/3 = 5Y_2/6 = 2Y_3/3 = Y_4 = Y_0$ is established in the design.

For simplification, the assumption $Y_1/3 = 5Y_2/6 = 2Y_3/3 = Y_4 = Y_0$ is established in the design. Therefore, the frequencies of first four resonant modes and two TZs versus L_1, L_2, L_3 and L_4 can be numerically derived as illustrated in Figure 11.29. Specifically, we can figure out that f_{2o} and f_{3e} could be controlled by L_1 while f_{2o} could be tuned by L_2. On the other hand, f_{1e} and TZ_1 could be controlled by L_3, while f_{3e}, f_{4o}, and TZ_2 could be adjusted by L_4. Therefore, these demonstrated properties are very useful for us to realize a balun bandpass filter with varied bandwidth and high selectivity.

According to the above analysis, a wideband balun bandpass filter can be formed up and its basic configuration is depicted in Part 1 of Figure 11.30. The corresponding coupling scheme is illustrated in Figure 11.31. Intuitively, the input signal is firstly launched by an unbalanced port (Port#1), then coupled to two quadruple-mode resonators with identical coupling magnitude with out-of-phase, and finally converted into a pair of balanced signals to be delivered to a balanced port with two terminals, i.e., Port#2 and Port#3, respectively.

For demonstration, a wideband balun filter with the central frequency of 1.65 GHz, 3-dB bandwidth of 550 MHz (fractional bandwidth of 33%), 18 dB return loss and two TZs at 1.34 and 2.07 GHz is designed. All the geometric parameters of the designed balun filter are denoted in Part 1 of Figure 11.30. Simulated and measured results of this balun filter, i.e., scattering parameters, magnitude imbalance and phase imbalance, are all plotted in Figure 11.32, and they are found in good agreement with each other. The wideband balun filter operates at the central frequency of 1.65 GHz with 3-dB FBW of about 32%. The measured minimum insertion loss is 0.6 dB while the return loss is more than 12.0 dB. Inside the desired passband, the balun filter exhibits good balanced performance with less than 0.49 dB amplitude imbalance and better than 7.2° phase imbalance. Meanwhile, we can

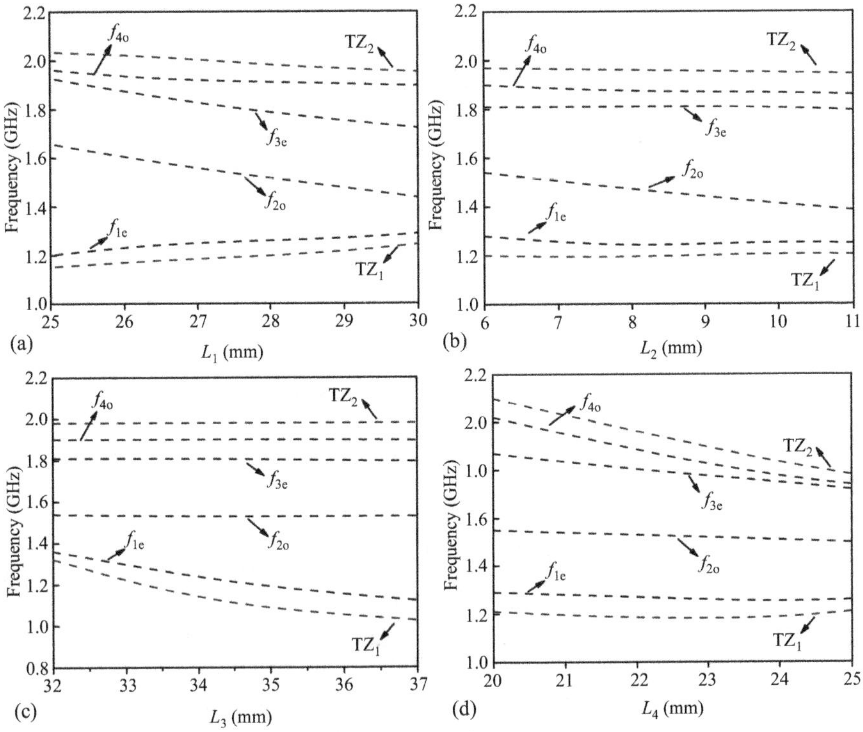

Figure 11.29 Variation of first four resonant frequencies and TZs as a function of: (a) L_1 ($L_2 = 6$, $L_3 = 34$, $L_4 = 21.6$); (b) L_2 ($L_1 = 27.4$, $L_3 = 34$, $L_4 = 21.6$); and (c) L_3 ($L_1 = 27.4$, $L_2 = 6$, $L_4 = 21.6$). (d) L_4 ($L_1 = 27.4$, $L_2 = 6$, $L_3 = 34$) (units: mm). Reprinted with permission from Reference [22]; © 2016 IEEE

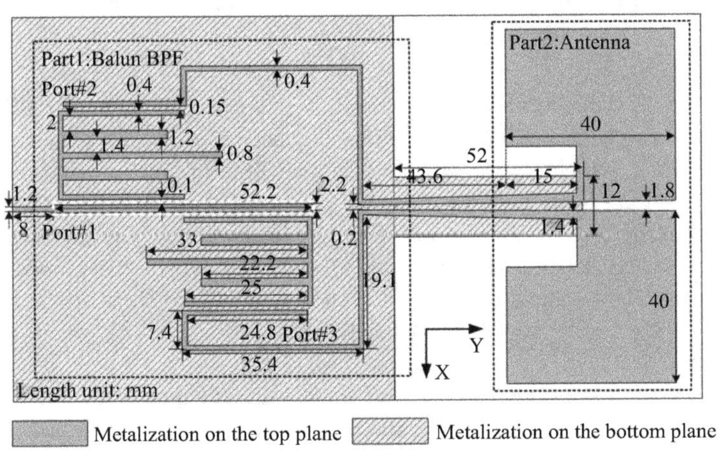

Figure 11.30 Configuration of the wideband balun filter for feeding a balanced fat dipole antenna. Reprinted with permission from Reference [22]; © 2016 IEEE

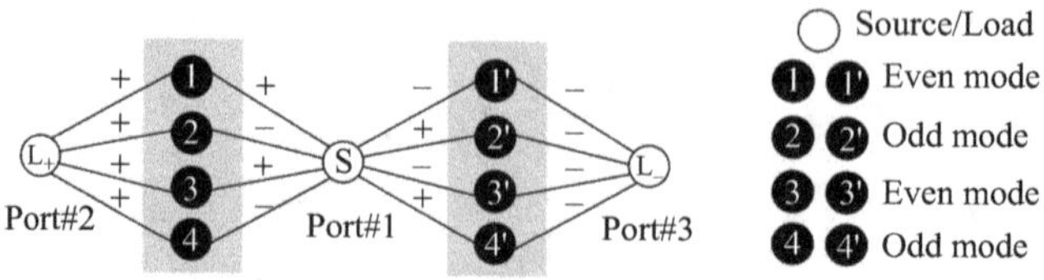

Figure 11.31 Coupling scheme for design of the quad-mode wideband balun bandpass filter. Reprinted with permission from Reference [22]; © 2016 IEEE

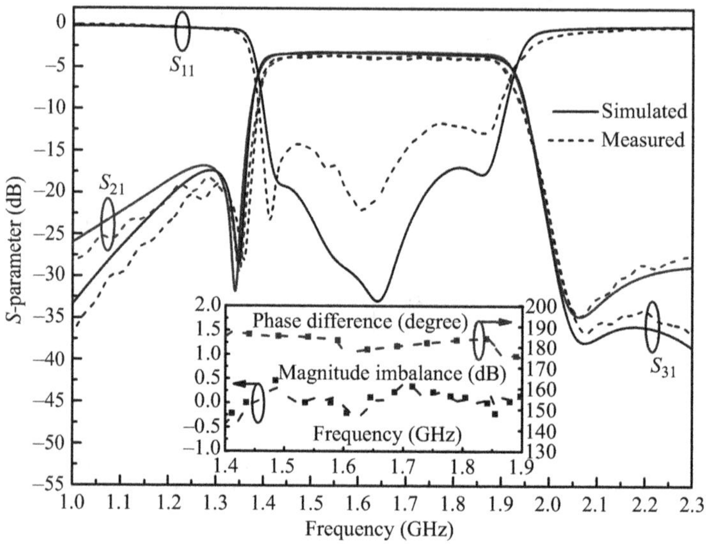

Figure 11.32 Simulated and measured results of the designed quad-mode wideband balun bandpass filter. Reprinted with permission from Reference [22]; © 2016 IEEE

see that two TZs outside the passband are located at 1.35 and 2.07 GHz, thus ensuring that the wideband balun filter has enhanced out-of-band selectivity.

Now, let us employ this wideband balun filter to feed a balanced dipole antenna with flat shape as shown in part 2 of Figure 11.30, aiming to achieve wideband radiation performance and high selectivity out of the desired band of radiation. Two sectional microstrip lines are adopted to link the proposed balun filter with the dipole antenna. The first section is a folded line with the length of about $\lambda/2$ utilized to transmit the balanced signal while keeping the 50-Ω output impedance of the balun filter unchanged. The second one is a tapered line utilized to transform the 50-Ω output impedance in the 120-Ω input balanced impedance of the flat dipole antenna. The detailed geometrical parameters are listed in Figure 11.30. The top- and bottom-view photographs of the fabricated antenna are

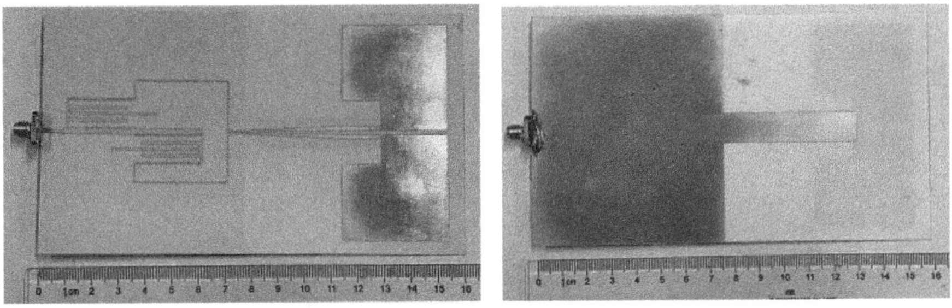

(a) (b)

Figure 11.33 Photograph of the fabricated antenna: (a) top side and (b) bottom side. Reprinted with permission from Reference [22]; © 2016 IEEE

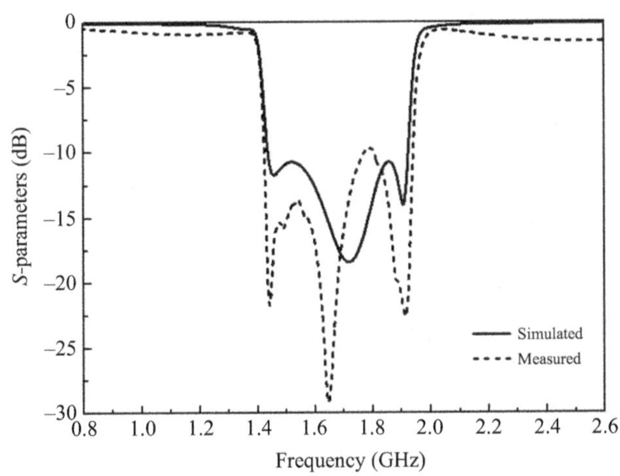

Figure 11.34 Simulated and measured reflection coefficient of balun BPF integrated dipole antenna. Reprinted with permission from Reference [22]; © 2016 IEEE

depicted in Figure 11.33. Simulated and measured reflection coefficients of the antenna are presented in Figure 11.34, and they are found to agree well with each other. The designed antenna covers the operating band from 1.4 to 1.9 GHz or the fractional bandwidth of about 30%, over which the reflection coefficient is less than −10 dB. The radiation patterns, including co- and cross-polarization, are measured in the two main planes, i.e., *E*-plane in the *XY*-plane and *H*-plane in the *YZ*-plane, as labeled in Figure 11.30. Figure 11.35 indicates that the measured radiation patterns at 1.5 and 1.8 GHz, respectively. They are exhibited to behave like quasi-omnidirectional H-plane and "8-shape" E-plane, similarly to those of an ideal dipole. The maximum radiation gain achieves 2.4 dBi in the band of interest.

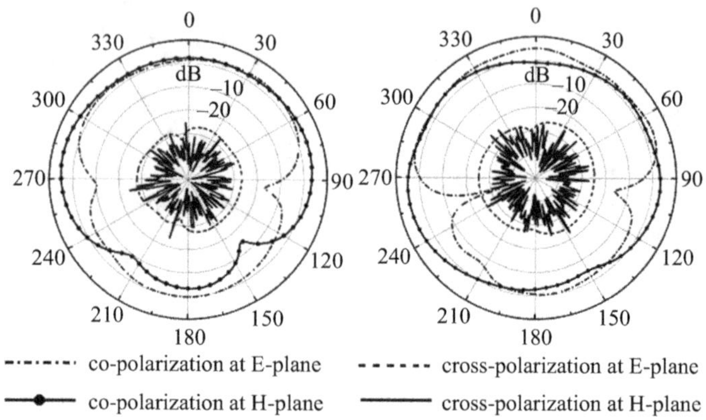

Figure 11.35 Measured radiation patterns of wideband balun filter integrated dipole antenna. Reprinted with permission from Reference [22]; © 2016 IEEE

11.6 Dual-mode balun diplexer

The balun is an essential RF module that has been broadly used in modern wireless systems while the diplexer serves as a good candidate in many multi-band and multi-service applications. When these two individual components coexist in the same front-end, the overall area of the system is bound to increase excessively. To tackle this problem, one effective approach is to integrate the balun and the diplexer into a single device, i.e., the balun diplexer. In this section, a planar balun diplexer based on stub-loaded dual-mode resonators is presented [27]. The proposed balun diplexer is a highly integrated design, which is able to provide the functions of unbalanced-to-balanced signal conversion, frequency selectivity, dual-band operation, port-to-port isolation, and common-mode suppression simultaneously.

Figure 11.36 shows the physical layout of the proposed balun diplexer. Basically, the circuit consists of two balun filter channels with a common input feed line, and each balun filter channel is implemented based on a pair of stub-loaded resonators (A and A′ or B and B′). As is shown in Figure 11.36, the employed each pair of resonators (A and A′ or B and B′) is rotationally symmetrical with regard to the center of the common input feed line and has to operate simultaneously at the same frequency. Note that the offset S is designed to reduce the capacitive coupling between the two adjacent open ends. Besides, owing to the direct connection of two dual-mode balun filters, no additional matching networks are demanded at the common input port, leading to a significant size reduction. In design, we can tune the length of the loaded stub of the resonator to properly adjust the even-mode frequency f_{even} while maintaining the odd-mode frequency f_{odd} fixed. This property can be utilized to achieve a filtering circuit, such as a balun diplexer to be discussed herein, with a certain bandwidth and two transmission poles in each operating band.

Figure 11.37 depicts the coupling scheme of this balun diplexer, where Port 1 is the unbalanced input port while Port $i+$ and $i-$ ($i = 2, 3$) are two pairs of

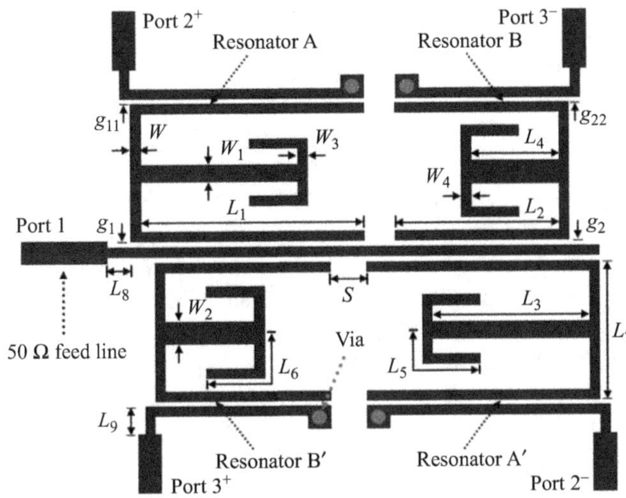

Figure 11.36 Configuration of the proposed balun diplexer. Reprinted with permission from Reference [27]; © 2016 IEEE

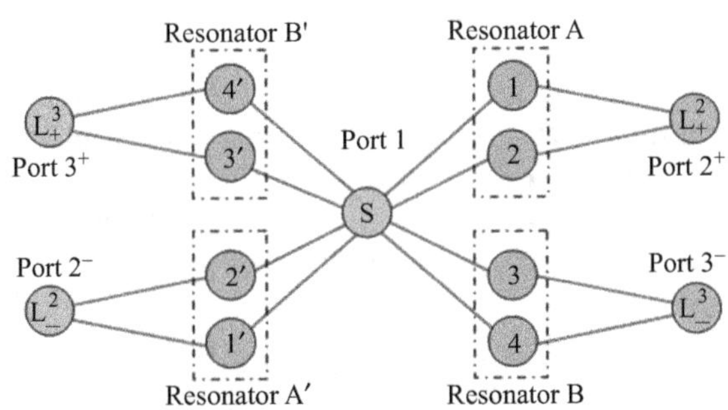

Figure 11.37 Coupling and routing scheme of proposed balun diplexer. Reprinted with permission from Reference [27]; © 2016 IEEE

balanced output ports. As illustrated, nodes 1 and 2 (1′ and 2′) represent the even and odd resonant modes of the resonator A (resonator A′), and they are utilized to form up the lower channel. Meanwhile, nodes 3 and 4 (3′ and 4′) denote those resonant modes of the resonator B (resonator B′), and they make up the upper channel. When the input signal is launched by the unbalanced Port 1, it will be simultaneously coupled to the resonators A and A′ (B and B′) with identical magnitude and contrary phase, and then as expected, translated into differential signals exported by two pairs of balanced output ports.

According to the design equation (11.13), the operation frequency of each channel can be individually controlled firstly. On the other hand, the input/output (I/O) coupling strength can be determined by fine tuning the gap between feed lines and resonators, thus to achieve the desired bandwidth in each balun filter channel design. In addition, final optimization of the entire circuit layout is executed. For verification, a prototype balun diplexer operating at 2.30 and 2.73 GHz is designed and fabricated on Rogers RO4003C substrate with a dielectric constant of 3.55, a loss tangent of 0.0027, and a thickness of 0.508 mm. Realizable physical dimensions are determined as: $L_1 = 17$, $L_2 = 13.5$, $L_3 = 12.35$, $L_4 = 7.7$, $L_5 = 5.8$, $L_6 = 7.25$, $L_7 = 8.9$, $L_8 = 1.6$, $L_9 = 0.95$, $W = 0.45$, $W_1 = 1$, $W_2 = 1.4$, $W_3 = 0.5$, $W_4 = 0.7$, $g_1 = 0.3$, $g_2 = 0.3$, $g_{11} = 0.2$, $g_{22} = 0.2$, $S = 1.6$, all in millimeters. The photograph of the fabricated circuit is portrayed in Figure 11.38, and the effective overall size is about $0.47\lambda_g \times 0.27\lambda_g$, where λ_g is the guided wavelength at 2.3 GHz.

Following the work in [28,29], the conversion between mixed-mode S-parameters and single-ended S-parameters is given as (11.20) and it is applied herein to characterize the balance of the balun diplexer.

$$S_{ss_{11}} = S_{11} \tag{11.20a}$$

$$S_{ds_{21}} = \frac{S_{21} - S_{31}}{\sqrt{2}} \tag{11.20b}$$

$$S_{cs_{21}} = \frac{S_{21} + S_{31}}{\sqrt{2}} \tag{11.20c}$$

where $S_{ss_{11}}$ is the return loss at unbalanced port 1, $S_{ds_{21}}$ is the two port S-parameter from unbalanced port 1 to differential-mode balanced port 2, and $S_{cs_{21}}$ is the two port S-parameter from unbalanced port 1 to common-mode balanced port 2.

Figure 11.39 displays final results of the proposed balun diplexer from both simulation and experiment. For the differential-mode operation, the tested lower channel passband is centered at 2.3 GHz with a 1 dB fractional bandwidth of 6.1%, while the upper one is located at 2.72 GHz with a 1 dB fractional bandwidth of

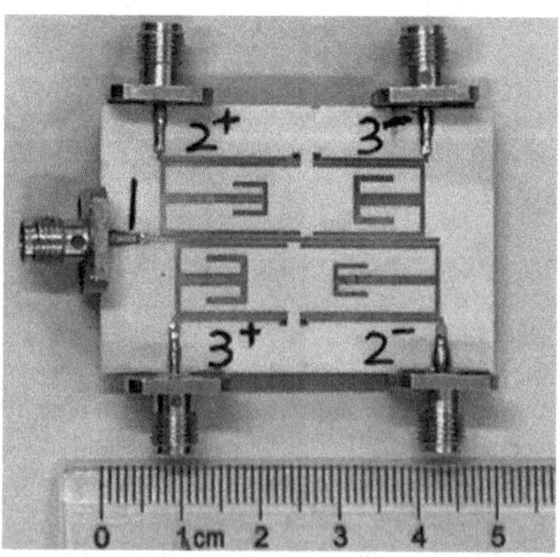

Figure 11.38 Photograph of the fabricated dual-mode balun diplexer. Reprinted with permission from Reference [27]; © 2016 IEEE

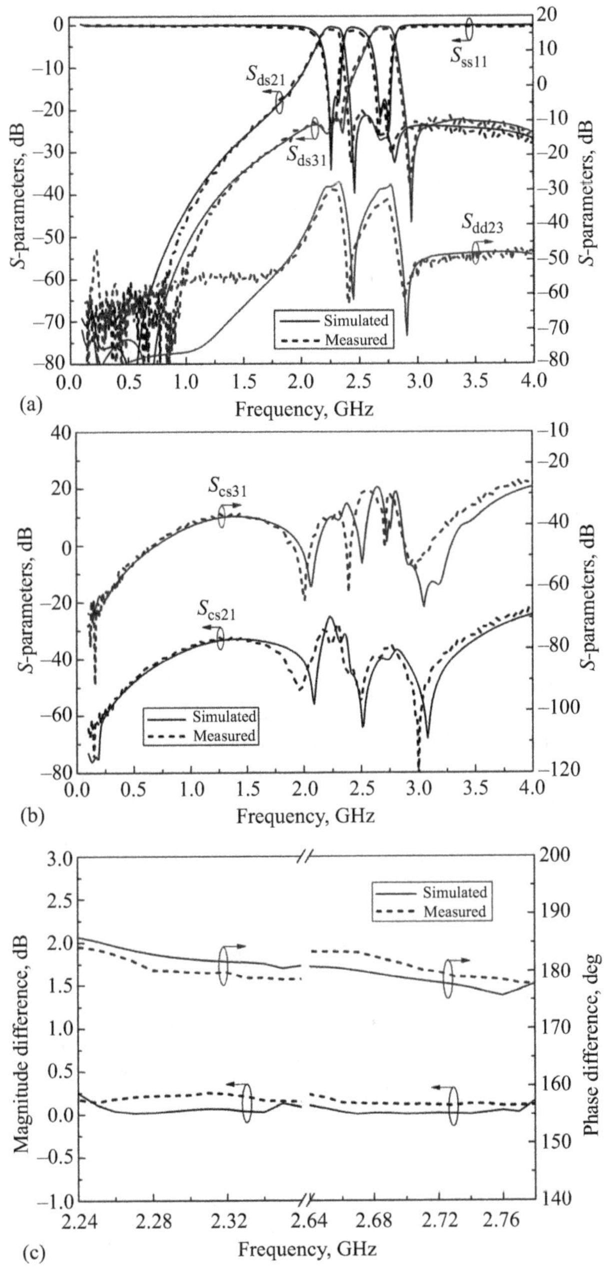

Figure 11.39 *Simulated and measured performance of the balun diplexer: (a) differential-mode response; (b) common-mode suppression; and (c) imbalance parameters in both lower/upper channel bands. Reprinted with permission from Reference [27]; © 2016 IEEE*

5.8%. Inside each passband, two transmission poles can be clearly observed. The measured insertion losses are smaller than 1.0 and 0.9 dB in the lower and upper passbands, and both return losses are better than 20 dB. Besides, the differential-mode channel isolation is better than 30 dB form 0.1 to 4.0 GHz. As shown in Figure 11.39(b), the common-mode suppression is larger than 30 dB for both differential-mode operation bands. Figure 11.39(c) showcases the phase and magnitude difference at each operating band. The displayed results indicate that both lower and upper passbands exhibit good balance performance with the amplitude imbalance of less than 0.3 dB and phase imbalance of better than 4.5°.

11.7 Transition structure-based balun bandpass filters

Different from above introduced balun bandpass filters, in this section several balun filters based on different transition structures have been introduced.

11.7.1 Balun filter using CPW-microstrip and CPS-microstrip transitions

Figure 11.40 shows the layout of the proposed transition structure based balun bandpass filter [30]. It is mainly comprised of two parts. Part I consists of two same step-impedance resonators coupling with both the unbalanced and balanced ports in the coupled-line format, which provides two outputs with equal amplitude and phase from the input. Part II is composed of two different types of transition structures, i.e., the transition structures I and II. Transition I is a microstrip-coplanar waveguide (CPW)-microstrip transition and transition II is a microstrip-coplanar stripline (CPS)-microstrip transition. Figure 11.41 shows the coupling scheme of the balun filter. We can see that when an incident signal is fed at port 1, it will be divided into two same signals by symmetrical coupling to two identical multimode

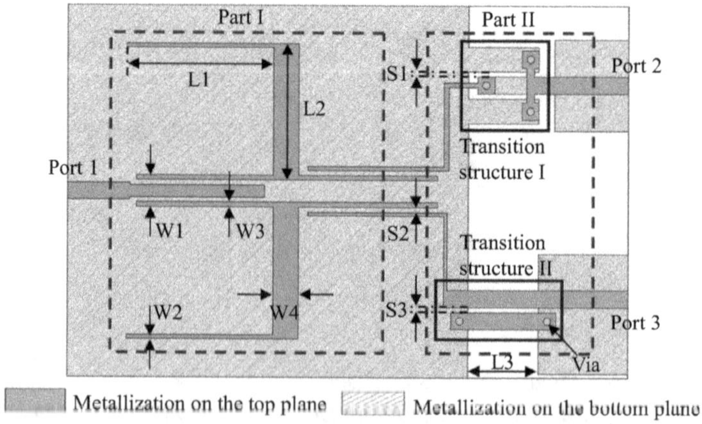

Figure 11.40 Layout of the proposed balun bandpass filter based on two transition structures. Reprinted with permission from Reference [30]; © 2015 IEEE

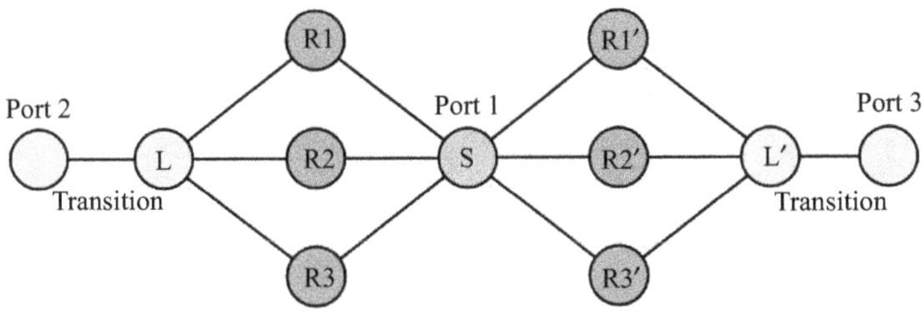

Figure 11.41 Coupling scheme of the balun bandpass filter. Reprinted with permission from Reference [30]; © 2015 IEEE

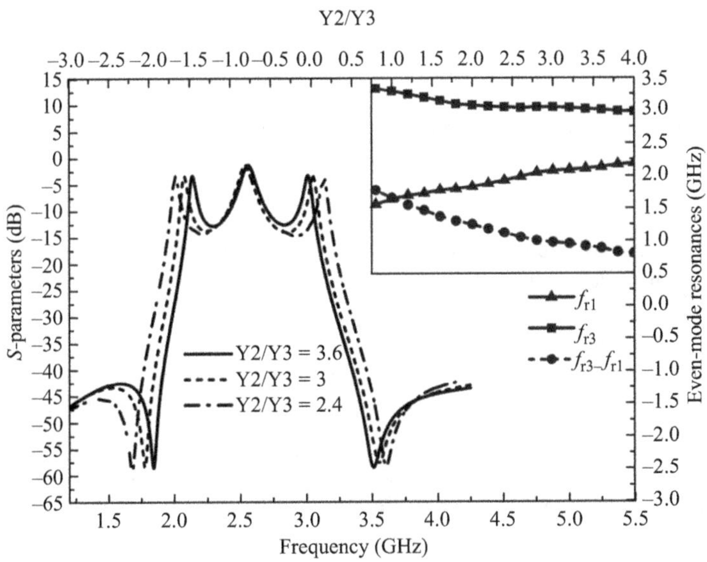

Figure 11.42 Effect of the admittance ratio $R = Y_2/Y_3$ on the resonance characteristic of the resonator under $K = 1/2$. Reprinted with permission from Reference [30]; © 2015 IEEE

stepped-impedance resonators. Then, by proper exciting of the first three resonant modes of the resonators (R1, R2, R3 and R1′, R2′, R3′ in Figure 11.41) and coupling them to the balanced ports, a tri-mode bandpass filter is realized. Finally, a balanced performance is achieved between outputs 2 and 3 when the two-way signals transmit through the transition structure I and structure II simultaneously.

The utilized multimode step-impedance resonator has been discussed in Section 11.5.1, and in this design K is readily chosen as 1/2. In this context, Figure 11.42 shows frequency response of the resonator with varied $R = Y_2/Y_3$ under $K = 1/2$. As we can see, the distance between the first two even-mode resonant frequencies f_{r1} and f_{r3} can be controlled by adjusting the admittance ratio R. This property can be utilized to achieve a proper bandwidth of the balun filter.

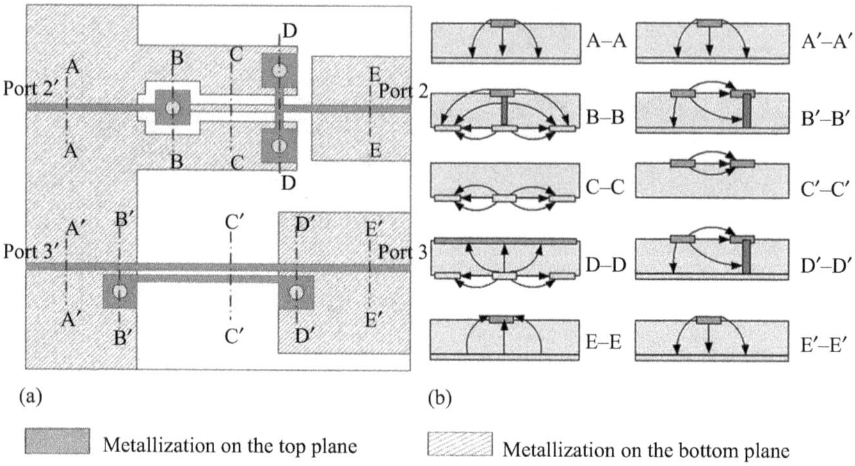

Figure 11.43 Working principle of two transition structures. Reprinted with permission from Reference [30]; © 2015 IEEE

To achieve a wideband phase and magnitude balance performance, two different types of wideband transition structures are introduced in the balun bandpass filter design. The working principles of these two transition structures are explained from the viewpoint of electric filed distributions in different cross-sections as described in Figure 11.43. When the signals in two ways are excited from the port 2′ and port 3′ with equal amplitude and phase, they will go through the transition I and the transition II, respectively. Comparing the electric field distributions in different cross-sections, we can find that the electric field lines at the A–A and A′–A′ sections are the same as the amplitude and phase of the two signals at these sections. After transmitting through different transition structures, the electric field distributions will undergo different variations as shown in Figure 11.43(b). Finally, the electric field lines at the E–E and E′–E′ sections are opposite, which means an 180° phase difference between the E–E section and E′–E′ section. Therefore, when the signals go through the two transitions, the phase difference between the signals at the two outputs will be out of phase. It should be mentioned that, due to the inherent filed characteristic of the CPW/CPS and the wideband specialty of the two transitions, the proposed scheme is able to achieve excellent phase and amplitude balance performance in the wide band.

To verify the design concept, a tri-mode balun BPF characterized with a central frequency of 2.6 GHz, bandwidth of 800 MHz and two transmission zeros at 1.86/3.55 GHz was designed on a single layer PCB. The substrate used here has a dielectric constant of 3.38, a loss tangent of 0.0027 and a thickness of 0.635 mm. The fabricated circuit is shown in Figure 11.44. The geometric parameters were determined as: $W_1 = 1.54$ mm, $W_2 = 0.2$ mm, $W_3 = 0.1$ mm, $W_4 = 2.5$ mm, $L_1 = 18.75$ mm, $L_2 = 16$ mm, $L_3 = 5.2$ mm, $S_1 = 0.26$ mm, $S_2 = 0.17$ mm, $S_3 = 0.1$ mm and the diameter of the via-hole is 0.6 mm.

Figure 11.44 Photograph of the fabricated balun bandpass filter. Reprinted with permission from Reference [30]; © 2015 IEEE

The simulated and measured results of this balun filter, i.e., scattering parameters, phase imbalance and magnitude imbalance, are plotted in Figure 11.45. As can be seen from Figure 11.45(a), the balun filter operates at the central frequency of 2.6 GHz with a fractional bandwidth of about 31%. Inside the passband, three transmission poles that are caused by the first three resonance modes of the multimode resonator can be clearly observed. The measured in-band insertion loss is smaller than 5.8 dB with the return loss better than 14 dB. Outside the passband, two transmission zeros emerge at 1.7 and 3.7 GHz for both S_{21} and S_{31}, which ensure the proposed balun filter has good frequency selectivity. As for the balance performance between the two outputs depicted in Figure 11.45(b), it can be seen that better than 0.5 dB amplitude imbalance and $180° \pm 5°$ phase difference are achieved. It should be mentioned that the small discrepancy between measured and simulated results are mainly due to the fabrication error and the minor SMA connection difference between port 2 and port 3. Overall, the proposed balun filter has good performance in phase imbalance, amplitude imbalance and frequency selectivity.

11.7.2 Balun filter using microstrip-to-slotline transitions

In this design, two different forms of microstrip-to-slotline transition structures, as shown in Figure 11.46(a), are adopted to achieve the desired balun function [31]. Transition structure I is formed by connecting the upper microstrip to the bottom slotline through shorting via. While for transition structure II, it is constructed by the upper quarter wavelength microstrip ($\lambda_m/4$) and the bottom $\lambda_s/4$ slotline on the ground for realizing maximum electromagnetic coupling. Specifically, the two upper microstrips are rotationally symmetrically arranged with respect to the center of the slotline, thus to ensure good balanced performance between the two outputs. To demonstrate the working mechanism, Figure 11.46(b) and Figure 11.46(c) depicts the electric field distributions in transition structure II. As expected, we can clearly find that when a signal is launched into the slotline from port 1, it will be split into two equal signals. Then, owing to the inherent out-of-phase electric field distributions on the adjacent grounds of slotline, signals with identical magnitude and out-of-phase can be easily attained at two output ports (port 2 and port 3).

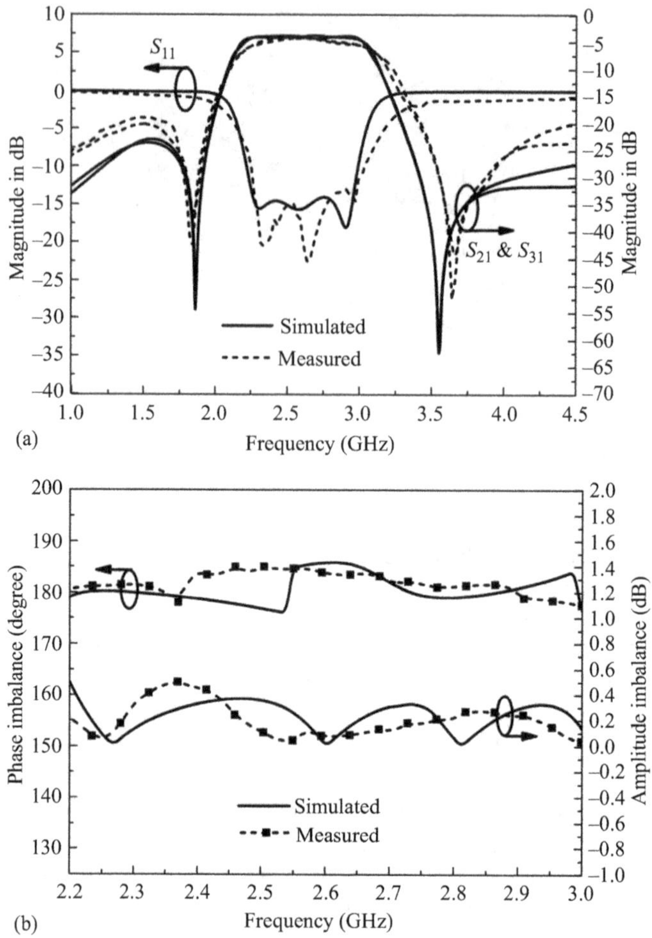

Figure 11.45 Measured and simulated S-parameters and amplitude and phase imbalance: (a) S-parameters; and (b) amplitude and phase imbalance. Reprinted with permission from Reference [30]; © 2015 IEEE

Based on aforementioned microstrip-to-slotline transition structures, a balun filter is subsequently designed by using dual-mode stub-loaded resonators. Figure 11.47 shows the configuration of the proposed balun filter. Similarly, the applied dual-mode stub-loaded resonators are bent into E-shape to reduce the circuit size. Figure 11.48 depicts the coupling scheme of the balun filter, where S, L+ or L−, and E_i or E_i' ($i = 1, 2$) stand for source, load, and two resonant modes, respectively. Note that the source S in Figure 11.48 includes two different types of microstrip-to-slotline transition structures. Port 1 is the unbalanced input while ports 2 and 3 are the balanced outputs. It is worth mentioning that once the two resonators dimensions are determined, next the input/output (I/O) coupling should

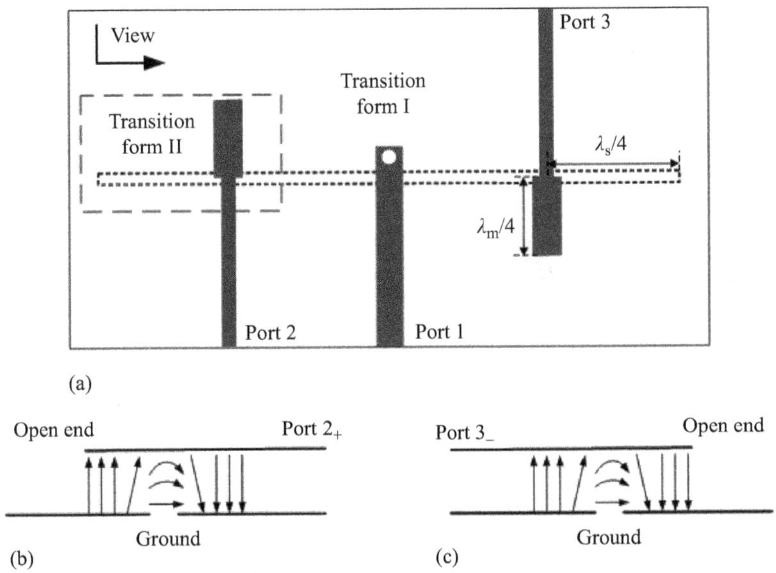

Figure 11.46 *(a) Layout of the microstrip-to-slotline transition structures; (b) and (c) left view of electric field distributions in the transition structure II. Reprinted with permission from Reference [31]; © 2018 IEEE*

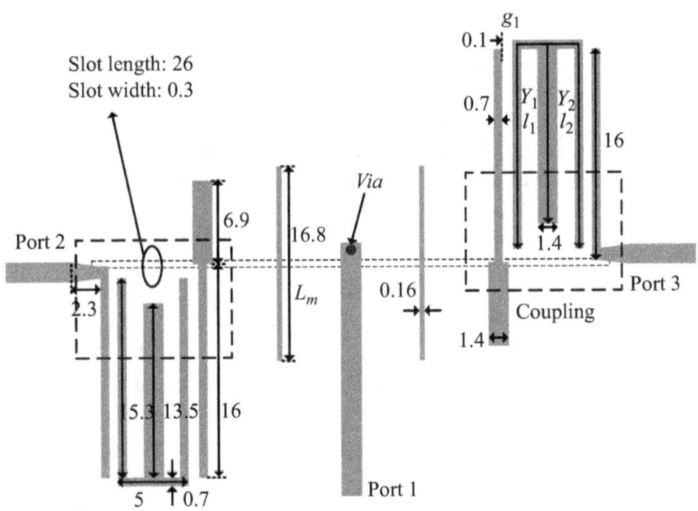

Figure 11.47 *Sketch of the proposed balun bandpass filter (unit: mm). Reprinted with permission from Reference [31]; © 2018 IEEE*

be taken into consideration to achieve the desired bandwidth. As we know, the I/O coupling is related to the external Q-factors. In the design the external Q-factors can be controlled by fine-tuning the gap (g_1) between feed lines and resonators, thus to meet the given bandwidth.

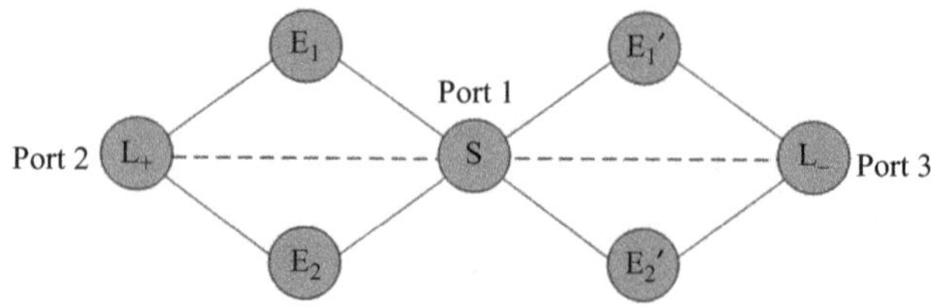

Figure 11.48 Coupling scheme of the balun filter. Reprinted with permission from Reference [31]; © 2018 IEEE

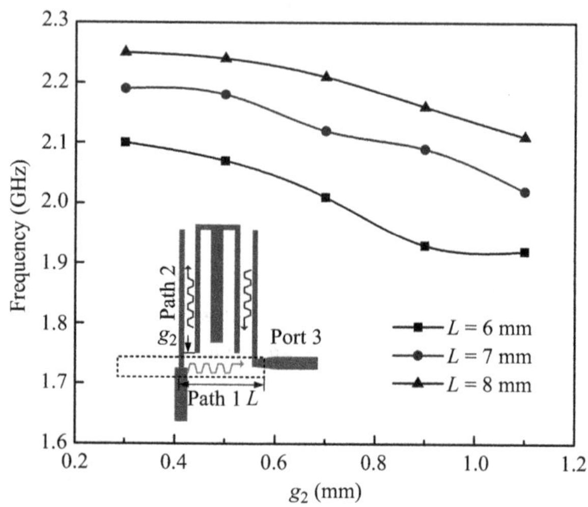

Figure 11.49 Relationship between frequency locations of TZ_1 and the coupling gap (g_2) with different slot lengths L. Reprinted with permission from Reference [31]; © 2018 IEEE

To realize high selectivity for the proposed balun filter, four TZs are introduced in the design according to different mechanisms. Above all, by wisely taking advantages of the slotline structure we can readily construct the source and load coupling scheme in the balun filter represented as path 1 in the inset of Figure 11.49. On the other side, at lower frequency outside the passband, signals from path 2 will be transmitted to port 3 in the odd mode form of the dual-mode resonator with respect to proper structure values of l_1 and l_2. Accordingly, at port 3 the phase difference of signals between path 1 and path 2 is approximately to 180°, thereby introducing the first TZ denoted by TZ_1 at lower frequency outside the passband. It should be mentioned that the frequency location of TZ_1 can be properly adjusted by changing the values of g_2 and L. For illustration, Figure 11.49 depicts the location of TZ_1 under different coupling gaps (g_2) and slotline lengths (L). In the next place, considering the utilized E-shaped dual-mode stub-loaded resonator, it will introduce the second TZ denoted by TZ_2 when the stub loaded on the central plane is at quarter-wavelength resonance. Furthermore, to produce another TZ denoted by TZ_3 outside the passband for harmonic suppression, a pair of open-circuited half-wavelength resonators is elaborately adopted in symmetrical location from the center of the slotline. As shown in Figure 11.49, the upper two identical resonators are arranged orthogonally with respect to the bottom slotline. And the resonant frequency can be derived as

$$f_{Z_3} = \frac{nc}{2L_{\mathrm{m}}\sqrt{\varepsilon_e}} \tag{11.21}$$

where c represents the speed of light in free space, ε_{e} is the effective dielectric constant, and L_{m} means the physical length of the half-wavelength resonator denoted in Figure 11.49. Thus, TZ_3 can be engendered as our expectation by changing the length of L_{m} to suppress the harmonic response. Finally, to further extend the stopband, an extra TZ represented as TZ_4 will also be achieved attributing to the inherent resonant properties of the adopted coupling topology between the feed lines and the dual-mode resonators [24]. In summary, with these introduced four TZs the proposed balun BPF are bound to exhibit high frequency selectivity performance.

For demonstration, the proposed balun filter described in Figure 11.47 is fabricated on the Rogers RO4003C substrate with a dielectric constant of 3.55, a loss tangent of 0.0027, and a thickness of 0.508 mm. The diameter of the via-hole is set as 0.3 mm. Figure 11.50 illustrate the fabricated circuit. Figure 11.51 shows the simulated and measured results, which indicate that the designed balun filter operates at the center frequency of 2.78 GHz, with a fractional bandwidth of 23.4% and two transmission poles inside the passband. The measured in-band insertion is smaller than 1.2 dB, and the return loss is better than 18 dB. Four TZs are

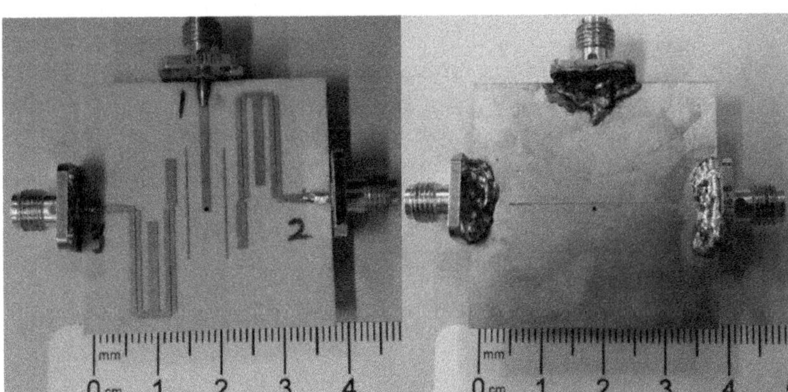

(a) (b)

Figure 11.50 Photograph of the fabricated balun bandpass filter: (a) top side and (b) bottom side. Reprinted with permission from Reference [31]; © 2018 IEEE

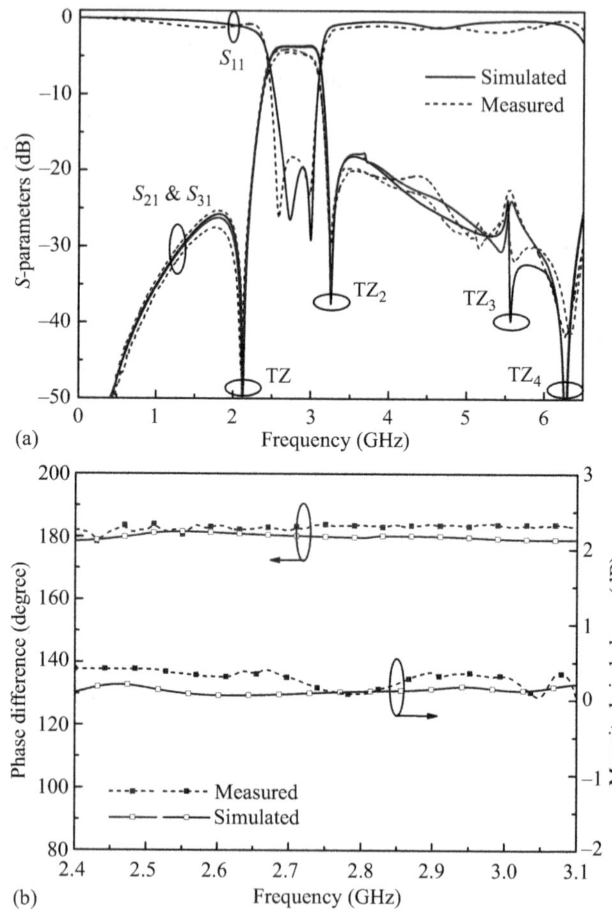

Figure 11.51 Measured and simulation results of the balun filter: (a) magnitude responses and (b) amplitude and phase differences inside the passband. Reprinted with permission from Reference [31]; © 2018 IEEE

generated, respectively, at 2.12, 3.25, 5.57, and 6.28 GHz, engendering sharp roll-off skirt and out-of-band harmonic suppression. As shown in Figure 11.51(b), the phase difference between two balanced ports is within $180 \pm 5°$ and the magnitude imbalance is less than 0.5 dB.

11.7.3 Ultra wideband balun filter using broadside-coupled structure

By using broadside-coupled structure, a new balun bandpass filter is introduced here for ultra wideband operation based on the conventional Marchand-type balun prototype [32]. Figure 11.52 shows the topology of the traditional Marchand balun.

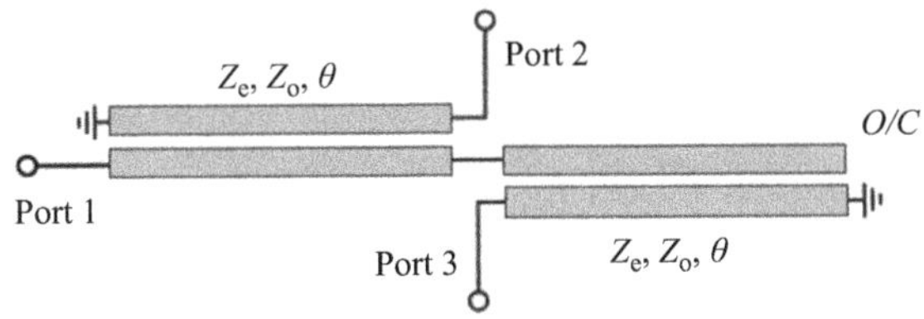

Figure 11.52 Conventional Marchand balun with two parallel microstrip coupled line sections. Reprinted with permission from Reference [32]; © 2017 IEEE

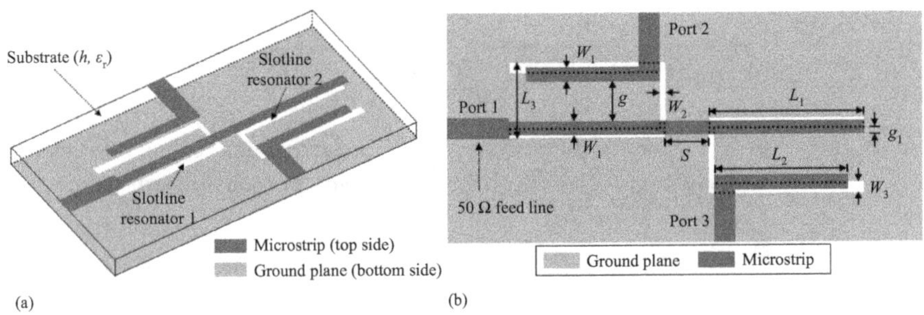

Figure 11.53 Configuration of the proposed wideband balun filter: (a) 3-D view and (b) top view. Reprinted with permission from Reference [32]; © 2017 IEEE

It is composed of two parallel microstrip coupled-line sections. As stated in [33], in order to achieve wide bandwidth, the coupling coefficient of the coupled-line should be set as a high value which means that the tight coupling gap is required. However, such strong coupling can hardly be realized between the parallel coupled microstrip lines in conventional planar PCB fabrication process.

In order to solve the problem, the broadside coupling structure between microstrip line and slotline resonator is proposed herein to achieve the desired coupling strength for wideband operation. In this context, a new balun bandpass filter with wideband performance is presented. Figure 11.53 depicts the 3-D view and top view of the proposed balun filter. As we can observe, the wideband balun filter is constructed on a single dielectric substrate with a thickness of h and a dielectric constant of ε_r. Two identical half-wavelength short-circuited slotline resonators are etched in the bottom ground plane of the substrate. Note that these two slotline resonators are both folded into a U-shaped configuration for size compactness with a horizontal distance of S. Besides, on the top side of the substrate, three microstrip structures serve as single-ended input feed line and balanced output feed lines through the broadside-coupling to the slotline resonators. More intuitively, when the unbalanced signal is transmitted into port 1, it will be simultaneously coupled to the slotline resonators 1 and 2 with identical magnitude and out-of-phase through the dielectric substrate. Therefore, the desired differential

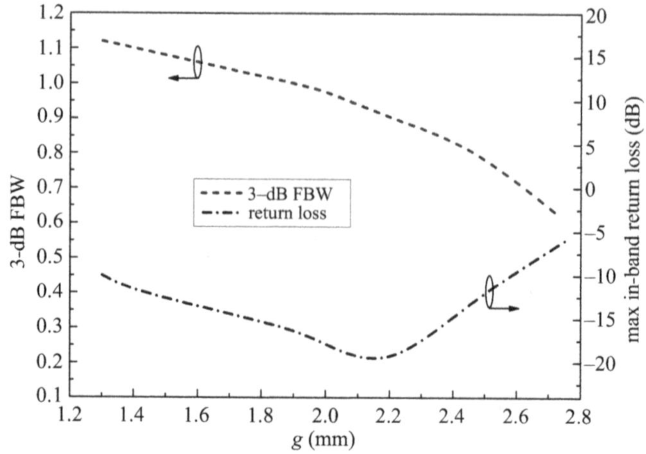

Figure 11.54 3-dB fractional bandwidth (FBW) and maximum in-band return loss versus the gap spacing g. Reprinted with permission from Reference [32]; © 2017 IEEE

signal would be achieved through the broadside-coupling again from slotline resonators to two balanced output ports.

Figure 11.54 shows the simulated 3-dB fractional bandwidth (FBW) and the maximum in-band return loss of the balun BPF against the parameter g as illustrated in Figure 11.53(b). The investigations are carried out on Rogers RO4003C substrate ($\varepsilon_r = 3.55$, tan $\delta = 0.0027$) with $h = 0.508$ mm. As we can observe, the performance of FBW tends to diminish when the distance g gets increased. Besides, we can also conclude that the circuit performs a low in-band return loss near $g =$ 2.1 mm. It should also be mentioned that the FBW and passband return loss can be further optimized by adjusting the other parameters, such as W_1, W_2, W_3, S, and g_1.

Based on the discussions mentioned above, a prototype wideband balun filter operating at 3.75 GHz is designed and fabricated. Final dimensions noted in Figure 11.53(b) are as: $L_1 = 13.5$, $L_2 = 12$, $L_3 = 4$, $W_1 = 0.7$, $W_2 = 0.1$, $W_3 = 0.5$, $g_1 =$ 0.35, $g = 2.15$, $S = 1.8$, all in millimeter. The photographs of the fabricated circuit are displayed in Figure 11.55, and its overall size is about $0.3\lambda_g \times 0.75\lambda_g$, where λ_g is the guided wavelength at the central frequency.

Figure 11.56 delineates the simulated and measured results of the designed wideband balun filter including the S-parameters, phase difference, and magnitude imbalance. The balun filter operates at 3.75 GHz with a 3-dB FBW of 93%. Inside the passband, three transmission poles which can be attributed to the half-wavelength slotline resonator and input/output microstrip transmission lines are clearly observed. The measured minimum in-band insertion loss is about 0.9 dB, while the return loss is better than 24 dB across a wideband frequency range. In addition, transmission zeros outside the passband are due to the cross-coupling between input and output microstrip feed lines. We can observe from Figure 11.56(b) that

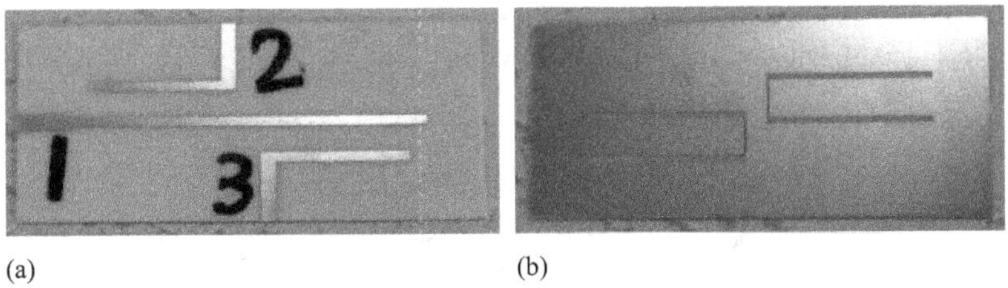

Figure 11.55 *Photograph of the fabricated wideband balun filter: (a) top side and (b) bottom side. Reprinted with permission from Reference [32]; © 2017 IEEE*

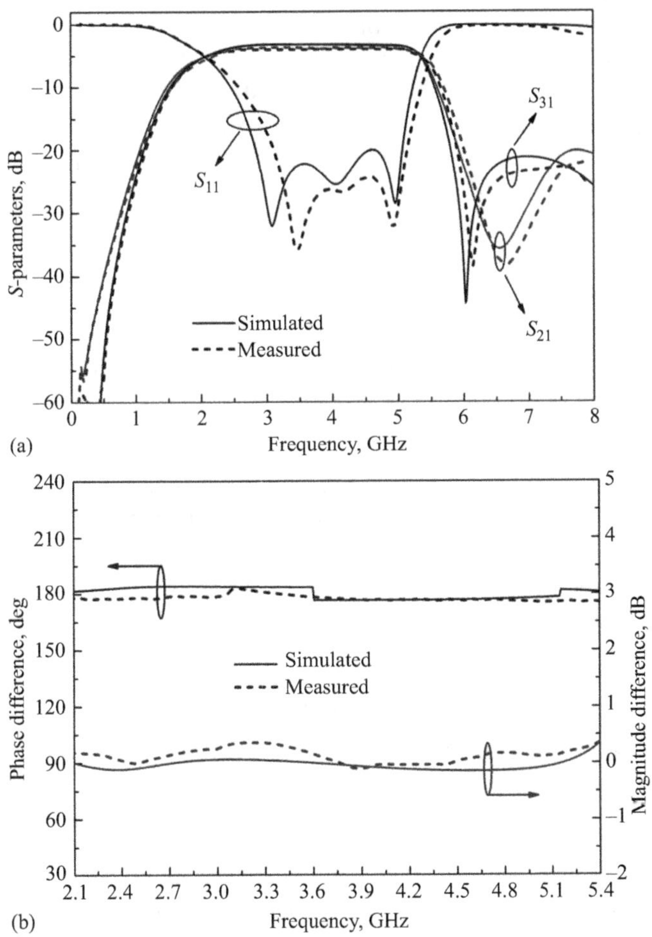

Figure 11.56 *Simulated and measured performance of the ultra-wideband balun filter: (a) S-parameters and (b) phase and magnitude differences. Reprinted with permission from Reference [32]; © 2017 IEEE*

the phase difference is better than 3.5°, and the magnitude imbalance is less than 0.34 dB over the entire operating band from 2.1 to 5.4 GHz.

11.8 Summary

In this chapter, a new coupled line balun topology is presented by analyzing the standing wave pattern around the open-circuited half-wavelength transmission line. Then, a class of planar-type balun bandpass filters with single-/dual-band and wideband responses is introduced based on this balun topology. This method to design balun filters is simple and effective with widely used multi-mode resonators. To further extend the application of the design approach, a balun diplexer is briefly introduced by using stub-loaded dual-mode resonators. What is more, some other planar balun bandpass filters based on different transition structures are also discussed in the end. All these designed circuits are verified by both simulation and experiment, and measured results indicate that the balun filters exhibit not only good balance performance, but also show high frequency selectivity. In a nutshell, the design method in this chapter provides us some new concepts for the implementation of balun bandpass filters.

References

[1] Y. C. Leong, K. S. Ang, and C. H. Lee, "A derivation of a class of 3-port baluns from symmetrical 4-port networks," in *IEEE MTT-S Int. Microw. Symp. Dig.*, Jun. 2002, pp. 1165–1168.

[2] L. K. Yeung and K.-L. Wu, "A dual band coupled-line balun filter," *IEEE Trans. Microw. Theory Techn.*, vol. 55, no. 11, pp. 2406–2411, 2007.

[3] C.-Y. Liou and S.-G. Mao, "Triple-band Marchand balun filter using coupled-line admittance inverter technique," *IEEE Trans. Microw. Theory Techn.*, vol. 61, no. 11, pp. 3846–3852, 2013.

[4] G.-S. Huang, C.-H. Wu, and C. H. Chen, "LTCC balun bandpass filters using dual-response resonators," *IEEE Microw. Wireless Compon. Lett.*, vol. 21, no. 9, pp. 483–485, 2011.

[5] C. L. Tsai and Y. S. Lin, "Analysis and design of new single-to- balanced multi coupled line bandpass filters using low-temperature co-fired ceramic technology," *IEEE Trans. Microw. Theory Techn.*, vol. 56, no. 12, pp. 2902–2912, 2008.

[6] T. Yang, P.-L. Chi, and T. Itoh, "Lumped isolation circuits for improvement of matching and isolation in three-port balun band-pass filter," in *IEEE MTT-S Int. Microw. Symp. Dig.*, May 2010, pp. 584–587.

[7] C.-H. Wu, C.-H. Wang, S.-Y. Chen, and C. H. Chen, "Balanced-to-unbalanced bandpass filters and the antenna application," *IEEE Trans. Microw. Theory Techn.*, vol. 56, no. 11, pp. 2474–2482, 2008.

[8] W. J. Feng and W. Q. Che, "Wideband balun bandpass filter based on a differential circuit," in *IEEE MTT-S Int. Microw. Symp. Dig.*, June 2012, pp. 1–3.

[9] G.-S. Huang and C. H. Chen, "Dual-band balun bandpass filter with hybrid structure," *IEEE Microw. Wireless Compon. Lett.*, vol. 21, no. 7, pp. 356–358, 2011.

[10] E.-Y. Jung and H.-Y. Hwang, "A balun-BPF using a dual-mode ring resonator," *IEEE Microw. Wireless Compon. Lett.*, vol. 17, no. 9, pp. 652–654, 2007.

[11] P. Cheong, T.-S. Lv, W.-W. Choi, and K.-W. Tam, "A compact microstrip square-loop dual-mode balun-bandpass filter with simultaneous spurious response suppression and differential performance improvement," *IEEE Microw. Wireless Compon. Lett.*, vol. 21, no. 2, pp. 77–79, 2011.

[12] S. S. Gao and S. Sun, "Compact dual-mode balun bandpass filter with improved upper stopband performance," *Electron. Lett.*, vol. 47, no. 23, pp. 1281–1283, 2011.

[13] S. Sun and W. Menzel, "Novel dual-mode balun bandpass filters using single cross-slotted patch resonator," *IEEE Microw. Wireless Compon. Lett.*, vol. 21, no. 8, pp. 415–417, 2011.

[14] G.-Q. Zhang, J.-X. Chen, J. Shi, H. Tang, H. Chu, and Z.-H. Bao, "Design of multilayer balun filter with independently controllable dual passbands," *IEEE Microw. Wireless Compon. Lett.*, vol. 25, no. 1, pp. 10–12, 2015.

[15] Y. L. Chen and H. H. Lin, "Novel broadband planar balun using multiple coupled lines," in *IEEE MTT-S Int. Microw. Symp. Dig.*, June 2006, pp. 1571–1574.

[16] D. Chen, L. Zhu, H. Bu, and C. Cheng, "A wideband balun filter on a triple-mode slotline resonator with controllable bandwidth," *IEEE Microw. Wireless Compon. Lett.*, vol. 27, no. 6, pp. 569–571, 2017.

[17] L. P. Feng and L. Zhu, "Wideband filtering balun on a novel hybrid multi-mode resonator with the functionality of vertical transition," *IEEE Trans. Compon., Packag. Manuf. Technol.*, vol. 7, no. 8, pp. 1324–1330, 2017.

[18] L. S. Wu, Y. X. Guo, J. F. Mao, and W. Y. Yin, "Design of a substrate integrated waveguide balun filter based on three-port coupled-resonator circuit model," *IEEE Microw. Wireless Compon. Lett.*, vol. 21, no. 5, pp. 252–254, May 2011.

[19] F. Huang, J. Wang, and L. Zhu, "A new approach to design a microstrip dual-mode balun bandpass filter," *IEEE Microw. Wireless Compon. Lett.*, vol. 26, no. 4, pp. 252–254, 2016.

[20] J. Wang, F. Huang, L. Zhu, C. Cai, and W. Wu, "Study of a new planar-type balun topology for application in the design of balun bandpass filters," *IEEE Trans. Microw. Theory Techn.*, vol. 64, no. 9, pp. 2824–2832, 2016.

[21] C. Cai, J. Wang, L. Zhu, and W. Wu, "A new approach to design microstrip wideband balun bandpass filter," *IEEE Microw. Wireless Compon. Lett.*, vol. 26, no. 2, pp. 116–118, 2016.

[22] G. Zhang, J. Wang, and W. Wu, "Wideband balun bandpass filter explored for a balanced dipole antenna with high selectivity," *Electron. Lett.*, vol. 52, no. 13, pp. 1153–1155, 2016.

[23] R. K. Mongia, I. J. Bahl, P. Bhartia, and J. Hong, *RF and Microwave Coupled-Line Circuits*. Norwood, MA: Artech House, 1999.

[24] D. M. Pozar, *Microwave Engineering*, Third ed. New York, NY, USA: Wiley, 2005.

[25] X. H. Guan, F. Q. Yang, H. W. Liu, and L. Zhu, "Compact and high-isolation diplexer using dual-mode stub-loaded resonators," *IEEE Microw. Wireless Compon. Lett.*, vol. 24, no. 6, pp. 385–387, 2014.

[26] J.-S. Hong and M. J. Lancaster, *Microstrip Filters for RF/Microwave Applications*. New York, NJ, USA: Wiley, 2001.

[27] F. Huang, J. Wang, L. Zhu, and W. Wu, "Compact microstrip balun diplexer using stub-loaded dual-mode resonators," *Electron. Lett.*, vol. 52, no. 24, pp. 1994–1996, 2016.

[28] Q. Xue, J. Shi, and J.-X. Chen, "Unbalanced-to-balanced and balancedto-unbalanced diplexer with high selectivity and common-mode suppression," *IEEE Trans. Microw. Theory Tech.*, vol. 59, no. 11, pp. 2848–2855, 2011.

[29] C.-M. Chen, S.-J. Chang, C.-F. Yang, and C.-Y. Chen, "A simple and effective method for designing frequency adjustable balun diplexer with high common-mode suppression," *IEEE Microw. Wireless Compon. Lett.*, vol. 25, no. 7, pp. 433–435, 2015.

[30] S. Lin, J. Wang, G. Zhang, and J. Hong, "Design of microstrip tri-mode balun bandpass filter with high selectivity," *Electronics Letters*, vol. 51, no. 13, pp. 998–999, 2015.

[31] H. Xu, J. Wang, L. Zhu, F. Huang, and W. Wu, "Design of a dual-mode balun bandpass filter with high selectivity," *IEEE Microw. Wireless Compon. Lett.*, vol. 28, no. 1, pp. 22–24, 2018.

[32] F. Huang, J. Wang, J. Hong, and W. Wu, "Wideband balun bandpass filter with broadside-coupled microstrip/slotline resonator structure," *Electron. Lett.*, vol. 53, no. 19, pp. 1320–1321, 2017.

[33] K. S. Ang, Y. C. Leong, and C. H. Lee, "Multisection impedance transforming coupled-line baluns," *IEEE Trans. Microw. Theory Tech.*, vol. 51, no. 2, pp. 536–541, 2003.

Chapter 12

Millimetre wave SIW diplexer with relaxed fabrication tolerances

J. Ross Aitken[1,2] *and Jiasheng Hong*[1]

12.1 Introduction

There is a lot of information regarding substrate-integrated waveguide (SIW) components such as filters and diplexers below millimetre wave frequencies in the literature [1]. However, when considering millimetre wave SIW diplexers, there is only a limited amount of information available. Above 30 GHz, component sizes are in the order of a few millimetres which makes fabrication very challenging, especially when considering that most SIW circuits use a substrate with a dielectric constant of 2 or more, making the structure even smaller when compared to the equivalent waveguide using air. In papers where there is a comparison between the measured and simulated results of millimetre wave SIW diplexers, there is a notable shift in the scattering parameters as a result of fabrication tolerances. For example, in [2,3] diplexers in the K-band part of the spectrum are presented where the shift is cited as a result of a +6 μm variation in via diameter by the authors. In [4], a diplexer operating in the V-band part of the spectrum is presented where the authors also cite fabrication tolerances as the explanation for the shift. From these discussions, it is apparent that there is a problem with fabrication tolerances in the manufacture of millimetre wave SIW diplexers and relaxing the circuit's sensitivity to these tolerances would be advantageous when considering volume production.

The proof of concept Ka-band diplexer that was presented in [5,6] provided some insight into how the use of a hybrid coupler type diplexer topology could be used to help relax sensitivity to fabrication errors in waveguide diplexers. Building on this work, an SIW equivalent diplexer was presented in [7] that suggested the use of this type of diplexer could help to relax the sensitivity of millimetre wave SIW diplexer circuits to fabrication errors. The design of this diplexer is the focus of this chapter where further information is provided that complements the work in [7].

The structure of this chapter is as follows. The equivalence between SIW and dielectric-filled waveguide (DWG) is outlined in Section 12.2 where a method of

[1]Institute of Sensors, Signals and Systems, School of Engineering and Physical Sciences, Heriot-Watt University, Edinburgh, UK
[2]Airborne & Space Systems Division, Leonardo MW Ltd, Edinburgh, UK

designing SIW components entirely in DWG and translating the final design to SIW is presented in Section 12.2.1. In Section 12.3, an overview of diplexer circuits is given, where a discussion on the operation of the hybrid coupler type diplexer is given in Section 12.3.1. In Section 12.4, the design of the Ka-band SIW diplexer in [7] is presented. Some discussions on the design of the diplexer's components are given in Sections 12.4.1–12.4.3, where simulated results of the complete SIW diplexer are presented in Section 12.4.4. To investigate the diplexer's handling of fabrication errors, simulated results of a tolerance analysis are discussed in Section 12.5 where changes to the via position, dielectric constant and via diameter of the substrate are considered in detail in Sections 12.5.1, 12.5.2, and 12.5.3, respectively. In Section 12.6, the design of the waveguide-to-SIW transition presented in [8] is discussed. Using this transition, some promising measured results of the SIW diplexer are observed in Section 12.7.

12.2 Equivalence between SIW and DWG

Analogous to DWG, SIW guides electromagnetic waves between four conducting walls. However, in SIW, the two side walls are replaced by an array of periodic vias that are plated using a conductive material such as copper as is shown in Figure 12.1. This means that SIW can be constructed using standard PCB substrates with top and bottom ground layers, where a PCB milling/laser-drilling fabrication process can be used to construct the vias. When compared to the CNC fabrication process used to make the waveguide diplexer in [5,6], fabricating SIW circuits is significantly cheaper in terms of the material cost and the time required to fabricate the circuit [9].

The fundamental propagating mode of SIW can be considered as quasi TE_{10} as the field pattern is similar to the TE_{10} mode in DWG. However, due to the currents induced by the H field of the TM mode set being broken by the periodic vias, excited modes of this type will radiate from the guide [11]. Thus, only TE_{m0} mode

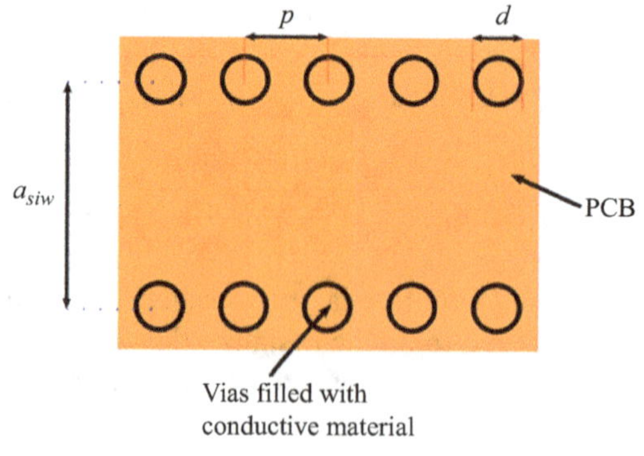

Figure 12.1 Schematic showing the geometry of SIW, where p represents the separation between the centre of the vias and d represents the diameter of the vias. The width of the SIW is represented by a_{siw}. Reprinted with permission from [10]

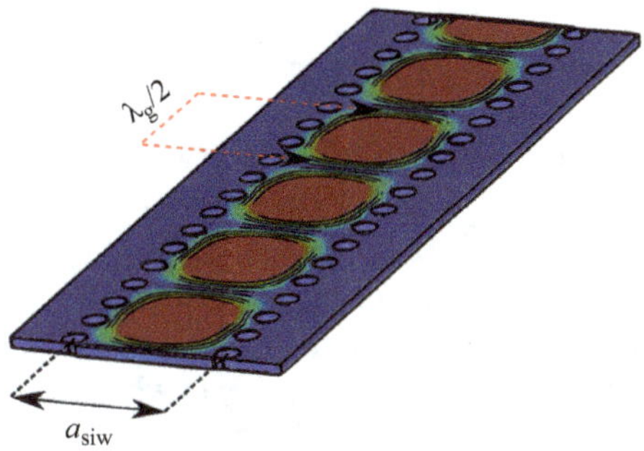

Figure 12.2 Quasi TE_{10} mode propagating in a section of SIW. The field distribution corresponds to the absolute value of the electric field where blue shading represents low field strength through to red shading which represents high field strength. Reprinted with permission from [10]

sets can be contained within SIW. An example of a quasi TE_{10} mode propagating in SIW can be seen in Figure 12.2.

So long as p and d are chosen correctly, there will be negligible radiation leaked from the vias [1]. There are some guidelines presented in [9,11,12] to avoid the leakage of radiation from the SIW and to prevent a bandgap in the operating band of the circuit. These are summarised below, where:

$$\frac{p}{d} \leq 2 \tag{12.1}$$

is to avoid leakage of radiation;

$$p > d \tag{12.2}$$

is to stop overlapping of the vias;

$$\frac{p}{\lambda_c} < 0.25 \tag{12.3}$$

is to push any bandgap generated as a result of the periodicity of the vias out of the operating passband of the circuit, and

$$\frac{p}{\lambda_c} > 0.05 \tag{12.4}$$

is to maintain the structural rigidity of the circuit by restricting the number of via holes used to realise the circuit. Here, λ_c represents the cut-off wavelength of the equivalent DWG of width a_{dwg}, which can be calculated using (12.5) for the fundamental mode.

$$\lambda_c = 2a_{\text{dwg}}\sqrt{\varepsilon_r} \tag{12.5}$$

In recent years, there has been a lot of work done to relate the cut-off frequency of DWG to that of SIW, with a number of design equations being proposed in

[11,13–15] and more recently [16] demonstrating equivalence between both cases. Of these, the equations in [13,14] are currently cited the most in the literature. To gain an insight into any differences associated with the published equations, a comparison between the formulas in [13,14,16] was carried out using the commercial solver CST Microwave Studio for identical sections of SIW calculated to have a cut-off frequency of 15 GHz. The equation in [11] was neglected from the analysis as it was shown in the same paper to only provide a marginal increase in accuracy when compared to the equation in [13]. Additionally, the equation in [15] was neglected as it has been shown to be inaccurate when compared to the equation in [13]. The three equations used in the analysis can be seen below, where (12.6) is the equation presented in [13], (12.7) is the equation presented in [14] and (12.9) is the equation presented in [16], where a_{dwg} is the equivalent width of a section of DWG having a cut-off frequency of 15 GHz in this case.

$$a_{\mathrm{dwg}} = a_{\mathrm{siw}} - \frac{d^2}{0.95p} \tag{12.6}$$

$$a_{\mathrm{dwg}} = a_{\mathrm{siw}} \cdot \left[\xi_1 \cdot \left(\frac{\xi_2}{\frac{p}{d} + \left(\frac{\xi_1 + \xi_2 - \xi_3}{\xi_3 - \xi_1} \right)} \right) \right] \tag{12.7}$$

where ξ_1, ξ_2, and ξ_3 are given in (12.8).

$$\begin{aligned} \xi_1 &= 1.0198 + \frac{0.3465}{\frac{a_{\mathrm{siw}}}{p} - 1.0684} \\ \xi_2 &= -1.1183 - \frac{1.2729}{\frac{a_{\mathrm{siw}}}{p} - 1.201} \\ \xi_3 &= 1.0082 - \frac{0.9163}{\frac{a_{\mathrm{siw}}}{p} + 0.2152} \end{aligned} \tag{12.8}$$

$$a_{\mathrm{siw}} = a_{\mathrm{dwg}} + p\left(0.766e^{\frac{0.4482d}{p}} - 1.176e^{\frac{-1.214d}{p}}\right) \tag{12.9}$$

By referring to Figure 12.3, the simulated comparison between (12.6), (12.7), and (12.9) can be seen. None of the equations meet the desired 15 GHz cut-off frequency, with (12.7) and (12.9) deviating by around 200 MHz and (12.6) deviating by 400 MHz. As the difference between (12.7) and (12.9) can be considered negligible, the use of (12.9) is recommended over (12.7) due to the simplicity of the equation. Further to this, the use of (12.6) is not recommended as the simulated cut-off frequency is significantly higher than the desired cut-off frequency when compared to (12.7) or (12.9).

It should be noted that the guide wavelength λ_g of SIW can be characterised by using the equivalent equation for a DWG. For example, the guide wavelength for the

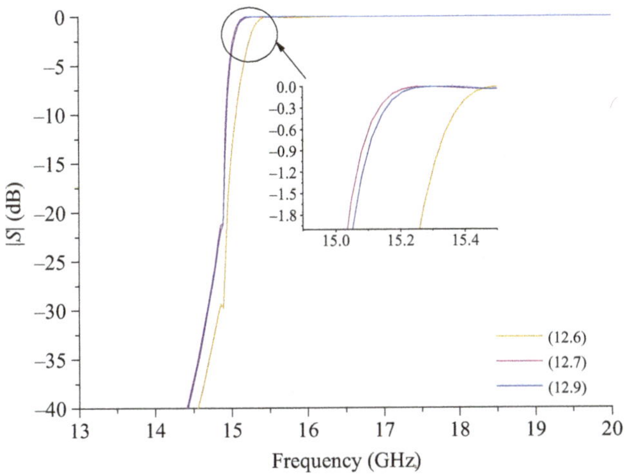

Figure 12.3 A comparison of the simulated cut-off frequency of sections of SIW designed using (12.6), (12.7) and (12.9). Reprinted with permission from [10]

fundamental mode can be calculated using (12.10), where λ_0 is the free space wavelength at a frequency f and $f_{c_{10}}$ is the cut-off frequency of the fundamental mode.

$$\lambda_g = \frac{\lambda_0}{\sqrt{\varepsilon_r}\sqrt{1-\left(\frac{f_{c_{10}}}{f}\right)^2}} \tag{12.10}$$

12.2.1 Design of SIW components

Due to the equivalence between SIW and DWG, the design of SIW components can be carried out by designing the structure entirely in DWG and then translating a final optimised design to SIW [7,17]. This method results in a faster design process, as the complete structure is optimised in DWG rather than in SIW where the mesh is denser due to the vias in the model when using a commercial solver such as CST Microwave Studio or HFSS. In practice, the translation stage is carried out by splitting the component into small sections of DWG of width a_{dwg} and length z and using an equation such as that in (12.9) to obtain the equivalent width a_{siw} of the SIW section for a particular via separation p, and via diameter, d [16]. As the length of the section is known, the required number of vias to construct the SIW side wall can be obtained from (12.11), forming the new SIW component. This is summarised in Figure 12.4. So long as the design guidelines in Section 12.2 are met, the diameter and pitch of the via can be chosen arbitrarily by the designer to suit a particular component. In this way, smaller diameter vias can be selected to ensure fine details such as septums in hybrid couplers can be realised, whereas larger diameter vias can be selected for parts of the circuit where small feature sizes are

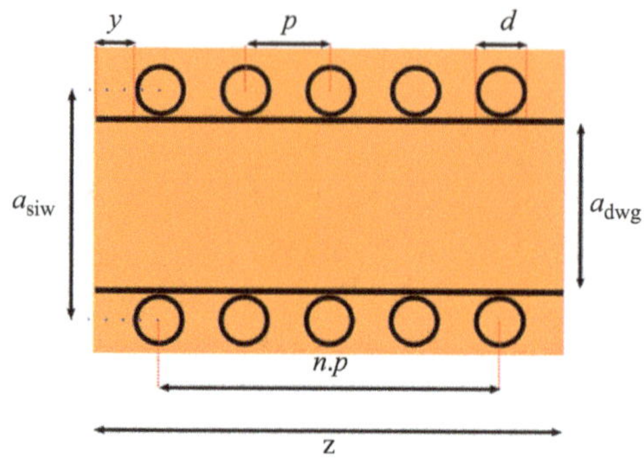

Figure 12.4 A diagram showing how the length of an SIW section, z, is related to the number of vias, n. Here, y is the distance to the edge of the section and can be chosen arbitrarily by the designer to make sure the optimum number of vias is chosen for a given p and d. Reprinted with permission from [10]

not required. Translating components in this way ensures that the SIW component matches the optimised DWG equivalent almost exactly and eliminates the need to optimise the translated SIW component entirely.

$$n = \frac{z - 2y - d}{p} \tag{12.11}$$

12.3 Diplexer overview

Diplexers are essential components that are used in communications systems to allow an antenna to transmit (TX) and receive (RX) simultaneously [18]. A block diagram of a generalised diplexer can be seen in Figure 12.5, where two frequency filters are matched to a power divider. The diplexing action is achieved through the use of two filters, where each filter provides a separate TX and RX signal path. In general, diplexers have very stringent passband and stopband requirements imposed on them to ensure the TX and RX channels do not interact with each other [18]. In this case, half-wavelength bandpass filters are commonly used to construct the diplexer as they provide the best trade-off with size and channel performance. As the diplexer is connected directly to the antenna, the insertion loss of the diplexer contributes to the overall noise figure of the receiver system and can impact the sensitivity of the antenna. This in practice requires diplexers to have high Q filters in order to keep insertion losses to a minimum and as such waveguide diplexers are preferred. It should be noted that the term duplexer is also used in the literature to describe diplexers.

Another type of diplexer configuration consists of the use of two identical hybrid couplers and two identical channel filters as can be seen in Figure 12.6. In its current form, the diplexer will only show the response of the channel filter when looking at S_{31}, S_{13}, S_{24} or S_{42}. Therefore, a second filter is usually required at one of

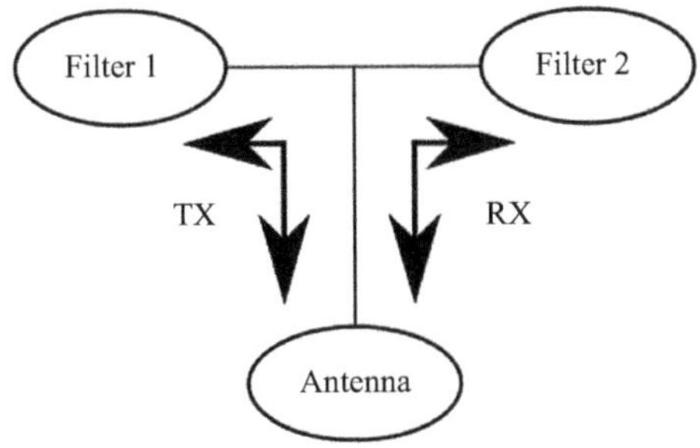

Figure 12.5 A block diagram of a generalised T-junction-based diplexer. This topology can also be used as a frequency combiner. Reprinted with permission from [10]

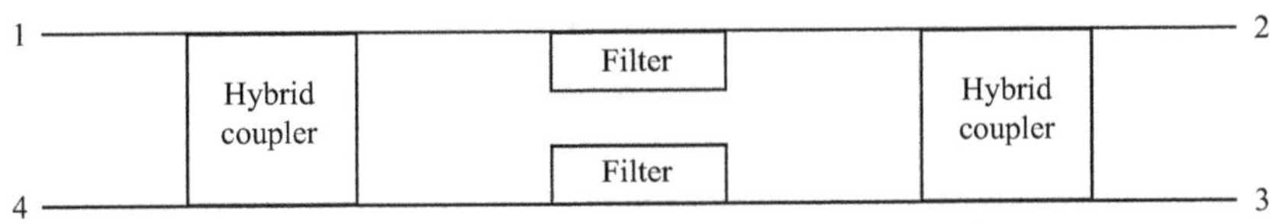

Figure 12.6 Block diagram of a hybrid coupler-based diplexer topology. The hybrid couplers in this case have a phase difference of 90° at their output ports

the ports to improve the rejection on the second channel [19]. For example, if ports 1 and 3 are selected to construct the TX channel, a second filter could be added to port 2 to construct the RX channel. It was shown in [5,6] and subsequently in [7] that a diplexer of this type is advantageous when fabrication tolerance is considered due to the reduced number of resonators contained within the structure. When compared to a T-junction-based diplexer operating over a similar fractional bandwidth, the hybrid coupler-based diplexers of [5–7] contain half the number of resonating structures as a result of using only one bandpass filter which subsequently relaxes the diplexer's sensitivity to fabrication errors. An additional benefit of the hybrid coupler-based diplexer topology is the increased power handling ability due to the hybrid couplers only supplying half the input power to one of the diplexing channels. For example, this type of diplexer is well suited for future satellite ground stations where the required power handling of Q/V band waveguide diplexers is in excess of 500W [20]. It should be noted that when compared to diplexers constructed using a T-junction, the hybrid coupler-based diplexers are of a larger size and have a slightly higher insertion loss [18].

12.3.1 Hybrid coupler type diplexer operation

An overview of the operation of the hybrid coupler type diplexer presented in [5–7] is given here. The RX filter has been removed from the analysis to simplify the discussion. Consider the schematic in Figure 12.7. For a signal passed from the

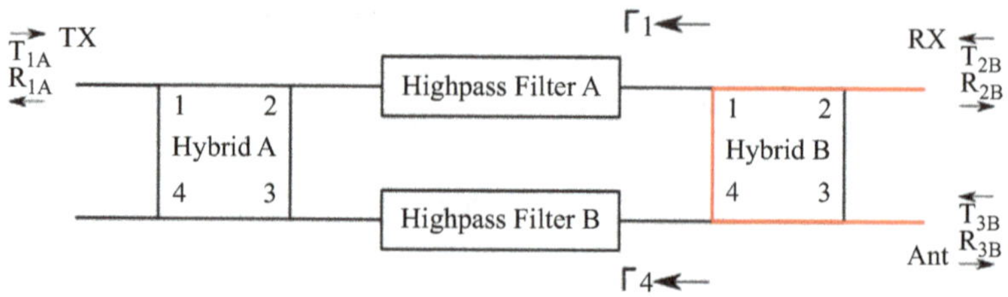

Figure 12.7 Schematic diagram showing a signal being routed to the RX port from the antenna port. The highpass filters act as a load when viewed from the antenna and reflect an incoming signal from the antenna to the RX port. Reprinted with permission from [10]

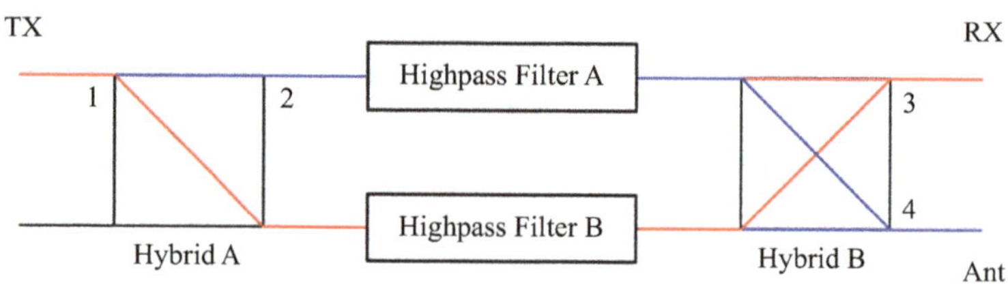

Figure 12.8 Schematic diagram showing the TX–Antenna path of the diplexer. At point 3 the signals are combined with a 180° difference in phase and cancel due to destructive interference. At point 4 the signals meet in phase and are passed to the antenna. Reprinted with permission from [10]

antenna to the RX port, the following scattering matrix is valid for identical hybrid couplers and highpass filters.

$$\begin{bmatrix} R_{2B} \\ R_{3B} \end{bmatrix} = \begin{bmatrix} \frac{1}{2}(\Gamma_4 - \Gamma_1) & \frac{j}{2}(\Gamma_1 + \Gamma_4) \\ \frac{j}{2}(\Gamma_1 + \Gamma_4) & \frac{1}{2}(\Gamma_1 - \Gamma_4) \end{bmatrix} \cdot \begin{bmatrix} T_{2B} \\ T_{3B} \end{bmatrix} \tag{12.12}$$

where T and R represent the transmitted and reflected waves, respectively [21,22]. Looking from the antenna port, the highpass filters act like a load with a reflection coefficient $\Gamma = \Gamma_1 = \Gamma_4$. In this case, (12.12) can be reduced to (12.13).

$$S_{\text{Ant-RX}} = \frac{R_{2B}}{T_{3B}} = j\Gamma \tag{12.13}$$

If $\Gamma = 1$, a signal detected at the antenna will be diverted entirely to the RX port without any additional loss. This requires the highpass filters to have sufficient rejection below the cut-off frequency to ensure there is a large reflection in the RX frequency band.

Now consider a signal incident at the TX port in Figure 12.8. At point 1, the signal is split into two components after passing through Hybrid A that has a 90°

phase difference when viewed from point 2. The signals pass through the identical highpass filters which act as phase shifters, and then are combined in Hybrid B. At point 3, the signals meet with a 180° phase difference and cancel due to destructive interference. However, at point 4 the signals meet in phase and are passed to the antenna. For lossless, ideal components, the isolation between the RX and antenna port is ideal [22].

12.4 SIW diplexer with relaxed fabrication tolerances

An overview of the design of the diplexer presented in [7] is given in this section, where a description of the method used to design the DWG bandpass and highpass filters, in addition to the DWG hybrid coupler is given in Sections 12.4.1–12.4.3, respectively. The design requirements of the diplexer can be found in Table 12.1 and are similar to the requirements presented in [5,6] for backhaul communications. The diplexer is constructed by joining together the optimised DWG components of Sections 12.4.1–12.4.3 and then subjecting the complete structure to an optimisation cycle. Once the requirements in Table 12.1 are met, the optimised DWG diplexer is translated to SIW using the methods in Section 12.2.1 without a further optimisation cycle being required. A diagram of the translated SIW design can be seen in Figure 12.9 with the addition of elliptic bends at ports 1, 3 and 4 to facilitate the use of a waveguide-to-SIW transition for measurement purposes. Other types of bends, such as the mitred corner bend in [23] were considered. However, due to the small septum size used to create the hybrid coupler, this type of bend would be situated too close to the bandpass filter which could result in vias overlapping during a fabrication cycle and is undesirable.

All simulated frequency responses presented in this Section were created using the commercial solver CST Microwave Studio with DWG ports and omit losses. Square vias were used in place of circular vias as they offer a reduced mesh density and simulation time without compromising the accuracy of the model. Moreover, the equivalence between circular and square vias has been validated experimentally in [24], [25] for filters and couplers, respectively, with good agreement. To ensure

Table 12.1 Design specifications for the SIW diplexer circuit. The diplexer is designed for use with RT 5880, which has a dielectric constant of 2.2 and a height 1.574 mm

	Channel 1	**Channel 2**
Passband	31.00–33.00 GHz	35.32–37.50 GHz
Stopband rejection	−55 dB between 35.32–37.5 GHz	−60 dB between 31.00–33.00 GHz
Return loss	−14 dB	−14 dB
Insertion loss	−2 dB	−2 dB

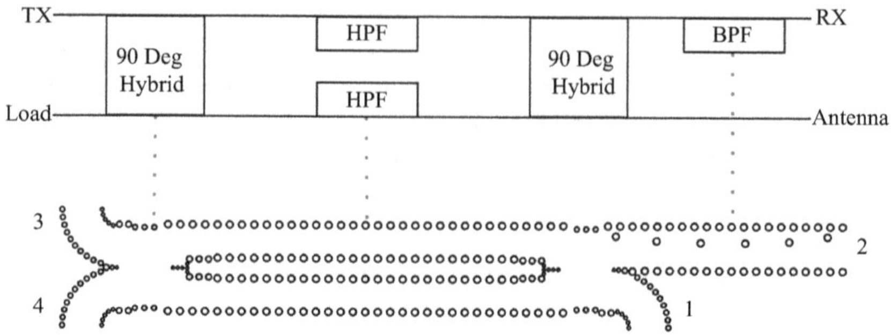

Figure 12.9 Schematic diagram of the SIW diplexer circuit. To facilitate the use of a waveguide-to-SIW transition, elliptic bends were added to ports 1, 3 and 4

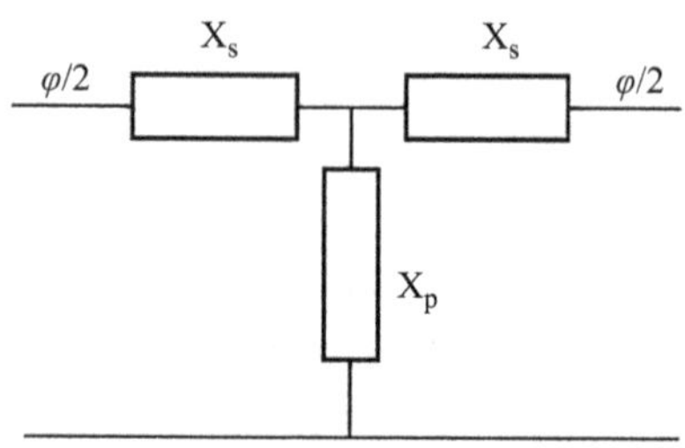

Figure 12.10 Equivalent circuit for inductive iris and inductive post type inverter discontinuities. Reprinted with permission from [10]

the correct equivalence between the square and circular vias, the length of the square via, L, was determined from (12.14) which is based on a geometric average, where d_{circle} is the diameter of the circular via [26].

$$L = d_{circle} \frac{\frac{1}{\sqrt{2}} + 1}{2} \tag{12.14}$$

12.4.1 DWG bandpass filter design

Typically, impedance inverters are realised in SIW bandpass filters using either inductive iris or inductive post type discontinuities due to their ease of fabrication in a single layer substrate. These discontinuities can be represented using the equivalent circuit shown in Figure 12.10 where X_s and X_p are the series and parallel reactances of the inverter, respectively. For both the inductive iris and inductive post cases, the electrical length of the inverter φ is negative. As negative lengths cannot be realised physically, they are absorbed into the resonator [27].

In terms of filter synthesis, the scattering parameters of a 3D impedance inverter model such as that in Figure 12.11 can be used to determine the physical dimensions of a narrow-band bandpass filter using a full-wave solver. The goal here is to produce a

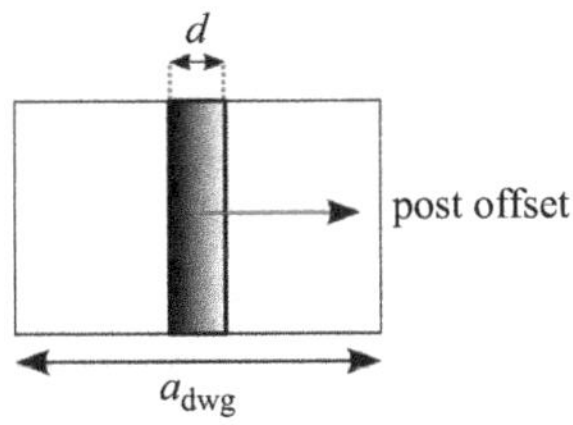

Figure 12.11 3D impedance inverter model used to extract the scattering parameters from an inductive post type discontinuity. The model is run over a range of post offset positions to produce a design curve showing how the normalised inverter impedance varies as a function of post position. The ports are embedded to the edge of the post discontinuity

design curve showing how the normalised inverter impedance varies against known physical dimensions from the 3D model. From this curve, the geometries that correspond to the theoretical normalised impedance inverter values as calculated using the equations in [28] for a given bandwidth and passband return loss can then be interpolated and used to construct an initial filter. Typically, a small optimisation cycle is then required to ensure the filter meets any specified design criteria.

A summary of the scattering parameter extraction process will now be discussed. It was shown in [29] that the S_{11} and S_{21} responses (both real and imaginary parts) can be used to determine the normalised series and parallel reactances of the circuit in Figure 12.10 by using (12.15).

$$j\frac{X_s}{Z_0} = \frac{1 - S_{21} + S_{11}}{1 - S_{11} + S_{21}}$$
$$j\frac{X_p}{Z_0} = \frac{2S_{21}}{(1 - S_{11})^2 - S_{21}^2} \tag{12.15}$$

The scattering parameters are extracted over a range of different inverter geometries at the centre frequency of the filters passband and used to solve (12.15) during a post-processing step. Using these values, the electrical length of the inverter can be calculated using (12.16) for the entire parameter range. These values are then used to determine the normalised inverter impedance using (12.17). A chart showing how the normalised inverter impedance varies against the known inverter dimensions from the 3D model can then be produced and used to determine the initial inverter dimensions of the filter [18].

$$\varphi = -\tan^{-1}\left(2\frac{X_p}{Z_0} + \frac{X_s}{Z_0}\right) - \tan^{-1}\frac{X_s}{Z_0} \tag{12.16}$$

$$\frac{K}{Z_0} = \left|\tan\left(\frac{\varphi}{2} + \tan^{-1}\frac{X_s}{Z_0}\right)\right| \tag{12.17}$$

Moreover, with knowledge of the required normalised impedance inverter values, the resonator lengths can be found from (12.18), where λ_{g0} is the guide wavelength at the centre frequency f_0 of the filter's passband.

$$l_i = \frac{\lambda_{g0}}{2\pi}\left(\pi + \frac{1}{2}(\varphi_i + \varphi_{i+1})\right) \quad i = 1 \text{ to } n \tag{12.18}$$

An example of this design method can be found in Figure 12.12, where the final optimised response of a DWG inductive post bandpass filter and translated SIW equivalent can be seen. The filter was designed to operate over the 31–33 GHz range with a minimum passband return loss of −15 dB. Figure 12.13 shows the design curve used to determine the initial positions of the inductive post inverters in the filter. Final optimised dimensions of the filter can be found in Table 12.2 in addition to the normalised impedance inverter values used to extract the initial filter dimensions from Figure 12.13. These are presented for reference and can be compared to the schematic in Figure 12.14. Good agreement between the DWG and translated SIW case can be observed using the method in Section 12.2.1 where additional optimisation of the SIW circuit is not required.

12.4.2 DWG highpass filter design

The design of highpass filters in DWG can be achieved by equating the cut-off frequency of the DWG to the cut-off frequency of the highpass filter [30]. In this case, knowledge of the traditional prototype circuit for highpass filters is not required as DWG naturally acts as a highpass filter: only knowledge of the propagation constant for a particular mode is required. Below the cut-off frequency, the wave attenuates exponentially following $e^{-\alpha z}$, where α is the attenuation constant and z is the distance travelled by the wave. In terms of highpass filter design, z corresponds to the length of the highpass filter which can be determined for a

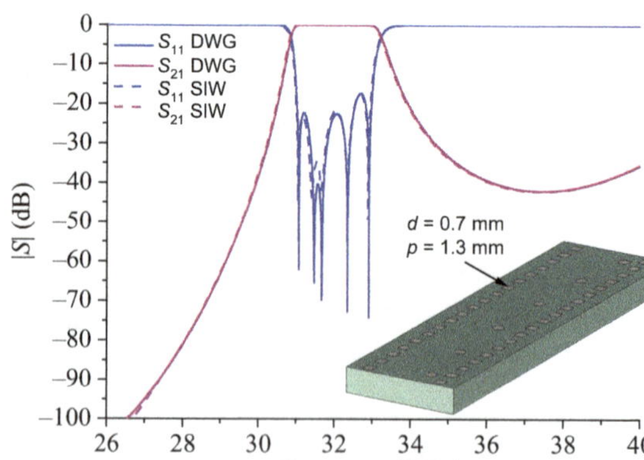

Figure 12.12 Simulated response of a DWG bandpass filter and the translated SIW equivalent used to construct the diplexer. The filter was constructed using inductive post inverters and has a minimum passband return loss of −15 dB

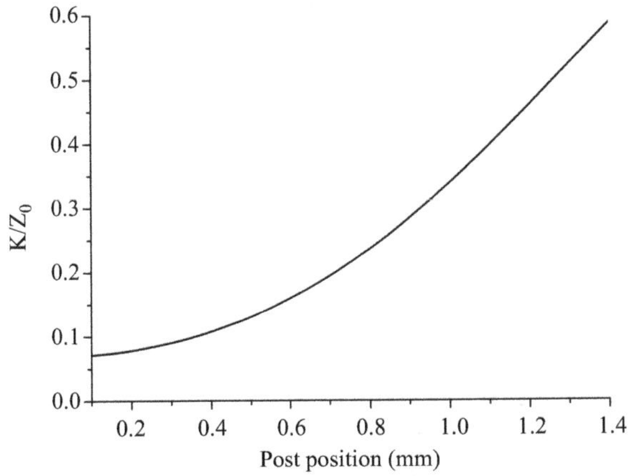

Figure 12.13 *Normalised inverter impedance of an inductive post discontinuity against the offset position of the post from the centre of a section of DWG. The diameter of the post is 0.7 mm. Reprinted with permission from [10]*

Table 12.2 *Optimised dimensions of the bandpass filter. The normalised impedance inverter values used to extract the initial filter dimensions from Figure 12.13 have been added for reference*

$\frac{K_{01}}{Z_0}$	$\frac{K_{12}}{Z_0}$	$\frac{K_{23}}{Z_0}$	$\frac{K_{34}}{Z_0}$	$\frac{K_{45}}{Z_0}$	$\frac{K_{56}}{Z_0}$
0.4904	0.2025	0.1488	0.1488	0.2025	0.4904
d_{off1}	d_{off2}	d_{off3}	d_{off3}	d_{off2}	d_{off1}
1.26 mm	0.73 mm	0.59 mm	0.59 mm	0.73 mm	1.26 mm
l_1	l_2	l_3	l_4	l_5	
3.35 mm	3.90 mm	4.00 mm	3.90 mm	3.35 mm	

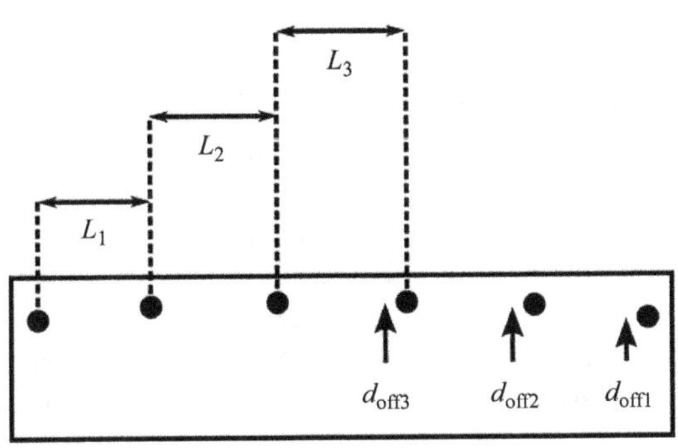

Figure 12.14 *Schematic relating the dimensions of the filter in Table 12.2 to the filter topology. The width of the DWG is 4.15 mm in this case*

particular stopband rejection using (12.19), where *SB* is the stopband rejection of the highpass filter in dB.

$$z = \frac{-\ln\left(10^{\frac{-SB}{20}}\right)}{\alpha} \tag{12.19}$$

Additionally, the attenuation constant is given by

$$\alpha = \sqrt{k_c^2 - k^2} \tag{12.20}$$

where k_c is the cut-off wavenumber, which is $\frac{\pi}{a_{\mathrm{dwg}}}$ for the fundamental mode, and k is the phase constant which is given by $\frac{2\pi f_{SB}\sqrt{\varepsilon_r}}{c}$, where f_{SB} is the desired stopband edge frequency of the highpass filter [31]. The length of the cut-off section in this case is 32.5 mm and has a width of 2.99 mm, which corresponds to a cut-off frequency of 33.8 GHz. As the cut-off section of the DWG will have a narrower width than the input port of the hybrid coupler, an inhomogeneous transformer is required to match the highpass filter to the hybrid couplers. Inhomogeneous transformers are characterised by cascaded sections of having different widths [28]. It was shown in [32] that these transformers can be designed entirely by using computer optimisation techniques, rather as relying on approximate design equations such as those summarised in [28] for rectangular waveguides. To design the transformer, the following optimisation constraints are enforced to ensure that only the fundamental mode is propagating in the transformer band:

$$\begin{aligned} &\frac{\lambda_l}{2} < a < \lambda_u \\ &0 < b < \frac{\lambda_u}{2} \end{aligned} \tag{12.21}$$

where λ_l and λ_u are wavelengths of the lower and upper band edges, respectively. Additionally, it is assumed that the maximum transformer section length will not exceed half the guide wavelength of the upper band edge λ_{gu}:

$$0 < l < \frac{\lambda_{gu}}{2} \tag{12.22}$$

where l is the length of the transformer section. As with conventional multi-section transformer theory, increasing the number of waveguide sections in the transformer improves the bandwidth performance of the matching network. In this case, it was found that a second-order transformer was required to ensure a sufficient match was obtained over the operating range of the highpass filter, which is 35.32–37.5 GHz. The simulated frequency response of the highpass filter used to construct the diplexer (including a two-section inhomogeneous transformer) can be seen in Figure 12.15. A return loss of −11.5 dB is obtained over the filters operating range and is sufficient to demonstrate the operation of the diplexer. Good agreement between the DWG and translated SIW case can be observed using the method in Section 12.2.1, where additional optimisation of the SIW circuit is not required.

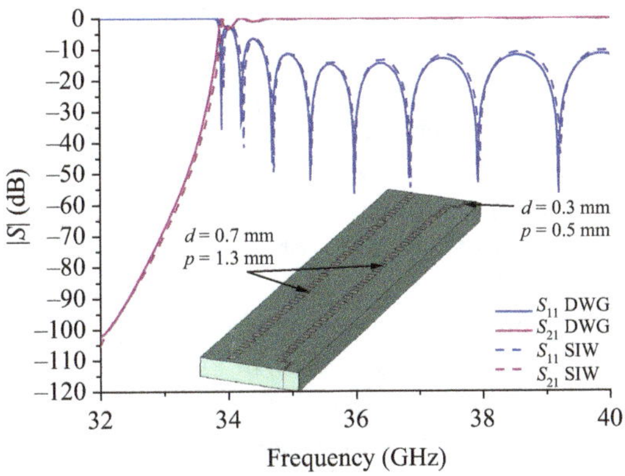

Figure 12.15 Simulated response of a DWG highpass filter and the translated SIW equivalent used to construct the diplexer. A return loss of −11.5 dB is obtained over the operating range of the highpass filter

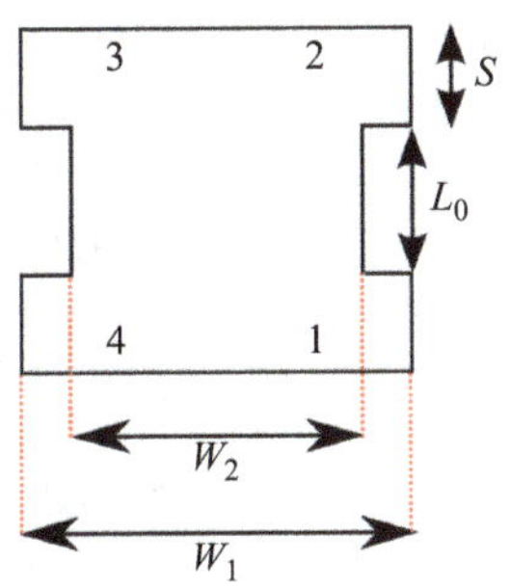

Figure 12.16 Schematic showing the coupling region of a Riblet type DWG hybrid coupler when looking directly down on the structure. Port numbers have been added for reference. The TE_{30} mode is evanescent in the centre part of the coupling region labelled L_0: this is essential for broadband performance [34,35]

12.4.3 DWG hybrid coupler design

When tolerance is considered, the short-slot (Riblet) coupler has benefits over other types of hybrid couplers, such as the branch-line coupler. This is due to the Riblet coupler possessing a large coupling region in place of tight coupling slots which at millimetre wave frequencies can be difficult to realise [33]. Additionally, due to the Riblet coupler's ability to be fabricated in the H-plane of a single layer section of SIW, their use has become popular in the literature. In this case the Riblet coupler was chosen to design the hybrid coupler for this design. A schematic diagram showing the coupling region of a Riblet coupler can be found in Figure 12.16.

The Riblet coupler works by exciting predominantly the TE_{10} and TE_{20} modes in the coupling region. Due to the phase relationships between these modes, isolation is achieved at port 4 as a consequence of destructive interference, whereas the input signal is passed to ports 2 and 3 with a −3 dB split and 90° difference in phase due to constructive interference. For this diplexer, the coupler is required to operate over both diplexing channels, resulting in a 19.1 per cent bandwidth requirement. It has been shown in [34,35] that the interaction with the TE_{30} mode (both propagating and evanescent) can improve the isolation and matching properties of the coupler. Thus, to achieve this broadband performance, the coupler was designed as a series of steps where the TE_{10}, TE_{20} and TE_{30} modes are excited: the TE_{30} mode is allowed to propagate in the outer sections of the coupling region, but is evanescent in the centre region labelled L_0 in Figure 12.16. With the addition of more steps in the circuit, the bandwidth of the coupler can be increased [36]. However, the required bandwidth performance for this design was achieved by using only three steps. A design procedure for broadband Riblet couplers can be found in [34], where the initial length of the coupling region can be calculated using (12.23) [37]. From this, the width and length of the steps in the coupler are optimised until a −3 dB split is obtained over the required bandwidth.

$$L = \frac{\lambda_{odd}\lambda_{even}}{4(\lambda_{odd} - \lambda_{even})} = L_0 + 2S \tag{12.23}$$

Here, λ_{even} and λ_{odd} represent the guide wavelengths of the TE_{10} and TE_{20} modes, respectively. The simulated frequency response of the coupling region used in the diplexer after an optimisation cycle can be seen in Figure 12.17 where a return loss of −20 dB in addition to a ±0.84 dB variation in S_{21} and S_{31} is obtained over the 27–38.8 GHz range. Final optimised dimensions of the coupling region can be found in Table 12.3 for reference. Good agreement between the DWG and

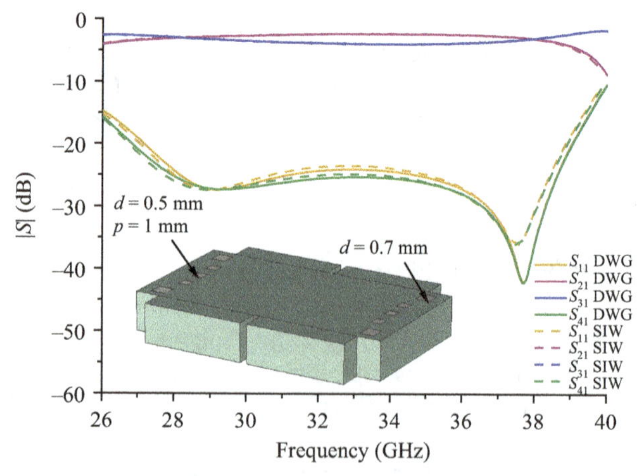

Figure 12.17 Simulated response of a DWG hybrid coupler and the translated SIW equivalent used to construct the diplexer. A return loss of −20 dB in addition to a ±0.84 dB variation in S_{21} and S_{31} is obtained over the 27–38.8 GHz range

Table 12.3 Optimised dimensions of the coupling region. The width of the septum used to create the ports is 0.29 mm

W_1	W_2	L	S	L_0
8.59 mm	8.11 mm	5.62 mm	1.25 mm	3.12

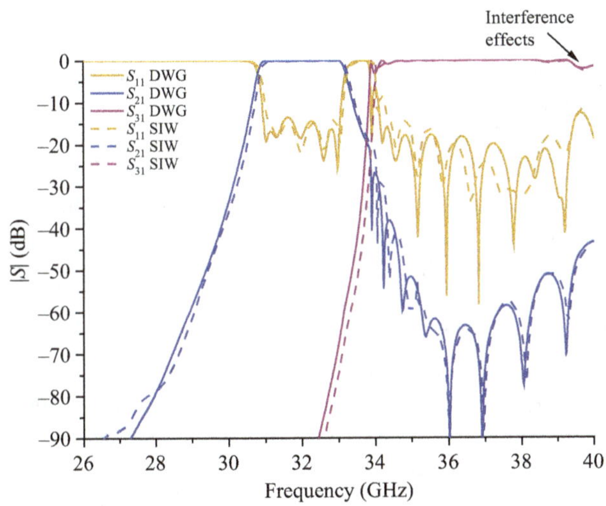

Figure 12.18 Simulated response of a DWG diplexer and the translated SIW equivalent. There is good agreement between the DWG and SIW cases without requiring additional optimisation of the SIW circuit

translated SIW case can be observed using the method in Section 12.2.1, where additional optimisation of the SIW circuit is not required.

12.4.4 Simulated diplexer response

The simulated frequency response of the complete diplexer can be seen in Figure 12.18 where a comparison between the optimised DWG case and translated SIW case without additional optimisation are shown. When comparing the S_{21} and S_{31} responses of the DWG and SIW cases, a small shift in frequency can be observed. This is more significant in the highpass filter where the cut-off frequency of the SIW case has shifted up in frequency by 140 MHz. This corresponds to a 0.41 per cent shift in frequency when considering it as a percentage and can be considered negligible. The shift in frequency can be attributed to the use of square vias in the simulation where their use can only be considered as a good approximation to circular vias [26]. However, even when considering this, there is good agreement between the simulated DWG and SIW frequency responses in both channels which gives further evidence to support the design method in Section 12.2.1.

The out-of-band performance of the diplexer is limited to the bandwidth of the hybrid coupler as can be seen in Figure 12.18, where interference effects are visible beyond the band edge of the highpass filter channel. Extending the useable bandwidth of the coupler will help to reduce this, improving the out-of-band performance of the diplexer for wide-band applications.

12.5 Tolerance analysis

To gain an understanding of how well the diplexer handles fabrication errors, a tolerance study was conducted where it was assumed the vias would be drilled using a PCB drilling machine and then plated using a conductive material. The tolerance study was carried out using the inbuilt parameter sweep macro in CST Microwave Studio to vary key circuit parameters over a specified range. Three separate cases are investigated: an error in the positioning of the vias; an error in the dielectric constant of the substrate; and an error in the via diameter. A discussion of each case will now follow.

12.5.1 Error in via position

The CNC drilling process is a popular fabrication method for drilling vias in SIW due to the quick production turnaround that can be achieved. In addition, extremely small via holes are achievable with a reasonable positional accuracy. For example, in [38] vias of 0.1 mm in diameter were demonstrated on a 0.1 mm thickness glass-fibre laminate substrate. To ensure high-precision drilling with low-drill breakage, the CNC drilling process drives the drill bit into the substrate with a high rotational speed which can be in the order of 300 krpm [39]. In this case, centripetal forces rotate the drill in such a way that the positional error in the entry point of the drill on the substrate can be as large as ± 0.04 mm [38]. This can result in overlapping of the vias which is undesirable, and ultimately limits the via separation that can be used to construct SIW side walls. To investigate how this affects the diplexer response, the position of the vias was altered by ± 0.04 mm over 20 samples for both variations in the x- and y-planes. The simulated frequency response can be seen in Figures 12.19 and 12.20 for both cases, respectively.

By comparing Figure 12.19 with Figure 12.20, changes to the via position in the y-axis appear to affect the diplexer response less than changes to the via position in the x-axis. This is expected, as the cut-off frequency of the SIW and subsequently the guide wavelength of the SIW is related to changes in the width for the fundamental mode. Moreover, this sensitivity is further demonstrated in Figure 12.19, where the cut-off frequency of the highpass filter has a shift of $^{+340}_{-470}$ MHz. However, the return loss in the highpass filter channel appears to be stable and does not fall below -15 dB in the passband of the highpass filter. This is not true for the bandpass filter, where the return loss varies more significantly as a consequence of the bandwidth changing. This suggests that the highpass filter is less sensitive to variations in the via position than the bandpass filter. Designing the

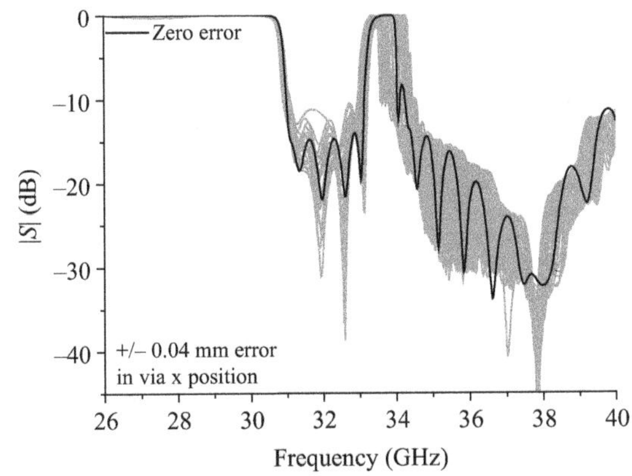

Figure 12.19 Simulated S_{11} frequency response of the diplexer with a ±0.04 mm variation in via x position over 20 samples

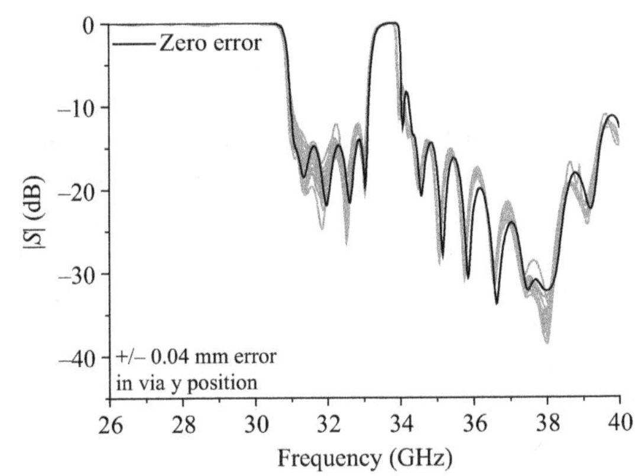

Figure 12.20 Simulated S_{11} frequency response of the diplexer with a ±0.04 mm variation in via y position over 20 samples

highpass filter so that the cut-off frequency is sufficiently far enough away from the passband of channel two would compensate for the shift in cut-off frequency [6].

12.5.2 Error in dielectric constant

Changes to the dielectric constant affect the cut-off frequency and guide wavelength of the SIW. In this case, choosing a substrate that has a low variation in the dielectric constant is essential. Some discussions are given in [40] where it is suggested that Rogers RT 5880 has the lowest variation in dielectric constant of ±0.02, in addition to having the lowest dielectric loss which is essential for millimetre wave applications. Moreover, this substrate is isotropic and the variation in dielectric constant is quoted up to 40 GHz which justifies the use of RT 5880 for this design [41]. To investigate how a change to the dielectric constant affects the

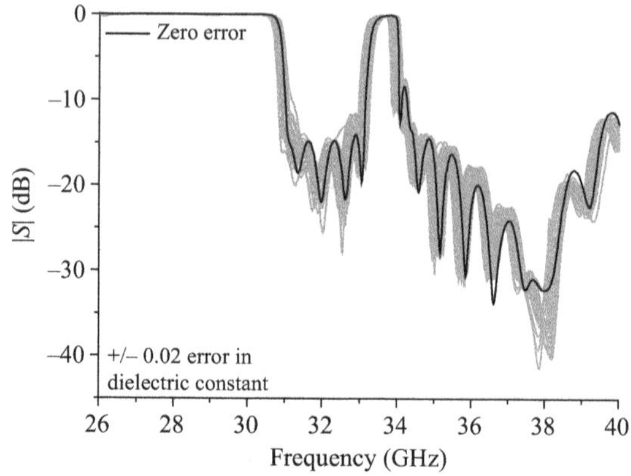

Figure 12.21 Simulated S_{11} frequency response of the diplexer with a ±0.02 variation in the dielectric constant over 20 samples

diplexer's frequency response, the dielectric constant was varied by ±0.02 over 20 samples. The simulated frequency response can be seen in Figure 12.21 where a shift in frequency can be observed. The bandpass filter appears to be more sensitive to variations in dielectric constant than the highpass filter due to the degraded bandwidth and return loss. This suggests that the highpass filter is less sensitive to variations in the dielectric constant than the bandpass filter as the return loss of the highpass filter remains stable. This is expected as the length of each resonator in the bandpass filter is a function of the guide wavelength. Therefore, any change in the dielectric constant will significantly alter the frequency response of the bandpass filter. These observations signify there is a need for the continued development of high-frequency substrates: further reducing the ±0.02 variation in the dielectric constant will reduce the observed shift in frequency.

12.5.3 Error in via diameter

Drilling through PCB substrates ultimately wears out the drill cutters. In [38], it was shown that this wear is related to the position on the drill cutter, where more wear is experienced closer to the tip of the drill. In this case, it was reported that the tip of the drill can be −0.025 mm smaller than new drills over a production cycle. This is undesirable as reducing the via diameter has the effect of shifting the diplexing response down in frequency as the width and length of each SIW component is made larger [40]. Additionally, due to the high rotational speed of the drill, loose substrate laminate material can heat up and bond to the drill which further affects the diameter of the via hole [39]. In this case, replacing worn drills would help to ensure a clean via is produced. Another factor in the diameter of via holes is the aspect ratio. For example, due to the centripetal forces acting on the drill, the entry hole will be larger than the exit hole as the drill enters the substrate at a slight angle [38]. This is more significant on smaller diameter drills and is related to the thickness of the substrate being processed. Ultimately, this limits the

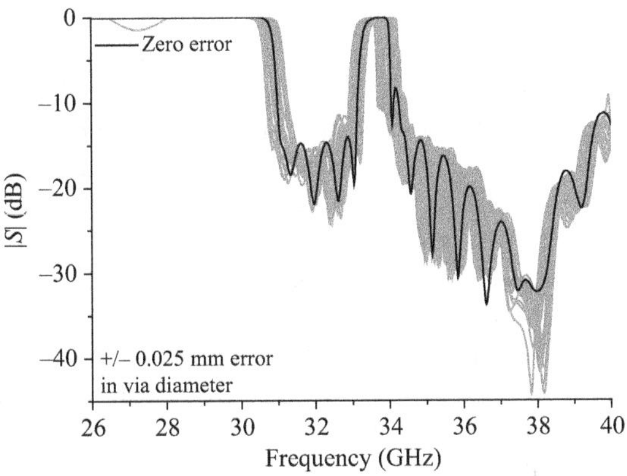

Figure 12.22 Simulated S_{11} frequency response of the diplexer with up to a ±0.025 mm reduction in via diameter over 20 samples

substrate thicknesses which can be processed to ensure acceptable aspect ratios are achieved. Taking into account all variables related to the diameter of the via hole, predicting how the circuit will react due to a change in via diameter is difficult, as the model is inherently complex. To investigate how changes to the via diameter affect the diplexer, a simplified model was used where the diameter of the entry and exit holes were changed systematically. Although a simplification, this model takes into account the reduction in via diameter due to the wearing of the drill cutters, and the larger entry hole of the via due to the drill entering at an angle. The simulated results can be seen in Figure 12.22 where the via diameter was varied by ±0.025 mm over 20 samples. As changes to the via diameter affect the equivalent width of the SIW, a shift in the cut-off frequency of the highpass filter and bandpass filter can be observed. The shift appears to be more significant in the bandpass filter, where a change to the filter's bandwidth and return loss can be seen. In contrast to this, the return loss of the highpass filter is stable in the passband of channel 2 and does not reduce below −15 dB. This suggests that the highpass filter is less sensitive to variations in the via diameter than the bandpass filter.

12.6 Waveguide-to-SIW transition

In order to measure the diplexer circuit, a transition to a suitable transmission line needs to be designed. Microstrip-to-SIW transitions have proved themselves to be a popular choice for designers due to their simple construction and easy integration with other planar components [42]. However, at millimetre wave frequencies microstrip is lossy and prone to radiating energy which can affect the stopband performance of the diplexer. This can be seen in [2,3] where the measured diplexer stopband rejection is reduced as a result of interference at the microstrip ports. Some discussions are given in [43] where the use of a conductor-backed coplanar waveguide-to-SIW transition is shown to reduce interference effects when compared to microstrip-to-SIW transitions. However, in [44], the measured insertion loss of a back-to-back coplanar waveguide-to-SIW transition is quoted as −1 dB over the 18–28 GHz range which is unsuitable for millimetre wave applications. One solution could be the use of a waveguide-to-SIW transition. For example, in [45] the insertion loss of a back-to-back waveguide-to-SIW transition is quoted as −0.68 dB over the 32–50 GHz range which is more suitable for millimetre wave applications.

At present, there are two main waveguide-to-SIW transitions discussed in the literature. The first is the aperture-coupled type, which works by essentially radiating an SIW signal through coupling apertures or equivalent into a waveguide flange [46–49]. However transitions of this type exhibit high losses and are not ideal for use on millimetre wave systems as low insertion loss is a key requirement. For example, in [47] and [49] the measured insertion loss of an aperture-coupled transition at Ka-band frequencies were found to be −1.5 dB and −1.4 dB, respectively, which is unsuitable. The second kind consists of designing a multi-section transformer to match the SIW circuit to a waveguide flange [45,50,51]. More promising insertion loss values have been reported in [45] using this type of transition, where an insertion loss of −0.68 dB is quoted over the 30–50 GHz range. In this case, the multi-section transformer type was selected for use on this design to match the SIW diplexer to a standard WG 22 flange [8].

Due to the equivalence between SIW and DWG, the transition was designed by considering it entirely as an inhomogeneous multi-section waveguide transformer where a DWG feedline is matched to a standard WG 22 flange. In this way, inhomogeneous waveguide transformer theory such as that discussed in Section 12.4.2 can be adapted to design the transition using computer optimisation. The transition is required to operate over the diplexer's operating range which is 31–37.5 GHz. In this case, a three-section inhomogeneous waveguide transformer was required. When compared to the four-section design in [45] where only two of the four sections of the transition are treated inhomogeneously, and the designs in [50] and [51] where all sections of the transition are treated homogeneously, this method offers more degrees of freedom for finding the optimum design. The simulated frequency response of the transition can be seen in Figure 12.23 where a return loss of −23.8 dB is obtained over the operating band of the diplexer. One factor limiting the bandwidth of the transition is the cut-off frequency of the SIW feedline, as close to cut-off there is a large change in impedance which can be difficult to match. In this case, properly selecting the cut-off frequency for the feed line will help to improve the match in the lower-frequency range of the transition which will increase the useable bandwidth. In addition, increasing the order of the transition with more sections would also help to improve the bandwidth performance at the expense of a larger footprint.

To investigate how sensitive the transition is to fabrication errors, a tolerance analysis was carried out where it was assumed the CNC milling process would systematically overcut key circuit dimensions in the transition by up to +0.03 mm. This includes variations to the height, width and length of each transformer section and includes all parameters. Additionally, to simulate misalignment between the SIW circuit and the transition, an air gap of up to +0.05 mm was included in the analysis. The simulated results can be seen in Figure 12.24 for 26 samples and

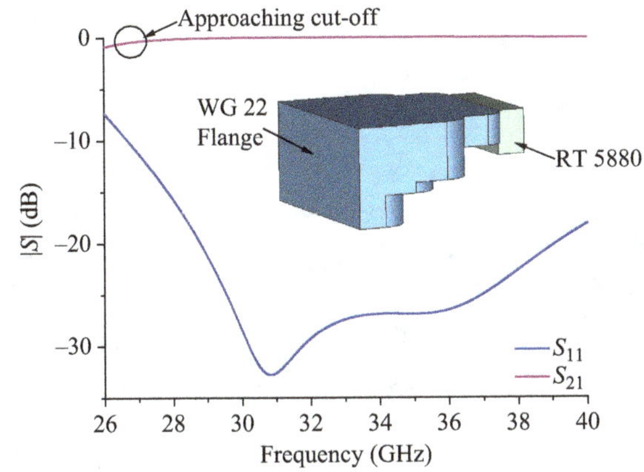

Figure 12.23 *Simulated response of a three-section waveguide-to-SIW transition. A return loss of −23.8 dB is achieved over the 31–37.5 GHz frequency range. Reprinted with permission from [8]*

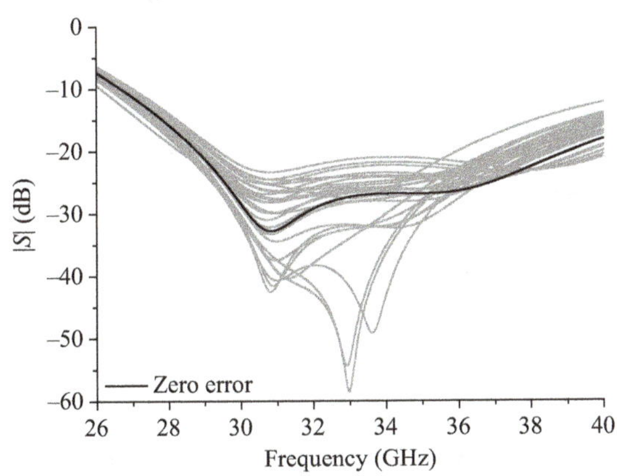

Figure 12.24 *Simulated S_{11} frequency response of the transition being exposed to fabrication errors. The return loss does not drop below −16 dB in the required 31–37.5 GHz frequency range. Reprinted with permission from [8]*

demonstrates how robust the transition is, as the return loss does not drop below −16 dB in the passband of the diplexer.

To validate the waveguide-to-SIW transition, a SIW through line was fabricated from Rogers RT 5880 using 0.6 mm diameter copper-plated vias and placed between ports 3 and 4 in the waveguide housing to represent a back-to-back connection. A photograph of this can be seen in Figure 12.25, where the port numbers correspond to the diplexer port numbers as laid out in Figure 12.9. The transition was measured using an Agilent N5225A PNA network analyser that was calibrated

Figure 12.25 Photograph of the fabricated SIW through line connected between ports 3 and 4 inside an aluminium alloy waveguide housing. A flat lid is used to complete the circuit. The waveguide housing was fabricated in house at Heriot-Watt University. However, fabrication of the SIW through line was outsourced to Southeast University in China due to limitations with Heriot-Watt's PCB fabrication process. Reprinted with permission from [10]

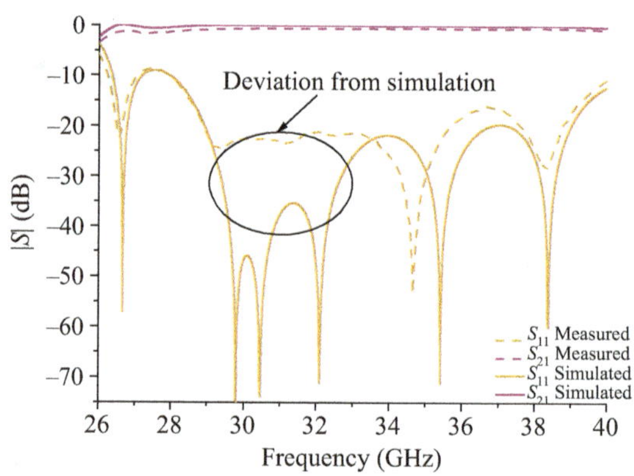

Figure 12.26 Simulated lossless and measured results of the back-to-back waveguide-to-SIW transition. The insertion loss of the transition is −0.48 dB at 34.09 GHz, which is the centre frequency of the diplexer's operating range. Additionally, the minimum return loss over this operating range is −16.07 dB. Reprinted with permission from [8]

using through-reflect-match standards. Simulated lossless and measured results for the back-to-back transition can be seen in Figure 12.26, where very good agreement can be observed. An insertion loss of −0.48 dB was measured at 34.09 GHz, which is the centre frequency of the required operating range. In comparison to the design in [45], the insertion loss here is significantly lower over a similar frequency range. Additionally, a minimum return loss of −16.07 dB was measured over the

28.6–39.1 GHz range which is larger than the required bandwidth. However, there is some deviation from the simulation between 29.2 and 33.42 GHz. This could be due to the through line being misaligned in the waveguide housing, resulting in destructive interference from reflected incident waves. As the deviation is below −20 dB, it is unlikely that it will affect the performance of the diplexer.

12.7 Measured performance of the SIW diplexer

A photograph of the fabricated diplexer can be seen in Figure 12.27 along with a comparison between measured and simulated results in Figure 12.28. The diplexer

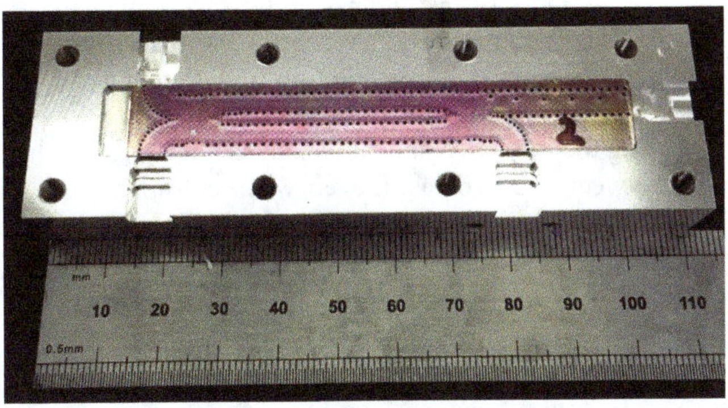

Figure 12.27 Photograph of the fabricated SIW diplexer situated inside an aluminium alloy waveguide housing. A flat lid is used to complete the circuit. Fabrication of the SIW diplexer was outsourced to Southeast University in China due to limitations with Heriot-Watt's PCB fabrication process

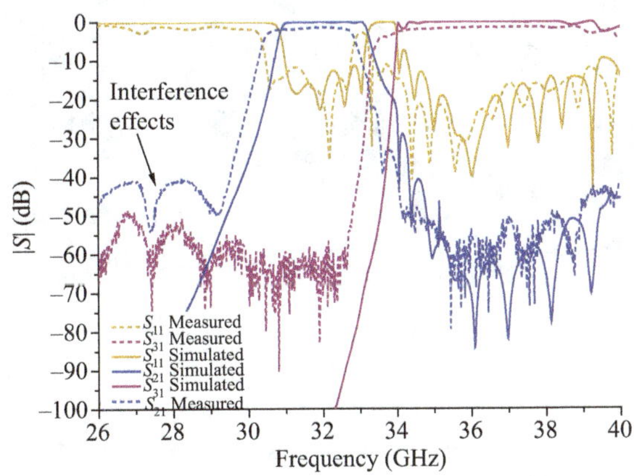

Figure 12.28 Simulated lossless and measured performance of the SIW diplexer using a waveguide-to-SIW transition

was measured using an Agilent N5225A PNA network analyser that was calibrated using through-reflect-match standards. The measured passband of the bandpass channel was found to be 30.62–32.72 GHz, whereas the measured cut-off frequency of the highpass channel was found to be 33.63 GHz. This corresponds to a shift of −350 MHz and −430 MHz, respectively. It should be noted that the measured bandwidth of the bandpass filter is 6.63 per cent which is marginally larger than the desired 6.25 per cent. In this case, more energy is passing through the filter which can only be attributed to a positive error in the via x position; a reduction in via diameter; or a combination of both. Assuming that these errors are systematic during the diplexer's fabrication, the shift in cut-off frequency of the highpass filter can also be attributed to these errors. To confirm this, the average via diameter, pitch and width of the highpass filter section was measured using a digital microscope. The measured results can be found in Figure 12.29 and are in agreement with the discussion above in addition to the predicted range of the tolerance analysis in Section 12.5.

When comparing the measured and simulated return loss of channel 2, there is good agreement between the two cases, even when considering the downshift in frequency. However, this is not true for the return loss of channel 1 where the shift

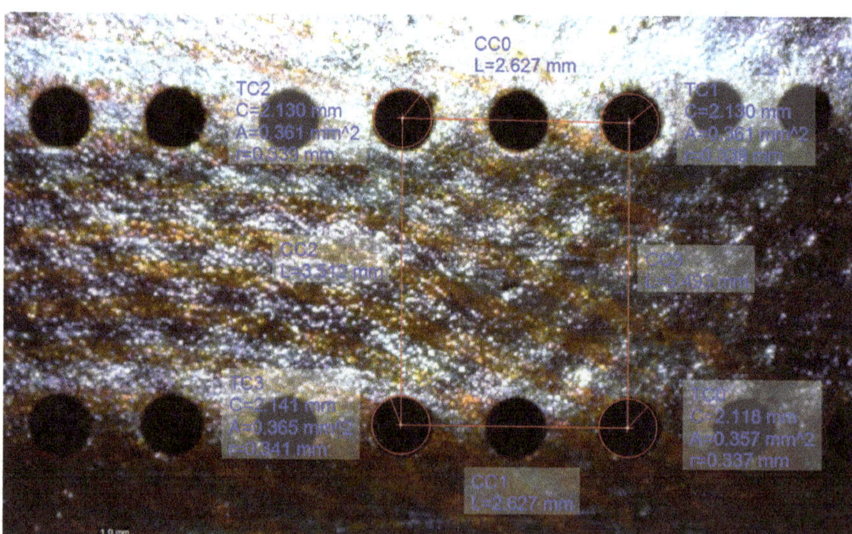

Figure 12.29 The average via diameter was measured to be 0.68 mm which is smaller than the 0.7 mm design value. Conversely, the via pitch was measured to be 1.32 mm which is larger than the 1.3 mm design value. The average width of the highpass filter section was found to be 3.5 mm which is larger than the 3.46 mm design value. These values are in line with the tolerance analysis in Section 12.5 and validate the method

Table 12.4 Summary of the measured diplexer's performance. Maximum insertion loss (IL) is given at the centre frequency of the shifted passbands; and the return loss (RL) and stopband (SB) performance are given as the minimum value in the shifted passbands

Channel 1			Channel 2		
IL (dB)	**RL (dB)**	**SB (dB)**	**IL (dB)**	**RL (dB)**	**SB (dB)**
−1.63	−11.8	−50.7	−1.60	−16.2	−56.8

in frequency degrades the passband return loss too significantly. This suggests that the highpass filter is less sensitive to fabrication errors than the bandpass filter, and justifies its use in helping to relax fabrication tolerances within the diplexer.

A summary of the measured diplexer's performance neglecting the downshift in frequency can be found in Table 12.4 for ease of reading. When comparing Table 12.4 to the desired specifications in Table 12.1, it can be seen that the diplexer would narrowly fail to meet the requirements for return loss in channel 1, and stopband performance in both channel 1 and channel 2 if there was no shift in frequency. The differences between the measured and simulated stopbands can be attributed to interference in the coupling region, where reflected TX and RX signals can interfere with the stopband. For example, consider the deviations between the measured and simulated S_{21} and S_{31} responses over the 35–40 GHz and 26–32 GHz ranges, respectively, where effects of destructive and constructive interference can be observed. In this case, increasing the number of steps in the coupling region will improve the coupler's isolation which will subsequently help to improve the stopbands of the bandpass and highpass filters [6]. Additionally, during measurement, the lower stopband edge of the bandpass filter appeared sensitive to contact between the SIW circuit and the waveguide housing. In this case, the screws connecting the lid of the waveguide to the body had to be closed with a large amount of torque to ensure good contact was made between the waveguide and the SIW, which is undesirable as it can compress the PCB. As a proof of concept design, the results are promising. For example, by considering the diplexers in [2–4], the measured insertion loss for this design is considerably less than the current state of the art. However, further work is required to relax the sensitivity of SIW bandpass filters to fabrication errors if SIW diplexers are to be used for commercial millimetre wave applications.

12.8 Summary

This chapter has concerned itself with the design of millimetre wave SIW diplexer circuits with an aim of improving their tolerance handling ability. A method of designing SIW components entirely in DWG and then translating a final optimised design to SIW is presented. Due to the equivalence between SIW and DWG, the translated SIW components do not require a further optimisation which is a significant advantage as it speeds up the design process. This method was demonstrated by showing simulated results of a Ka-band bandpass filter, highpass filter and hybrid coupler where good agreement is observed between the original DWG response and the translated SIW response. Using these components, the design of a Ka-band SIW diplexer based on the topology proposed in [5,6] and subsequently [7] is presented. To gain an understanding of how fabrication errors will affect the diplexer's frequency response, a tolerance analysis where errors in the via x and y position, dielectric constant, and via diameter were studied in detail. These observations suggest that the highpass filter in the diplexer is less sensitive to fabrication errors than the bandpass filter and its use in the diplexer helps to improve the tolerance handling ability of the circuit. To validate the simulations, a Ka-band diplexer was fabricated using RT 5880 in addition to an aluminium alloy waveguide-to-SIW transition where good agreement is observed between the measured and simulated frequency responses in both cases. However, the results suggest that further work is required in relaxing the fabrication tolerance of SIW bandpass filters if SIW diplexers are to be used for commercial millimetre wave applications, such as backhaul.

References

[1] M. Bozzi, A. Georgiadis, and K. Wu, "Review of substrate-integrated waveguide circuits and antennas," *IET Microwaves, Antennas Propag.*, vol. 5, no. 8, pp. 909–920, 2011.

[2] Z. Kordiboroujeni and J. Bornemann, "Mode matching design of substrate integrated waveguide diplexers," in *Proc. MTT-S International Microwave Symposium Digest, Seattle, Washington*, 2013, pp. 1–3.

[3] Z. Kordiboroujeni, J. Bornemann, and T. Sieverding, "K-band substrate integrated waveguide T-junction diplexer design by mode-matching techniques," in *Proc. Asia-Pacific Microwave Conference, Sendai, Japan*, 2014, pp. 1297–1299.

[4] N. Athanasopoulos, D. Makris, and K. Voudouris, "A 60 GHz planar diplexer based on substrate integrated waveguide technology," *Act. Passiv. Electron. Components*, vol. 2013, Article ID 948217, 6 pages, 2013.

[5] J. R. Aitken and J. Hong, "Tolerance considerations for wireless backhaul diplexer circuits," in *Proc. 44th European Microwave Conference, Rome*, 2014, pp. 620–623.

[6] J. R. Aitken and J. Hong, "Design of millimetre wave diplexers with relaxed fabrication tolerances," *IET Microwaves, Antennas Propag.*, vol. 9, no. 8, pp. 802–807, 2015.

[7] J. R. Aitken, J. Hong, and Z. C. Hao, "Millimetre wave SIW diplexer circuits with relaxed fabrication tolerances," *IET Microwaves, Antennas Propag.*, vol. 11, no. 8, pp. 1133–1138, 2017.

[8] J. R. Aitken and J. Hong, "Millimetre wave wideband low-loss waveguide-to-substrate integrated waveguide transition," *Microw. Opt. Technol. Lett.*, vol. 59, no. 1, pp. 10–12, 2017.

[9] X. P. Chen and K. Wu, "Substrate integrated waveguide filter: Basic design rules and fundamental structure features," *IEEE Microw. Mag.*, vol. 15, no. 5, pp. 108–116, 2014.

[10] J. R. Aitken, "Improving tolerance aspects of millimetre wave diplexer circuits," EngD Thesis, Heriot-Watt University, May 2017.

[11] F. Xu and K. Wu, "Guided-wave and leakage characteristics of substrate integrated waveguide," *IEEE Trans. Microw. Theory Tech.*, vol. 53, no. 1, pp. 66–73, 2005.

[12] D. Deslandes and K. Wu, "Accurate modeling, wave mechanisms, and design considerations of a substrate integrated waveguide," *IEEE Trans. Microw. Theory Tech.*, vol. 54, no. 6, pp. 2516–2526, 2006.

[13] Y. Cassivi, L. Perregrini, P. Arcioni, M. Bressan, K. Wu, and G. Conciauro, "Dispersion characteristics of substrate integrated rectangular waveguide," *IEEE Microw. Wirel. Components Lett.*, vol. 12, no. 9, pp. 333–335, 2002.

[14] L. Yan, W. Hong, K. Wu, and T. J. Cui, "Investigations on the propagation characteristics of the substrate integrated waveguide based on the method of lines," *IEE Proc. - Microwaves, Antennas Propag.*, vol. 152, no. 1, pp. 35–42, 2005.

[15] W. Che, K. Deng, D. Wang, and Y. L. Chow, "Analytical equivalence between substrate-integrated waveguide and rectangular waveguide," *IET Microwaves, Antennas Propag.*, vol. 2, no. 1, pp. 35–41, 2008.

[16] Z. Kordiboroujeni and J. Bornemann, "Designing the width of substrate integrated waveguide structures," *IEEE Microw. Wirel. Components Lett.*, vol. 23, no. 10, pp. 518–520, 2013.

[17] J. R. Aitken and J. Hong, "Design of millimetre wave substrate integrated waveguide diplexers," in *Proc. 1st URSI Atlantic Radio Science Conference (URSI AT-RASC), Gran Canaria, Spain*, 2015, pp. 1–1.

[18] R. J. Cameron, C. M. Kudsia, and R. R. Mansour, "Microwave filters for communication systems," *John Wiley Sons, INC.*, New Jersey, 2007.

[19] K. K. Belostotskay, C. N. Klimov, B. V. Sestroretskiy, V. A. Ruchenkov, and T. V. Kamishev, "Waveguide diplexer for feed reflector-type antenna for deep-space communication," in *Proc. 5th International Conference on Antenna Theory and Techniques, Ukraine*, 2005, pp. 457–459.

[20] European Space Agency, "High Power Q/V-Band Diplexers for Ground Stations," *ARTES 5.1 6B.025*, October, 2015.

[21] J. Hong, M. Lancaster, R. Greed, and D. Jedamzik, "A high-temperature superconducting duplexer for cellular base-station applications," *IEEE Trans. Microw. Theory Tech.*, vol. 48, no. 8, pp. 1336–1343, 2000.

[22] J. Hong and M. Lancaster, "Wiley: Microstrip Filters for RF / Microwave Applications," *John Wiley & Sons, INC.*, 2001.

[23] Z. Kordiboroujeni and J. Bornemann, "Mode-matching analysis and design of substrate integrated waveguide T-junction diplexer and corner filter," *Int. J. Numer. Model. Electron. Networks, Devices Fields*, vol. 28, no. 5, pp. 497–507, 2015.

[24] Z. Kordiboroujeni, F. Taringou, and J. Bornemann, "Efficient mode-matching design of substrate-integrated waveguide filters," in *Proc. 42nd European Microwave Conference, Amsterdam*, 2012, pp. 253–256.

[25] Z. Kordiboroujeni, J. Bornemann, and T. Sieverding, "Mode-matching design of substrate-integrated waveguide couplers," in *Proc. Asia-Pacific Symposium on Electromagnetic Compatibility, Singapore*, 2012, pp. 701–704.

[26] M. Buchta and W. Heinrich, "On the equivalence between cylindrical and rectangular via-holes in electromagnetic modeling," in *Proc. 37th European Microwave Conference, Munich*, 2007, pp. 142–145.

[27] I. Hunter, "Theory and Design of Microwave Filters," The Institution of Engineering and Technology, London, 2001.

[28] E. G. Matthaei, Young, and L. Jones, "Microwave Filters, Impedance-Matching Networks, and Coupling Structures", Artech House, INC., Massachusetts, 1964.

[29] Y. Leviatan, P. G. Li, A. T. Adams, and J. Perini, "Single-post inductive obstacle in rectangular waveguide," *IEEE Trans. Microw. Theory Tech.*, vol. 31, no. 10, pp. 806–812, 1983.

[30] W. Hauth, R. Keller, and U. Rosenberg, "The corrugated-waveguide band-pass filter – A new type of waveguide filter," in *Proc. 18th European Microwave Conference, Stockholm*, 1988, pp. 945–949.

[31] D. Pozar, "Microwave Engineering," *4th Edition, John Wiley & Sons, INC.*, New Jersey, 2012.

[32] J. W. Bandler, "Computer optimization of inhomogeneous waveguide transformers," *IEEE Trans. Microw. Theory Tech.*, vol. 17, no. 8, pp. 563–571, 1969.

[33] Y. Fang and X. Yan, "Design of waveguide narrow-wall 3dB coupler for 3mm-wave frequency band," in *Proc. 5th Global Symposium on Millimeter-Waves, Harbin*, 2012, pp. 166–169.

[34] U. Rosenberg and K. Beis, "Improved narrow-wall short slot coupler design exhibiting significant increased bandwidth and low cost production," in *Proc. 31st European Microwave Conference, London*, 2001, pp. 1–4.

[35] L. W. Hendrick and R. Levy, "Design of waveguide narrow-wall short-slot couplers," *IEEE Trans. Microw. Theory Tech.*, vol. 48, no. 10, pp. 1771–1774, 2000.

[36] L. T. Hildebrand, "Results for a simple compact narrow-wall directional coupler," *IEEE Microw. Guid. Wave Lett.*, vol. 10, no. 6, pp. 231–232, 2000.

[37] H. Riblet, "The short-slot hybrid junction," *Proc. IRE*, vol. 40, no. 2, pp. 180–184, 1952.

[38] H. Watanabe, H. Tsuzaka, and M. Masuda, "Microdrilling for printed circuit boards (PCBs)—Influence of radial run-out of microdrills on hole quality," *Precis. Eng.*, vol. 32, no. 4, pp. 329–335, 2008.

[39] L. Zheng, C. Wang, L. Yang, Y. Song, and L. Fu, "Characteristics of chip formation in the micro-drilling of multi-material sheets," *Int. J. Mach. Tools Manuf.*, vol. 52, no. 1, pp. 40–49, 2012.

[40] X. P. Chen and K. Wu, "Substrate integrated waveguide filters: Practical aspects and design considerations," *IEEE Microw. Mag.*, vol. 15, no. 7, pp. 75–83, 2014.

[41] *https://www.rogerscorp.com*, "Accessed 02/04/2018."

[42] D. Deslandes, "Design equations for tapered microstrip-to-substrate integrated waveguide transitions," in *Proc. IEEE MTT-S International Microwave Symposium, California*, 2010, pp. 704–707.

[43] X. P. Chen and K. Wu, "Self-packaged millimeter-wave substrate integrated waveguide filter with asymmetric frequency response," *IEEE Trans. Components, Packag. Manuf. Technol.*, vol. 2, no. 5, pp. 775–782, 2012.

[44] F. Taringou, J. Bornemann, K. Wu, and T. Weiland, "Broadband interconnects between coplanar waveguide and substrate integrated waveguide for dense packaging and integration," in *Proc. IEEE MTT-S International Microwave Symposium (IMS2014), Florida*, 2014, pp. 1–3.

[45] J. L. Cano, A. Mediavilla, and A. R. Perez, "Full-band air-filled waveguide-to-substrate integrated waveguide (SIW) direct transition," *IEEE Microw. Wirel. Components Lett.*, vol. 25, no. 2, pp. 79–81, 2015.

[46] Y. Li and K. M. Luk, "A broadband V-band rectangular waveguide to substrate integrated waveguide transition," *IEEE Microw. Wirel. Components Lett.*, vol. 24, no. 9, pp. 590–592, 2014.

[47] T. Li and W. Dou, "Broadband right-angle transition from substrate-integrated waveguide to rectangular waveguide," *Electron. Lett.*, vol. 50, no. 19, pp. 1355–1356, 2014.

[48] R. Głogowski, J. F. Zürcher, C. Peixeiro, and J. R. Mosig, "Broadband Ka-band rectangular waveguide to substrate integrated waveguide transition," *Electron. Lett.*, vol. 49, no. 9, pp. 602–604, 2013.

[49] J. Dong, Y. Liu, H. Lin, and T. Yang, "Low loss and broadband transition between substrate integrated waveguide and rectangular waveguide," *Int. J. RF Microw. Comput. Eng.*, vol. 26, no. 1, pp. 54–61, 2016.

[50] E. Moldovan, R. G. Bosisio, and K. Wu, "W-band multiport substrate-integrated waveguide circuits," *IEEE Trans. Microw. Theory Tech.*, vol. 54, no. 2, pp. 625–632, 2006.

[51] Z. Li, X. P. Chen, and K. Wu, "A surface mountable pyramidal horn antenna and transition to substrate integrated waveguide," in *Proc. International Symposium on Signals, Systems and Electronics, Montreal, Quebec*, 2007, pp. 607–610.

Index

www.ingramcontent.com/pod-product-compliance
Lightning Source LLC
LaVergne TN
LVHW011829200926
848484LV00002B/342